CHROMATOGRAPHY AND ISOLATION OF INSECT HORMONES AND PHEROMONES

CHROMATOGRAPHIC SOCIETY SYMPOSIUM SERIES

CHIRAL SEPARATIONS
Edited by D. Stevenson and I. D. Wilson

CHROMATOGRAPHY AND ISOLATION OF INSECT HORMONES AND PHEROMONES
Edited by A. R. McCaffery and I. D. Wilson

RECENT ADVANCES IN CHIRAL SEPARATIONS
Edited by D. Stevenson and I. D. Wilson

RECENT ADVANCES IN THIN-LAYER CHROMATOGRAPHY
Edited by F. A. A. Dallas, H. Read, R. J. Ruane, and I. D. Wilson

CHROMATOGRAPHY AND ISOLATION OF INSECT HORMONES AND PHEROMONES

Edited by

A. R. McCaffery
University of Reading
Reading, United Kingdom

and

I. D. Wilson
ICI Pharmaceuticals
Macclesfield, Cheshire, United Kingdom

PLENUM PRESS • NEW YORK AND LONDON

Library of Congress Cataloging-in-Publication Data

International Symposium on Chromatography and Isolation of Insect Hormones, Pheromones, and Related Compounds (1st : 1989 : Reading, England)
Chromatography and isolation of insect hormones and pheromones : [proceedings of the First International Symposium (jointly organized by the Chromatographic Society and the Royal Entomological Society of London) on Chromatography and Isolation of Insect Hormones, Pheromones, and Related Compounds, held March 21-23, 1989, in Reading, United Kingdom] / edited by A.R. McCaffery and I.D. Wilson.
p. cm. -- (Chromatographic Society symposium)
Includes bibliographical references and indexes.

1. Insect hormones--Analysis--Congresses. 2. Pheromones--Analysis--Congresses. 3. Chromatographic analysis--Congresses. I. McCaffery, A. R. II. Wilson, Ian D. III. Chromatographic Society (Great Britain) IV. Royal Entomological Society of London. V. Title. VI. Series.
QL495.I548 1989
595.7'01927--dc20 90-46480
CIP

Proceedings of the First International Symposium (jointly organized by the Chromatographic Society and the Royal Entomological Society of London) on Chromatography and Isolation of Insect Hormones, Pheromones, and Related Compounds, held March 21–23, 1989, in Reading, United Kingdom

ISBN 978-1-4684-8064-1 ISBN 978-1-4684-8062-7 (eBook)
DOI 10.1007/978-1-4684-8062-7

Softcover reprint of the hardcover 1st edition 1990

A Division of Plenum Publishing Corporation
233 Spring Street, New York, N.Y. 10013

PREFACE

The connection between the study of insects, their development, behaviour and biochemistry, and chromatography is perhaps not immediately obvious. However, this connection exists and it is of fundamental importance to our understanding of many areas of insect physiology. Insects range in size from small to minute and consequently the amounts of hormones or pheromones they produce are equally minute. Ultimately any attempt at understanding the processes which control development, social behaviour or the biochemistry of insects requires some means of isolating the tiny quantities of the hormones and pheromones responsible in sufficient quantity and purity for identification. The ability to devise novel techniques to separate these materials from frequently complex biological mixtures including precursors and metabolites and to devise detection systems for them is vital. Methods for the quantification of these substances at different stages in the life cycle, or in response to environmental change or stress, are then essential. Chromatography, both as a means for isolation and as a method for quantitative analysis, has been an essential tool in these studies.

This volume represents the outcome of a joint international symposium organized by the Chromatographic and Royal Entomological Societies at the University of Reading between the 21st and 23rd March 1989 aimed specifically at discussing the chromatography and isolation of insect hormones, pheromones and related substances.The papers presented at that meeting, and collected together here, covered many aspects of the subject including the chromatography of juvenile hormones, ecdysteroids, peptides, pheromones and semiochemicals.

Almost every area of chromatography was discussed, including GLC, HPLC, TLC, SFC, CZE and electrophoresis, with the separations described often of an impressively high standard. Many authors have devised novel combinations of separation and detection or coupled separative techniques to improve resolution. Together with modern immunological, molecular and sequencing techniques the science of insect hormone and pheromone isolation has identified many active materials. As shown in this volume, the continued development of sensitive bioassays in combination with chromatographic techniques has been an important factor in this progress. The editors believe that these proceedings represent a body of knowledge and experience which will be of immense value to any researcher in this field for many years to come.

A.R. McCaffery and I.D. Wilson
April 1990

CONTENTS

PHEROMONES

PATTERN RECOGNITION

THE CHROMATOGRAPHIC SOCIETY

The Chromatographic Society is the only international organization devoted to the promotion of, and the exchange of information on, all aspects of chromatography and related techniques.

With the introduction of gas chromatography in 1952, the Hydrocarbon Chemistry Panel of the Hydrocarbon Research Group of the Institute of Petroleum, recognizing the potential of this new technique, set up a Committee under Dr. S. F. Birch to organize a Symposium on 'Vapor Phase Chromatography' which was held in London in June 1956. Almost 400 delegates attended this meeting and success exceeded all expectation. It was immediately apparent that there was a need for an organized forum to afford discussion of development and application of the method and, by the end of the year, the Gas Chromatography Discussion Group had been formed under the Chairmanship of Dr. A. T. James with D. H. Desty as Secretary. Membership of this Group was originally by invitation only but, in deference to popular demand, the Group was opened to all willing to pay the modest subscription of one guinea and in 1957 A.J.P. Martin, Nobel Laureate, was elected inaugural Chairman of the newly-expanded Discussion Group.

In 1958 a second Symposium was organized, this time in conjunction with the Dutch Chemical Society, and since that memorable meeting in Amsterdam the Group, now Society, has maintained close contact with kindred bodies in other countries, particularly France (Groupement pour l'Avancement des Methodes Spectroscopiques et Physico-chimiques d'Analyse) and Germany (Arbeitskreis Chromatographie der Gesellschaft Deutscher Chemiker) as well as interested parties in Eire, Italy, The Netherlands, Scandinavia, Spain and Switzerland. As a result Chromatography Symposia, in association with Instrument Exhibitions, have been held biennially in Amsterdam, Edinburgh, Hamburg, Brighton, Rome, Copenhagen, Dublin, Montreux, Barcelona, Birmingham, Baden-Baden, Cannes, London, Nurnburg, Paris and Vienna.

In 1958 'Gas Chromatography Abstracts' was introduced in journal format under the Editorship of C.E.H. Knapman; first published by Butterworths, then by the Institute of Petroleum, it now appears as 'Gas and Liquid Chromatography Abstracts' produced by Elsevier Applied Science Publishers and is of international status — abstracts, covering all aspects of chromatography, are collected by Members from over 200 sources and collated by the Editor Mr. E. R. Adlard assisted by Dr. P. S. Sewell.

Links with the Institute of Petroleum were severed at the end of 1972 and the Group established a Secretariat at Trent Polytechnic in Nottingham, Professor Ralph Stock playing a prominent part in the establishment of the Group as an independent body. At the same time, in recognition of expanding horizons, the name of the organization was changed to the Chromatography Discussion Group.

In 1978, the 'Father' of Partition Chromatography, Professor A. J. P. Martin was both honoured and commemorated by the institution of the Martin Award which is designed as testimony of distinguished contribution to the advancement of chromatography. Recipients of the award include:

E.R. Adlard
Professor E. Bayer
Professor U.A.Th. Brinkman
Dr. L.S. Ettre
Professor J.C. Giddings
Professor G. Guiochon
Professor J.F.K. Huber
Dr. C.E.R. Jones
Dr. R.E. Kaiser
C.E.H. Knapman
Professor J.H. Knox
Dr. E. Kovats
Professor A. Liberti
Dr. C.S.G. Phillips
Dr. G. Schomburg
Dr. R.P.W. Scott
Dr. L. Snyder
Professor R. Stock
and Dr. G.A.P. Tuey

The Group celebrated its Silver Jubilee in 1982 with the 14th International Symposium held, appropriately, in London. To commemorate that event the Jubilee Medal was struck as means of recognizing significant contributions by younger workers in the field. Recipients of the Jubilee Medal include: Dr. J. Berridge, Dr. H. Colin, Dr. K. Grob Jr., Dr. J. Hermannson, Dr. E.D. Morgan, Dr. P.G. Simmonds, Dr. P. Schoenmachers and Dr. R. Tijssen.

In 1984 the name was once again changed, this time to The Chromagraphic Society, which title was believed to be more in keeping with the role of a learned society having an international membership of some 1000 scientists drawn from more than 40 countries. At that time, the Executive Committee instituted Conference and Travel Bursaries in order to assist Members wishing to contribute to, or attend major meetings throughout the world.

The Society is run by an Executive Committee elected by its Members, in addition to the international symposia, seven or eight one-day meetings covering a wide range of subjects are organized annually. One of these meetings, the Spring Symposium, is coupled with the Society's Annual General Meeting when, in addition to electing the Society's Executive Committee, Members have the opportunity to express their views on the Society's activities and offer suggestions for future policy.

Regular training courses in all aspects of chromatography are run in conjunction with the Robens Institute of the University of Surrey and it is hoped that this particular service will eventually include advanced and highly specialized instruction.

Reports of the Society's activities, in addition to other items of interest to its members (including detailed summaries of all papers presented at its meetings), are given in the Chromatographic Society Bulletin which is produced quarterly under the editorship of I.W. Davies.

At the time of writing three grades of membership are offered: Membership with Abstracts, Membership, and Student Membership (includes Abstracts). Members receive the Bulletin free of charge, benefit from concessionary Registration Fees for all Meetings and Training Courses and are, of course, eligible to apply for Travel and/or Conference Bursaries.

For further information and details of subscription rates please write to:

Mrs. J. Challis
Executive Secretary
THE CHROMATOGRAPHIC SOCIETY
Trent Polytechnic
Burton Street
Nottingham NG1 4BU
United Kingdom

THE ROYAL ENTOMOLOGICAL SOCIETY OF LONDON

The objects of the Society are 'the improvement and diffusion of entomological science'. These objects are pursued by holding meetings on all aspects of entomology; publishing the results of entomological research; maintaining a large entomological library; supporting entomological expeditions; and more informally by generating discourse between entomologists, particularly at the Society's Rooms.

The Royal Entomological Society of London was founded in 1833 as the Entomological Society of London. It was the successor to a number of short-lived societies dating back as far as 1745. The first meetings were held in the Thatched House Tavern, St. James's Street. Various other places in their turn became the scene of the Society's activities before the freehold of the present headquarters at 41 Queen's Gate was bought in 1920. In 1885 a Royal Charter was granted to the Entomological Society by Queen Victoria, and the privilege of adding the word 'Royal' to the title was granted by King George V in 1933, the Centenary of the Society's foundation. Many eminent scientists of the past, Darwin and Wallace to mention but two, have been Fellows of the Society. Through the years down to the sesquicentennial anniversary (1983) of the Society, most internationally recognized entomologists have been and are, numbered among the Fellowship.

The Society is governed by its Royal Charter and Bye-Laws; it is administered by a Council of 16 persons which includes the President, Honorary Treasurer, Honorary Secretary and Honorary Editorial Officer. All Fellows of three years' standing are eligible for election as Officers of the Society or as Members of Council, such elections being held at the Annual Meeting in June. Standing Committees appointed by Council oversee the Society's financial and scientific affairs. The day-to-day running of the Society's business is carried out by the Registrar and staff at 41 Queen's Gate, London SW7 5HU.

Regular monthly evening meetings are held at the Society's Rooms, and twice a year there are similar meetings in the Regions. At these meetings a paper is presented by an invited speaker or speakers with a discussion afterwards. Informal discussion takes place during tea before the meetings and afterwards, when refreshments are provided. The year's programme is formulated to cover all fields of entomology. One- and two-day Workshops organized by the Special Interest Groups that operate within the Society are held three or four times a year. Symposia take place biennially on specific topics when leading entomologists from all over the world give comprehensive and up-to-date accounts of research in their fields. Joint meetings are organized from time to time with other scientific and learned societies.

The elegant premises at Queen's Gate are open to Fellows and their guests Monday to Friday between 9.30 am and 5.00 pm. The Library may be used for private study, and light refreshments are obtainable. The General Office is open for the sale of publications and enquiries. There are other rooms that may be hired for private meetings.

The Society possesses one of the finest entomological libraries in the world, with a superb collection of eighteenth and nineteenth century works on insects. There are in addition collections of the manuscripts, diaries, drawings and correspondence of entomologists. The Library acts as a 'special source' for the British Library. In 1983 it contained some 9000 monographs and works of reference, 55,000 separates and holdings of over 600 journal titles, of which 270 are taken currently. A loan and photocopying service is available to Fellows. The Library is open from 9.30-5.00 pm on all weekdays, and until 7.00 pm on Meeting Days. The Librarian is in the Library on Tuesdays, Wednesdays and Thursdays.

The Society publishes four scientific journals: *Ecological Entomology, Physiological Entomology, Systematic Entomology* and *Medical and Veterinary Entomology*. The Society's house journal *Antenna* contains reports on all Society Meetings, Workshops, Symposia and other activities; there are also articles, readers' correspondence, an important Diary Section, and many other items of interest to all entomologists. The Handbooks for the Identification of British Insects is an expanding series of important works. The highly successful Symposia Volumes are published biennially following each Symposium. Details of these and other occasional publications are available from the Registrar's Office.

Fellows are elected to the Society by the Council on the proposal of three Fellows, after consideration by the Fellowship Committee and on the recommendation of the Officers. All interested in entomology are welcome to apply for a Fellowship. However, the activities of the Society are at the graduate level and while academic qualifications are not a prerequisite, it is possible that only entomologists (whether professional or amateur) whose activities and interests are of that standard will benefit from a Fellowship. At the time of the Society's sesquicentennial anniversary there were some 2100 Fellows, of whom about a quarter were living and working overseas. The privileges of the Fellowship include: the right to attend Ordinary Meetings of the Society and to introduce guests; the use of the Library, including loans (UK only), and facilities for photocopying; the receipt of the quarterly bulletin *Antenna* free of charge; the right to buy any of the Society's publications at greatly reduced prices; the right to attend, at a greatly reduced fee, those Society activities where a registration fee is charged; the right to propose Candidates; the right to make use of the title 'Fellow of the Royal Entomological Society' and suffix "FRES'; and the right to attend and vote at Annual Meetings. A certificate of Fellowship is issued to Fellows on election to the Society.

For further information and details of subscription rates please write to:

Mr. G. Bentley
Registrar
ROYAL ENTOMOLOGICAL SOCIETY OF LONDON
41 Queen's Gate
London SW7 5HU
United Kingdom

JUVENILE HORMONES

ENDOGENOUS JUVENILE HORMONE III TITRES AND *IN VITRO* RATES OF HORMONE BIOSYNTHESIS BY CORPORA ALLATA DURING THE REPRODUCTIVE CYCLE OF ADULT FEMALE *PERIPLANETA AMERICANA*

J. P. Edwards, J. Chambers, J. E. Short, N. R. Price,
R. J. Weaver, L. Abraham and C. M. Walter

ADAS Central Science Laboratory
Ministry of Agriculture, Fisheries and Food
London Road, Slough, Berkshire SL3 7HJ, UK

SUMMARY

The endogenous titres of juvenile hormone III (JH III), and the *in vitro* rates of juvenile hormone biosynthesis by isolated corpora allata have been measured at intervals during the reproductive cycle of adult female *Periplaneta americana*. There was a close correlation between the rates of JH III biosynthesis and endogenous hormone titres. Only JH III was detected in whole body extracts, and hormone titres reached a peak (mean approximately 29 ng/g) at about the mid-point of the ootheca production cycle.

INTRODUCTION

Juvenile hormones (JHs) are known to play an important role in regulating both metamorphosis and reproduction in insects. In those species where endogenous hormone levels have been measured, JH titres show fluctuations related to changes in developmental or reproductive physiology. In some cockroach species, *in vitro* rates of JH biosynthesis by isolated corpora allata show distinct peaks which correlate well with each reproductive cycle. In *Leucophaea maderae* [1] and *Nauphoeta cinerea* [2] high titres of JH in the haemolymph coincided with the times of maximum oocyte growth and vitellogenin synthesis. In adult female *Diploptera punctata* [3] and *Blattella germanica* [4], endogenous JH III titres were also closely linked to oocyte development, and were reflected by *in vitro* rates of JH biosynthesis by isolated corpora allata.

In mated adult female *Periplaneta americana*, oothecae are produced regularly at 3 to 5 day intervals. Previous studies [5,6] have shown that *in vitro* rates of JH III biosynthesis by isolated corpora allata change substantially during this cycle. Maximum rates (approximately 25 pmole/gland pair/h) occuring just prior to the mid-point of each gonadal cycle [6,7].

The existence of marked fluctuations in endogenous JH titres, and in the biosynthetic activity of the endocrine glands *in vitro*, suggests that corpus allatum activity and endogenous hormone levels are intimately linked. Thus, since the rate of biosynthesis of JH by corpora allata is likely to be a major factor contributing to the final levels of hormone in the insect body, we have investigated the relationship between endogenous JH III levels in intact insects and the rates of JH biosynthesis by isolated corpora allata, at precisely-timed points during the female reproductive cycle.

MATERIALS AND METHODS

Cockroaches were from a standard laboratory strain of *Periplaneta americana* maintained at 27°C and 45% rh as previously described [8]. Adult females of this strain produce oothecae regularly at intervals of about 4 days (mean interval between successive oothecae = 4.26 $\pm$ 0.11 days, n = 72). Mated adult females were selected for experimentation

Chromatography and Isolation of Insect Hormones and Pheromones
Edited by A. R. McCaffery and I. D. Wilson
Plenum Press, New York, 1990

at the start of an ootheca production cycle. Such females show the first portion (1-2 mm) of the untanned ootheca protruding from the genital pouch. We have used the term "extrusion" to denote this stage in the ootheca production cycle, and arbitrarily designate this time point as the zero-hour (0 h) point in the gonadal cycle. Adult females selected at extrusion were either used immediately for experimentation, or kept in groups of 5-15 animals with food and water for periods of 12, 24, 36, 48, 60, 72 or 96 h before use. Oocyte development was measured using the method of Weaver [9]. From batches of 10 adult females aged 0, 24, 48, 72 and 96 h post-extrusion, ovaries were removed in sterile Ringer solution, mounted on glass microscope slides and the lengths of the terminal and penultimate oocytes (T and T1 oocytes) measured with a calibrated eyepiece graticule.

Rates of juvenile hormone III biosynthesis by isolated pairs of corpora allata were measured using the *in vitro* radiochemical method described previously by Pratt and Tobe [10], Pratt et al. [7] and Pratt and Weaver [11], incorporating the rapid partition technique described by Feyeriesen and Tobe [12]. A minimum of 5 gland pairs were used for determination of biosynthetic rates at each time point during the female reproductive cycle.

Determination of endogenous JH levels was accomplished using the method of Bergot et al. [13]. In these experiments, timed adult female cockroaches were combined in batches of 4 (approximately 5 g wet weight) prior to extraction and JH titre determination by gas chromatography-mass spectrometry (GC-MS). A minimum of 4 replicate batches were used to determine JH titre at each time point during the adult female reproductive cycle. Juvenile hormone, extracted from cockroaches homogenized in HPLC grade acetonitrile, was subjected to various purification steps and derivatized to a d_{-3} methoxyhydrin prior to further clean-up (HPLC) and analysis by GC-MS [13]. In each set of experiments, a procedure blank (no insect material) was included, and recoveries (typically 60 to 80% after HPLC) were measured using an internal standard (^{3}H-iso-juvenile hormone II). GC-MS was performed on a VG 7035 mass spectrometer linked to a Dani 3800 HR gas chromatograph equipped with a 1 m packed column containing 3% OV 101 on Ultrabond II (100-120 mesh). Separation of juvenile hormone methoxyhydrins and internal standard was performed at 250°C, and the retention time for JH III was approximately 2 min. Identification and quantification of the hormone was accomplished by the use of GC-MS in the selective ion monitoring mode set to measure ions at m/z 76. Using this method, the limit of detection for JH III in cockroach extracts was about 0.01 ng/g insect material.

In some of the 36, 48 and 60 h post-extrusion samples the appropriate peaks were confirmed as being wholly due to juvenile hormone III methoxyhydrin by measuring the peak areas at m/z 76, 193 and 225. In all cases, the relative proportions were close to the 100:7:7 ratio obtained with an authentic sample of JH III methoxyhydrin in a standard solution containing known quantities of the methoxyhydrins of juvenile hormones I, II, III and 0. In addition, some of the 36, 48 and 60 h post-extrusion samples were also analyzed by selective ion monitoring at m/z 90. Although some small peaks were observed during these analyses, their retention times were different from those of the methoxyhydrins of authentic juvenile hormones I, II and 0 in the standard solution. Since similar peaks were also observed in the procedure blank samples, we conclude that these homologues are absent in adult females, and that only juvenile hormone III was present in this stage.

RESULTS

The length of the terminal and penultimate oocytes in the ovarioles of adult female *P. americana* at various times during the reproductive cycle is shown in Table 1. Both the terminal and the penultimate oocytes show a pattern of continuous growth throughout the period. By 96 h after extrusion, the majority of the terminal oocytes had reached maximum length (approximately 3.6 mm). However, in some of the insects examined at this time point, ovulation of the terminal oocytes (and extrusion of the ootheca) had already occurred.

The *in vitro* rates of JH III biosynthesis by isolated pairs of corpora allata removed from females at different time points during the reproductive cycle, and the corresponding endogenous titres of juvenile hormone III measured in whole body extracts, are shown in Table 2. Mean rates of JH III biosynthesis were highest (21.7 pmol/gland pair/h) 48 h after extrusion, and were markedly lower at extrusion (0 h) and at the end of the 4 day cycle (96 h). In addition, corpora allata removed from 48 h post-extrusion females produced the highest rate of JH III biosynthesis for an individual gland pair (36.2 pmol/gland pair/h).

Table 1. Oocyte development in ovarioles during the ootheca production cycle of adult female *Periplaneta americana* (n = 10)

Time after extrusion (hr)	Mean oocyte length (mm) ± SEM	
	Terminal oocyte	Penultimate oocyte
0[a]	1.97 ± 0.06	0.75 ± 0.02
24	2.62 ± 0.06	0.88 ± 0.04
48	2.87 ± 0.07	1.05 ± 0.05
72	3.37 ± 0.08	1.46 ± 0.07
96[a]	3.59 ± 0.07	1.85 ± 0.12

[a] Times of ootheca extrusion

Table 2. Juvenile hormone III titres and *in vitro* rates of JH III biosynthesis by isolated corpora allata, during the ootheca production cycle of adult female *Periplaneta americana*

Time after extrusion (hr)	(n =)	JH III titre (ng/g insect)	(n =)	JH III biosynthesis (pmol/gland pair/hr)
0[a]	4	0.74 ± 0.16	8	7.06 ± 1.56
12	4	2.52 ± 0.43	5	10.08 ± 1.26
24	5	2.37 ± 0.33	12	12.15 ± 2.01
36	4	7.34 ± 1.30	10	14.60 ± 2.12
48	8	24.89 ± 5.61	21	21.71 ± 1.71
60	4	29.10 ± 5.63	8	11.78 ± 3.28
72	5	4.74 ± 1.65	9	10.14 ± 1.91
96[a]	4	0.78 ± 0.24	7	6.06 ± 1.10

[a] Times of ootheca extrusion

Endogenous titres of juvenile hormone III were low (< 3 ng/g) for the first 24 h after extrusion. By 36 h post-extrusion, mean JH III levels had increased, and high levels were found in females at 48 h post-extrusion (24.9 ng/g) and at 60 h post-extrusion (29.1 ng/g). By 72 h post-extrusion, the endogenous JH III titre had dropped markedly to 4.7 ng/g and, by 96 h post-extrusion, had declined further to less than 1.0 ng/g. Although the highest mean JH titre occurred at 60 h post-extrusion, the highest individual batch titre (55.3 ng/g) was recorded from females aged 48 h after extrusion.

The relationship between JH III biosynthesis by isolated corpora allata and endogenous JH III titres in whole insects is illustrated in Figure 1.

DISCUSSION

In these studies we have detected only juvenile hormone III (methyl 10,11-epoxy 3,7,11 trimethyl 2,6 dodecadienoate) in extracts of adult female *Periplaneta americana*. In addition, we have demonstrated a clear relationship between the *in vitro* rate of juvenile hormone III biosynthesis by isolated corpora allata and endogenous levels of this hormone in adult females during the 4 day ootheca production cycle. The results of the *in vitro* studies confirm the findings of previous workers [5,6] that rates of biosynthesis are maximum just prior to the mid-point in the gonadal cycle. In the present experiments we have shown that these maximum rates of biosynthesis result in a marked peak of JH III which occurs at about the mid-point of the cycle. Highest *in vitro* rates of biosynthesis were recorded in corpora allata removed from females timed at 48 h post-extrusion, and

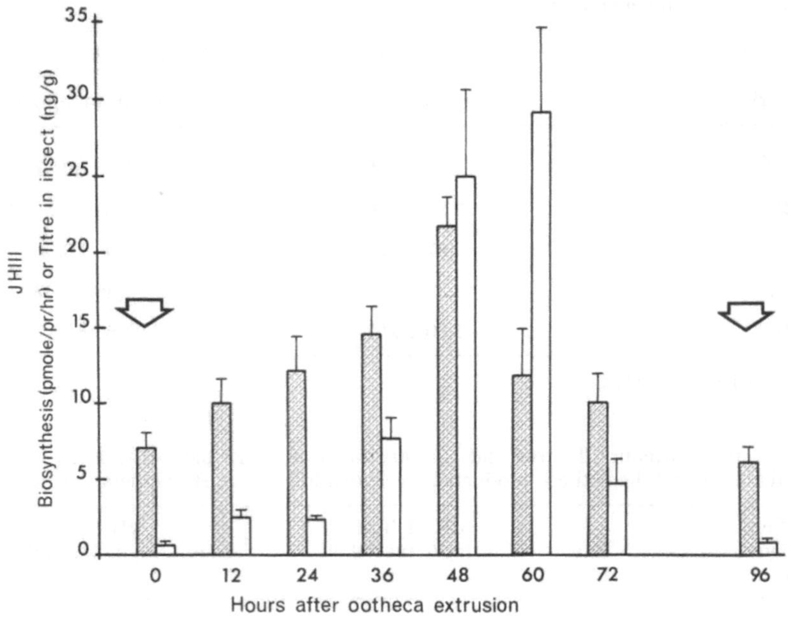

Fig. 1. Rates of *in vitro* juvenile hormone III biosynthesis (shaded histograms) and endogenous levels of juvenile hormone III (clear histograms) during the ootheca production cycle of adult female *Periplaneta americana*. Arrows indicate the times at which oothecae are produced.

although highest mean titre in whole animals was found at 60 h post-extrusion (i.e., some 12 h later), the highest individual batch titre (55 ng/g) was also recorded at 48 h post-extrusion.

Thus, it seems likely that both corpus allatum activity and endogenous juvenile hormone titres in individual insects reach their maxima close to 48 h after the extrusion of the previous ootheca. There was considerable variation in the mean juvenile hormone III titres measured at 48 and 60 h after extrusion. The range of titres measured was 10.1 to 55.3 ng/g at 48 h and 17.5 to 39.2 ng/g at 60 h. Some of this variability may be simply a reflection of differences in the duration of the ootheca production cycle in different batches of insects. In this strain of *P. americana*, the average interval between successive oothecae is 4.2 days, but the range is 3 to 6 days, so some variation in the timing of the peak of hormone is to be expected. In addition, this inherent variability in the insect material, coupled with the wide range of values obtained for JH III titres at the mid-point of the cycle, suggests that the peak of hormone in individual insects may be much sharper and much higher than is indicated by the mean titre values based on batches of insects. The lowest levels of endogenous hormone (< 1.0 ng/g) were found at the time of extrusion of the first ootheca (0 h) and 96 h later, at the extrusion of the subsequent ootheca. Similarly, lowest rates of hormone biosynthesis by isolated corpora allata were obtained from glands removed from females at the time of extrusion.

Most other determinations of juvenile hormone levels in cockroaches have been made on haemolymph samples only, and most have used indirect (i.e., non physico-chemical) methods to quantify levels of hormone. The most detailed study has been made by Tobe and co-workers in *Diploptera punctata*. Using GC-MS techniques, they found very high quantities (up to 1500 ng/ml) of juvenile hormone III in haemolymph samples taken from females at mid-cycle [3]. In addition, they reported that at some points during the reproductive cycle, the amounts of hormone in whole body extracts (up to 740 ng/g) were substantially higher than was expected on the basis of quantities in haemolymph, and that ovaries in particular contained higher concentrations than other tissues. In other species Lanzrein et al. [2] recorded the equivalent of 1.7 ng/ml juvenile hormone III in female *Nauphoeta cinerea*, using simple bioassay of haemolymph samples (*Galleria* wax test). Koeppe et al. [1], using the *Manduca* black mutant assay, found up to 15 ng/ml in adult female *Leucophea maderae*, and in *Blattella germanica*, Camps et al. [4], using direct chemical (GC-MS) methods, found fluctuating titres (3-11 ng/ml) in haemolymph samples taken from adult females. In all these studies, the authors have reported that fluctuations in quantities of endogenous hormone coincided with high rates of JH biosynthesis *in vitro*, and/or with high rates of oocyte growth. In *P. americana* Shaaya [14] used an indirect biochemical technique (measurement of protocatechuic acid glucoside in colleterial glands) to investigate juvenile hormone titres in 6th instar nymphs. This resulted in an estimate of 30-35 ng/g juvenile hormone equivalents at this developmental stage - a titre that is similar to the quantities of JH III that we have found in adult females at the mid-point of the ootheca production cycle.

In adult female *Periplaneta americana* there is a continuous vitellogenic cycle in which the terminal and penultimate oocytes sequester yolk protein concurrently [15]. When the terminal oocytes reach maximum size, ovulation occurs, and the penultimate oocytes become the next wave of terminal oocytes. In the strain of *P. americana* used in these experiments, where the mean interval between the production of successive oothecae is 4.2 days (Weaver and Edwards, unpublished results), this concurrent vitellogenic cycle is clearly visible in the measurements of oocyte length (Table 1). Thus, the penultimate oocytes increased in length from about 0.75 mm to about 1.9 mm during the first 4 day cycle, and then continued to grow to about 3.6 mm during the subsequent 4 day period. In *P. americana*, the presence of juvenile hormone is necessary both for the synthesis of vitellogenins by the fat body and for their uptake by the oocytes [16]. However, the apparent rate of growth (as reflected by changes in oocyte length) of terminal and penultimate oocytes seemed to be relatively constant throughout the 4 day cycle, and oocyte length did not appear to change substantially in response to the peak titre of endogenous hormone that occurred around the mid-point of the cycle. Similarly, only minor changes in rates of increase in oocyte growth (as measured by increases in oocyte volume) were observed by Bell [16] during the ootheca production cycle. Thus, the function (if any) of this peak of juvenile hormone III does not seem to be associated with changes in the rate of oocyte growth. It is possible, therefore, that the frequency with which the peak of hormone occurs may serve only to regulate the frequency of ootheca production. Some evidence that this may be the case comes from previous experiments in this species [17]. Treatment of adult females with a juvenile hormone analogue just after the time of extrusion accelerated the rate of production of the subsequent ootheca, and treatment with an anti-juvenile hormone agent (fluoromevalonate) extended the interval between ootheca production.

It is clear from our results that the production of juvenile hormone by the corpora allata changes substantially over the period of the gonadal cycle, and that such changes are reflected by the titre of hormone in the intact animal. However, there was no time when the glands were completely inactive, or when juvenile hormone was completely absent in the insects. Nevertheless, the pattern of biosynthesis and endogenous titre, strongly suggest that the activity of the glands is modulated in the intact insect. In previous studies [8,17] we have found evidence that endogenous juvenile hormone titres in *P. americana* are regulated by negative feedback. Thus, high levels of hormone in the insect might inhibit the production of more hormone by the corpora allata. Alternatively, high endogenous titres might induce a more rapid degradation of hormone by metabolizing enzymes. A system of negative feedback control could clearly result in the pattern of endogenous JH III titres observed in the present experiments, and would be consistent with the cyclical changes in corpus allatum activity reported here and elsewhere [5,7]. However, although we cannot exclude the possibility that changes in the rate of metabolism may play a part in altering endogenous hormone titre, the changes in the spontaneous activity of the corpora allata *in vitro* suggest that fluctuations in hormone levels in female *P. americana* are largely a result of differential rates of juvenile hormone biosynthesis by the corpora allata.

ACKNOWLEDGEMENTS

The authors would like to thank Dr F. C. Baker and Dr D. A. Schooley (Sandoz Crop Protection, Zoecon Research Institute, Palo Alto, USA) for the gift of several chemicals, and Mr J. P. G Wilkins (ADAS Harpenden Laboratory) for Mass Spectrometry facilities.

REFERENCES

1. J. K. Koeppe, S. G. Brantley and M. M. Nijhout, Structure, haemolymph titre and regulatory effects of juvenile hormone during ovarian maturation in *Leucophaea madeerae*, *J. Insect Physiol.*, 26:749-753 (1980).

2. B. Lanzrein, V. Gentinetta, R. Fehr and M. M. Luscher, Correlation between haemolymph juvenile hormone titre, corpus allatum volume, and corpus allatum *in vivo* and *in vitro* activity during oocyte maturation in a cockroach (*Nauphoeta cinerea*), *Gen. Comp. Endocr.*, 36:339-345 (1978).
3. S. S. Tobe, R. P. Ruegg, B. B. Stay, F. C. Baker, C. A. Miller and D. A. Schooley, Juvenile hormone titre and regulation in the cockroach *Diploptera punctata*, *Experientia*, 41:1028-1034 (1985).
4. F. Camps, J. Casas, F.-J. Sanchez and A. Messeguer, Identification of juvenile hormone III in the haemolymph of *Blattella germanica* adult female by gas chromatography-mass spectrometry, *Arch. Insect Biochem. Physiol.*, 6:b181-189 (1987).
5. R. J. Weaver, G. E. Pratt and J. R. Finney, Cyclic activity of the corpus allatum related to gonadotrophic cycles in adult female *Periplaneta americana*, *Experientia*, 31:597-598 (1975).
6. R. J. Weaver and G. E. Pratt, The effect of enforced virginity and subsequent mating on the activity of the corpus allatum of *Periplaneta americana* measured *in vitro* as related to changes in the rate of ovarian maturation, *Physiol. Ent.*, 2:59-76 (1977).
7. G. E. Pratt, S. S. Tobe, R. J. Weaver and J. R. Finney, Spontaneous synthesis and release of C-16 juvenile hormone by isolated corpora allata of female locust *Schistocerca gregaria* and female cockroach *Periplaneta americana*, *Gen. Comp. Endocr.*, 26:478-484 (1975).
8. J. P. Edwards, J. Chambers, N. R. Price and J. P. G. Wilkins, Action of a juvenile hormone analogue on the activity of *Periplaneta americana* corpora allata *in vitro* and on juvenile hormone III levels *in vivo*, *Insect Biochem.*, 17:1115-1118 (1987).
9. R. J. Weaver, Effects of food and water availability, and of NCA-1 section upon juvenile hormone biosynthesis and oocyte development in adult female *Periplaneta americana*, *J. Insect Physiol.*, 30:831-838 (1984).
10. G. E. Pratt and S. S. Tobe, Juvenile hormones radiobiosynthesized by corpora allata of adult female locusts *in vitro*, *Life Sci.*, 14:575-586 (1974).
11. G. E. Pratt and R. J. Weaver, Juvenile hormone biosynthesis by cultured cockroach corpora allata, *J. Endocr.*, 64:67 (1975).
12. R. Feyereisen and S. S. Tobe, A rapid partition assay for routine analysis of juvenile hormone released by insect corpora allata, *Analyt. Biochem.*, III:372-375 (1981).
13. B. J. Bergot, M. Ratcliff and D. A. Schooley, Method for the quantitative determination of the four known juvenile hormones in insects tissue using gas chromatography-mass spectroscopy, *J. Chromatog.*, 204:231-244 (1981).
14. E. Shaaya, Ecdysone and juvenile hormone activity in the larvae of the cockroach *Periplaneta americana*, *Insect Biochem.*, 8:193-195 (1978).
15. W. J. Bell and K. G. Adiyodi, Reproduction, *in*: "The American Cockroach", W. J. Bell and K. G. Adiyodi, eds., Chapman and Hall, London, 529 pp (1981).
16. W. J. Bell, Dual role of juvenile hormone in the control of yolk formation in *Periplaneta americana*, *J. Insect Physiol.*, 15:1279-1290 (1969).
17. J. P. Edwards, D. C. Cerf and G. B. Staal, Inhibition of ootheca production in *Periplaneta americana* with the anti-juvenile hormone fluoromevalonate, *J. Insect Physiol.*, 31:723-728 (1985).

USE OF THIN-LAYER CHROMATOGRAPHY, HIGH PERFORMANCE LIQUID CHROMATOGRAPHY AND GAS CHROMATOGRAPHY-MASS SPECTROMETRY TO INVESTIGATE THE RELATIONSHIP BETWEEN JUVENILE HORMONE TITRE AND CORPUS ALLATUM ACTIVITY IN ADULT MALE *PERIPLANETA AMERICANA*

R. J. Weaver, J. P. G. Wilkins,
A. T. Roberts and J. P. Edwards

ADAS Central Science Laboratory
Ministry of Agriculture, Fisheries and Food
London Road, Slough, Berkshire SL3 7HJ, UK

SUMMARY

Estimates of juvenile hormone (JH) biosynthesis by isolated corpora allata from adult male *Periplaneta americana* using a non-discriminatory partition assay suggested glandular activity levels that were comparable with the moderate to highly active rates observed in mid-gonotrophic cycle females. Parallel measurements of endogenous whole body JH titres using micro-derivatization, HPLC and GC-MS showed that male cockroaches contained less than 10% of the amount of hormone found in mid-cycle females. Radio-TLC measurements of the short-term *in vitro* synthetic performance of male glands failed to reveal any major differences from female corpora allata. Follow-up JH titre studies combined with surgical sectioning of the allatal nerves seemed to indicate that the corpora allata of adult male cockroaches are subject to neural inhibition *in vivo*.

INTRODUCTION

The essential gonadotrophic role of the corpora allata in females of many insect species is a well-established phenomenon [1]. In contrast, the situation in male insects remains somewhat questionable [2,3]. Isolated corpora allata from reproductively active female *Periplaneta americana* biosynthesize and release juvenile hormone when incubated or superfused in a defined tissue culture medium. Incorporation of the methyl moiety of methyl-[^{14}C]methionine or methyl-[^{3}H]methionine permitted indentification of the hormone by radio-liquid chromatography (LC) and radio-gas chromatography (GC), and its measurement following radio-thin-layer chromatography (TLC) by scintillation counting [4-7]. The primary biosynthetic product of the corpora allata in adult female *P. americana* has been unequivocably identified as juvenile hormone III [4,7,8]. Analysis of the incubation products of adult male corpora allata by high performance liquid-chromatography (HPLC) only has indicated that the glands of males release the same hormone [9].

Quantitative radiochemical assays of corpus allatum activity in mated adult females revealed substantial peaks in juvenile hormone biosynthesis which appeared to be closely correlated with the ongoing reproductive cycles. These large and predictable pulses of juvenile hormone biosynthesis observed *in vitro* have been interpreted as affording convincing evidence that corpus allatum activity is a precisely regulated phenomenon *in vivo* and, as such, of prime importance in the regulation of juvenile hormone titre in the whole animal [5,10]. This view has been greatly substantiated by the recent demonstration of a very close relationship between *in vitro* rates of juvenile hormone III biosynthesis in isolated corpora allata of accurately timed adult female *P. americana* and the endogenous whole body titres of juvenile hormone in equivalently-staged animals throughout the 4-day ootheca production cycle (Edwards et al., this volume).

The exact means by which insect corpora allata are physiologically regulated is largely unknown. Trophic and inhibitory substances seem to be important components in

Chromatography and Isolation of Insect Hormones and Pheromones
Edited by A. R. McCaffery and I. D. Wilson
Plenum Press, New York, 1990

corpus allatum regulation in many species of Dictyoptera. These controlling factors appear to manifest their effects by way of both neural and neurohumoral pathways [1,11-14]. The same may well be true for insects of other orders [1]. Until quite recently none of these factors had been isolated or fully characterized and their chemical nature was completely unknown [15].

A comprehensive understanding of the biochemical processes which govern the regulation of corpus allatum activity and reproductive development in adult insects might lead to the development of safer or more effective pesticides. To this end we have been using a specific and sensitive gas chromatographic-mass spectrometric (GC-MS) method [16] to quantitatively monitor changes in the juvenile hormone titre in *P. americana* in response to injections of various tissue extracts which we anticipate might contain the putative allatotactive compounds. During part of this study we had occasion to measure the endogenous hormone levels in a small number of sexually mature adult male cockroaches and we were somewhat surprised to discover that the titres in these animals were always quite low. Our earlier determinations of corpus allatum activity *in vitro* having previously indicated that male glands might be moderately active *in vivo*. A non-discriminatory rapid partition assay had been used for these gland activity measurements [17], and this left us in some doubt as to the true nature of the radiolabeled product released from male glands. With this in mind, and because so little seemed to be known about the role of juvenile hormone in adult male insects, we decided to investigate the performance of male corpora allata in the radiochemical assay more thoroughly and to then compare the activities of glands from males of known ages with those in the adult female. Here we describe these results and in addition the results of a nerve-severance experiment which seems to indicate that the corpora allata of adult male *P. americana* may be restrained by way of intact nerves from the brain. This finding would seem to suggest that the *in vitro* assay might prove to be an unreliable indicator of endogenous juvenile hormone biosynthesis under certain circumstances.

MATERIALS AND METHODS

Animals

P. americana were maintained in a stock colony at 27°C and 45% r.h. [10,18]. Newly emerged adult males were transferred from the stock colony 0 to 12 h after the final ecdysis, and held thereafter in 2 kg Kilner jars under similar culture conditions, as groups of 8 to 12 individuals. Mated adult females were selected from the stock colony at the time of commencement of an ootheca production cycle (= early extrusion stage) and were used either 36 h or 48 h later (Edwards et al., this volume).

Juvenile Hormone Titres

Juvenile hormone titres were measured according to the GC-MS method of Bergot et al. [16] using the electron impact mode with selected ion-monitoring (SIM) for detection. For each determination in these experiments we pooled 4 equally-staged male or female insects (approx. 5 g wet weight). Juvenile hormone was extracted from whole animals by homogenization (Virtis S-23) in HPLC grade acetonitrile. Then 20 μg (nominal) of [10-^{3}H]iso-JH II ethyl ester (75,000 dpm) was added to each homogenizer flask as an internal standard. The extracts were then subjected to various purification steps (pentane/saline partition, Sep-Pak, Alumina mini-column) prior to micro-derivatization, by acid-catalyzed methan-d_3-olysis, of both endogenous hormone and internal standard to the respective 11-methyoxy-d_3-10-hydroxy (methoxyhydrin) derivatives. This was followed by a further clean-up stage involving reversed-phase HPLC on a 25 cm x 0.45 cm i.d. stainless steel column packed with Spherisorb S5W (PhaseSep) and eluted with 3% diethyl ether in dichloromethane (50% water saturated). Only that fraction eluting between two non-radioactive, but strongly UV-absorbing, phenyl-urea marker compounds was collected for further analysis. Post-HPLC recoveries were typically 60 to 80%. GC-MS was performed on a VG 7035 mass spectrometer coupled to a Dani 3800 HR gas chromatograph. The GC column was 1.0 m x 2 mm i.d. packed with 3% OV 101 on Ultrabond II (100-120 mesh). Identification and quantification were accomplished by means of SIM with the machine set to measure ions at m/z 76. The detection limit for juvenile hormone III using this method was approximately 10 pg per gram of insect tissue. Appropriate GC peaks from certain male extracts were confirmed as being wholly due to juvenile hormone III methoxyhydrin by comparing the peak areas at m/z 76, 193 and 225. Furthermore, in order to confirm the absence of other known juvenile hormones (e.g., JH I and JH II) from extracts of males, many of the samples were additionally analyzed at m/z 90.

Radiolabeled Incubations

Juvenile hormone biosynthetic activities of freshly excised male and female corpora allata were determined from the incorporation into juvenile hormone III of methyl-[^{14}C]methionine according to standard procedures [10,19]. In many experiments we measured the incorporation into methyl farnesoate as well. This substance is the immediate juvenile hormone precursor in *P. americana* [20,21]. Pairs of corpora allata were incubated individually in 100 ml sterile-filtered TC-199 (Flow Laboratories) as previously described [5,10]. Radiolabeled methionine was routinely diluted to provide final L-methionine concentrations within the range 0.25 to 0.28 mM. Final specific activities ranged, in different experiments, between 1.29 and 1.40 GBq/mmol. Incubations were carried out at 30°C with gentle shaking for 3 h. Biosynthesis was terminated by the addition to each tube of 200 µl methanol, 100 µl EDTA (1% aq. soln.) and 1 ml chloroform. The products of the incubations (both glandular and released) were separately extracted in the presence of non-radioactive markers (20 µg each of methyl farnesoate and juvenile hormone III) then analyzed by silica TLC essentially as described elsewhere [19,21]. Products on the TLC plates were visualized by beta-imaging (Birchover 450/4) prior to liquid scintillation counting of appropriate zones. The combined amounts of radiolabeled juvenile hormone III and methyl farnesoate were measured in certain experiments (total ester determination) using a minor modification of the isooctane partition assay described by Feyereisen and Tobe [17].

Allatectomy and Sectioning of Allatal Nerves

Insects were prepared for surgery by cold anaesthesia. The retrocerebral complex was exposed under sterile Ringer [19] and the nervi corporis allati (NCA-1, NCA-2) were carefully sectioned using ultrafine watchmakers forceps [13]. Allatectomy necessitated the added severance of the smaller post-allatal nerves. The wounds were sealed with molten wax. Antibiotics were not employed. Sham-operated animals were subjected to similar manipulations without cutting or damaging the nerves.

Reagents

Methyl farnesoate was synthesized in this laboratory (gift of Dr J. Chambers). Juvenile hormone III was purchased from Sigma. Methoxy-d_3-hydrin standards and [10-^{3}H]iso-juvenile hormone II ethyl ester were a gift of Drs F. C. Baker and D. A. Schooley (Sandoz Crop Protection, Zoecon Research Institute, Palo Alto, USA). Methanol-d_4 (99.9%) and trifluoroacetic acid were purchased from Aldrich. Methyl-[^{14}C]methionine (2.23 Gbq/mmol) was supplied by Amersham International. All other chemicals were of the highest grade commercially available.

RESULTS

Corpus Allatum Activity Versus Endogenous Juvenile Hormone Titres in Selected Males and Females

The mean titre of juvenile hormone in adult male *P. americana* 14 to 18 days after ecdysis is shown in Table 1. The average titre in mid-gonadotrophic cycle adult females has been included for comparative purposes. This latter result has been taken from Edwards et al. (this volume). Also shown are estimates of corpus allatum activity in both sexes for the same stages of development. Glandular activities were calculated from the incorporation of methionine radiolabel into isooctane-extractable incubation products in this preliminary experiment and we observed that the biosynthetic activities of freshly excised corpora allata from randomly selected males were seemingly commensurate with the moderately active glands from reproductively active females. This assumes that the bulk of the extracted radioactivity can be equated with juvenile hormone III production. By contrast however, though the endogenous juvenile hormone III titres in adult females were found to be generally very high, the equivalent titres in males were invariably very low. The highest titre recorded in any individual batch of adult males was 2.30 ng/g but titres in excess of 20 ng/g were relatively common in adult females. It is perhaps appropriate to mention at this point, that in all of these studies we detected only juvenile hormone III in the derivatized extracts of both male and female *P. americana*. There was no evidence for any of the higher juvenile hormone homologues in extracts of either sex.

Table 1. Comparison of endogenous JH III titres in adult male and female *Periplaneta americana* with the mean rates of JH and methyl farnesoate synthesized by their corpora allata *in vitro*

Sex	Stage	JH III Titre (ng/g insect) ±S.E.	(n)	Ester synthesis (pmol/gland pr/h) ±S.E.	(n)
Adult female	36-48 h into cycle	19.04 ± 4.44	(12)	19.41 ± 1.46	(31)
Adult male	14-18 days old	1.56 ± 0.12	(12)	15.72 ± 1.04	(45)

Juvenile Hormone Synthesis and Release

It was next established, using corpora allata from males of different ages, that there was a linear relationship between the rate of synthesis and the rate of release of juvenile hormone III from male glands over a wide range of spontaneous activities. In these experiments single pairs of corpora allata were incubated in radiolabeled medium for a period of 3 h. The glands were then separated from the medium by means of a platinum wire loop, and the total radioactive juvenile hormone III present was estimated both in the glands and in the incubation.

Despite the inherent limitations of this procedure [20], we found a remarkably clear relationship between the rate of synthesis and the rate of release of hormone (Figure 1). If we compare, moreover, these present results with those obtained previously, using corpora allata from adult female *P. americana* [22], it can be seen that in glands of both sexes there was a strictly linear relationship between rate of release and rate of synthesis of hormone over the full range of spontaneous activities investigated. More than 95% of the radiolabeled hormone synthesized by male corpora allata during a 3 h incubation could be recovered from the incubation medium at the end of the incubation, a figure which is slightly higher than the 85% quoted for glands from adult females [22] and is exactly commensurate with the higher figure reported later [10]. The hourly progress of juvenile hormone release, on an accumulative basis, for seven separate pairs of male corpora allata having different spontaneous activities is shown in Figure 2. The rate of release of newly-biosynthesized hormone was relatively constant for individual pairs of glands up to the maximum 11 h period investigated.

In Figure 3 the relationship between total glandular content of methyl farnesoate at the end of the 3 hr incubation period and the rate of juvenile hormone synthesis during the same period for a collection of 46 pairs of corpora allata taken from adult male *P. americana* of different ages is shown. The available data appear to show a fundamental correlation between intraglandular methyl farnesoate levels and the rate of its epoxidative conversion to juvenile hormone. This finding is consistent with similar observations obtained before with corpora allata from adult female *P. americana*. The likely significance of such data in terms of enzyme kinetics has been discussed [21]. The important points for our present purposes are that methyl farnesoate levels in adult male corpora allata rarely exceed 14% of the juvenile hormone synthesized, and that high levels of intraglandular methyl farnesoate (> 6.0 pmol/3 h per pair) are hardly ever observed.

In Table 2 we have compared the mean rates of juvenile hormone and methyl farnesoate synthesis in corpora allata from 14 to 18 day old adult males with those of mid-gonadotrophic cycle (48 h post-extrusion) adult females. It is clear from these results, and from the data described both above and in Pratt et al. [21] that the glands of adult males have an epoxidative capacity which is certainly not lower than observed in female corpora allata.

Corpus Allatum Activity During the Course of Sexual Maturation

Using the radiochemical assay with TLC analysis, corpus allatum activity was monitored in males during the first 22 days of adult life (Figure 4). The rate of juvenile hormone synthesis was found to be low (< 1.0 pmol/h per pair) on the first day after the final moult, but then increased rapidly over the next 8 days to reach a peak in excess of 20 pmol/h per pair. Hormone biosynthetic activity was seen to decline thereafter, at first quite sharply, then more gradually, until it appeared to reach a plateau of approximately 12 pmol/h per pair from day 18 onwards. These observed rates are not too dissimilar to those exhibited by corpora allata from mated adult females [10]. With males, however, we failed to detect the marked fluctuations in activity which in females always appear to accompany

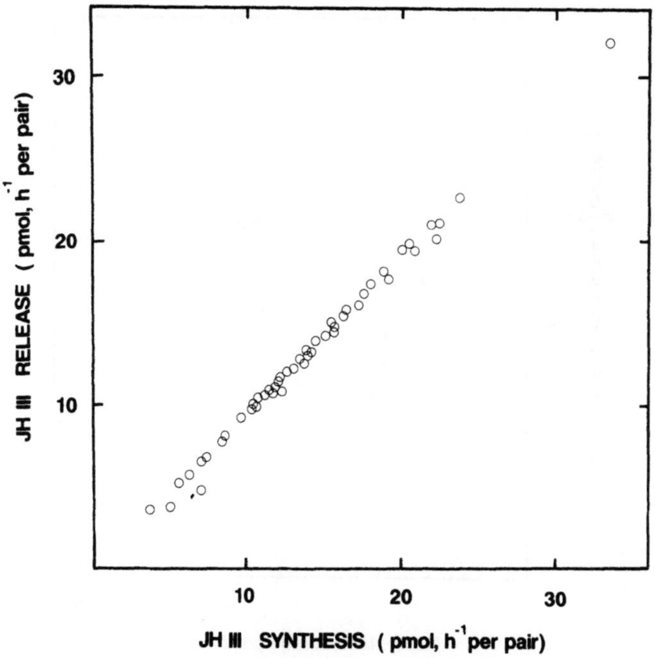

Fig. 1. Relationship between rates of synthesis and release of juvenile hormone III by individual pairs of corpora allata from adult male *Periplaneta americana*. Glands were incubated for 3 h in culture medium containing methyl-[^{14}C]methionine. The rate of release of freshly synthesized hormone was directly proportional to the measured rate of synthesis of hormone for a wide range of intrinsic glandular activities.

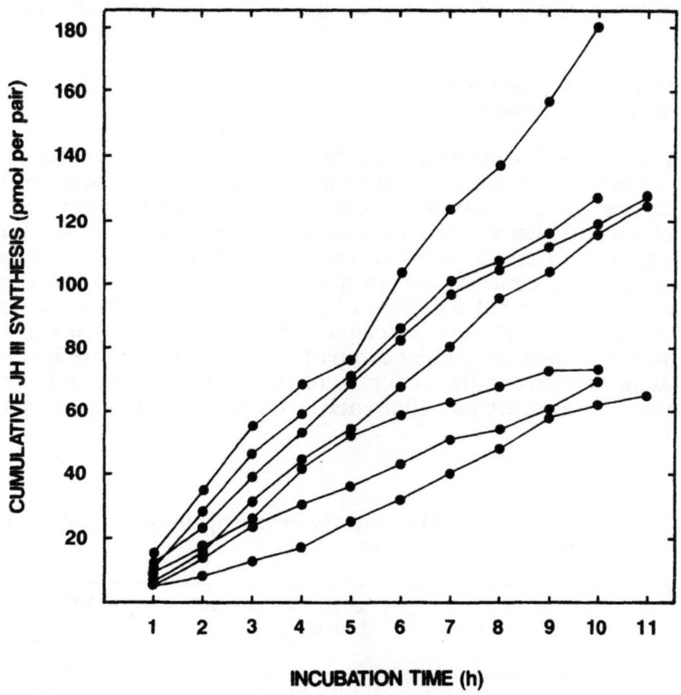

Fig. 2. Progress of release of radiolabeled juvenile hormone III from single pairs of corpora allata from adult male *Periplaneta americana* of different ages. Rates of release of hormone were essentially constant for at least 11 hr.

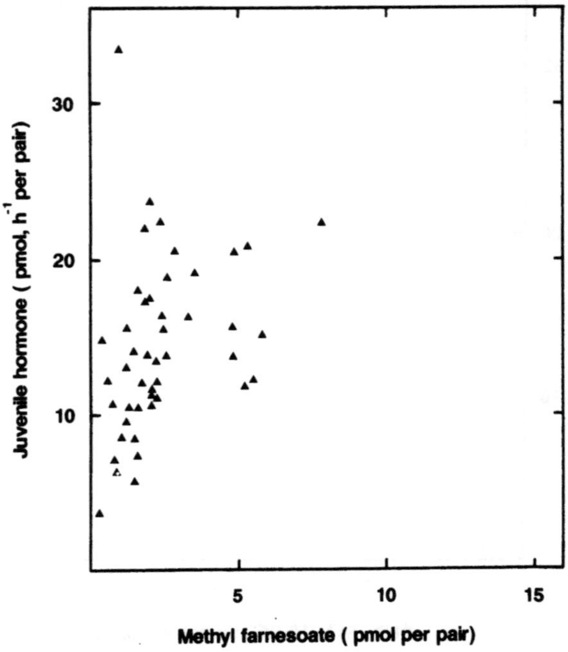

Fig. 3. Relationship between intraglandular levels of methyl farnesoate and rates of juvenile hormone synthesis for individual pairs of corpora allata taken from adult male *Periplaneta americana* of various ages. Levels and rates were determined from methyl-[^{14}C]methionine incorporation after 3 h incubations. Each point represents one gland pair from one male.

the 4 day reproductive cycles. From the two sets of data it is apparent that the corpora allata of adult males become competent for hormone biosythesis somewhat earlier than do the glands of females. We are unable to rule out the possibility, however, that the use of two different laboratory strains of cockroach might be responsible for this temporal divergence.

Effect of Nerve Sectioning upon Endogenous Juvenile Hormone Titres in Adult Males

The NCA-1 and NCA-2 were severed bilaterally in a small number of mature males some 72 h prior to measuring their endogenous hormone levels in order to investigate any possible role of the allatal nerves in the apparent discrepancy between *in vitro* measurements of corpus allatum activity and whole-body juvenile hormone titres. The results are summarized in Table 3. Low titres of juvenile hormone III (< 2.0 ng/g wet weight) were found in unoperated controls and in sham-operated insects but those animals in which the allatal nerves were cut had much higher titres of juvenile hormone III, up to 4 times as much as in control groups. Precision was assured in these experiments by the inclusion of an internal standard and by comparison with a procedure blank. We are thus reasonably confident concerning the accuracy of these results despite the relatively small number of observations and the fact that each determination was performed on a pooled extract of 4 males.

Table 2. Rates of juvenile hormone III and methyl farnesoate synthesis by corpora allata from adult male and female *Periplaneta americana*

Age & Sex	(n)	JH III synthesis (pmol/gland pr/h) ±S.E.	MF synthesis (pmol/gland pr/h) ±S.E.
14-18 day males	(13)	17.42 ± 1.38	2.35 ± 0.45
Mid-cycle females (48 h)	(21)	18.51 ± 2.30	3.71 ± 0.50

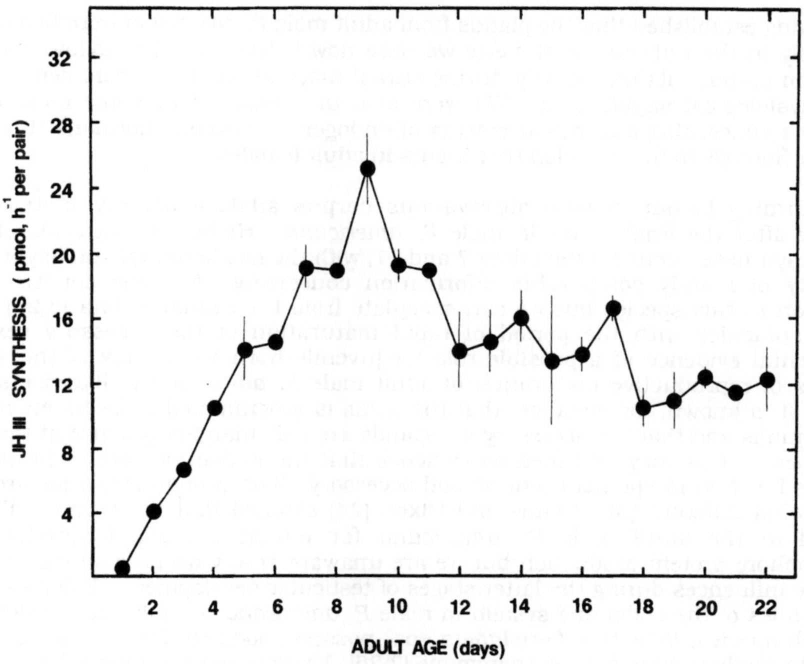

Fig. 4. Rates of synthesis of juvenile hormone III by corpora allata from adult male *Periplaneta americana* during the first 22 days of adult life. Each point is a mean value for a minimum of five determinations. Vertical bars depict standard errors.

DISCUSSION

The simplest explanation for the observation that low endogenous juvenile hormone titres in adult male *P. americana* were reflected by low rates of hormone synthesis in isolated corpora allata was that the non-discriminatory isooctane partition assay had been providing us with misleading information. There were a number of ways in which this could have happened. The most obvious entailed a hypothetical overproduction of methyl farnesoate at the expense of juvenile hormone III. According to this theory it was conceivable that the corpora allata of males were either accruing or discharging much larger quantities of this putative precursor than had ever been observed in incubations of female glands in the absence of exogenous farnesoic acid. A compound such as methyl farnesoate would be readily detected in the partition assay by virtue of its methyl ester group and its inherent lipophilicity, but since it lacks an epoxide moiety it would not derivatize to a methoxyhydrin and would be unlikely to generate the characteristic m/z 76 and m/z 90 fragments which we were monitoring by mass spectrometry. By reverting to the technique of using gland transfers and TLC we were soon able to dismiss the possibility of methyl farnesoate excess but our experiences should serve as a precautionary reminder to those contemplating the investigation of corpus allatum activity by means of simplified methodology such as the isooctane partition assay. This procedure should only be used as the authors specifically intended, i.e., in cases where the identity of the released products are known, and with the glands first removed from the medium [17].

Table 3. Effect of bilateral NCA-I and NCA-II section upon endogenous juvenile hormone titre in adult male *Periplaneta americana*

Treatment	No. (groups of 4)	Juvenile hormone III titre (ng/g wet weight)
Control (14-18 days old)	6	1.45 ± 0.20
Control (14 days old)	3	1.87 ± 0.21
Control (18 days old)	3	1.85 ± 0.55
Sham-operated (18 day old)	3	1.44 ± 0.12
Bilateral NCA-I/NCA-II (18 day)	3	7.52 ± 1.23

Having established that the glands from adult male *P. americana* were behaving quite predictably in the radiochemical assay we were now better placed to follow any possible changes in corpus allatum activity during sexual maturation and to comment upon their likely physiological significance. We were also in a position to make more confident statements concerning our measurements of endogenous juvenile hormone titres, and to relate our findings to the situation that occurs in adult females.

According to our *in vitro* observations corpus allatum activity increases quite markedly after the final moult in male *P. americana*. Highest mean rates of juvenile hormone synthesis occur between days 7 and 11, with the maximum rate on day 9. There is a scarcity of readily comparable information concerning the time course of sexual maturation in this species but we can speculate from the available data [3,23] that this interval coincides with the period of rapid maturation of the accessory sex glands. Experimental evidence of a possible role for juvenile hormone in any of the additional processes of reproductive maturation of adult male *P. americana* is likewise somewhat meagre. It is known, for instance, that the testis is programmed to be functional in last instar nymphs and that the accessory sex glands are only marginally active at this stage of development. This may be taken as evidence that the endocrine control (if any) of the processes involved in spermatogenesis and accessory gland protein synthesis are likely to be somewhat different [3]. Blaine and Dixon [24] claimed that the corpora allata were essential in the adult male *P. americana* for mushroom gland activity and for spermatophore protein production but we are unaware of any decisive studies concerning endocrine influences during the latter stages of testicular development. Evidence for other possible roles of the endocrine system in male *P. americana* is equivocal. Males of many cockroach species, including *Periplaneta* spp. possess modified dermal glands which are believed to produce an arrestant pheromone [2,25]. There is some evidence that the corpora allata may be involved in stimulating the production of this short-range sex attractant [2,26]. There are also conflicting observations which argue strongly for direct neuronal modulation [27].

Data on juvenile hormone biosynthetic rates for both sexes are available for only a very limited number of species. These include *Schistocerca gregaria* [28,29], *Diploptera punctata* [30,31], *Melanoplus sanguinipes* [32,33] and *Locusta migratoria* [34]. The average rates of hormone synthesis, at times when glands are active, tend to be much lower in the males than in females of the first three species. The same does not appear to be true for either *L. migratoria* or *P. americana* for radiochemical assays revealed no readily discernible quantitative differences between the sexes in these latter species.

Comparisons such as those mentioned above have become problematic in the light of our present observations which clearly indicate that in male *P. americana* the corpus allatum activities measured *in vitro* may bear little or no relation to their juvenile hormone biosynthetic activity immediately prior to being removed from the animal. The possibility that these discrepancies are being generated by a higher rate of juvenile hormone degradation in males has been largely discounted now that we have established that haemolymph juvenile hormone esterase levels in adult male and female *P. americana* are much the same (Dr T. S. Corbitt, personal communication). We are thus left with the alternative hypothesis, partially supported by our experimental observations with nerve-sectioned males, that male corpora allata are in some way inhibited *in vivo*, and that this inhibition of juvenile hormone synthesis is nullified *in vitro*.

Finally, we had hoped at the outset to show that male corpora allata could be more favorable than female glands for testing putative allatoactive factors. We reasoned that the activities of adult male corpora allata might be far less variable than those of females since males have no obvious requirement for a rapidly fluctuating titre of juvenile hormone. We conclude that, whilst this series of experiments has apparently served to demonstrate that the theory is at least partly well-founded, it has by coincidence also uncovered a further complication to the problem of corpus allatum control in cockroaches.

ACKNOWLEDGEMENTS

We would like to thank Dr T. S. Corbitt for juvenile hormone esterase determinations. We are also grateful to Dr J. Chambers for advice concerning HPLC separations.

REFERENCES

1. F. Engelmann, "The Physiology of Insect Reproduction", Pergamon, Oxford (1970).
2. R. H. Barth, The comparative physiology of reproductive processes in cockroaches. Part I. Mating behavior and its endocrine control, *in*: "Advances in Reproductive Physiology", A. McLaren, ed., Academic Press, NY, pp 167-207 (1968).
3. W. J. Bell and K. G. Adiyodi, Reproduction, *in*: "The American Cockroach", W. J. Bell and K. G. Adiyodi, eds., Chapman and Hall, London, pp 343-370 (1981).
4. G. E. Pratt and R. J. Weaver, Juvenile hormone biosynthesis by cultured cockroach corpora allata, *J. Endocrinol.*, 64:67 (1975).
5. R. J. Weaver, G. E. Pratt and J. R. Finney, Cyclic activity of the corpus allatum related to gonadotrophic cycles in adult female *Periplaneta americana*, *Experientia*, 31:597-598 (1975).
6. G. E. Pratt, R. J. Weaver and A. F. Hamnett, Continuous monitoring of juvenile hormone release by superfused corpora allata of *Periplaneta americana*, *in*: "The Juvenile Hormones", L. I. Gilbert, ed., Plenum, NY, pp 164-178 (1976).
7. A. F. Hamnett and G. E. Pratt, Use of automated capillary column radio gas chromatography in the identification of insect juvenile hormones, *J. Chromatog.*, 158:387-399 (1978).
8. P. J. Muller, P. Masner, K. H. Trautmann, M. Suchy and H. K. Wipf, The isolation and identification of juvenile hormone from cockroach corpora allata *in vitro*, *Life Sci.*, 15:915-921 (1975).
9. K. H. Dahm, G. Bhaskaran, M. G. Peter, P. D. Shirk, K. R. Seshan and H. Roller, On the identity of the juvenile hormone in insects, *in*: "The Juvenile Hormones", L. I. Gilbert, ed., Plenum, NY, pp 19-47 (1976).
10. R. J. Weaver and G. E. Pratt, The effect of enforced virginity and subsequent mating on the activity of the corpus allatum of *Periplaneta americana* measured *in vitro*, as related to changes in the rate of ovarian maturation, *Physiol. Entomol.*, 2:59-76 (1977).
11. B. Scharrer, Section of the nervi corporis cardiaci in *Leucophaea maderae* (orthoptera) *Anat.Rec.* 96:577 (1946).
12. R. J. Weaver, Comparable activities of left and right corpora allata consistent with humoral control of juvenile hormone biosynthesis in the cockroach, *Can. J. Zool.*, 57:343-345 (1979).
13. R. J. Weaver, Effects of food and water availability, and of NCA-1 section, upon juvenile hormone biosynthesis and oocyte development in adult female *Periplaneta americana*, *J. Insect Physiol.*, 30:831-838 (1984).
14. S. S. Tobe and B. Stay, Structure and regulation of the corpus allatum, *Adv. Insect Physiol.*, 18:305-432 (1985).
15. H. Kataoka, A. Toschi, J. P. Li, R. L. Carney, D. A. Schooley and S. J. Kramer, Identification of an allatotrophin from adult *Manduca sexta*, *Science*, 243:1481-1483 (1989).
16. B. J. Bergot, M. Ratcliff and D. A. Schooley, Method for quantitative determination of the four known juvenile hormones in insect tissue using gas chromatography-mass spectroscopy, *J. Chromatog.*, 204:321-244 (1981).
17. R. Feyereisen and S. S. Tobe, A rapid partition assay for routine analysis of juvenile hormone release by insect corpora allata, *Anal. Biochem.*, 111:372-375 (1981).
18. J. P. Edwards, J. Chambers, N. R. Price and J. P. G. Wilkins, Action of a juvenile hormone analogue on the activity of *Periplaneta americana* corpora allata *in vitro* and on juvenile hormone III levels *in vivo*, *Insect Biochem.*, 17:1115-1118 (1987).
19. G. E. Pratt and S. S. Tobe, Juvenile hormones radio-biosynthesized by corpora allata of adult female locusts *in vitro*, *Life Sci.*, 14:575-586 (1974).
20. G. E. Pratt, S. S. Tobe and R. J. Weaver, Relative oxygenase activities in juvenile hormone biosynthesis of corpora allata of an African locust (*Schistocerca gregaria*) and American cockroach (*Periplaneta americana*), *Experientia*, 31:120-122 (1975).
21. G. E. Pratt, R. C. Jennings and R. J. Weaver, The influence of a P_{450} inhibitor on methyl farnesoate levels in cockroach corpora allata, *in vitro*, *Insect Biochem.*, 14:609-614 (1984).
22. G. E. Pratt, S. S. Tobe, R. J. Weaver and J. R. Finney, Spontaneous synthesis and release of C_{16} juvenile hormone by isolated corpora allata of female locust *Schistocerca gregaria* and female cockroach *Periplaneta americana*, *Gen. Comp. Endocrinol.*, 26:478-484 (1975).

23. V. R. Vijayalekshmi and K. G. Adiyodi, Accessory sex glands and their secretions in male *Periplaneta americana* (L.).II. Secretory behavior and maturation of the mushroom gland complex, *Ind. J. Exp. Biol.*, 11:515-520 (1973).
24. W. D. Blaine and S. E. Dixon, The effect of juvenile hormone on the function of the accessory gland of the adult male cockroach *Periplaneta americana* (Orthoptera: Blattidae), *Can. Entomol.*, 105:1275-1280 (1973).
25. L. M. Roth and G. P. Dateo, A sex pheromone produced by males of the cockroach *Nauphoeta cinerea*, *J. Insect Physiol.*, 12:255-265 (1966).
26. R. H. Barth, The endocrine control of mating behavior in the cockroach *Byrsotria fumigata* (Guerin), *Gen. Comp. Endocrinol.*, 2:53-69 (1962).
27. H. B. Hartman and M. Suda, Pheromone production and mating behavior by allatectomized males of the cockroach, *Nauphoeta cinerea*, *J. Insect Physiol.*, 19:1417-1422 (1973).
28. S. S. Tobe and G. E. Pratt, Corpus allatum activity *in vitro* during ovarian maturation in the desert locust, *Schistocerca gregaria*, *J. Exp. Biol.*, 62:611-627 (1975).
29. L. I. Avruch and S. S. Tobe, Juvenile hormone biosynthesis by the corpora allata of the male desert locust *Schistocerca gregaria* during sexual maturation, *Can. J. Zool*, 56:2097-2102 (1978).
30. S. S. Tobe and B. Stay, Corpus allatum activity *in vitro* during the reproductive cycle of the viviparous cockroach, *Diploptera punctata* (Eschscholtz), *Gen. Comp. Endocrinol.*, 31:138-147 (1977).
31. S. S. Tobe, A. Musters and B. Stay, Corpus allatum function during sexual maturation of male *Diploptera punctata*, *Physiol. Entomol.*, 4:79-86 (1979).
32. A. R. McCaffery and V. A. McCaffery, Corpus allatum activity during overlapping cycles of oocyte growth in adult female *Melanoplus sanguinipes*, *J. Insect Physiol.*, 29:259-266 (1983).
33. G. A. Couche, C. Gillott, S. S. Tobe and R. Feyereisen, Juvenile hormone biosynthesis during sexual maturation and after mating in the adult male migratory grasshopper, *Melanoplus sanguinipes*, *Can. J. Zool.*, 63:2789-2792 (1985).
34. F. Couillaud, Influence of sexual organs on corpora allata biosynthetic activity in *Locusta migratoria*, *Physiol. Entomol.*, 11:397-403 (1986).

USE OF MICRODERIVATIZATION TECHNIQUES IN COMBINATION WITH THIN-LAYER CHROMATOGRAPHY, LIQUID CHROMATOGRAPHY AND GAS CHROMATOGRAPHY-MASS SPECTROMETRY FOR INVESTIGATION OF JUVENILE HORMONES AND RELATED COMPOUNDS

F. C. Baker, C. C. Reuter, L. W. Tsai, P. A. Brindle†, D. S. Richard*, S. S. Tobe** and D. A. Schooley††

Sandoz Crop Protection, Research Division
Palo Alto, CA 94304, USA
*Department of Biology, University of North Carolina, Chapel Hill, NC 27599-3280, USA, and
**Department of Zoology, University of Toronto
Toronto, Ontario M5S 1A1, Canada
†Present address: American Cyanamid
PO Box 400, Princeton, NJ 08540, USA
††Present address: Department of Biochemistry
University of Nevada, Reno, NV 89557, USA

SUMMARY

The qualitative juvenile hormone (JH) profile in lepidopteran species, as determined by both *in vivo* and *in vitro* techniques, is markedly different from that in insects from other orders, the majority of which contain only JH III. A gas chromatographic-mass spectrometric (GC-MS) method for quantification of JHs *in vivo* was adapted to study the mode of JH biosynthesis by corpora allata incubated *in vitro* with selected ^{14}C-radiolabeled substrates of high specific activity. We analyzed derivatives of JH using selected ion monitoring to measure ratios of the fragment ion and corresponding ions 2,4, and 6 mass units higher, allowing determination of dilution of the propionate unit which is incorporated as an intermediate. The results showed conclusively that the ethyl branch of JH II from *Manduca sexta* derives ultimately from metabolism of the branched-chain amino acids isoleucine and valine. A purification procedure for isolation of JHs from biological tissue was modified to allow isolation of JH acids and compounds structurally related to JH, including JH diols, methyl farnesoate (MF), farnesoic acid (FA), and farnesol. To allow analysis, JH acids were methylated and the resulting JHs were detected and quantified, as their d_3-methoxyhydrin derivatives, by GC-MS with selected ion monitoring (SIM). A detailed study of fifth stadium *Manduca sexta* showed that JH I and II acids predominate over the corresponding esters at the prepupal stage. GC-MS methods were devised for quantification of JH diols, MF, FA, and farnesol from arthropods. None of the latter compounds, nor any of the known JHs, could be detected in the hemipteran *Oncopeltus fasciatus*. Ring glands from a few dipteran species appear to secrete a new JH *in vitro*, and details are given for the isolation and analysis of this compound. Haemolymph from several crustaceans contain detectable levels of MF, and mandibular organs from crustaceans secrete MF and/or FA *in vitro*.

INTRODUCTION

A rigorous purification procedure for isolation of juvenile hormones (JHs) from insect tissue, in combination with a gas chromatographic-mass spectrometric (GC-MS) method for identification and quantification of JHs [1], has been utilized for investigation of JHs from numerous insect and several crustacean species (see [2,3] for reviews). Results from these studies plus those of other researchers using alternative physico-chemical methods, or *in vitro* JH biosynthesis techniques, have generally confirmed that the lepidopterans appear unique in their biogenesis of ethyl branched JHs (JH O, I, II, and 4-methyl JH I, see Figure 1

Chromatography and Isolation of Insect Hormones and Pheromones
Edited by A. R. McCaffery and I. D. Wilson
Plenum Press, New York, 1990

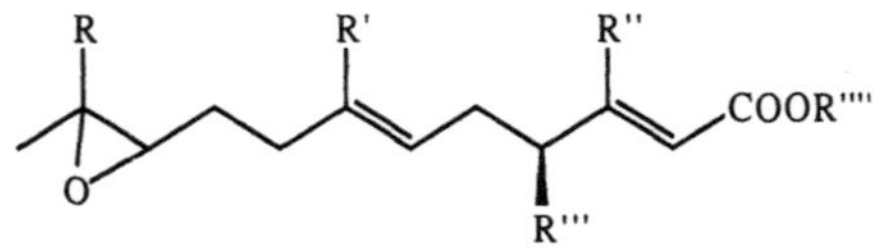

R=R'=R''=Me, R'''=H, (10*R*)-JH III
R=Et, R'=R''=Me, R'''=H, JH II
R=R'=Et, R''=Me, R'''=H, (10*R*,11*S*)-JH I
R=R'=R''=Et, R'''=H, JH O
R=R'=Et, R''=Me, R'''=Me, 4-MeJH I

R''''= Me for JH, R''''= H for JH acids

Fig. 1. Structures of known JHs and JH acids.

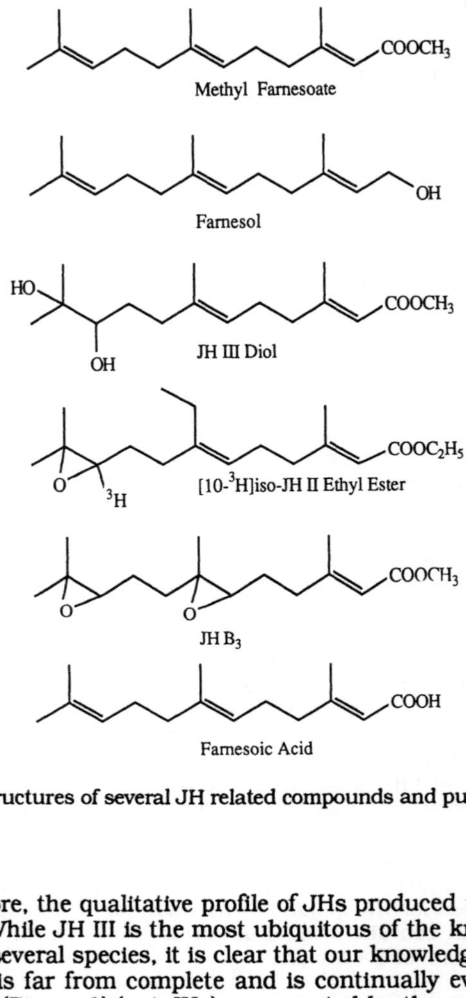

Fig. 2. Structures of several JH related compounds and putative JHs.

and Ref. [2]). Furthermore, the qualitative profile of JHs produced *in vitro* generally appears to reflect that *in vivo*. While JH III is the most ubiquitous of the known JHs, and occurs in high concentrations in several species, it is clear that our knowledge concerning JH identity in many insect orders is far from complete and is continually evolving. Thus, it is now apparent that JH acids (Figure 1) (not JHs) are secreted by the corpora allata of *Manduca sexta* at the prepupal stage of development [4,5]. None of the known JHs could be detected in *Oncopeltus fasciatus* [6] despite reports that this species contains high levels of a JH active component [7,8]. Similarly, several fly species have been investigated and found to contain none of the known JHs [2]. However, recently Richard et al. [9] have observed that ring glands from larval *Drosophila melanogaster* and other dipterans secrete a putative new JH (JHB$_3$; Figure 2). Other studies have indicated that corpora allata (CA) secrete JH III diol [10]. Several crustacean species have also been investigated with respect to JH content. To date, results suggest that the closely related compounds methyl farnesoate (MF) or farnesoic acid (FA) (Figure 2), but not JH, may perform a role in crustaceans similar to that of JH in insects [11-14].

We have developed physico-chemical methods to isolate and quantify the expanding family of JHs and related compounds, and details on the methodology are presented here. Furthermore, we have used our JH titre procedure to investigate the mode of JH biosynthesis, and showed that the branched-chain amino acids isoleucine and valine are the important precursors of propionate units crucial to ethyl-branched JH biosynthesis [15].

MATERIALS AND METHODS

Sources of arthropod tissues, rearing information, etc. are provided in references within the text. Various radiolabeled internal standards used for the studies were prepared earlier, as described in the reference below. The corresponding acids were prepared from [10-^{3}H]JH III and [10-^{3}H]MF [16] as described by Goodman and Adams [17] for JH III acid, except that acetic acid, rather than hydrochloric acid, was used to adjust the pH to 5.0 after the hydrolysis was complete (thus, avoiding potential hydrolysis of the acid-labile oxirane ring). Other chemicals and radiochemicals were obtained from commercial sources or prepared as described in the following methods.

JH titres in haemolymph or whole-body samples were determined by the method of Bergot et al. [1] with modifications described by Sliter et al. [18]. JH acids were isolated as described by Baker et al. [5,6], and after methylation with diazomethane, were processed as JHs. Procedures for *in vitro* biosynthesis of radiolabeled JHs from insect corpora allata incubated in medium 199 containing [U-^{14}C]branched-chain amino acids of high specific activity (87-95 atom % ^{14}C), are described by Brindle et al. [15,19]. The resulting JHs were purified by LC and converted to d_3-methoxyhydrin derivatives; the latter were purified by LC and analyzed by GC-MS with selected ion monitoring [15]. To detect methyl farnesoate,

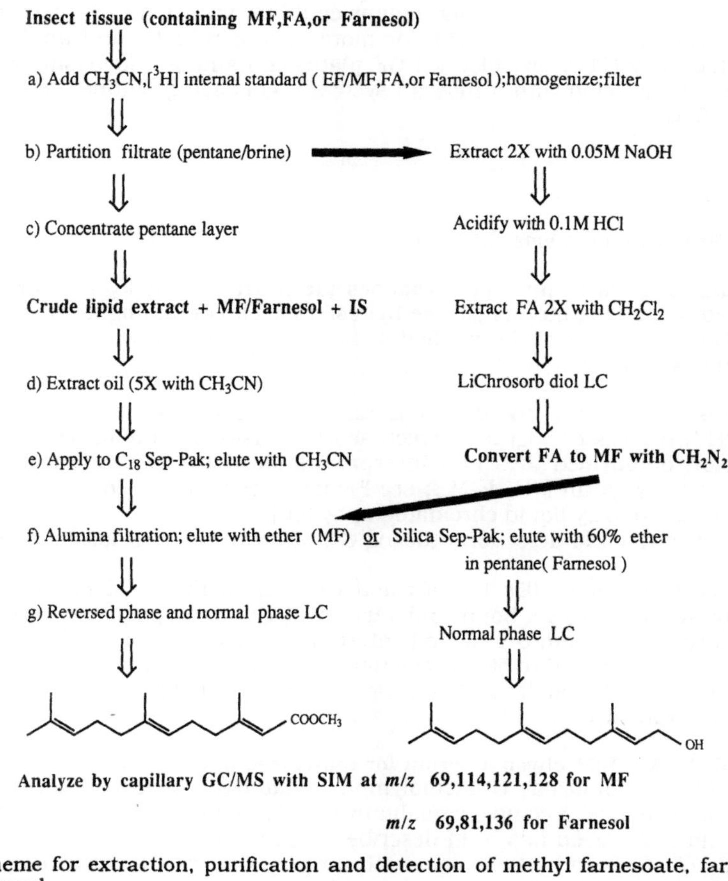

Fig. 3. Scheme for extraction, purification and detection of methyl farnesoate, farnesoic acid or farnesol.

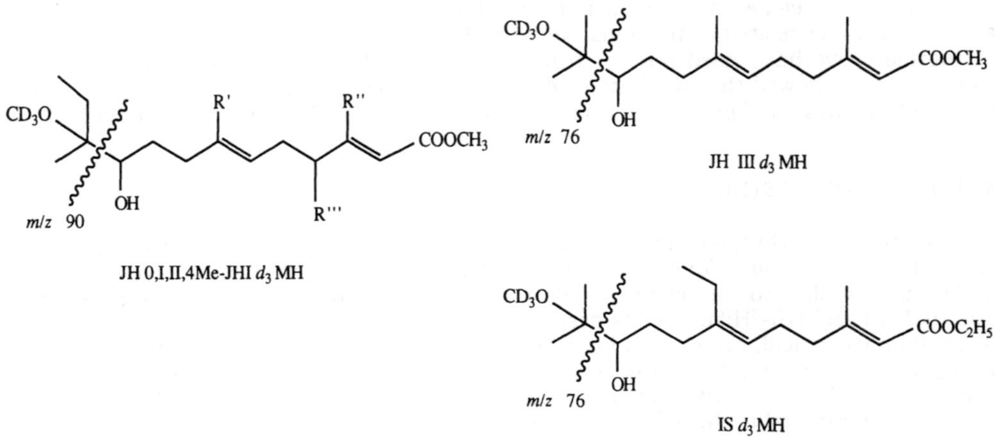

Fig. 4. Major electron impact mass spectrometric fragmentation pattern of JH deuteromethoxyhydrins (JH d3-MH).

farnesoic acid or farnesol (Figure 2) in insect or crustacean tissue, samples were processed essentially as described in Figure 3. Additional details are described by Laufer et al. [11] and Borst et al. [13] for MF; by Baker et al. [6] for farnesol and MF; by Tobe et al. [14] for MF and FA, and by Baker and Schooley [16] for preparation of radiolabelled samples. JH diols were extracted from insect tissue, converted to bis-trimethylsilyl (TMS) ethers and quantified by GC-MS [6]. Methyl *trans*-6,7;10,11-bisepoxy-3,7,11-trimethyl-(2*E*)-dodecenoate (JHB_3) was isolated from *in vitro* cultures of larval *D. melanogaster* ring glands stimulated with farnesoic acid [9]. JH III and related products from *in vitro* biosynthesis studies using insect corpora allata or crustacean mandibular organs could be extracted from medium by simple partition into iso-octane (or chloroform for more polar products) and analyzed by thin-layer chromatography (TLC) on silica gel GF plates containing a focussing zone. Radiolabeled zones were detected with a Bioscan system 200 imaging scanner equipped with an auto analyzer 1000.

RESULTS

JH/JH Acid Titre Determinations In Vivo

The recognized insect juvenile hormones (JHs) are a homologous series of acyclic sesquiterpenoid/homosesquiterpenoid methyl esters each containing a 10,11-epoxy group (Figure 1). Salient features of the method of Bergot et al. [1] for quantification of JHs in biological samples include:

a) A rigorous purification procedure and use of a radiolabeled internal standard ([10-^{3}H]iso-JH II ethyl ester, Figure 2) which allows measurement of recovery.
b) Conversion of purified JHs and internal standard to their relatively more polar deuteromethoxyhydrin (JH d_3-MH; see Figure 4) derivatives which are subsequently purified as a group by liquid chromatography (LC).
c) Final purification and detection utilizing GC-MS with selected ion monitoring (SIM).

During electron-impact (EI) MS of the JH deuteromethoxyhydrins, scission occurs predominantly at the C-10,11 bond and results in the base peak ion, m/z 90, for all derivatives except those from JH III and internal standard which produce m/z 76 ions (Figure 4). These ions carry 30 to 50% of the total ion current, and account for the sensitivity of the technique (typically, the limit of detection = 0.01 to 0.02 ng/g using ~5 g sample and ~5 ng of internal standard).

A typical GC-MS SIM chromatogram for standards and d_3-MH derivatives from JHs isolated from fifth stadium day 0 haemolymph of *Manduca sexta* is shown in Figure 5. Using the above technique, a complete qualitative and quantitative study of JH titres during the fifth stadium of *M. sexta* has been described [5], and has also been conducted for the entire life cycle of *M. sexta* by Schooley et al. (in preparation). Furthermore, by a slight modification of the extraction procedure, it was also possible to isolate JH acids which were

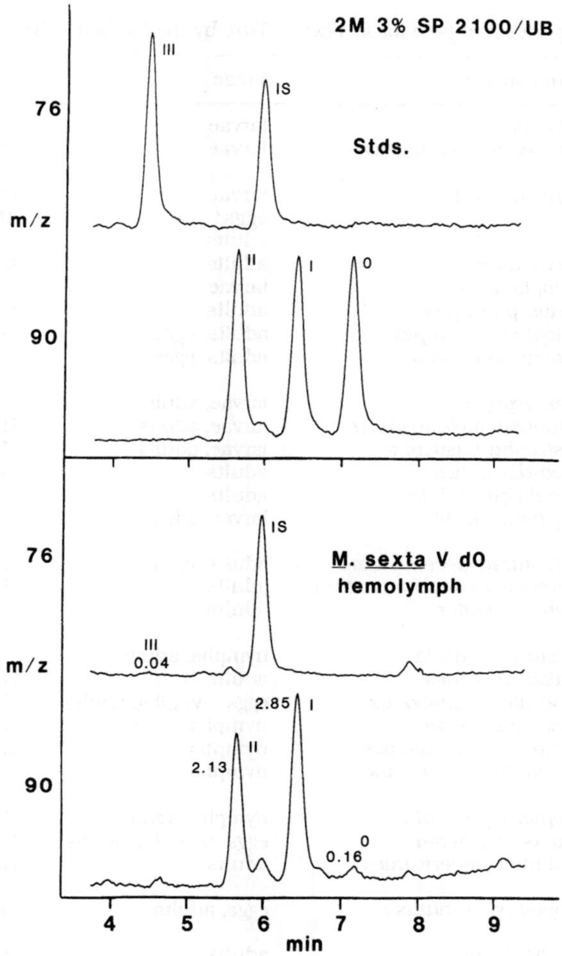

Fig. 5. Detection of JHs as d_3-MH derivatives by GC-MS with selected ion monitoring; above = JH d_3-MH standards; below = products from fifth stadium *M. sexta* day 0 haemolymph.

subsequently converted to their methyl esters (JHs) for purification, derivatization and detection [5,6]. JH I acid and JH II acid appeared relatively more predominant than the corresponding JHs during the prepupal stage of development of *M. sexta* [5].

The method of Bergot et al. [1] has been applied to JH titre analysis of no less than 2000 samples at Zoecon/Sandoz over the last decade. Although it is impossible here to list all of the important features of these studies, the qualitative results are summarized in Table 1. These clearly indicate that the ethyl branched JH homologs (JH 0, I,II, 4-Me JH I) are found only in lepidopteran species, whereas JH III is found in other insect orders and Lepidoptera.

In Vitro *JH Biosynthesis Studies: Source of Propionate Units for Biogenesis of Ethyl-Branched JHs*

Much of our interest in JH biosynthesis has been directed toward the origin of the unusual ethyl-branched JHs, and a rationale to explain their occurrence only in lepidopteran species. Earlier studies [20] proved that the C-11, 12 and 13 carbon atoms of JH II are derived from propionate (see Figure 6A). Recently, we attempted to identify the metabolic precursors of propionate important for ethyl-branched JH biosynthesis [15,19]. We were able to apply the method of Bergot et al. [1] to analyze the JHs secreted *in vitro* by corpora allata incubated with various radiolabeled precursors. The origin of each of the carbon atoms in JH II and III (shown as their d_3-MH derivatives) is depicted in Figure 6. Clearly, if ^{14}C-labeled propionate from a highly enriched precursor (specifically or uniformly labeled) was incorporated into the JH II skeleton, an increase in mass of the m/z 90 fragment would be observed upon analysis. Using this technique, we were able to show

Table 1. Species Analyzed for *In Vivo* JH Titre by the Method of Bergot et al (ref. 1)

Order	Genus/Species	Stage	JHs detected
Hymenoptera	*Chelonus sp.*	larvae	III
"	*Biosteres longicaudatus*	larvae	III
Lepidoptera	*Manduca sexta*	larvae	O, I, II, III (and acids)
"	" "	eggs	4Me-JH I, O, I, II
"	" "	adults	I, II, III
"	*Attacus atlas*	adults	I, II
"	*Trichoplusia ni*	larvae	II
"	*Danaus plexippus*	adults	I, II, III (and acids)
"	*Hyalophora cecropia*	adults, eggs	O, I, II
"	*Heliothis virescens*	adults, eggs	O, I, II
Diptera	*Aedes aegypti*	larvae, adults	III
"	*Drosophila melanogaster*	larvae, adults	III
"	*Anastrepha suspensa*	larvae, adults	III
"	*Musca domestica*	adults	none
"	*Sarcophaga bullata*	adults	none
"	*Calliphora vicina*	larvae, adults	none
Coleoptera	*Leptinotarsa decemlineata*	adults, eggs	III
"	*Diabrotica undecempunctata*	adults	III
"	*Tenebrio molitor*	adults	III
Orthoptera	*Locusta migratoria*	nymphs, adults	III
"	*Taeniopoda eques*	adults	III
"	*Teleogryllus commodus*	eggs, nymphs, adults	III
"	*Schistocerca vaga*	nymphs	III
"	*Melanoplus sanguinipes*	nymphs	III
"	*Gastrimargus africanus*	nymphs	III
Dictyoptera	*Diploptera punctata*	nymphs, adults	III
"	*Nauphoeta cinerea*	eggs, nymphs, adults	III
"	*Periplaneta americana*	adults	III
Hemiptera	*Oncopeltus fasciatus*	eggs, adults	none
Homoptera	*Megoura viciae*	adults	III
"	*Aphis fabae*	adults	III
Thysanura	*Thermobia domestica*	eggs, adults	III

* For references to individual studies see reference 2.

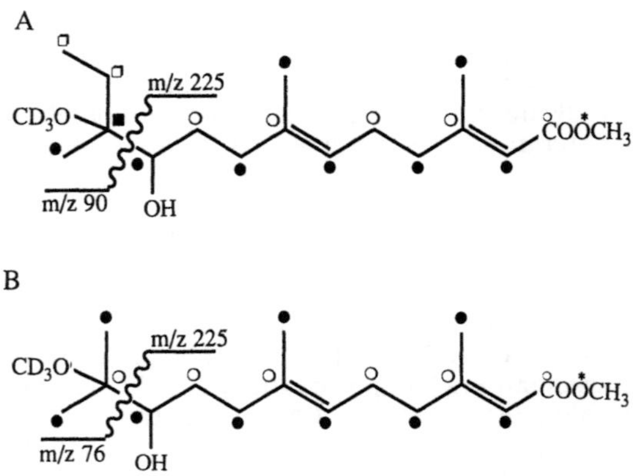

Fig. 6. Labeling pattern of A) JH II-d_3-MH and B) JH III-d_3-MH from JH biosynthesis using: [1-^{14}C]acetate (○); [2-^{14}C]acetate (●); ;1- ^{14}C]propionate (■); C-2, C-3 of propionate (□); L-[*methyl*-^{3}H]methionine (*). Mass/charge ratios m/z 76 and m/z 90 refer to unlabeled fragments from EI mass spectrometry (see Ref. [15]).

that [U-^{14}C]isoleucine and [U-^{14}C]valine of very high specific activity (> 59 mCi/milliatom of carbon) were metabolized to [U-^{14}C]propionate which subsequently labelled the C-11, 12 and 13 carbons of the secreted JH II (m/z 96, Figure 7). The extent of dilution of the [U-^{14}C]isoleucine, [U-^{14}C]valine, or their metabolites could be determined from the ratios of abundancies of the m/z 90, 92, 94 and 96 ions. These were monitored by SIM rather than from a mass spectrum due to the small amounts of samples.

Some enrichment of JH III was observed due to incorporation of acetate units, also derived from metabolism of isoleucine. However, the acetate was much more diluted than propionate, because of the naturally large number of acetate precursors and the high concentration of unlabeled acetate in the medium. Subsequent experiments showed that corpora allata from several lepidopteran species used branched-chain amino acids (BCAA) for JH biosynthesis while non-lepidopteran corpora allata failed to catabolize BCAA and consequently did not utilize these substrates for JH biosynthesis [19].

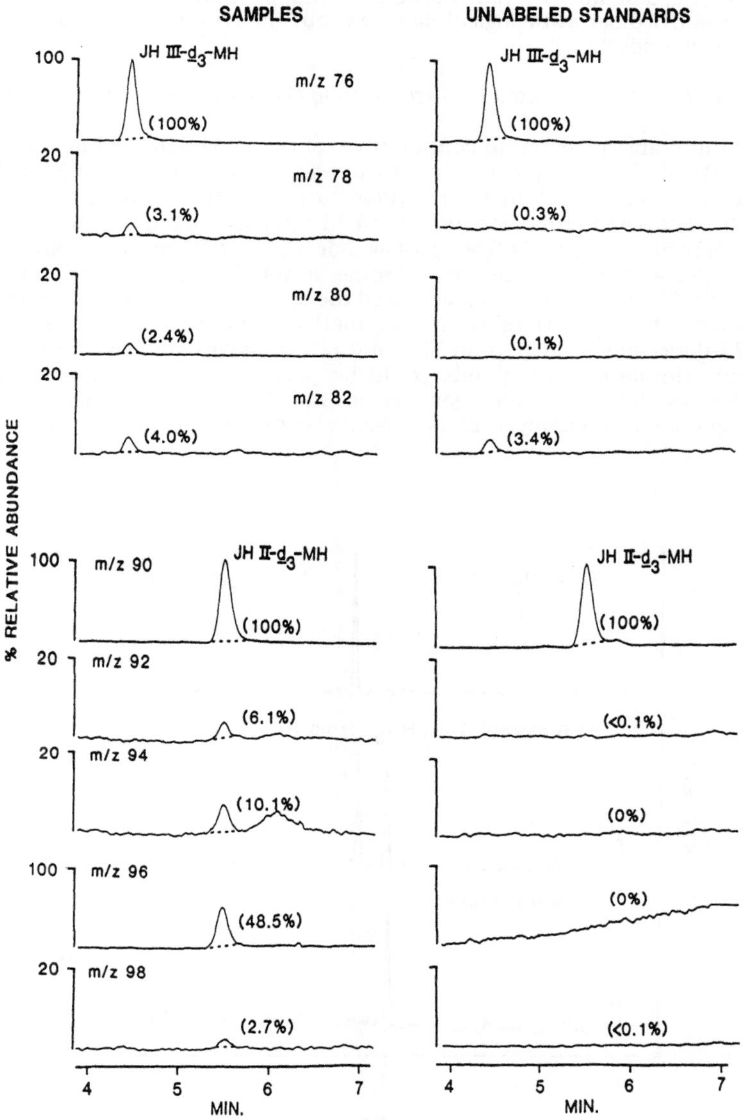

Fig. 7. GC-MS selected ion monitoring chromatograms of JH III-d_3-MH and JH II-d_3-MHfrom JHs secreted by CC-CA from adult female *M. sexta* incubated in medium 199 containing [U-^{14}C]isoleucine (87-95 atom % ^{14}C).

Search for JHs and Related Compounds in Oncopeltus fasciatus

As indicated in Table 1, several species appeared to contain none of the known JHs when analyzed. Data for the hemipteran, *Oncopeltus fasciatus*, were particularly intriguing since very high titres of JH (via bioassay) had previously been reported [7,8]. We examined eggs and adults from three different strains of *Oncopeltus fasciatus*, but detected none of the known JHs or JH acids at significant levels [6]. We considered that this species might instead contain one of several materials related to JH with various degrees of biological activity including farnesol, MF, or even JH III diol. Subsequently, we modified the isolation procedure for JHs [1] to allow purification of these JH related compounds (Figure 3; see Ref. [6] for the JH diol purification method, and for additional steps such as low temperature precipitation of lipids in oily samples, e.g., *O. fasciatus* eggs). GC-MS/SIM methods employing radioisotope dilution techniques were used to detect and quantify MF (m/z 69, 114, 121) or farnesol (m/z 69, 81, 136) in *O. fasciatus*, but none of these compounds were observed above 1 ng/ml and could not account for the high level of JH biological activity reported by other groups. We quantified JH diols as their 10,11 *bis*-TMS ether derivatives, again using GC-MS with SIM (m/z 131, 145); however, we used [10-^{3}H]iso-JH II ethyl ester 10,11 diol as internal standard. This procedure appears satisfactory for quantification of JH diols in biological samples, but no JH diols could be detected (< 0.1 ng/g) in *O. fasciatus* adults.

Identification of a New JH Secreted In Vitro *by Ring Glands from* Drosophila melanogaster

Several dipterans appeared to contain none of the known JHs while others contained detectable levels of JH III (Table 1). In *Drosophila melanogaster*, adults contained the highest concentrations of JH III while larvae (pre-wandering) contained only very low amounts [18]. Post-wandering larvae contained intermediate quantities of JH III (~0.2 ng/g). Surprisingly, ring glands from post-wandering *D. melanogaster* larave or puparia, incubated *in vitro* with radiolabeled methionine secreted a radiolabeled compound more polar than JH III [9]. Farnesoic acid stimulated the amounts of this secreted product, which had chromatographic properties similar to methyl *trans*-6,7;10,11-bis-epoxy-3,7,11-trimethyl-(2*E*)-dodecenoate (JHB_3) on TLC and LC. Biosynthetic, radiolabeled JHB_3 from farnesoic acid stimulated ring glands could be partially purified by sequential normal phase LC using two different solvent systems (see Ref. [9]), and had the same k' as synthetic JHB_3. Subsequent GC-MS analysis showed several components (Figure 8), one of which had

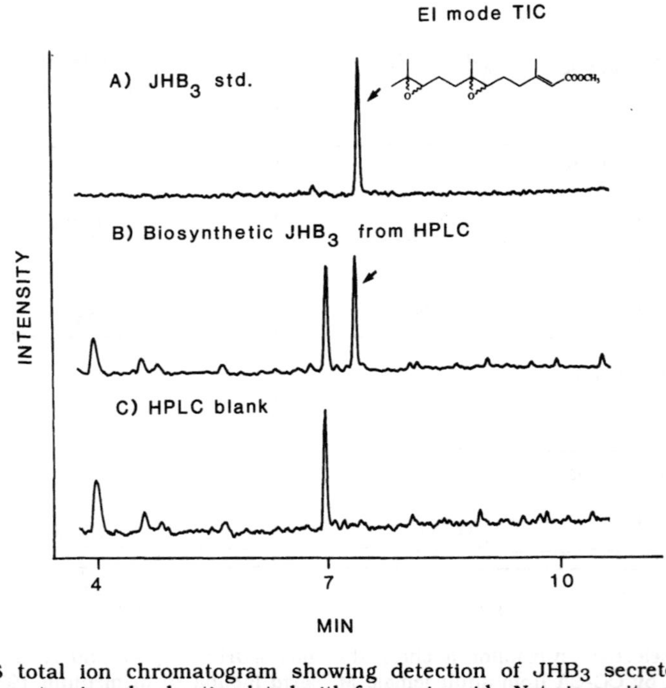

Fig. 8. GC-MS total ion chromatogram showing detection of JHB_3 secreted *in vitro* by *D. melanogaster* ring glands stimulated with farnesoic acid. Note impurity peaks also present in processed blank (see Ref. [9]).

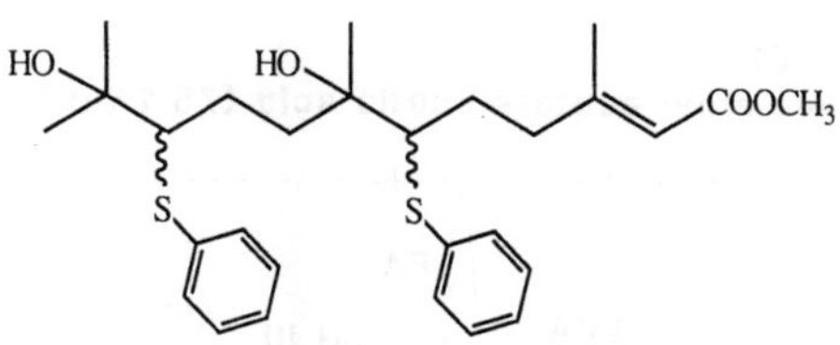

Fig. 9. Structure of methyl 6,10-bis(thiophenyl)-7,11-dihydroxyfarnesoate.

the same retention time (R_t 7.41 min) and same electron impact mass spectrum (not shown) as synthetic JHB_3. A closely eluting peak was an impurity which was also present in a blank processed through the same LC purification procedure (see Figure 8). Conversion of the biosynthetic product to several derivatives, including methyl 6,10-bis(thiophenyl)-7,11-dihydroxyfarnesoate (Figure 9), and a mixture of acid hydrolysis products, further confirmed the identity of JHB_3 [9]. Methyl 6,10-bis(thiophenyl)-7,11-dihydroxyfarnesoate prepared from synthetic 10(*RS*)-JHB_3 was separable by normal phase LC into two diastereomers, each of which is a DL pair. When the [^{3}H]-radiolabeled biosynthetic derivative was similarly analyzed, tritium was detected only in the fast eluting peak, indicating that a maximum of two of the four possible JHB_3 isomers were biosynthesized. Larval ring glands from other cyclorrhaphous dipterans also appeared to secrete JHB_3. However, adult *D. melanogaster* corpora allata produced both JHB_3 and JH III, the former predominating [9].

Identification of JH Related Compounds in Crustaceans

The JH titre method of Bergot et al. [1] and the modified method for JH acid quantification [6] was applied to analysis of haemolymph from the spider crab *Libinia emarginata* [11]. Low titres of JH III and JH III acid were detected. However, further investigation suggested that these low concentrations of JH III/JH III acid were artifacts of the work-up and analytical procedures and probably were derived from methyl farnesoate which was present at much greater levels (100-1000 x higher than corresponding JH acids/JHs). Methyl (2*E*,6*E*)-farnesoate was detected in haemolymph samples (processed essentially as described in Figure 3) using methyl [10-^{3}H]farnesoate as internal standard and analysis by GC-MS with SIM at m/z 69, 114 and 121. *L. emarginata* mandibular organs, incubated *in vitro* with [*methyl*-^{3}H]methionine, secreted radiolabeled methyl farnesoate and JH III (analyzed by normal phase and reversed-phase LC, and GC-MS) in a ratio of ~400:1. Analysis of *in vitro* biosynthesized JH III for chirality, by conversion to the 10,11-diol and subsequently to the 10-(+)-α-methoxy-α-trifluoromethylphenyl acetate (MTP) derivative, indicated that only 59% of the radiolabeled material had the 10*R* configuration while 41% was 10*S*. This absence of stereochemical purity again implies that the JH III detected is perhaps an artifact formed from non-biological oxidation of MF.

Significant quantities of methyl farnesoate have been detected in other crustacean species, either *in vivo*, or secreted *in vitro* by mandibular organs [13]. A lesser amount of MF was secreted *in vitro* by mandibular organs of the lobster *Homarus americanus* compared to those of *L. emarginata*, and the concentration of MF in the haemolymph from *H. americanus* was also relatively much lower. To facilitate accurate quantification of MF at low concentrations *in vivo*, the analytical method was modified to use ethyl [10-^{3}H]farnesoate as internal standard [13]. Only 0.1 ng MF/g was detected in second or third stage *H. americanus* larvae. Interestingly, *H. americanus* larvae exposed to seawater containing 10^{-6} M MF took a small, but significant, increase in the time for metamorphosis compared to controls (22.0 ± 1.3 vs. 20.8 ± 1.5 days, respectively). Furthermore, rates of MF secretion by mandibular organs from *L. emarginata* appeared to vary according to stages of ovarian growth, with highest rates found in adult females containing brown embryos when oocyte development and vitellogenesis were occurring [11].

Taken together, the latter results are consistent with MF performing a JH-like role in crustacean species. However, *in vitro* studies [13] indicated that several other radiolabeled products were also secreted by mandibular organs incubated in culture medium containing [*methyl*-labelled]methionine. While these products presumably contain a methyl ester moiety, their further identification was not pursued. It is also possible that additional important products are secreted by the mandibular organs. If these products lack a methyl ester group, they would remain undetected in methods relying on labelling by [*methyl*-

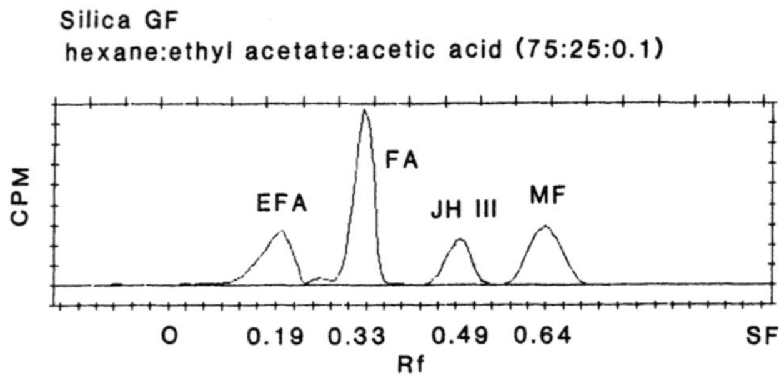

Fig. 10. Radiochromatogram scan of epoxyfarnesoic acid (JH III acid), farnesoic acid, JH III and methyl farnesoate separated by silica GF, TLC.

radiolabeled]methionine. Subsequently, Tobe et al. [21] found that on an hourly basis < 4% of the MF formed by mandibular organs of the mud crab, *Scylla serrata*, is actually secreted. By means of double labeled studies with [2-^{14}C]acetate and [*methyl*-^{3}H]methionine, Tobe et al. [14] showed that ^{14}C-labeled farnesoic acid was a major secretory product of the mandibular organs. The product was identified by TLC, HPLC and GC-MS (following methylation with diazomethane). Thin-layer chromatography is still perhaps the most useful method for preliminary analysis of a mixture comprising radiolabeled JH-related compounds of widely different polarity. Results from analysis of a mixture of radiolabeled epoxyfarnesoic acid, farnesoic acid, JH III, and MF are shown in Figure 10.

To analyze for farnesoic acid *in vivo* (e.g., in haemolymph samples), it was necessary to further modify the extraction and purification procedure for MF to include aqueous base extraction and subsequent acidification, to separate FA from MF (see Figure 3). [10-^{3}H]Farnesoic acid was used as internal standard in what was essentially an isotope-dilution method. For the end point analysis in this method, the corresponding methyl ester was analyzed using GC-MS with SIM [11]. In analyses of haemolymph from male *S. serrata* we detected MF at 2.1 ng/ml while neither JH III nor farnesoic acid were detectable [14].

DISCUSSION

Several physico-chemical methods are currently available for the extraction and quantification of JHs from miscellaneous insect tissues (see, for example, [1,22,23]; for a Review see Ref. [2]). The method of Bergot et al. [1] has proved convenient for analysis of a wide range and large number of insect species. From these studies and others (e.g., [24]), a pattern has emerged which suggests that only lepidopterans contain the rather unusual ethyl-branched JHs in addition to JH III. Insects of other orders contain only JH III or, in some cases, none of the known JHs. Besides measurement of physiological JH titres, this method has been used to study the effect of miscellaneous treatments on insect JH titre. These include treatment with anti-juvenile hormones or juvenoids [25,26], parasitization [27,28], and effects of surgical manipulation [29]. Furthermore, the technique has proved useful for confirming that the source of propionate units, crucial for biogenesis of ethyl-branched JHs, stems from the branched-chain amino acids isoleucine and valine [15].

Using only simple modifications of the JH extraction procedure of Bergot et al. [1], and applying similar principles of GC-MS with selected ion monitoring to analyze and quantify products, it has been possible to detect miscellaneous JH precursors or metabolites in arthropod tissue (*in vivo* or *in vitro*). An additional extraction step using a polar solvent allowed us to isolate JH acids after initial extraction of JHs. We have used this technique to quantify JHs and JH acids during the entire fifth stadium of *M. sexta*, and were able to show the appearance of JH acids prior to JHs at the prepupal stage of development [5]. The latter results confirmed the observations of Bhaskaran et al. [4] that only JH acids are secreted by corpora allata of prepupal *M. sexta* due to absence of the methyl transferase enzyme, and that the JH acids are subsequently methylated in the imaginal discs. Earlier, we had demonstrated that CC-CA from prepupal *M. sexta* secrete substantial amounts of JH acids *in vitro* (F. C. Baker, S. J. Kramer, D. A. Schooley and L. W. Tsai, unpublished results). Taken together, these results suggest that the reported *in vitro* secretion of JH by CC-CA from prepupal *M. sexta* measured by an RIA method [30] is actually due to detection of JH acids.

The absence of known JHs in several strains of *O. fasciatus* [6] was perplexing in the light of the reported high level of JH biological activity in this species [7,8]. This prompted us to investigate JH acids in *O. fasciatus*, but we were unable to consistently detect levels above 0.01 to 0.02 ng/g. We developed techniques to identify and quantify additional JH related compounds, including JH 10,11 diol, farnesol or methyl farnesoate. While the techniques appear feasible for detection of the latter compounds, we were unable to detect them in *O. fasciatus*, and our conclusions favour the presence of an as yet undiscovered JH or minute levels of the known JHs in this species. Bowers' group has claimed that CA from several hemipteran species secrete MF [31] or JH III [32]. However, we investigated *in vitro* JH biosynthesis by *O. fasciatus*, but were unable to detect JH III or MF [6].

As indicated in Table 1, JH III has been detected in several dipteran species, including *D. melanogaster*, while others did not appear to contain any of the known JHs. Surprisingly, *in vitro* organ culture of ring glands from *D. melanogaster* resulted in secretion of a lipophilic methyl ester which had chromatographic properties different from known JHs. Secretion of this product was greatly stimulated by addition of farnesoic acid to the medium, a result which gave a strong clue as to potential identity of the unknown and which facilitated biosynthesis of sufficient product for elucidation of the JHB_3 structure [9]. This use of the *in vitro* organ culture technique in the isolation of a previously undiscovered JH is reminiscent of the discovery of JH III by Judy et al. [33]. Additional confirmation of the ubiquity of JHB_3 will await development of methods to quantify this product *in vivo* which may be a challenge because of the instability and reactivity of the 6,7;10,11 bisepoxide. Meanwhile, JHB_3 appears to be secreted by ring glands from other cyclorrhaphous dipteran larvae [9] and *Phormia regina* corpora allata secrete an unidentified product with chromatographic properties similar to JHB_3 [34].

The many similarities between insect and crustacean development have prompted comparison of their endocrinology. The moderate titres of methyl farnesoate detected in several crustaceans are particularly intriguing since this product requires only a single epoxidation step for conversion to JH III. However, the observed JH III detected at a low level in *L. emarginata* (*in vivo* or *in vitro*) was not enantiomerically pure and was probably an artifact of the isolation procedure [11]. The possibility exists that MF is a crustacean juvenile hormone in its own right and there is some modest evidence for this [11-13]. Furthermore, very high titres of methyl farnesoate are known to occur in some insect species (e.g., *Nauphoeta cinerea* embryos) [34], although it is doubtful that this compound *per se* plays an important regulatory role in insects. While the number of crustaceans in which MF has been detected continues to expand, it is apparent that the picture is not quite as simple as first perceived. The mandibular organ of the crab *S. serrata* appears to sequester methyl farnesoate and instead secretes predominantly farnesoic acid [14,21]. However, very little farnesoic acid, but substantial MF, could be detected in *S. serrata* haemolymph [14]. Additional studies will be required to determine what the important secretory product is and whether farnesoic acid can be methylated in other tissues.

ACKNOWLEDGEMENTS

We thank the National Science Foundation (grant DMB-8518299) for partial financial support of F. C. Baker, C. C. Reuter, L. W. Tsai, P. A. Brindle and D. A. Schooley; the National Institute of Health (grants DK35347 and DK30118 to Prof. L. I. Gilbert, UNC) for partial financial support of D. S. Richard, and we thank our collaborators involved in many of the studies listed in Table 1.

REFERENCES

1. B. J. Bergot, M. Ratcliff and D. A. Schooley, Method for quantitative determination of the four known juvenile hormones in insect tissue using gas chromatography-mass spectroscopy, *J. Chromatog.*, 204:231-244 (1981).
2. F. C. Baker, Techniques for identification and quantification of juvenile hormones and related compounds in arthropods, *in*: "Morphogenetic Hormones of Arthropods/Recent Advances in Comparative Morphology, Physiology and Development of Arthropods", A. P. Gupta, ed., Rutgers University Press, New Brunswick, pp. 389-453, in press (1990).
3. D. A. Schooley, F. C.Baker, L. W. Tsai, C. A. Miller and G. C. Jamieson, Juvenile hormones, 0, I and II exist only in Lepidoptera, *in*: "Biosynthesis, Metabolism and Mode of Action of Invertebrate Hormones", J. Hoffmann and M. Porchet, eds., Springer-Verlag, Berlin, pp. 373-383 (1984).

4. G. Bhaskaran, S. P. Sparagana, P. Barrera and K. H. Dahm, Change in corpus allatum function during metamorphosis of the tobacco hornworm *Manduca sexta*: regulation at the terminal step in juvenile hormone biosynthesis, *Arch. Insect Biochem Physiol.*, 3:321-338 (1986).
5. F. C. Baker, L. W. Tsai, C. C. Reuter and D. A. Schooley, *In vivo* fluctuation of JH, JH acid, and ecdysteroid titer, and JH esterase activity, during development of fifth stadium *Manduca sexta*, *Insect Biochem.*, 17:989-996 (1987).
6. F. C. Baker, L. W. Tsai, C. C. Reuter and D. A. Schooley, The absence of significant levels of the known juvenile hormones and related compounds in the milkweed bug, *Oncopeltus fasciatus*, *Insect Biochem.*, 18:453-462 (1988).
7. A. Dorn, Structure and function of the embryonic corpus allatum of *Oncopeltus fasciatus* Dallas (Insecta, Heteroptera), *Verhandlungen der Deutschen Zoologischen Gesellschaft*, pp. 85-89 (1975).
8. M. A. Rankin and L. M. Riddiford, Significance of hemolymph juvenile hormone titer changes in timing of migration and reproduction in adult *Oncopeltus fasciatus*, *J. Insect Physiol.*, 24:31-38 (1978).
9. D. S. Richard, S. W. Applebaum, T. J. Sliter, F. C. Baker, D. A. Schooley, C. C. Reuter, V. C. Henrich and L. I. Gilbert, Juvenile hormone bisepoxide biosynthesis *in vitro* by the ring of *Drosophila melanogaster*: a putative juvenile hormone in the higher Diptera, *Proc. Nat. Acad. Sci. USA*, 86:1421-1425 (1989).
10. M. Gadot, A. Goldman, M. Cojocaru and S. W. Applebaum, The intrinsic synthesis of juvenile hormone-III diol by locust corpora allata *in vitro*, *Mol. Cell. Endocrinol.*, 49:99-107 (1987).
11. H. Laufer, D. Borst, F. C. Baker, C. Carrasco, M. Sinkus, C. C. Reuter, L. W. Tsai and D. A. Schooley, Identification of a juvenile hormone-like compound in a crustacean, *Science*, 235:202-205 (1987).
12. H. Laufer, M. Landau, E. Homola and D. W. Borst, Methyl farnesoate: its site of synthesis and regulation of secretion in a juvenile crustacean, *Insect Biochem*, 17:1129-1131 (1987).
13. D. W. Borst, H. Laufer, M. Landau, E. S. Chang, W. A. Hertz, F. C. Baker and D. A. Schooley, Methyl farnesoate and its role in crustacean reproduction and development, *Insect Biochem.*, 17:1123-1127 (1987).
14. S. S. Tobe, D. A. Young, H. W. Khoo and F. C. Baker, Farnesoic acid as a major product of release from crustacean mandibular organs *in vitro*, *J. Exp. Zool.*, 249:165-171 (1989).
15. P. A. Brindle, F. C. Baker, L. W. Tsai, C. C. Reuter and D. A. Schooley, Sources of propionate for the biogenesis of ethyl-branched insect juvenile hormones: role of isoleucine and valine, *Proc. Nat. Acad. Sci. USA*, 84:7906-7910 (1987).
16. F. C. Baker and D. A. Schooley, Preparation of (10*R*,S)-[10-^{3}H]juvenile hormone III and (10*R*,*S*,11*S*,*R*)-[10-^{3}H]juvenile hormone 0; conversion of [10-^{3}H]juvenile hormone III to methyl (2*E*, 6*E*)-[10-^{3}H]farnesoate and (2*E*,6*E*)-[10-^{3}H]farnesol, *J. Labeled Comp. Radiopharmaceut.*, 5:533-543 (1986).
17. W. G. Goodman and B. Adams, Semipreparative synthesis and purification of juvenile hormone acids by high-performance liquid chromatography, *J. Chromatog.*, 294:447-451 (1984).
18. T. J. Sliter, B. J. Sedlak, F. C. Baker and D. A. Schooley, Juvenile hormone in *Drosophila melanogaster*.Identification and titer determination during development, *Insect Biochem.*, 17:161-165 (1987).
19. P. A. Brindle, D. A. Schooley, L. W. Tsai and F. C. Baker, Comparative metabolism of branched-chain amino acids to precursors of juvenile hormone biogenesis in corpora allata of lepidopterous versus non-lepidopterous insects, *J. Biol. Chem.*, 263:10653-10657 (1988).
20. D. A. Schooley, K. J. Judy, B. J. Bergot, M. S. Hall and J. Siddall, Biosythesis of the juvenile hormones of *Manduca sexta*: labeling pattern from mevalonate, propionate and acetate, *Proc. Nat. Acad. Sci. USA*, 70:2921-2925 (1973).
21. S. S. Tobe, D. A. Young and H. W. Khoo, Production of methyl farnesoate by the mandibular organs of the mud crab, *Scylla serrata*: validation of a radiochemical assay, *Gen. Comp. Endocrinol.*, 73:342-353 (1989).
22. H. Rembold and B. Lackner, Convenient method for the determination of picomole amounts of juvenile hormone, *J. Chromatog.*, 323:355-361 (1985).
23. B. Mauchamp, R. Lafont, M. Hardy and D. Jourdain, Analysis of insect juvenile hormones by gas chromatography mass spectrometry: problems of sample preparation and choice of detection procedure, *Biomed. Mass Spec.*, 6:276-281 (1979).
24. K. H. Trautmann, P. Masner, A. Schuler, M. Suchy and H.-K. Wipf, Evidence of the juvenile hormone methyl(2*E*,6*E*)-10,11-epoxy-3,7,11-trimethyl-2,6-dodecadieno-

ate JH-3) in insects of four orders, *Zeitschrift fur Naturforschung.*, 29c:757-759 (1974).

25. J. P. Edward, B. J. Bergot and G. B. Staal, Effects of three compounds with anti-juvenile hormone levels in the tobacco hornworm, *Manduca sexta, Insect Physiol.*, 29:83-89 (1983).
26. F. C. Baker, C. A. Miller, L. W. Tsai, G. C. Jamieson, D. C. Cerf and D. A. Schooley, The effect of juvenoids, anti juvenile hormone agents, and several intermediates of juvenile hormone biosynthesis on the *in vivo* juvenile hormone levels in *Manduca sexta* larvae, *Insect Biochem.*, 16:741-747 (1986).
27. P. O. Lawrence, F. C. Baker, L. W. Tsai, C. A. Miller, D. A. Schooley and L. Geddes, JH III levels in superparasitized pharate pupae of *Anastrepha suspensa* (Diptera: tephritidae) and in larvae of the parasitic wasp *Biosteres longicaudatus* (Hymenoptera: braconidae), *Arch. Insect Biochem. Physiol.*, in press (1990).
28. G. Jones, T. Hanzlik, B. D. Hammock, D. A. Schooley, C. A. Miller, L. W. Tsai and F. C. Baker, The juvenile hormone titer during the penultimate and ultimate larval stadia of *Trichoplusia ni.*, *J. Insect. Physiol.*, in press (1990).
29. W. Loher, D. A. Schooley and F. C. Baker, Influence of the ovaries on JH titer in *Teleogryllus commodus, Insect Biochem.*, 17:1099-1102 (1987).
30. N. A. Granger, S. M. Niemiec, L. I. Gilbert and W. E. Bollenbacher, Juvenile hormone synthesis *in vitro* by larval and pupal corpora allata of *Manduca sexta, Mol. Cell. Endocrinol.*, 28:587-604 (1982).
31. M. F. Feldlaufer, W. S. Bowers, D. M. Soderlund and P. H. Evans, Biosynthesis of the sesquiterpenoid skeleton of juvenile hormone 3 by *Dysdercus fasciatus* corpora allata *in vitro, J. Exp. Zool.*, 223:295-298 (1982).
32. W. S. Bowers, P. A. Marsella and P. H. Evans, Identification of an hemipteran juvenile hormone: *in vitro* biosynthesis of JH III by *Dysdercus fasciatus, J. Exp. Zool.*, 228:555-559 (1983).
33. K. J. Judy, D. A. Schooley, L. L. Dunham, M. S. Hall, B. J. Bergot and J. Siddall, Isolation, structure and absolute configuration of a new natural insect juvenile hormone from *Manduca sexta, Proc. Nat. Acad. Sci. USA*, 70:1509-1513 (1973).
34. M.-A. Liu, G. L. Jones, J. G. Stoffolano, Jr. and C.-M. Yin, Conditions for estimation of corpus allatum activity in the blowfly, *Phormia regina, in vitro, Physiol Entomol.*, 13:69-79 (1988).
35. B. Lanzrein, H. Imboden, C. Bürgin, E. Brüning and H. Gfeller, On titers, origin and functions of juvenile hormone III, methyl farnesoate, and ecdysteroids in embryonic development of the ovoviviparous cockroach *Nauphoeta cinerea, in*: "Biosynthesis, Metaboism and Mode of Action of Invertebrate Hormones", J. Hoffmann and M. Porchet, eds., Springer-Verlag, Berlin, pp. 454-465 (1984).

THE PRECOCENE ANTIJUVENILE HORMONES (ALLATOTOXINS): A CASE HISTORY IN INSECT TOXICOLOGY

G. T. Brooks and A. R. McCaffery

School of Animal and Microbial Sciences
University of Reading, Whiteknights
PO Box 228, Reading RG6 2AJ, UK

SUMMARY

The precocenes (ageratochromenes) have attracted attention as the first examples of simple molecules, superficially dissimilar in structure to the natural juvenile hormones (JHs) yet able to elicit remarkable JH-antagonistic effects in some species of insects. Precocenes appear to exert a cytotoxic action on the corpora allata (CA), mediated by an oxidative bioactivation *in vivo*. Reactive epoxide formation in the CA is the mechanism that currently best explains the results of several radiochromatographically based metabolism studies in precocene-sensitive insects. Other oxidatively derived reactive intermediates such as quinone-methides or oxyradicals have been suggested but these mechanisms await substantiation. The limited pharmacokinetic studies on precocenes conducted so far have not revealed differences that convincingly explain their species selectivity but methods must be found to maintain the bioavailability of parent precocenes to the CA for longer periods, especially in the seemingly insensitive larvae of lepidoptera.

INTRODUCTION

The possibility that simple chemicals with structures unlike those of the natural insect juvenile hormones (JH) might antagonize JH-action was realized when Bowers, in 1976, discovered such activity in two simple chromenes (precocenes 1 and 2; P1 and P2, Figure 1) isolated from a plant, *Ageratum houstonianum* [1]. In sensitive insects treated with these compounds (anti-JHs), JH-deprivation is signalled by precocious metamorphosis in immature stages and sterility in adult females. An associated necrotic degeneration of the corpora allata (CA), together with the production of P2-3,4-dihydrodiol (P2-DHD; Figure 1) as a significant insect metabolite from P2 [2] suggested [3] an action on the CA analogous to the cytotoxic effects produced by compounds such as bromobenzene in vertebrate liver and believed to be mediated by the formation of a reactive epoxide. Mode of action studies on precocenes have depended heavily on chromatographic separation techniques. The precise molecular mechanism of action of these compounds and the basis for their remarkable species selectivity remain unknown and continue to challenge those interested in the insect endocrine system as a target for insect control by chemicals.

Early Indicators of Oxidative Bioactivation of Precocenes

P2 at rather high concentrations (ca. 5 x 10^{-4}M) was shown to inhibit JH-biosynthesis by intact CA of the insensitive cockroach, *Periplaneta americana in vitro* [4] and structure-activity studies [5-7] established the requirement for a 3,4-double bond and a 7-alkoxy-substituent to produce the overt signs of JH-deficiency seen *in vivo* in sensitive insects. Early metabolism studies on nine, mainly insensitive, insect species, implicated mixed function oxidase enzymes (MFO) in 3,4-dihydrodiol (DHD) formation from [^{14}C]-labeled P2 in whole bodies and tissues [2] but the formation of a 3,4-epoxide intermediate was not substantiated, even in the presence of epoxide hydrolase inhibitors [2,8,9]. Brooks et al. [6] showed that P2

Chromatography and Isolation of Insect Hormones and Pheromones
Edited by A. R. McCaffery and I. D. Wilson
Plenum Press, New York, 1990

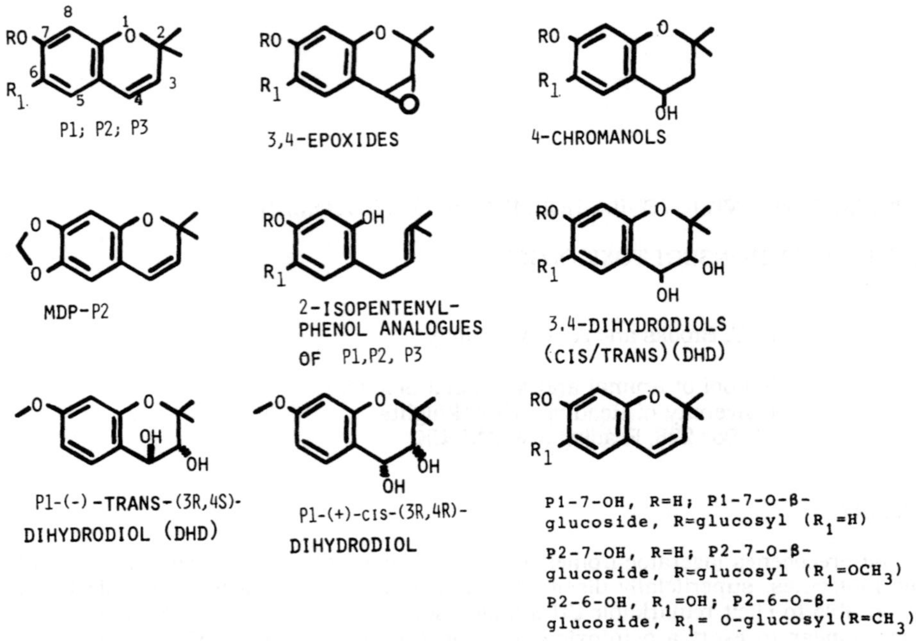

Fig. 1. Key to chemical structures mentioned in the text. Except where otherwise indicated, R = CH_3 for P1, P2, and C_2H_5 for P3 related structures; R_1 = H for P1, CH_3O for P2 and P3 related structures.

action was antagonized by the structurally related MFO-inhibitor MDP-P2 (Figure 1) *in vivo*, which supported the requirement for an oxidative bioactivation.

Thin-layer adsorption chromatography (TLC) on silica gel provided the basis for analytical separations of radiolabeled compounds in these investigations, which were continued by Soderlund et al. [9], using two-dimensional silica TLC, with and without metabolite derivatization, together with both normal and reversed-phase (RP)-high pressure liquid chromatography (HPLC) and gas-liquid chromatography (GLC) for separations (Table 1). These authors also reported the first synthesis of P2-epoxide (Figure 1) which had proved elusive due to its high instability and was said to be detectable without decomposition only by GLC. Their method was used to prepare P1-epoxide (Figure 1) [19], which proved to be highly reactive, as expected. At this time, the already evident species selectivity of precocenes emphasized the need for (1) information about precocene metabolism in the CA of sensitive insects and (2) the measurement of precocene levels in haemolymph following an effective dose, as an indication of the level of exposure of the CA required *in vivo*.

Metabolism in the Corpora Allata

Pratt et al. [11] examined the metabolism of [4-^{3}H]-P1 in intact CA isolated from *Locusta migratoria migratorioides* (R&F), which is equally sensitive to both P1 and P2 and responds with precocious metamorphosis in immature stages and retarded oocyte development in adult females [10]. When CA from reproductively mature adult females were incubated with P1 in tissue culture medium for 1 to 5h, 99% of the radiolabel was recoverable by chloroform extraction and 96% of this co-chromatographed with authentic *trans*/*cis* P1-3,4-dihydrodiol (P1-DHD; Figure 1), unchanged P1 accounting for the remainder. The diol mixture was not resolved by the RP- or normal phase (NP-) TLC systems used (Table 1) but of the radiolabel recovered from the P1-DHD zone after RP-TLC, 96% co-chromatographed with the authentic marker *trans*- and *cis*-diols separated by analytical HPLC on Zorbax-sil (Table 1). The significant *cis*-DHD content of the mixture (*trans*:*cis*, 62:38) indicated a non-enzymic hydration of a reactive intermediate such as the 3,4-epoxide by an S_N1 mechanism, proceeding via protonation of the epoxide ring and a C-4 carbonium ion, which might also be expected to alkylate cellular macromolecules in the vicinity of its site of formation. Thus, the primary metabolism of P1 to DHD in *Locusta* corpora allata occurred without any evident subsequent conjugation but, in support of the proposed mechanism, non-extractable radiolabel was found to be associated with CA macromolecules to a much greater extent (80-fold) than that found in rat liver following an hepatotoxic dose of bromobenzene.

In another investigation, Soderlund et al. [18] showed that homogenates of CA from the grasshopper, *Gastrimargus africanus*, when fortified with NADPH, gave the corresponding DHD as major metabolites from [^{3}H]-P2, and again CA macromolecules became radiolabeled. The metabolism of [^{3}H]-P2 was inhibited by piperonyl butoxide (PB), indicating MFO involvement in DHD formation. Investigations on the chemical hydration of synthetic P1 and P2 epoxides using a variety of hydrolytic conditions [9,20] gave *trans:cis*-DHD ratios (ca. 7:3) generally similar to those found in precocene metabolism studies, in further support of epoxide formation as an intermediate step.

The results of these metabolism studies on CA suggest that P1 and P2 are alternative substrates for the glandular P450-type epoxidase that converts methyl farnesoate (MF) into its 10R,11-epoxide (JH-III; 10,11-epoxy methyl farnesoate) and that following epoxidation by this enzyme, precocenes act as general cytotoxins, rather than direct inactivators of the epoxidase. In support of the view that precocenes are alternative substrates, Hamnett and Pratt [14] showed that oxidative attack on P1 in *Locusta* CA is completely stereospecific and occurs in the same steric sense as the epoxidation of MF. To this end, [^{3}H]-DHD isolated from *Locusta* CA incubated with [^{3}H]-P1 [11] was mixed with authentic racemic *trans* and *cis*-DHD and converted into a mixture of *bis*-(-)-menthoxyacetates, which, following clean-up on TLC and RP-TLC, was completely resolved into *cis*- and *trans*-diastereoisomer pairs by HPLC on Zorbax-sil (Table 1 and Figure 2). The radiolabel was found to be entirely associated with the marker (-)-*trans* (3R,4S) and (+)-*cis* (3R,4R) diastereoisomers of the diol bis-(-)-menthoxyacetates previously characterized by Jennings [13], indicating that attack by P450 on the P1 double bond, expected to be at C-3, results in a 3R-hydroxy substituent. This corresponds with the direction of attack on MF to give its 10R,11-epoxide, JH-III. However, JH-III is much more stable than P1-epoxide and gives predominantly the corresponding *trans*-dihydrodiol on hydrolysis, by an S_N2-type mechanism; whereas JH-III is readily isolable from biological preparations, the formation of precocene epoxides has not been demonstrated directly.

PRECOCENE PHARMACOKINETICS AND SPECIES SELECTIVITY

Interspecific differences in sensitivity of the corpora allata and/or precocene pharmacokinetics are the most likely causes of differences in sensitivity to these compounds. To date, the number of pharmacokinetic investigations reported is rather limited. Pratt et al. [11] found that in contrast to the pattern in adult *Locusta* CA, [^{3}H]-P1 metabolism in Malpighian tubules, abdominal fat body and ligatured mid-guts resulted, according to RP-TLC co-chromatography with reference compounds, in the formation of very small amounts of P1-DHD and possibly the 7-hydroxy-analogue of P1 (i.e., the product of oxidative demethylation), with larger amounts of unknown, polar products, which were not found in the CA. The *trans:cis* ratio of DHD was similar to that found in adult CA metabolism but DHD did not accumulate in these tissues, suggesting further formation of conjugates. The radiolabel appeared in two polar zones migrating more rapidly than DHD on RP-TLC but the products were not identified, leaving open the question of the identity and quantity of the primary metabolites formed in these tissues.

Subsequent work showed that increasing doses of P1 injected in oil into 5 day old adult *Locusta* prior to vitellogenesis caused progressive reduction in JH-III biosynthesis by the CA and in CA volume but had little effect on oocyte development below 600 μg per female, which caused total necrosis of the CA and complete suppression of vitellogenesis [10]. When the haemolymph content of [^{3}H]-P1 was determined by RP-TLC (Table 1) after administration of 600 μg and compared with that in the haemolymph of 4th instars following topical application of a dose (70 μg per insect) causing complete precocious metamorphosis, the peak concentration occurred after about one hour and was 10-fold higher (2 x 10^{-4}M) in 4th instars than in the adults (Figure 3). Based on these haemolymph concentrations, a ten-fold greater sensitivity of adult females than 4th instars is indicated; a 7 μg dose of P1 to 4th instars is the maximum no-effect dose for precocious metamorphosis and produced the same haemolymph concentration of P1 as resulted from the 600 μg dose injected into adults (1 to 2 x 10^{-5}M). The difference in these administered doses may reflect a greater metabolic capacity of the adults for precocene.

Table 1. Summary of chromatographic methods used to separate precocenes, their metabolites and derivatives.

TLC	RP-TLC	HPLC	RP-HPLC	GC or GC/MS	Reference
Silicagel G(SiG); 0.5mm; Et/Hx (4:1) or Bz/Ea (93:7)				1.5m x 4mm glass: 3% OV-225 on 100/120 Chromosorb W-HP; 170-220°C; carrier gas methane (GC-MS)	(2)
SiG; Ea/Xy (1:3)	0.25mm Merck KC-18; Me/H_2O (7:3)	250 x 4.6mm; Zorbax-sil; Et/pentane/2-Pr (49.9:49.9:0.2), 1.5ml/min	Zorbax C8; An/H_2O (6:4), isocratic. An/H_2O, 25-75% gradient. LiChrosorb RP18; An/H_2O (1:3); 1.0-1.8ml/min		(10, 11)
SiG; 0.25mm; Ch/Mc/Ea/Me (10:10:1:1); 2-dimensional; 7 other solvent systems used for metabolite resolution.		250 x 4.6mm; Merck LiChrosorb Si 100; Ch/Pr/Ea (18:1:1): 2ml/min	250 x 4.6mm; ODS on Bio-sil A (Bio-Rad): Me/H_2O (1:1); 1ml/min	1.8m x 2mm glass; 3% OV-101 on Gas-Chrom Q, or 3% QF1 on Chromosorb W; 30-200°C; carrier N_2 (GC), flame ionization. Also GC-MS.	(9)
		250 x 4.6mm; LiChrosorb SI100; Et/Mc (4:96), 50% H_2O satd., 1.2ml/min	250 x 4.6mm; LiChrosorb RP-8; An/H_2O (1:3), then 3:2). 1.5ml/min	2.0m x 2mm; 3% OV-17 on Chromosorb W-AW-DMCS; 250-300°C programmed (GC-MS)	(12)
		300 x 3.9mm; Waters μPorasil; Et/Hx (7:93), 100% H_2O satd.	(separation of diastereoisomeric bis(-)menthoxyacetyl derivatives of precocene-1 *cis* and *trans*-diols and assignment of absolute configuration		(13)

TLC	RP-TLC	HPLC	RP-HPLC	GC or GC/MS	Reference
SiG; 0.25mm; Ea/Xy (35:65)	0.25mm, ODS; 100% Me	250 x 4.6mm; Zorbax-sil; Et/Hx (6:94)+0.01% propane-1,2-diol, 1.5ml/min	(absolute configuration of P1-diols from the CA of *Locusta migratoria*)		(14)
			250 x 4.6mm; Spherisorb S5 Octyl; linear gradient 25-75% An/H_2O; 1ml/min		(15)
SiG; 0.25mm or 0.5mm; Ch/Me (19:1)+ ten other solvent systems				1.8m x 2mm glass; 3% OV-101 on Gas-Chrom W-HP; 140-270°C; carrier N_2, flame ionization.	(16)
			Hypersil ODS; 80:20 buffer/solvent (H_2O/An/THF (3:1:1). 1.8ml/min. (separation of glutathione adducts from P2-epoxide)		(17)

An = acetonitrile, Bz = benzene, Ch = chloroform, Ea = ethyl acetate, Et = diethyl ether, Hx = hexane, Me = methanol, Mc = methylene dichloride, ODS = octadecylsilane, Pr = propanol, THF = tetrahydrofuran, Xy = xylene.

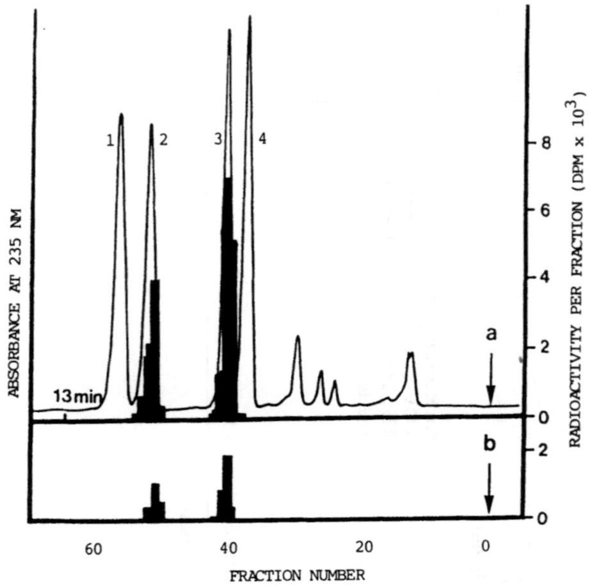

Fig. 2. Normal phase HPLC separation of bis(-)-menthoxyacetate derivatives of the [4-^{3}H] precocene-1 (P1) dihydrodiols isolated from *Locusta migratoria* corpora allata after a two hour incubation *in vitro* with [4-^{3}H]P1. (a) Radiolabel in incubation medium, (b) radiolabel subsequently released from the glands in incubation medium at 95°C (30 min). Shows concordance of radiolabeled P1-dihydrodiol metabolite derivatives (solid bars) with the synthetic 3R-diol derivatives located by their UV absorbance. (1) (-)*cis* (3S,4S); (2) (+)*cis* (3R,4R); (3) (-)*trans* (3R,4S); (4) (+)*trans* (3S,4R). Chromatography as in Table 1 [14].

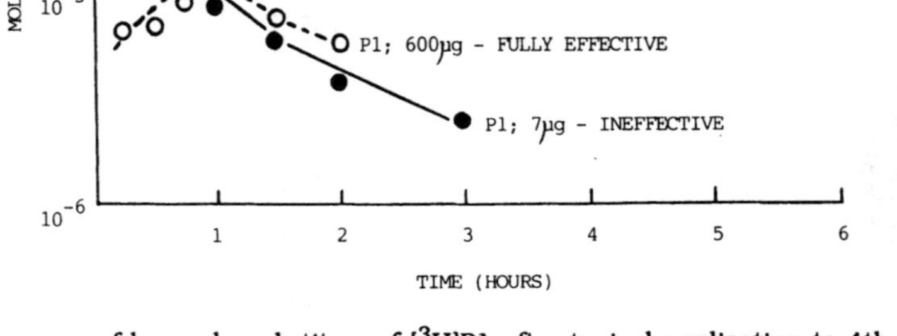

Fig. 3. Time course of haemolymph titres of [^{3}H]P1 after topical application to 4th instar *Locusta migrtoria* (●), oil injection into adult *Locusta* migratoria (○), and of [^{3}H]P2 after topical application to 5th instar *Spodoptera exempta* (▲) at the indicated doses [10,15].

At the effective concentration in adults, isolated CA metabolized P1 to DHD at up to 200 pmol/hr/gland pair and to macromolecular bound residues at about 40 pmol/hr/gland pair [11]. Solvent extraction of 4th instar carcasses [10] after haemolymph sampling and surface rinsing revealed, apart from unchanged [^{3}H]-P1, significant amounts of radiolabel which co-chromatographed on RP-HPLC (Table 1) with the 7-hydroxy-analogue of P1 (P1-7-

OH, Figure 1) and small amounts of DHD, barely resolved from a more polar metabolite 'G'. The latter reached a level of 10^{-4}M in the haemolymph within a few hours after treatment. Although a corresponding aglycone could not be obtained, it seems likely, considering its chromatographic behavior and the apparent presence of P1-7-OH, that this was a 7-O-β-glucoside corresponding to that isolated by Bergot et al. [12] and characterized as a metabolite of [^{3}H]-P2 in six species of insects, including the grasshoppers, *Gastrimargus africanus, Melanoplus sanguinipes* and *Schistocerca nitens*. In their RP-HPLC system (Table 1), the glucoside 'X' chromatographed between marker P2-DHD and P2-4-chromanol (Figure 1) and was subsequently resolved by normal phase HPLC as the tetra-acetyl derivative, into a mixture (1.5:1) of the P2-6- and P2-7-O-β–glucosides (Figure 1). In the grasshoppers, the haemolymph concentration of [^{3}H]-P2 reached 4 x 10^{-4}M one hour after topical application of a 250 μg dose; in contrast to the situation with P1 in Locusta [10], no free phenols were found in the insects examined. Only traces of P2-DHD were present but it may be noted that P2-DHD was the major metabolite produced from [^{3}H]-P2 by homogenates of *G. africanus* CA fortified with NADPH [18].

McCaffery and McDowell [15] (Figure 4) found a metabolite with similar behavior to glucoside 'X' on RP-HPLC (Table 1) following topical application of 50 μg of [^{3}H]-P2 to early 5th instar larvae of precocene insensitive *Spodoptera exempta*. The metabolite reached a haemolymph concentration of ca. 1 x 10^{-4}M four hours after treatment; similar maximal concentrations of DHD and P2 were found, the latter peaking at about two hours (Figure 3), together with 3-fold higher concentrations of more polar material, presumed to be other conjugates, peaking at 5 to 6 hours. The total radiolabel recovered from the haemolymph was much higher (6 to 8% of the applied dose) than was found in the insects studied by Bergot et al. [12] but in neither investigation was the haemolymph P2 level more than 1 to 2% of the applied dose. Tissue balance studies indicated that the radiolabel was fully recoverable several hours after application and may be held up in the cuticle-body wall matrix, largely as unchanged P2 [12]. Evidently, much higher concentrations of P2 were maintained for several hours in the sensitive grasshoppers (2 to 4 x 10^{-4}M) than in larvae of insensitive *Manduca sexta* [12], or *Heliothis armigera* [15]. However, the amount found in *S. exempta* larval haemolymph (Figure 3), although ineffective against this insect, should easily suffice to sterilize adult female *Locusta* and would be expected to cause precocious metamorphosis in 4th instars of this insect.

Haunerland and Bowers [21] found that insensitive *Heliothis zea* and sensitive *Oncopeltus fasciatus* absorbed topically applied [^{14}C]-P2 at equal rates. There was no retention in the cuticle and P2 was stored unchanged in the fat body of *O. fasciatus* from which it disappeared only slowly, whereas it rapidly reached the gut of *H. zea*, where large amounts of P2-DHD were formed. Thus, slow release of precocene from inert storage in cuticle or fat body might help to maintain an effective haemolymph concentration in sensitive insects. Although fat body and haemolymph concentrations remained nearly constant for 20 hours in *O. fasciatus*, blood concentrations in the sensitive grasshoppers appeared to decline more rapidly, despite any possible reservoir of unchanged P2 in cuticle or body wall. Part of the haemolymph content of P2 in *G. africanus* after injection appears to be bound to lipoprotein [18], which may protect it from metabolism and favor its delivery to the CA.

Any reactive metabolites produced by oxidative metabolism of precocenes in peripheral tissues or haemolymph are unlikely to reach the CA before being inactivated by water, conjugating enzymes or a variety of tissue macromolecules; the formation of precocene epoxides in biological milieu has not been demonstrated so far, DHD formation and its release from radiolabeled tissues being the only available indicator of such bioactivation. Full quantification of DHD formation requires complete analysis of the more polar metabolite fractions from such studies, since DHD conjugates may be present therein. The formation of P1-7-OH and P2-7-OH in insects treated with the respective parent precocenes is of some interest. Oxidative demethylation leading to their formation has not been demonstrated in CA but unlike the epoxides, these phenols are sufficiently stable to travel in the haemolymph. They are potential precursors of reactive p-quinone-methides that might produce cytotoxic effects in sensitive organs by alkylating critical macromolecules. However, they appear to be conjugated rapidly and P2-7-OH had no anti-JH action on *O. fasciatus* [12], nor had P1-7-OH on *O. fasciatus* or *L. migratoria* by topical application [22]. The CA of precocene insensitive insects may be intrinsically susceptible to precocene action [4,9]. If this is the situation with lepidopteran larvae, methods are needed to ensure adequate exposure of the CA for sustained time periods, which might be achieved by careful formulation to protect the chemicals from metabolic destruction in peripheral tissues.

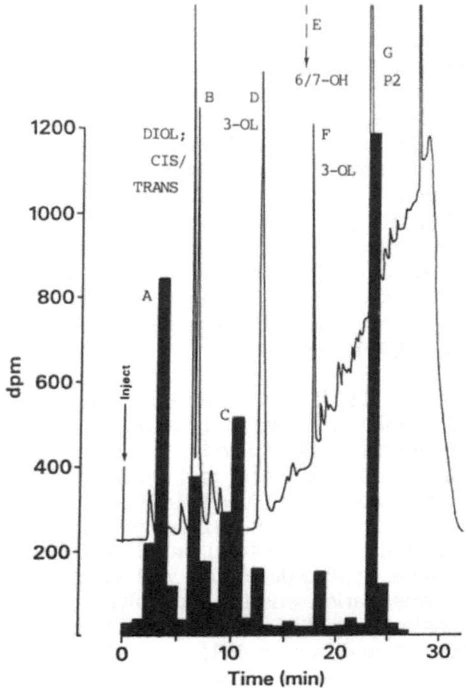

Fig. 4. Chromatogram obtained by HPLC of a haemolymph sample from a 5th instar larva of *Spodoptera exempta* two hours after topical treatment with [5-^{3}H] precocene-2 (P2), with added reference compounds. a. Presumed polar conjugates; b. *cis* and *trans* P2-3,4-dihydrodiols; c. unidentified metabolite; d. P2-4-chromanol; e. elution time of P2-7-OH (see Figure 1); f. P2-3-chromanol; g. P2. Chromatography as in Table 1 [15].

Molecular Mechanism of Precocene Action

Identification of the cellular entities to which precocenes become bound in the CA has not been achieved but might help to clarify their molecular mode of action. As expected, the synthetic precocene epoxides react readily with nucleophiles. Thus, Aizawa et al. [16] used normal phase TLC (Table 1) with eleven different solvent systems, combined with GLC, to explore P2-epoxide reactivity toward various simple nucleophiles in tetrahydrofuran as a solvent. Under their conditions, this epoxide reacted much less rapidly with water than with thiols, which in tissues may therefore compete successfully with water for any epoxide formed from the precocene *in situ*. Among the compounds tested, reaction was fastest with L-cysteine methyl ester and thiophenol.

As a well established protective nucleophile, glutathione (GSH) might be expected to participate in precocene detoxication and Bellés et al. [24] reported a synergistic effect of the GSH depletor, diethyl maleate (DEM) on the anti JH action of P1 and P2 on the seed bug, *Oxycarenus lavaterae* (F.). Camps et al. [17] have synthesized GSH-conjugates from P2-epoxide but GSH conjugates of precocenes have not been reported as insect metabolites so far. To account for the inhibition of JH-III biosynthesis by 3,4-dihydro-P2 in CA of the cockroach *Blattella germanica in vitro*, an oxidative activation involving a free radical rather than an epoxide intermediate has been suggested [24]. An oxyradical might be quenched by water to give DHD, or by GSH to give its oxidized form (GSSG), rather than GSH conjugates derived from precocenes. In that case, depletion of GSH in the CA by an agent such as DEM might potentiate the adverse effect of precocenes on CA function. The hepatocyte toxicity of the quinone, menadione, is believed to result from protein-thiol loss and disruption of intracellular calcium homeostasis, mainly through oxidative mechanisms following GSH depletion [25]. Similar mechanisms may be responsible for precocene cytotoxicity; quinone-methide formation has been discussed in relation to precocene action [22] and the precocene-like activity of certain 2-isopentenyl phenols that give the corresponding precocenes by chemical cyclization [22,26]. Paradoxically, P2 applied to early 5th instar *L. migratoria* behaves as a JH-mimic, although degeneration of the CA is still observed in the resulting juvenilized adults [27]. There appears to be a burst of JH synthesis/release which may reflect an early disturbance in calcium homeostasis, leading through neurohormone release to allatal stimulation, before the cytotoxic effects become manifest.

It is not known whether the "open chain" analogues of P1, P2 and P3 (Figure 1) convert into the corresponding precocenes *in vivo*. However, an interesting example of species selectivity arises in this group; whereas P1, P2 and P3 are all effective on *Locusta*, none of the "open" derivatives are active and only those related to P2 and P3 are effective on *O. fasciatus* [22]. One possible explanation is that the corresponding precocenes are not significant metabolic products in *Locusta*, which may deactivate the parent phenols by direct conjugation, or their putative quinone-methides by other routes that lead to detoxication. Since P1 is 10-fold less active than P2 or P3 toward *Oncopeltus*, its inadequate formation from the corresponding "open" precursor might explain the latter's inactivity. Detailed metabolism studies are needed to determine whether the other phenols do generate precocenes in this insect, or if the anti JH action results from some mechanism not requiring the intermediate formation of precocenes. It may be noted that some linear furochromenes related to P1 also show moderate anti JH activity toward *Locusta* but not toward *Oncopeltus* [28].

ACKNOWLEDGEMENT

Acknowledgement is made to former colleagues whose work is cited and discussed herein.

REFERENCES

1. W. S. Bowers, T. Ohta, J. S. Cleere and P. A. Marsella, Discovery of insect anti-juvenile hormones in plants, *Science*, 193:542-547 (1976).
2. T. Ohta, R. J. Kuhr and W. S. Bowers, Radiosynthesis and metabolism of the insect anti-juvenile hormone, precocene-II, *J. Agricul. Food Chem.*, 25:478-481 (1977).
3. G. T. Brooks, Epoxide hydratase as a modifier of biotransformation and biological activity, *Gen. Pharmacol.*, 8:221-226 (1977).
4. G. E. Pratt and W. S. Bowers, Precocene-II inhibits juvenile hormone biosynthesis by cockroach corpora allata *in vitro*, *Nature*, 265:548-550 (1977).
5. W. S. Bowers, Anti-juvenile hormones from plants: chemistry and biological activity, *Pontificiae Academiae Scientiarum Scripta Varia*, 41:129-156 (1977).
6. G. T. Brooks, G. E. Pratt and R. C. Jennings, The action of precocenes in milkweed bugs (*Oncopeltus fasciatus*) and locusts (*Locusta migratoria*), *Nature*, 281:570-572 (1979).
7. G. T. Brooks, A. F. Hamnett, R. C. Jennings, A. P. Ottridge and G. E. Pratt, Aspects of the mode of action of precocenes on milkweed bugs (*O. fasciatus*) and locusts (*L. migratoria*), Proceedings of the 1979 British Crop Protection Conference - Pests and Diseases, 1:273-279 (1979).
8. M. E. Burt, R. J. Kuhr and W. S. Bowers, Metabolism of precocene-II in the cabbage looper and European corn borer, *Pest. Biochem. Physiol.*, 9:300-303 (1978).
9. D. M. Soderlund, A. Messeguer and W. S. Bowers, Precocene-II metabolism in insects: synthesis of potential metabolites and identification of initial *in vitro* biotransformation products, *J. Agricul. Food Chem.*, 28:724-731 (1980).
10. G. E. Pratt, R. C. Jennings and A. F. Hamnett, The mode of action of precocenes, Scientific Papers of the Institute of Organic and Physical Chemistry of Wroclaw Technical University, No. 22, Part 1:347-356 (1981).
11. G. E. Pratt, R. C. Jennings, A. F. Hamnett and G. T. Brooks, Lethal metabolism of precocene-I to a reactive epoxide by locust corpora allata, *Nature*, 284:320-323 (1980).
12. B. J. Bergot, K. J. Judy, D. A. Schooley and L. W. Tsai, Precocene-II metabolism: comparative *in vivo* studies among several species of insects, and structure elucidation of two major metabolites, *Pest. Biochem. Physiol.*, 13:95-104 (1980).
13. R. C. Jennings, Synthesis, resolution and absolute configuration of the diol metabolites of precocene-I, *Tetrahedron Letters*, 23:2693-2696 (1982).
14. A. F. Hamnett and G. E. Pratt, The absolute configuration of precocene-I dihydrodiols produced by metabolism of precocene-I by corpora allata of *Locusta migratoria in vitro*, *Life Sciences*, 32:2747-2753 (1983).

15. A. R. McCaffery and P. G. McDowell, Titres of precocene-I and its metabolites in the haemolymph of larvae of the African armyworm *Spodoptera exempta* following topical treatment with the compound, *Pest. Sci.*, 19:185-196 (1987).
16. H. Aizawa, W. S. Bowers and D. M. Soderlund, Reactions of precocene-II epoxide with model nucelophiles, *J. Agricul. Food Chem.*, 33:406-411 (1985).
17. F. Camps, A. Conchillo and A. Messeguer, 3,4-Epoxyprecocene as a model of cytotoxic epoxides: synthesis of *trans* adducts occuring in the glutathione metabolic pathway, *Tetrahedron*, 43:3067-3074 (1987).
18. D. M. Soderlund, M. F. Feldlaufer, S. Y. Takahashi and W. S. Bowers, Mechanisms of selectivity in the action of precocenes, *in*: "Juvenile Hormone Biochemistry", G. E. Pratt and G. T. Brooks, eds., Elsevier/North Holland, Amsterdam, pp 353-362 (1981).
19. R. C. Jennings and A. P. Ottridge, The synthesis of precocene-I epoxide, *J. Chem. Soc. Chem. Comm.*, 920-921 (1979).
20. A. F. Hamnett, A. P. Ottridge, G. E. Pratt, R. C. Jennings and K. M. Stott, Kinetics and products of the hydrolysis of 3,4-dihydroprecocene-I-epoxide in aqueous organic solvents, *Pest. Sci.*, 12:245-254 (1981).
21. N. H. Haunerland and W. S. Bowers, Comparative studies on pharmacokinetics and metabolism of the anti-juvenile hormone precocene-II, *Archives Insect Biochem. Physiol.*, 2:55-63 (1985).
22. G. T. Brooks, A. P. Ottridge, R. C. Jennings, D. W. Mace and B. A. J. Alexander, The effect of 2,2-dimethylchromene derivatives and some other compounds on the development of *O. fasciatus* (Dallas) and *Locusta migratoria* (R&F), *Pest. Sci.*, 16:571-588 (1985).
23. X. Bellés, M. Baldellou and A. Messeguer, Synergistic action of diethyl maleate on the morphogenetic and antigonadotropic activity of precocenes on the seed bug *Oxycarenus lavaterae* (F.), *Archives Insect Biochem. Physiol.*, 4:107-112 (1987).
24. X. Bellés, F. Camps, J. Casas, A. Messeguer and M. D. Piulachs, *In vitro* inhibition of juvenile hormone III biosynthesis by precocene-II and 3,4-dihydroprecocene-II on *Blattella germanica*, *J. Insect Physiol.*, 34:457-461 (1988).
25. S. Orrenius, H. Thor, D. Dimonte, G. Bellomo, P. Nicotera, D. Ross and M. T. Smith, Mechanisms of oxidative cell injury studied in intact cells, *in*: "Microsomes and Drug Oxidations", A. R. Boobis, J. Caldwell, F. De Matteis and C. R. Elcombe, eds., Taylor & Francis, London, pp 238-247 (1985).
26. W. S. Bowers, P. H. Evans, P. A. Marsella, D. M. Soderlund and F. Bettarini, Natural and synthetic allatotoxins: suicide substrates for juvenile hormone biosynthesis, *Science*, 217:647-648 (1982).
27. R. C. Miall and W. Mordue, Precocene-II has juvenile hormone effects in 5th instar *Locusta migratoria*, *J. Insect Physiol.*, 26:361-364 (1980).
28. G. T. Brooks, A. P. Ottridge and D. W. Mace, The effect of some furochromene and benzochromene analogues of precocene-I and benzofuran precursors on *Oncopeltus fasciatus* (Dallas) and *Locusta migratoria migratorioides* (R&F), *Pest. Sci.*, 22:41-50 (1988).

A NOTE ON THE REGULATION OF THE CORPORA ALLATA BY THE BRAIN IN THE ADULT FEMALE AMERICAN COCKROACH, *PERIPLANETA AMERICANA*

I. Ghanem and A. R. McCaffery

Department of Pure and Applied Zoology
School of Animal and Microbial Sciences
The University of Reading, Whiteknights
PO Box 228, Reading RG6 2AJ, UK

INTRODUCTION

The corpora allata (CA), the source of juvenile hormones (JH) in insects, play a central role in the regulation of insect metamorphosis and reproduction [1]. The ability of CA to biosynthesize and release JH may be measured *in vitro* by incubating isolated CA in a defined tissue culture medium in the presence of [methyl-C^{14}]-methionine or [methyl-H^3]-methionine. The radiobiosynthesized juvenile hormone is separated on radio-TLC and measured by scintillation counting according to methods described previously [2-4].

In cockroaches (Dictyoptera), the activity of the CA varies depending on the stage and the reproductive cycle of the insect. Regulation of the CA is thought to involve both neural and neurohumoral mechanisms [1,5-7] and there is some evidence to support the presence of both stimulatory and inhibitory factors [8,9]. Recently, the presence of inhibitory factor(s) in brain extracts of the cockroach *Diploptera punctata* has been reported (for review see reference 10). Much of what is known about the regulation of CA in Dictyoptera is derived from studies carried out on the viviparous species *Diploptera punctata* [9,11] and the ovoviviparous species *Leucophaea maderae* [1] and *Nauphoeta cinerea* [12]. These species have long, clearly-defined pregnancy periods during which the females carry their embryos internally and their CA are thought to be inactive [1]. The American cockroach *Periplaneta americana*, however, is an oviparous species. The female does not have a pregnancy period and its ovarian cycles overlap in such a fashion that there are always vitellogenic oocytes present in the ovaries [13,14]. Moreover, the CA of this species are always active and capable of maintaining a minimum level of activity under normal conditions [14,15] and seem to be less dependent upon intact nervous connections for their regulation [6]. Because of these differences between other cockroaches and the American cockroach it is of interest to investigate the regulation of CA activity in the latter.

One of the problems encountered while attempting to investigate the regulation of CA activity in the adult female American cockroach, is the speed at which these glands change their rate of activity. The CA of this insect show a cyclic pattern of activity during which the rate of JH production changes from a minimum to a maximum and returning to a minimum again in a period of just three days [15]. This makes it difficult to define a long enough period during which the CA maintain a constant level of activity. One solution to this problem has been to impose physiological regime(s) which lead to constant rate of CA activity. Virginity and starvation reduce the spontaneous rate of activity of CA to extremely low levels [15,16]. The effect is reversible and the glands recover their original level and cyclic pattern of activity shortly after lifting the imposed regimes [15,16]. This suggests the presence of a regulatory mechanism involved in modulating the function of CA as a result of starvation and virginity [7].

In this work we have used starvation and virginity as a means of suppressing CA activity in the adult female cockroach [7], and subsequently investigated the possibility that this modulation of CA activity is controlled by factor(s) produced by the brain.

Chromatography and Isolation of Insect Hormones and Pheromones
Edited by A. R. McCaffery and I. D. Wilson
Plenum Press, New York, 1990

MATERIALS AND METHODS

Insects

Cockroaches *Periplaneta americana* were reared according to a previously described method [14]. To obtain starved, virgin females the insects were transferred from the stock colony 0 to 24 h after ecdysis, and kept thereafter in 33 x 22 x 19 cm tanks in groups of 15-25 animals. Drinking water was provided *ad lib.* Under these conditions the females did not produce any oothecae for the starvation period of 15 to 17 days.

Assay of Corpora Allata Activity

A radiochemical assay was used to determine the activity of the CA [3,4]. Single pairs of CA were incubated at 30°C in TC-199 (Gibco) medium to which [methyl-C^{14}]-methionine (Amersham, UK) was added (final specific activity 28.35 to 34.55 mCi/mmol in different experiments). Radiolabeled juvenile hormone (JH) released by the glands to the medium was extracted by routine methods, separated by thin layer chromatography (TLC), identified by a combination of UV light and TLC linear radioanalyzer, and finally measured by liquid scintillation counting [4].

Co-incubation of CA with Brains

Brains were dissected from starved, virgin adult female cockroaches (or from normal females in the case of control treatments) and kept on ice-cold TC-199 medium until used. Each brain was cut at the optic lobes and circumoesophageal connectives. CA were dissected from normal, fed females. The glands were pre-incubated for 1 h in radiolabeled medium (50 µl/pair) to achieve isotopic equilibrium [3]. The medium was then aspirated and fresh medium was added. A three-hour incubation was carried out to establish the initial rate of activity of the glands. The glands were then co-incubated either with brains from starved, virgin females or with brains from normal females (control). Brains were removed, rinsed with cold medium and each brain rinse was added to the corresponding incubation medium. Brains were analyzed for their JH content by disrupting the tissue by ultrasonication followed by extraction procedures identical to those used in the radiochemical assay. Glands were incubated in fresh labeled medium for a further two hours. The medium was aspirated and JH released was quantified as described above.

Hourly Monitoring of CA Activity

Single pairs of glands were dissected from normal fed female cockroaches 18 h after being collected carrying half-tanned oothecae. The glands were pre-incubated for 1 h in radiolabeled medium (50 µl/pair) to achieve isotopic equilibrium within the gland cells. A series of 1 h incubations was carried out using fresh radiolabeled medium at hourly intervals (75 µl/pair). At the end of every incubation the media were aspired. JH released to the medium was extracted and assayed as described above. Figure 1 represents the hourly monitoring of the activity of the CA from normal adult female cockroach *in vitro*. The glands under the conditions of this assay showed a significant increase above their initial level of activity after 5 h. This increased level was maintained, or slightly but not significantly increased, up to 9 h from where it returned to the initial level.

Brain Tissue Extraction

Brains and corpora cardiaca from 50 starved, virgin female cockroaches were dissected under ringer solution and stored separately at -20°C in 75% ethanol containing 0.2 M HCl and 0.1 mg/ml dithiothretol (DTT). The tissues were homogenized in an all glass homogenizer, placed in a water bath (90°C) for 10 min followed by centrifugation at 10,000 g for 20 min. The supernatant was recovered and the pellet was resuspended and centrifuged as before, the supernatant being recovered and pooled with that recovered earlier. Bovine serum albumin (BSA) was added (0.05 mg/ml) and the pH was adjusted to 7.2 using ammonia. Ethanol and HCl were removed using a vacuum centrifuge (Univap, Uniscience) and the dry extract was stored at -20°C. The sample was resuspended in 5 ml 0.2% trifluoracetic acid (TFA) loaded onto a pre-washed C18 Sep-pak cartridge (Millipore, UK). Before loading the sample Sep-pak cartridges were pre-washed with the following solutions and in the following order: 1) 10 ml of acetonitrile; 2) 10 ml of high performance liquid chromatography (HPLC) grade water; 3) 10 ml of 0.2% TFA; 4) 10 ml of 0.1% Poly Pep to fill any unspecific adsorption sites on the cartridge (P. J. Lowry, personal communication); and 5) 10 ml of 0.2% TFA. The salts were removed by eluting the sample with 20 ml of HPLC grade water (BDH). The sample was eluted from the cartridge in a step-wise fashion using

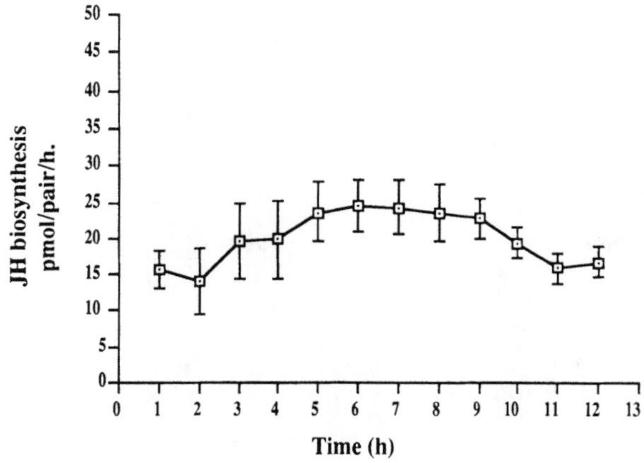

Fig. 1. Hourly monitoring of CA activity *in vitro*. The incubation medium was aspirated and fresh radiolabeled medium was added at hourly intervals. Each point is the mean value of 6 determinations. Vertical bars depict standard errors of the mean values.

increasing concentrations of acetonitrile in 0.2% TFA. The concentrations used were 10, 20, 30, 40 and 50% acetonitrile in 0.2% TFA. A fraction of 5 ml was collected for each concentration. To each fraction BSA and DTT was added and the pH adjusted to 7.2. TFA and acetonitrile were removed by vacuum centrifugation. For control treatments all the steps followed in the extraction of brain tissues were applied without the presence of any tissue. The fractions were stored at -20°C until tested.

Bioassays Used to Test the Effect of Brain Extract on CA Activity

The effect of each brain fraction on the CA activity was tested by resuspending each fraction in a known amount of radiolabeled TC-199, so as to keep the final concentration of BSA and DTT in the solution at 0.05 mg/ml and 0.1 mg/ml respectively, and then incubating the CA in a known amount of this medium. In one set of experiments, single pairs of CA were dissected in ringer solution and placed in cold TC-199. The cold medium was aspirated and the glands were pre-incubated for 1 h in radiolabeled TC-199 (50 µl/pair). The medium was aspirated and the glands were incubated for 3 h in fresh radiolabeled medium (100 µl/pair) to establish their initial rate of activity. After aspirating the medium the glands were incubated for 3 h in the presence of the brain extract. The medium was aspirated and the glands were finally incubated in radiolabeled TC-199 only. The medium was aspirated at the end of the final incubation. The JH released to the media collected from the different incubations was extracted and quantified as described above. The rate of activity of CA in the presence of the brain extract was then compared to their initial rate of activity without the extract. In a second experiment the hourly monitoring assay of CA activity was used (see above). In this experiment the glands were incubated in the presence of brain extract at hours 3, 4 and 7. Student t-test was used for statistical analysis of the results.

RESULTS

The results of co-incubating CA from normal fed female cockroaches with brains from starved, virgin females are shown in Figure 2. In the control treatment the glands were co-incubated with brains from normal females collected from the stock colony carrying half-tanned oothecae and left for 18 h before dissecting the brains out. Incubating CA with brains from starved, virgin females resulted in a highly significant ($p < 0.01$) inhibition in the rate of synthesis of JH by the glands as compared to their initial rate of activity (52% inhibition). This inhibition was reversible since removing the brains from the incubation medium resulted in a rapid recovery of the glands and their activity reached a level significantly higher ($p < 0.001$) than their initial level. There was a non-significant, reduction in the activity of the glands co-incubated with brains from normal fed females.

The effects of one fraction (eluted with 30% acetonitrile in 0.2% TFA) of brain extract and of a corpora cardiaca extract on the *in vitro* biosynthetic activity of the CA are shown

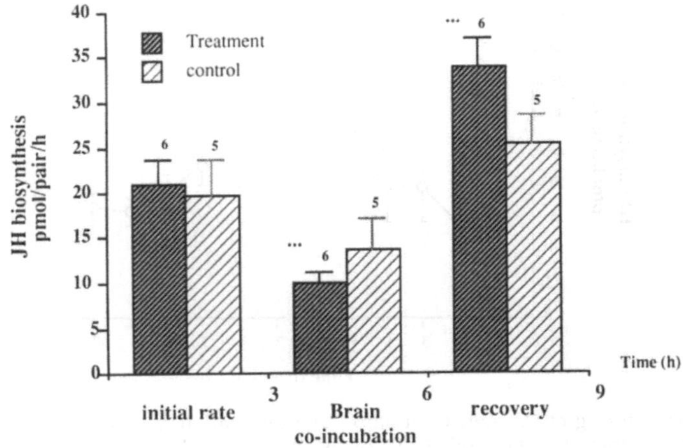

Fig. 2. The effect of *in vitro* co-incubation of CA from normal female cockroaches with brains from starved, virgin female cockroaches. Controls for this experiment were glands co-incubated with brains from normal females. All glands were first incubated in the absence of the brains to establish the glands initial rate of activity. Vertical bars depict standard errors of the mean values. Figures above bars indicate the number of determinations for each measurement.

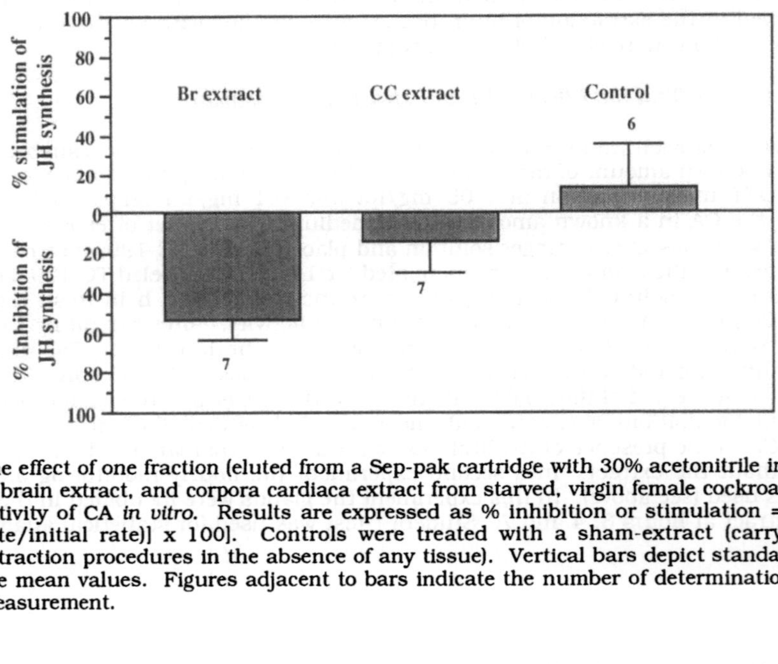

Fig. 3. The effect of one fraction (eluted from a Sep-pak cartridge with 30% acetonitrile in 0.2% TFA) of brain extract, and corpora cardiaca extract from starved, virgin female cockroaches on the activity of CA *in vitro*. Results are expressed as % inhibition or stimulation = [1-(treated rate/initial rate)] x 100]. Controls were treated with a sham-extract (carrying out all extraction procedures in the absence of any tissue). Vertical bars depict standard errors of the mean values. Figures adjacent to bars indicate the number of determinations for each measurement.

in Figure 3. The results are expressed as % inhibition of JH biosynthesis. There was a 54% inhibition in the biosynthetic activity of the CA incubated in the presence of the brain extracts (Figure 3), with JH biosynthesis reduced, on average, from 22 pmol/pair/h to 9 pmol/pair/h (n = 7). There was no significant inhibition in the biosynthetic activity of the CA incubated in the presence of CC extract. Likewise, there was no inhibition in the biosynthetic activity of the CA used as control (Figure 3). Average rates were 25.64 pmol/pair/h and 24.55 pmol/pair/h for the initial and the treatment incubations respectively.

To determine how quickly the activity of the CA declined in the presence of brain extract in the incubation medium and how quickly they recovered after substituting the

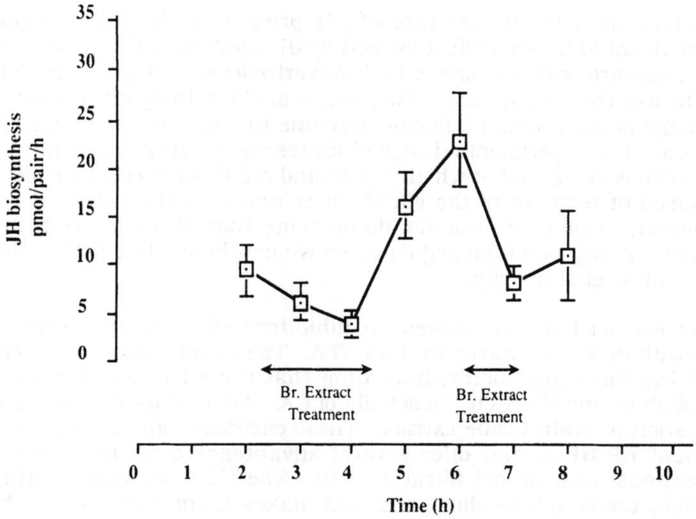

Fig. 4. Hourly monitoring of the effect of one brain extract fraction (eluted with 30% acetonitrile in 0.2% TFA) on the activity of CA *in vitro*. Arrows indicate the times when the glands were incubated in the presence of the brain extract-fraction. Each point represents the mean value of 6 determinations. Vertical bars depict standard errors of the mean values.

medium with fresh medium lacking the brain extract, we used the hourly monitoring assay (Figure 1). The results obtained from hourly monitoring of the effect of brain extracts (fraction 30% acetonitrile in 0.2% TFA) on the activity of the CA *in vitro* are shown in Figure 4. Brain extract was added to the incubation medium at hours 3, 4 and 7. The activity of the CA, as compared to their initial level of activity, was reduced by 36% ($p < 0.05$) during the first hour of incubating the glands in the presence of brain extract, and by 64% ($p < 0.01$) during the second hour of incubating the glands under the same conditions. The glands recovered very quickly after removing brain extracts from the incubation media. Their activity reached a rate 68% higher than their initial rate ($p < 0.01$) during the first hour (hour 5) of substituting the incubation medium with a fresh one free from brain extract. The glands more than doubled their initial rate ($p < 0.001$) during the second hour (hour 6) of removing the brain extract from the incubation medium. Reimposition of the brain extract into the medium at hour 7 medium again depressed the biosynthetic activity of the glands.

DISCUSSION

Co-incubation of CA with brains from starved, virgin female American cockroaches resulted in an inhibition of CA activity, whereas a slight but non-significant inhibition occurred when the glands were co-incubated with brains from normal female cockroaches. Inhibition of JH production occurred to some extent in the presence of brains from normal insects. This then raised a question as to whether part of this apparent inhibition resulted from JH degradation by the brain tissue and moreover whether the significant inhibition observed in the case of co-incubating CA with brains from starved, virgin insects was merely the result of an increase in the level of JH degrading enzymes. We carried out some preliminary experiments to test this hypothesis by incubating brains from starved virgin females and brains from normal females with authentic radiolabelled JH produced by the glands. The results indicated that there was no difference in the ability of both types of brain to degrade JH. The loss of JH in the medium caused by the presence of both types of brains was about 30%. This was assayed by quantifying the amount of radioactivity co-migrating with cold JH on the TLC system. We also observed an increase in the amount of radioactivity co-migrating with methyl farnesoate on the TLC system. These results do not account for the high inhibition of CA activity observed when CA were co-incubated with brains from starved virgin female cockroaches. Thus, we believe that brains from starved, virgin female cockroaches probably contain inhibitory factor(s) which, when released into the incubation medium, depress the biosynthetic activity of the CA *in vitro*. These results are consistent with the suggestions that CA of the adult female american cockroaches are less dependent on their nervous connections for controlling their activity *in vivo* and that they are controlled predominantly via the haemolymph [6,7].

The apparent increase in the rate of JH production by CA monitored at hourly intervals (Figure 3) could be an artifact caused by JH binding to the glassware since we did not treat our glassware with carbowax [17]. Nevertheless, it is possible, as evident from these results, to use the hourly-monitoring assay of CA activity since within an hour the glands are capable of producing sufficient hormone to measure their rate of activity. This protocol can be used in experiments designed for testing potential regulators of CA activity. It enabled us to detect the speed at which a potential regulator exerts its effect on the glands and also the speed of recovery of the glands after removing the potential regulator. It is important, however, to incubate the glands for more than three hours before starting an assay of the effect of any potential regulator so we can be confident that the glands have reached a constant level of activity.

One brain extract-fraction showed an inhibitory effect on the activity of CA. This fraction eluted with 30% acetonitrile in 0.2% TFA. The glands recovered very quickly upon the removal of the inhibitory factor, indicating that the effect was physiological. Other fractions did not show any effect on the activity of CA. We have used C18 Sep-pak cartridges for the fractionation of brain tissue extract. These cartridges are a low pressure version of the more efficient RP-HPLC and offer several advantages over more traditional sample preparation methods such as gel filtration [18]. The C18 Sep-pak cartridges contain a hydrophobic solid phase (ODS-silica) and this makes them suitable for the isolation of neuropeptides. Indeed, these cartridges were used for the purification of the cytochromogenic hormone from corpora cardiaca extract of the cockroach *Blaberus discoidalis* [18]. At this stage we can only speculate on the nature of the CA-inhibitory factor described in this study. The extraction and fractionation procedures used indicate that the factor(s) is probably a neuropeptide, and further work is needed to characterise it.

REFERENCES

1. F. Engelmann, "The Physiology of Insect Reproduction", Pergamon, Oxford (1970).
2. G. E. Pratt and S. S. Tobe, Juvenile hormones radiobiosynthesized by the corpora allata of adult female locusts *in vitro*, *Life Sciences*, 14:575-586 (1974).
3. G. E. Pratt, S. S. Tobe, R. J. Weaver and J. R. Finney, Spontaneous synthesis and release of C_{16} juvenile hormone by isolated corpora allata of female locust *Schistocerca gregaria* and female cockroach *Periplaneta americana*, *Gen. Comp. Endocrinol.*, 26:478-484 (1975).
4. R. J. Weaver, G. E. Pratt, A. F. Hamnett and R. C. Jennings, The influence of incubation conditions on the rates of juvenile hormone biosynthesis by corpora allata isolated from adult females of the beetle *Tenebrio molitor*, *Insect Biochem.*, 10:245-254 (1980).
5. S. S. Tobe and B. Stay, Structure and regulation of the corpus allatum, *Adv. Insect Physiol.*, 18:305-432 (1985).
6. R. J. Weaver, Comparable activities of left and right corpora allata consistent with humoral control of juvenile hormone biosynthesis in the cockroach, *Can.J. Zool.*, 57:243-245 (1979).
7. R. J. Weaver, Effects of food and water availability and of NCA-1 section, upon juvenile hormone biosynthesis and oocyte development in adult female *Periplaneta americana*, *J. Insect Physiol.*, 30:831-838 (1984).
8. S. S. Tobe and B. Stay, Control of juvenile hormone biosynthesis during the reproductive cycle of a viviparous cockroach: III Effects of denervation and age on compensation with unilateral allatectomy and supernumerary corpora allata, *Gen. Comp. Endocrinol.*, 40:89-98 (1980).
9. B. Stay and S. S. Tobe, Control of the corpora allata during a reproductive cycle in a viviparous cockroach, *Am. Zool.*, 21:663-674 (1981).
10. B. Stay and C. R. Paulson, Regulation of juvenile hormone sysnthesis by the corpora allata during metamorphosis, *in*: "The Molecular Regulation of Invertebrate Development", J. D. O'Connor, ed., Alan R. Liss, NewYork, pp 289-305 (1987).
11. S. S. Tobe, Regulation of the corpora allata in adult female insects, *in*: "Insect Biology in the Future", M. Locke and D. S. Smith, eds., Academic Press, pp 345-367 (1980).
12. B. Lanzrein, R. Wilhelm and J. Buschor, On the regulation of the corpora allata activity in adult females of the ovoviviparous cockroach *Nauphoeta cinerea*, *in*: "Juvenile Hormone Biochemistry", G. E. Pratt and G. T. Brooks, eds., Elsevier/North Holland, pp 147-160 (1981).
13. G. E. Pratt, Association of folin-positive material with the blood proteins of the American cockroach and the oviposition cycle, *Nature*, 214:1034-1035 (1967).

14. R. J. Weaver, G. E. Pratt and J. R. Finney, Cyclic activity of the corpus allatum related to gonotropic cycles in adult female *Periplaneta americana*, *Experientia*, 31:597-598 (1975).
15. R. J. Weaver and G. E. Pratt, The effect of enforced virginity and subsequent mating on the activity of the corpus allatum of *Periplaneta americana* measured *in vitro* as related to changes in the rate of ovarian maturation, *Physiol. Entomol.*, 2:59-76 (1977).
16. R. J. Weaver and G. E. Pratt, Effects of starvation and feeding upon corpus allatum activity and oocyte growth in adult female *Periplaneta americana*, *J. Insect Physiol.*, 27:75-83 (1981).
17. D. Schooley, Analysis of the naturally occurring juvenile hormones - their isolation, identification and titre determination at physiological levels, *in*: "Analytical Biochemistry of Insects", R. B. Turner, ed., Elsevier, pp 241-287 (1977).
18. T. K. Hayes and L. L. Keeley, The isolation of insect neuropeptides using reverse phase high performance liquid chromatography, *in*: "Insect Neurochemistry and Neurophysiology", A. B. Borkovec and T. J. Kelly, eds., Plenum Press, pp 223-250 (1984).

ECDYSTEROIDS

CHROMATOGRAPHIC SEPARATIONS OF ECDYSONE ACYL ESTERS AND THEIR APPLICATION TO THE DISTRIBUTION AND IDENTIFICATION OF ECDYSTEROIDS IN ADULT HOUSE CRICKETS, *ACHETA DOMESTICUS*

P. Whiting and L. Dinan

Department of Biological Sciences
University of Exeter
Washington Singer Laboratories
Perry Road, Exeter, Devon EX4 4QG, UK

SUMMARY

Apolar ecdysteroid conjugates (acetates and long-chain fatty acyl esters) are being identified from a wide variety of invertebrate sources. It is apparent that they form a major class of ecdysteroid conjugates which can serve either as inactivation products prior to excretion or as a storage form of ecdysteroid. We have assessed the suitability of reversed-phase (RP) high performance liquid chromatography (HPLC) on a C_{18}-stationary phase and normal phase (NP) HPLC on silica for the separation and identification of such conjugates. When used in conjunction, these systems provide adequate resolution and selectivity for the preliminary identification of ecdysteroid acetates and fatty acyl esters.

We have compared the quantities of ecdysteroids present in mature male, and both immature and mature female house crickets (*Acheta domesticus*) by differential radioimmunoassay (RIA). Ecdysteroids were detected in all three adult samples, regardless of which of three antisera was used. Titres in mature adult females were significantly higher than in immature females, which in turn were higher than in adult males. We have analyzed the distribution of ecdysteroids in mature adult female insects and find that they are present in all tissues in relatively high concentrations. Further, evidence is provided that hydrolysable apolar conjugates are present in all tissues. Much lower concentrations of hydrolysable polar conjugates are also present in some tissues (ovaries and haemolymph). The chromatographic behaviour of the free and apolar conjugated ecdysteroids has been analyzed by RP and NP HPLC/RIA. The major free ecdysteroid present in all extracts was ecdysone; no evidence has been found for the presence of 20-hydroxyecdysone in adult house crickets. The proportion of the total ecdysteroids found in the form of apolar conjugates varied considerably from organ to organ, and it is clear that several different apolar conjugates occurred in some of the extracts. In addition to the previously identified ecdysone 22-fatty acyl esters, which were present in all tissues of mature adult females, preliminary evidence is presented for the occurrence of ecdysone 2,22-diacetate in extracts of fat-body, gut and haemolymph. The nature and distribution of ecdysteroids found *in vivo* is compared to the previously determined metabolism of [^{3}H]ecdysone by organs from mature adult females *in vitro*.

INTRODUCTION

In the past fifteen years there has been considerable interest in the occurrence and function of ecdysteroids in adult insects (for reviews, see Refs. [1,2]). Ecdysteroids have been detected in both male and female adult insects, but the majority of published work to date has concerned female adults. No unifying pattern concerning the types of ecdysteroid molecules, their distribution, quantities or functions in adult female insects has yet

Chromatography and Isolation of Insect Hormones and Pheromones
Edited by A. R. McCaffery and I. D. Wilson
Plenum Press, New York, 1990

emerged from the handful of insect species which have been studied in any detail. Thus, in mosquitoes, ecdysone is secreted into the haemolymph by the ovary, is converted to 20-hydroxyecdysone and induces vitellogenin synthesis in the fat-body [1]. Whether ecdysteroids control vitellogenin synthesis in other Diptera is not so clear [2]. In locusts, the ovary is also a site of synthesis of ecdysteroids, but synthesis occurs after vitellogenin synthesis (which is known to be induced by juvenile hormone in these insects) and almost all (98%) of the synthesized ecdysteroids are incorporated into the developing oocytes and removed from the female on egg deposition [3]. Truly massive amounts of these hormonal molecules are found in newly-laid locust eggs (ca. 75 μg/g fresh wt.) [4]. Moreover, more than 95% of the ecdysteroid was in the form of polar conjugates. These conjugates have been identified as 22-phosphates of ecdysone, 2-deoxyecdysone, 20-hydroxyecdysone and 2-deoxy-20-hydroxyecdysone in the desert locust *Schistocerca gregaria* [5]. Ecdysteroid phosphates have also been identified in the ovaries of the silkworm *Bombyx mori* [6,7], whereas newly-laid eggs of *Periplaneta americana* [8,9] contained predominantly apolar ecdysteroid conjugates. Newly-laid eggs of some other species (e.g., stick insects) appear to contain no conjugated ecdysteroid and only low concentrations of free ecdysteroids ([10] and Dinan, unpublished observation). In the case of *Drosophila melanogaster* newly-laid eggs, ecdysteroids were only detectable after digestion of egg proteins with proteases in the presence of *Helix* hydrolases [11].

During embryogenesis of *L. migratoria* [12], *Blaberus craniifer* [13] and *Nauphoeta cinerea* [14], peaks of free ecdysteroids have been observed to correlate with embryological cuticular events. In *L. migratoria*, evidence has also been presented which suggests that the production of ecdysteroids in the follicle cells and their transfer into the oocyte is associated with meiotic reinitiation in the oocyte [15,16].

The occurrence and metabolism of ecdysteroids during embryogenesis of the lepidopteran *Manduca sexta* have been extensively investigated ([17,18] and references cited therein). Very high concentrations of ecdysteroids were found in newly-laid eggs (ca. 40 μg/g), predominantly as 26-hydroxyecdysone 26-phosphate, which was converted during embryogenesis into a series of other ecdysteroids. Correlation of the appearance of the various ecdysteroids with cuticular events is poor [17]. It does not appear that the embryo synthesizes ecdysteroids *de novo* in this species, as all newly-formed ecdysteroids can be explained by conversions of pre-existing ecdysteroids, all ultimately derived from the maternal 26-hydroxyecdysone 26-phosphate. Surprisingly, the prime ecdysteroid in this system is 26-hydroxyecdysone; titres of ecdysone and 20-hydroxyecdysone are very low throughout embryogenesis. This does not however appear to be a general feature of lepidoptera, as ovaries and eggs of *B. mori* [19] and *Galleria mellonella* [20] contain significant proportions of 2-deoxyecdysone, ecdysone, 20-hydroxyecdysone and 2-deoxy-20-hydroxyecdysone.

In contrast to the situation in locusts, only 68% of the total ecdysteroids present adult female *G. mellonella* are associated with the ovary [20]. Several other insects have also been shown to have significant quantities of ecdysteroids in tissues other than the ovary: *Calliphora vicina* [21], *A. domesticus* [22], *N. cinerea* [23] and *P. americana* [24].

Further roles which have been attributed to ecdysteroids during oogenesis and/or embryogenesis are the separation of follicles from the germarium (in *Tenebrio molitor, G. mellonella* and *A. aegypti*), inactivation of corpora allata (in the viviparous cockroaches *N. cinerea* and *Diploptera punctata*), loss of regenerative capacity in embryonic appendages (in *B. craniifer*), control of egg sclerotization (in *A. aegypti*) and control of embryonic diapause (in *B. mori*) (reviewed in Ref. [2]).

Clarification of the functions of the ecdysteroids during oogenesis and embryogenesis will only be achieved by extension of investigations on the already partially studied species and the study of a wider range of insects to determine if further patterns of ecdysteroid occurrence and distribution occur. Consequently, we have begun to investigate the role of ecdysteroids in reproduction in house cricket, *A. domesticus*. We have previously shown that significant quantities of ecdysteroids are present in newly-laid eggs, mainly in the form of apolar conjugates (732 ng ecdysone equivalents/g fresh wt.) [25], which we have identified as a mixture of long-chain fatty acyl esters at the C-22 position of ecdysone [26]. Ovaries from mature females are capable of converting [^{3}H]-ecdysone into the same mixture of fatty acyl esters *in vitro* [26]. We have also made a preliminary study of the metabolism of [^{3}H]-ecdysone *in vitro* by other organs of adult females [27]. Here we consider the distribution and identification of endogenous ecdysteroids in mature adult female insects and compare these with adult male and immature female insects.

MATERIALS AND METHODS

Ecdysteroids

Ecdysone and 20-hydroxyecdysone were purchased from Simes, Milan. 2-Deoxyecdysone and 2-deoxy-20-hydroxyecdysone were from Sigma. Ecdysone 22-fatty acyl esters were chemically synthesized as described previously [28] and ecdysone acetates were prepared according to the method of Galbraith and Horn [29].

Insects

House crickets were reared and dissected as described previously [25,27]. Dissected organs were washed in insect Ringer solution before extraction to remove any adhering haemolymph. Mature females were recognized by their distended abdomens and by the fact that they had begun egg laying, while immature females were used after the adult moult, but prior to extensive ovarian development.

Extraction Procedure

Insects and tissues were extracted as described previously [25]. Briefly, samples were homogenized and extracted with ethanol (5 ml/g fresh wt.). The pooled supernatants after centrifugation were taken to dryness, adsorbed onto Celite [4] and separated on silicic acid (2 g) to give three fractions: chloroform (apolar lipids), methanol-chloroform (3:7 v/v; free ecdysteroids and apolar ecdysteroid conjugates) and methanol (polar ecdysteroid conjugates and ecdysonoic acids).

Radioimmunoassay

The RIA procedure has been described previously [25]. Three antisera were used in this work: DBL-1 (at 3000-fold dilution), H-2 (at 300-fold dilution) and H-22 (at 1600-fold dilution). The DBL-1 antiserum was very kindly provided by Dr L. Reum and Prof. J. Koolman (Marburg, FRG), while the H-2 and H-22 antisera were generously provided by Prof. L. I. Gilbert and Prof. J. D. O'Connor (Chapel Hill, USA). The cross-reactivities of all three antisera have been described previously [30], but it should be mentioned that the RIA conditions used here are different and this may influence the cross-reactivities. All RIA results are expressed in ecdysone equivalents.

Enzymic Hydrolyses

Silica column fractions were tested for the presence of ecdysteroid conjugates by enzymic hydrolysis of RIA-negative conjugates for 5 days at 37°C with *Helix* enzymes (Sigma type I: 10 mg/ml 0.1 M acetate buffer [pH 5.4]), using conditions suitable for the hydrolysis of ecdysone fatty acyl esters [25]. Ecdysteroids released from the conjugates were determined by RIA.

High-Performance Liquid Chromatography (HPLC)

Portions of methanol-chloroform (3:7 v/v) fractions containing ca. 20 ng RIA-positive material were separated by reversed- and normal-phase (RP and NP) high-performance liquid chromatography (HPLC) on columns which had been extensively washed with mobile phase to remove any contaminating ecdysteroids. Samples for RP-HPLC were dissolved in methanol (15 µl), to which water(30 µl) was added just prior to separation. Any precipitate of lipid was removed by centrifugation before the sample was injected onto the column (Spherisorb ODS-2; 25 cm x 4.6 mm i.d.). The column was eluted at a constant flow-rate of 1 ml/min by linear gradient elution from methanol-water (3:7 v/v) to pure methanol over 30 min, followed by isocratic elution with pure methanol for a further 50 min. For NP-HPLC, samples were dissolved in mobile phase (50 µl), injected onto a Spherisorb SW-5 column (25 cm x 4.6 i.d.) and eluted isocratically with dichloromethane-propan-2-ol-water (125:25:2 v/v/v) at 1 ml/min.

The effluent from the HPLC separations was collected in fractions of 1 min (RP-HPLC) or 30 sec (NP-HPLC). Aliquots of each fraction were subjected to RIA to determine the elution profile of "free" (RIA-positive) ecdysteroids. The remainder of each fraction was subjected to enzymic hydrolysis followed by RIA, to determine additionally the retention times of any hydrolysable conjugates. Certain fractions containing hydrolyzed conjugates were resubjected to HPLC in order to determine the nature of the released ecdysteroid moiety.

RESULTS

HPLC of Ecdysteroid Acyl Esters

The retention times of various ecdysone acyl esters on a reversed-phase gradient system and an isocratic normal-phase (silica) system are given in Table 1 and compared to those of several free ecdysteroids in these same systems. The two systems in combination have sufficient resolution and selectivity for the preliminary identification of ecdysone acyl esters in biological extracts by HPLC/RIA prior to thorough identification by mass spectrometry (MS) and nuclear magnetic resonance spectroscopy (NMR) etc.

Preliminary Comparison of Ecdysteroid Levels in Adult Crickets

Mature adult males, immature adult females and mature adult females were extracted with ethanol and these extracts were assessed for ecdysteroid content by differential RIA using three antisera (Table 2). Owing to the significantly different cross-reactivities of these three antisera [30], it should be possible to detect almost all types of ecdysteroid with at least one of the antisera. Thus, the H-2 antiserum is sensitive to changes (including conjugation) in the side-chain, while the H-22 antiserum generally is less cross-reactive with ecdysteroids modified in the nucleus. RIA-positive material was detected in all extracts with all three antisera, but the amounts varied considerably between extracts. Only very low concentrations of ecdysteroids were detected in adult male insects, while moderate amounts were detected in immature females and the highest quantities in mature females.

The close similarity in the values obtained for adult males or immature female extracts with each of the antisera indicated that ecdysone (which was used as the reference ecdysteroid for all three RIAs) or a structurally very similar ecdysteroid was the major compound present in these extracts. The greater variation in the RIA responses for the mature female extract indicated that modified ecdysteroids may be present in this extract.

All further studies in this work were conducted using the DBL-1 antiserum only.

Polarity and Titre of Free Ecdysteroids and Hydrolysable Conjugates

The whole insect extracts were separated on silicic acid to give three broad fractions. Each of these fractions was assessed for RIA-positive material (Table 3) and this revealed that in each case the majority of the RIA-positive material eluted in the methanol-chloroform (3:7 v/v) fraction, although significant amounts were also detected in the methanol fraction in certain cases.

Table 1. Summary of retention times for ecdysteroids and ecdysone acyl esters on reversed-phase and normal-phase HPLC. HPLC conditions are given in the text

Ecdysteroid	Retention times (mins) normal-phase	reversed-phase
20-hydroxyecdysone	24.3	16.2
2-deoxy-20-hydroxyecdysone	10.8	20.7
2-deoxyecdysone	7.9	23.6
ecdysone	15.3	19.4
ecdysone 2-acetate	6.4	23.5
ecdysone 3-acetate	6.2	21.5
ecdysone 22-acetate	10.8	26.0
ecdysone 2,3-diacetate	3.9	26.5
ecdysone 2,22-diacetate	4.9	28.8
ecdysone 3,22-diacetate	4.4	27.2
ecdysone 2,3,22-triacetate	3.3	30.8
ecdysone 2,3,22,25-tetraacetate	3.0	32.2
ecdysone 22-myristate	8.8	42.2
ecdysone 22-palmitate	8.8	44.7
ecdysone 22-stearate	8.8	47.4
cholesterol	3.0	60.5
void volume	3.0	3.3

Table 2. Quantities of RIA-positive material in ethanolic extracts of adult crickets

Insect	RIA Response (ng ecdysone equivalents/insect) DBL-1	H-2	H-22
Mature male	2.1	1.8	2.4
Immature female	10.2	13.6	13.2
Mature female	37.5	48.0	52.5

Each of the fractions was subjected to enzymic hydrolysis with a crude mixture of *Helix* enzymes to detect the presence of hydrolysable conjugates releasing RIA-positive ecdysteroids. In view of the lengthy incubation period, an appropriate enzyme-free control was established for each fraction. In determining the levels of RIA-positive material in the hydrolysates and controls, a range of aliquot sizes was tested for each sample. If ecdysteroids alone were responsible for the RIA response, there should have been a linear relationship between RIA response and aliquot size (within the range of the RIA: 20 to 1000 pg ecdysone equivalents with DBL-1). An example of such plots is given in Figure 1 for the ovarian fractions. Some samples did not give a linear relationship at all and it is presumed that the RIA response is a consequence of interfering substances. Other samples gave a linear relationship at low aliquot size, but deviated at high aliquot size. In these cases, the quantity of ecdysteroid-like material was determined from the linear portion of the plot. Samples containing high concentrations of ecdysteroids typically gave linear plots over the entire range tested.

Table 3. Polarity of RIA-positive ecdysteroids and hydrolysable conjugates in extracts of whole adult insects and organs of adult female insects. Ethanolic extracts were separated on silicic acid eluted with increasing percentages of methanol in chloroform (0, 30 and 100%). These fractions were assessed for RIA-positive material with the DBL-1 antiserum and further aliquots were incubated with *Helix* enzymes in acetate buffer or as buffer controls before RIA

Sample (wt/insect)	% Methanol	RIA Response ng ecdysone equivalents/insect (ng ecdysone equivalents/g fresh wt.) Fraction	Control	Hydrolysate
mature male	0%	-	2.02 (7.83)	3.34 (13.0)
(0.258 g)	30%	2.61 (10.1)	4.54 (17.6)	6.56 (25.4)
	100%	?	5.05 (19.6)	6.06 (23.5)
immature female	0%	0.45 (2.11)	2.16 (10.1)	9.54 (44.8)
(0.213 g)	30%	10.1 (47.5)	15.3 (71.7)	24.7 (115.9)
	100%	3.05 (14.3)	2.67 (12.5)	6.80 (31.9)
mature female	0%	-	1.72 (3.38)	?
(0.509 g)	30%	51.8 (101.7)	59.1 (116.2)	93.6 (184.0)
	100%	4.93 (9.69)	7.64 (15.0)	15.5 (30.5)
haemolymph	0%	-	-	-
(0.017 g)	30%	2.83 (164.5)	4.16 (241.9)	6.02 (350.0)
	100%	0.57 (33.1)	0.56 (32.6)	2.17 (126.2)
fat-body	0%	-	-	-
(0.008 g)	30%	1.75 (218.8)	2.05 (256.3)	3.20 (400.0)
	100%	0.42 (52.5)	0.21 (26.3)	0.21 (26.3)
gut	0%	0.45 (9.0)	0.21 (4.2)	1.03 (20.6)
(0.050 g)	30%	2.46 (49.2)	2.05 (41.0)	3.76 (75.2)
	100%	0.52 (10.4)	1.42 (28.4)	1.86 (37.2)
ovaries	0%	-	-	?
(0.098 g)	30%	4.11 (41.9)	6.65 (67.9)	32.5 (331.6)
	100%	0.57 (5.82)	0.77 (7.86)	2.44 (24.9)
carcass	0%	0.44 (1.44)	0.39 (1.28)	0.82 (2.69)
(0.305 g)	30%	29.5 (96.7)	25.6 (83.9)	25.6 (83.9)
	100%	?	3.69 (12.1)	?

- : not detected
?: non-linearity of response owing to interfering substances

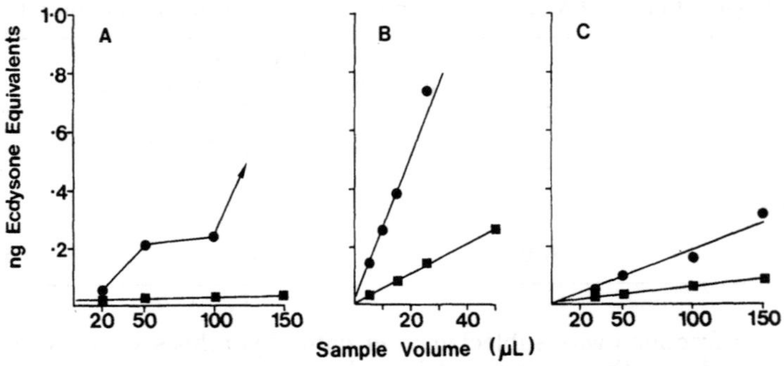

Fig. 1. Relationship between RIA response (DBL-1 antiserum) and aliquot size for (a) chloroform; (b) methanol-chloroform (3:7 v/v); and (c) methanol fractions from the silicic acid separation of the ethanolic extract of ovaries of mature adult female house crickets: ●, response after hydrolysis with *Helix* enzymes (10 mg/ml in 0.1 M acetate buffer, pH 5.4, for 5 days at 37°C): ■, response in enzyme-free control.

It is significant that the quantities of ecdysteroids present in the hydrolysis controls for the methanol-chloroform (3:7 v/v) fractions were higher than those found in the original fractions (Table 3). This indicates the presence of labile conjugates in these fractions which were being degraded during the incubation to produce a more RIA-positive substance.

Hydrolysis revealed the presence of significant quantities of hydrolysable conjugates in the methanol-chloroform (3:7 v/v) and methanol fractions of immature and mature female extracts. Low concentrations of conjugates may also be present in adult males and in the chloroform fraction of immature females.

Distribution of Ecdysteroids and Hydrolysable Conjugates in Mature Adult Females

Mature adult female crickets were bled and dissected. The pooled organs were weighed and then extracted and separated by the same procedure used for the whole insects. Fractions were monitored by RIA and also subjected to enzymic hydrolysis followed by RIA (Table 3). The haemolymph weight of 17 mg/insect given in Table 3 corresponds to the average obtained from each insect, not the total volume of haemolymph in an insect. The total volume of haemolymph in adult female *A. domesticus* has been previously determined to be 107 μl [22].

Ecdysteroid Concentrations in Testes and Male Accessory Glands

To determine whether the low titres of ecdysteroids present in adult males were associated with the male reproductive organs, extracts of the testes and accessory glands from 20 insects were assessed for their RIA-positive content both before and after treatment with *Helix* hydrolases. No RIA-positive material was detected in accessory glands, but low concentrations (54.5 ng/g tissue: 1.58 ng/testis pair) were detected in the testes, of which ca. 40% was composed of free ecdysteroids and 60% of (apolar) conjugates (data not shown). This accounted for between 1/3 and 1/2 of the total ecdysteroids found in adult male insects.

RP-HPLC of Methanol-Chloroform (3:7 v/v) Fractions

Appropriate portions of these fractions were evaporated to dryness and redissolved in methanol-water (1:2 v/v). In those samples which contained a large amount of lipid (in particular the ovary extract), this procedure caused a precipitate to form, which was removed by centrifugation before the rest of the sample was injected onto the HPLC column. Very apolar ecdysteroids, such as long-chain fatty acyl esters may have partially or totally co-precipitated with the lipid and been lost from the sample. The HPLC separation of each sample was fractionated and each of these fractions was assessed for their RIA-positive response both before and after enzymic hydrolysis. The results are summarized in Figure 2.

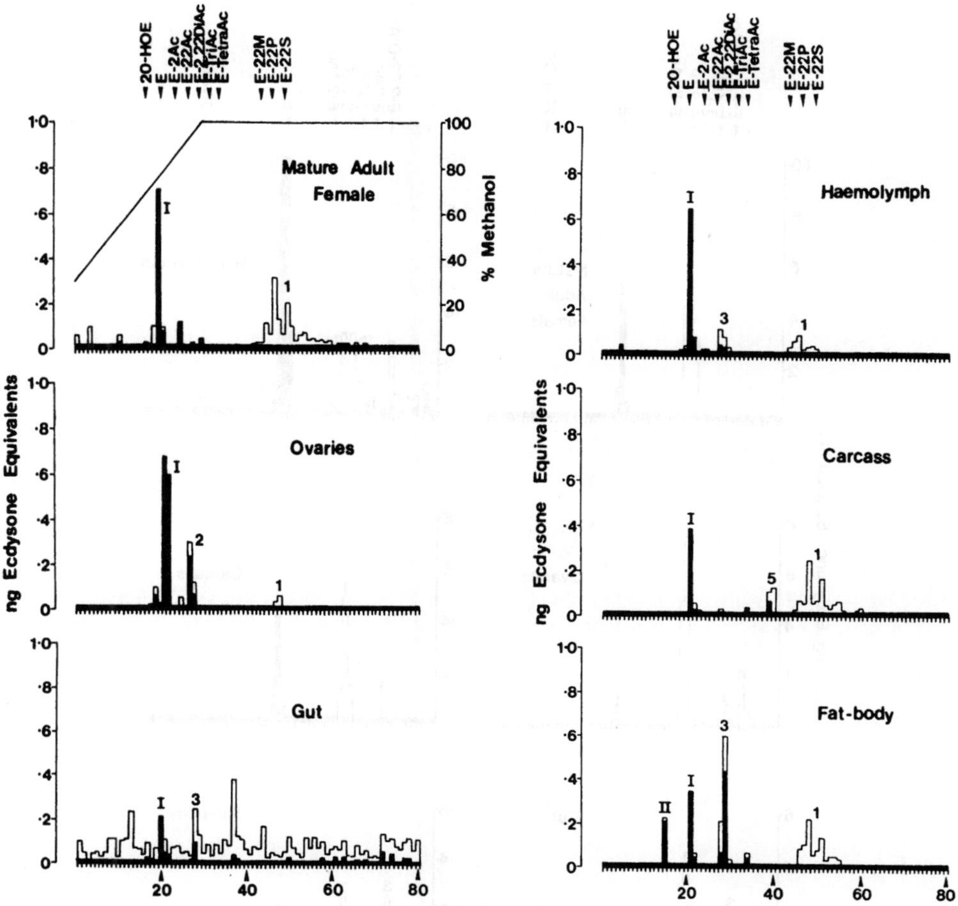

Fig. 2. RP-HPLC/RIA separations of free and hydrolysable apolar conjugated ecdysteroids in extracts of entire mature adult female house crickets and organs dissected from them. HPLC conditions are given in the text. Fractions (1 min duration) of the column effluent were collected and subjected to RIA (DBL-1 antiserum). Black columns give the RIA response of each fraction. White columns give the extra RIA-positive material detected after hydrolysis of fractions with *Helix* enzymes. The positions at which reference compounds elute are indicated by arrows: 20-HOE = 20-hydroxyecdysone, E = ecdysone, E-2Ac = ecdysone 2-acetate, E-22Ac = ecdysone 22-acetate, E-2,22diAc = ecdysone 2,22-diacetate, E-triAc = ecdysone 2,3,22-triacetate, E-tetraAc = ecdysone 2,3,22,25-tetraacetate, E-22M = ecdysone 22-myristate, E-22P = ecdysone 22-palmitate and E-22S = ecdysone 22-stearate. RIA peaks of the same chromatographic behaviour are numbered (Roman numerals: "free" ecdysteroids. Arabic numerals: conjugated ecdysteroids) to assist comparison of chromatograms.

NP-HPLC of Methanol-Chloroform (3:7 v/v) Fractions

No solubility problems were observed in dissolving samples in the mobile phase for NP-HPLC. Results are summarized in Figure 3.

Separation of Adult Male and Immature Female Extracts by NP-HPLC

In view of the problem of the co-precipitation of apolar ecdysteroid conjugates with lipid during the preparation of samples for RP-HPLC, it was thought that the RP-separation of the methanol-chloroform (3:7 v/v) fractions of the immature female and adult male extracts, which contained large amounts of lipid but only low concentrations of ecdysteroids, would be of dubious significance. Therefore, these samples were analyzed by NP-HPLC only (Figure 4).

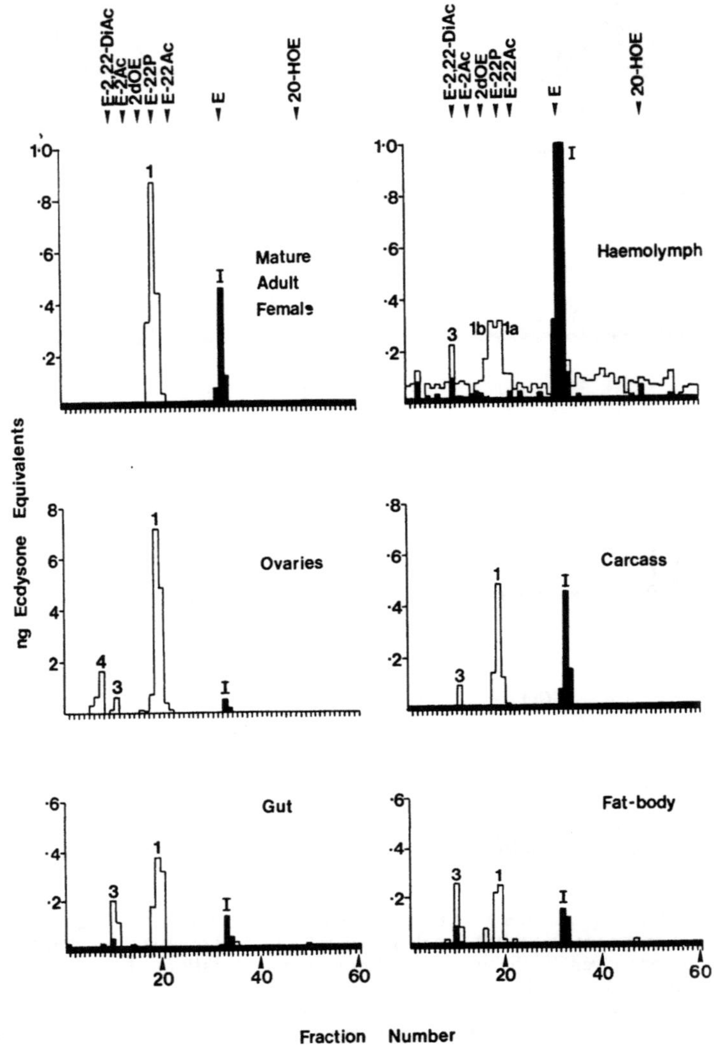

Fig. 3. NP-HPLC/RIA separations of free and hydrolysable apolar conjugated ecdysteroids in extracts of entire mature adult female house crickets and organs dissected from them. HPLC conditions are given in the text. Column fractions (30 sec duration) were subjected to RIA (DBL-1 antiserum) before (black columns) and after (black + white columns) enzymic hydrolysis with *Helix* enzymes. The abbreviations of the reference compounds are as given for Figure 2 with the addition of 2dOE = 2-deoxyecdysone.

DISCUSSION

HPLC Separation of Ecdysone Acyl Esters

A description of the chromatographic behaviour of ecdysone 22-fatty acyl esters on silica-HPLC and several RP-HPLC systems has been published recently [31]. This demonstrated that the ecdysone 22-long-chain fatty acyl esters co-migrated on silica HPLC, but separated on RP-HPLC. Increased length of the fatty acyl carbon chain length increased the retention time on RP-HPLC, whereas increasing unsaturation decreased the retention time. The shift in retention time caused by the addition of each double bond was dependent on the nature of the stationary phase (alkyl or nitrile), the chain length of alkyl stationary phases and on the mobile phase (methanol-water or acetonitrile-methanol). A non-aqueous RP-HPLC system has also been developed for chromatography of ecdysone mono-, di- and tri-fatty acyl esters [4]. The data in Table 1 relate the retention times of the ecdysone 22-fatty acyl esters to those for the ecdysone acetates and some free ecdysteroids. Normal-phase HPLC is useful for the recognition of the ecdysone mono (2/3- or 22-) fatty acyl esters

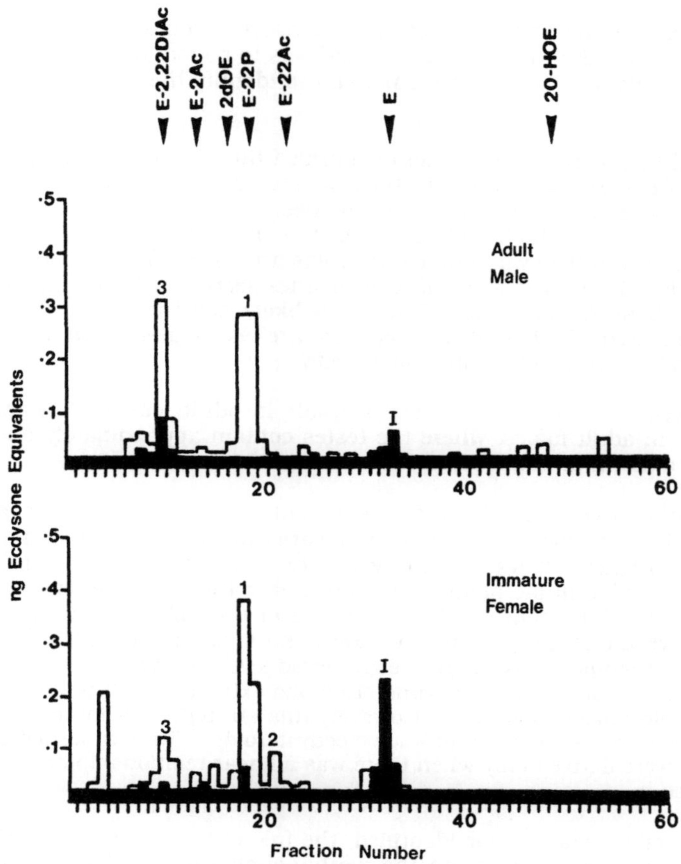

Fig. 4. NP-HPLC/RIA separations of free and hydrolysable apolar ecdysteroid conjugates in extracts of entire mature males and immature female house crickets. Details are as described in the text and in the legend to Figure 3.

as classes, whereas RP-HPLC allows the separation of the fatty acyl ester homologues. Application of several RP-HPLC systems in series permits the identification of the fatty acyl composition of the conjugates [27,31]. Ecdysteroid acetates show extensive separation on NP-HPLC, but the 2- and 3-monoacetate and the 2,22- and 3,22-diacetate pairs almost co-elute. There is complete resolution of all acetates of ecdysone on RP-HPLC and all are clearly resolved from the much later eluting 22-fatty acyl esters. As yet, 2- or 3-mono-, di- or tri-fatty acyl esters of ecdysteroids have not been detected in biological systems. However, as various mono- and di-acetates of ecdysteroids have been identified [27,32-38], it seems probable that the corresponding fatty acyl esters will also eventually be detected. The elution order for the various acetates on NP- or RP-HPLC was maintained for the corresponding positional isomers of the fatty acyl esters. Alteration of the esterified ecdysteroid caused the expected change in retention behaviour. Thus, 20-hydroxyecdysone 22-fatty acyl esters eluted before the corresponding ecdysone 22-fatty acyl esters on RP-HPLC and after them on NP-HPLC (Dinan, unpublished observation). The two HPLC systems presented here have considerable resolving power and, when used in conjunction, are suitable for the preliminary identification of ecdysteroid acyl esters. However, it is conceivable that biological extracts containing a complex mixture of ecdysteroid acyl esters which differ in the ecdysteroid moiety and the length, degree of unsaturation and the location of the acyl group may not allow the unambiguous identification of some components.

Ecdysteroids in Adult House Crickets

Renucci and Strambi [22] have previously investigated ecdysteroid titres in ovaries and haemolymph of adult female *A. domesticus* by RIA. They measured titres before and after enzymic hydrolysis with *Helix* enzymes, but it is not clear whether the conditions used would be suitable for the hydrolysis of acyl ecdysteroid conjugates [25]. However, they concluded that one third of the total ecdysteroid was to be found in the haemolymph. We wished to extend this study to other organs and to identify the nature of the RIA-positive material.

Differential RIA using three antisera confirmed the presence of ecdysteroids in adult house crickets (Table 1). Low concentrations were detected in mature adult males, while high titres were detected in mature adult females. Low to intermediate amounts were detected in adult females which had not yet begun egg laying. The ecdysteroids in adult females are produced in the ovary and ecdysteroids are absent from ovariectomized females [22]. We have also shown that *A. domesticus* ovaries secrete ecdysteroids *in vitro* (Whiting and Dinan, unpublished observation). Thus, it is likely that the intermediate quantities of ecdysteroids detected in "immature" females are associated with incipient ovarian maturation in some of these asynchronous insects.

If the ovaries are the source of ecdysteroids in adult females, a parallel situation appears to exist in adult males, where the testes contain approximately one third of the, albeit far lower, amount of total ecdysteroid.

Separation of the ethanolic extracts of entire insects on silicic acid columns permitted a preliminary consideration of the polarity of the RIA-positive material and the hydrolysable conjugates. It was found necessary to determine the level of RIA-positive material by means of multiple point measurements, rather than by assessment of single samples giving an RIA response within the range of the RIA. In certain cases, the RIA response was a result of the presence of interfering substances, which produced a false positive response or could mask the true ecdysteroid level. To a certain extent this problem is a consequence of working with a crude ethanolic extract, which is not further purified prior to silicic acid chromatography. However, this strategy was intentionally developed [25] so as to minimize losses of any unknown ecdysteroids. The ecdysteroid concentrations given in Table 3 were derived only when there was a linear relationship between aliquot size and RIA response.

Of the zooecdysteroids so far identified, the free ecdysteroids and apolar conjugates elute from the silicic acid column in the methanol-chloroform (3:7 v/v), whereas polar conjugates and ecdysonoic acids elute in the methanol fraction. Only in the extract of mature male insects was the quantity of RIA-positive material present in the methanol fraction comparable to that in the methanol-chloroform (3:7 v/v) fraction (before or after enzymic hydrolysis). In immature and mature female insects, the concentrations of RIA-positive material were 5- to 10-fold higher in the methanol/chloroform (3:7 v/v) fraction than in the corresponding methanol fraction. Further, these fractions showed a considerable increase in RIA response on enzymic hydrolysis. The data in Table 3 do provide some evidence for the occurrence of polar ecdysteroid conjugates in female insects. The significance of the apparent increase in RIA response on hydrolysis of the chloroform fraction from immature females is not at all clear at present, but this may be due to interference from fatty acids released from triacylglycerols by esterases present in the *Helix* enzymes.

Distribution of Ecdysteroids in Mature Adult Females

All organs of mature adult female insects contained ecdysteroids (Table 3). The RIA-positive material in all organ extracts was primarily recovered in the methanol-chloroform (3:7 v/v) fraction, but significant amounts of hydrolysable polar conjugates were detected in the ovary and haemolymph. Only in the case of the carcass extract did enzymic hydrolysis not reveal the presence of apolar ecdysteroid conjugates. However, HPLC/RIA results (see later) conclusively showed that this extract did also contain apolar hydrolysable conjugates. The reason for this contradiction is not clear at present. Concentrations of RIA-positive ecdysteroids and hydrolysable apolar ecdysteroid conjugates (expressed as ng ecdysone equivalents/g tissue) are not uniform throughout the insect. This may represent: (i) the presence of different ecdysteroids in different organs; (ii) the selective accumulation of ecdysteroids in some organs; (iii) the concentration correlating with a particular component of the tissue (e.g., water content) and not the total fresh weight; or (iv) a combination of the above.

The increase in RIA-positive material on enzymic hydrolysis varied from 1.5- to 2-fold in the haemolymph, fat-body and gut extracts to 5-fold in the ovarian extract. When the results for the ovarian extract are compared with those for a comparable extract of newly-laid eggs, it is apparent that the ratio of free ecdysteroids to apolar conjugates is higher and that the amount of apolar conjugates is lower. This is in agreement with the finding of Renucci and Strambi [22], who showed that the ratio of free to conjugated ecdysteroids in the ovary decreased as ovarian development proceeded and that the ratio was very low in newly-laid eggs.

The majority (54%) of "free" ecdysteroids in adult females were associated with the carcass. Considering that this consists predominantly of cuticle, muscle and epidermis, it would be revealing to determine the distribution of the ecdysteroid within this compartment. The sum of the "free" ecdysteroid in the methanol-chloroform (3:7 v/v) fractions from the organ extracts (having allowed for the amount of haemolymph lost) is 54.8 ng ecdysone equivalents/insect. This is very similar to the 51.8 ng ecdysone equivalents/insect found in the methanol-chloroform (3:7 v/v) fraction for the extract of entire females.

When one considers the total (free + conjugated) ecdysteroids, the highest amounts of RIA-positive material were found in the haemolymph (36% allowing for haemolymph losses on bleeding), ovary (32%) and carcass (25%, but see earlier caveat).

HPLC Separation of Free and Apolar Conjugated Ecdysteroids from Extracts of Mature Female Insects and Organs

Mature Female Insects: The major free ecdysteroid present in extracts of entire mature adult female insects was ecdysone (peak I), which is demonstrated by co-chromatography with reference ecdysone on RP-HPLC (Figure 2) and NP-HPLC (Figure 3). Hydrolysis of column fractions detected a family of apolar conjugates (peak 1) which were eluted late on RP-HPLC. These conjugates eluted as one peak on normal-phase. This behaviour is characteristic of the ecdysone 22-fatty acyl esters [31] which have been isolated from newly-laid eggs of *A. domesticus* and conclusively identified [26]. On both systems, the conjugates co-chromatographed with chemically synthesized ecdysone 22-long-chain fatty acyl esters. Thus, the major ecdysteroids present in mature adult female insects were ecdysone and ecdysone 22-fatty acyl esters in a ratio of 1:2:5.

Ovaries: In the ovarian extract, the major free ecdysteroid present was again ecdysone (peak I), but there was a smaller amount of material which co-chromatographed with ecdysone 22-acetate on reversed-phase (peak II). The RIA response of the putative ecdysone 22-acetate peak was slightly enhanced on enzymic hydrolysis, indicating that it was a conjugate of some sort. Ecdysteroid acetates are poorly hydrolyzed by *Helix* enzymes and the incubation conditions may not give complete hydrolysis. The major conjugates in the ovaries were the ecdysone 22-fatty acyl esters (peak 1), as shown in Figure 3. These conjugates were not apparent after RP separation presumably because of co-precipitation of these very apolar conjugates with other lipids during the preparation of the sample. Two other peaks (3 and 4) of RIA-positive material were detected after hydrolysis which eluted before the ecdysone 22-fatty acyl esters. The first of these (peak 4) eluted soon after the void volume of the column and may simply be due to interference in the RIA by fatty acids released by hydrolysis of lipids. The second (peak 3) eluted at a retention time similar to ecdysone 2,22-diacetate. The identification of these two peaks requires further investigation.

Haemolymph: In haemolymph, in addition to free ecdysone (peak I; 62% of the total), there were apparently 2 or 3 apolar conjugates which contributed 38% of the total ecdysteroids present. On RP-HPLC two groups of apolar conjugates were detected, one of which (peak 1) co-chromatographed with the ecdysone 22-fatty acyl esters and the other (peak 3) with ecdysone 2,22-diacetate. On NP-HPLC conjugates were detected in fraction 10, co-chromatographing with ecdysone 2,22-diacetate, and in fraction 15-22. This second peak appeared to be a double peak (1a and 1b) both because of the partial separation and because of the breadth of the peak. The portion with the greater retention time on NP-HPLC co-chromatographed with the ecdysone 22-palmitate reference. The identity of the earlier eluting portion is at present unknown. However, separation on RP-HPLC of the RIA-positive material released on hydrolysis of fractions 15 to 22 resulted in co-chromatography with ecdysone. Therefore, peak 1b is a conjugate of ecdysone.

Carcass: Ecdysone (peak I) and ecdysone 22-fatty acyl esters (peak 1) predominated in the carcass. An extra small conjugate peak (5) is indicated in fractions 39 and 40 on RP-HPLC. This peak was not detected in any other extract, nor did an equivalent peak appear in the NP separation of the carcass extract. Therefore, its validity must be considered dubious at the present.

Fat-Body: On RP-HPLC of fat-body extracts two peaks of free ecdysteroid (I and II) were detected. Peak I corresponded to ecdysone while peak II was more polar, eluting before the 20-hydroxyecdysone reference. Peak II was not detected on NP-HPLC, but this may have been because it may not have eluted within 30 min because of its increased polarity. Two major apolar conjugates were detected (1 and 3), corresponding to the ecdysone 22-fatty acyl esters and ecdysone 2,22-diacetate, respectively.

Gut: Owing to the low ratio of ecdysteroids to tissue mass in gut, a large amount of sample had to be injected onto the HPLC columns. The impurities resulted in a high background after enzyme hydrolysis of the fractions from RP-HPLC. However, it is apparent from the initial RIA on the RP-HPLC fractions and the RIA results before and after hydrolysis of the NP-HPLC fractions, that ecdysone, ecdysone 22-fatty acyl esters and putative ecdysone 2,22-diacetate were present. Again, the ecdysone 22-fatty acyl esters were not present on the RP chromatogram because of co-precipitation of these compounds with lipid on preparation of the sample.

Normal-Phase HPLC Separation of Free and Apolar Conjugated Ecdysteroids in Extracts of Mature Male and Immature Female Insects

Essentially the same ecdysteroids were detected in adult males and in immature females as in mature female insects. Although based on co-chromatography in only one system, one can presume that the free ecdysteroid was ecdysone, while the conjugated ecdysteroids were ecdysone 22-fatty acyl esters and ecdysone 2,22-diacetate.

Comparison of Endogenous Ecdysteroids with Metabolites of [^{3}H]ecdysone Produced In Vitro by Organs of A. Domesticus Mature Adult Females

We previously investigated the metabolism of [^{3}H]ecdysone by organs of *A. domesticus* mature adult females *in vitro* [27]. It is of interest to compare the results in this report with those previously reported.

The carcass tissues did not metabolize [^{3}H]ecdysone significantly and therefore one can conclude that the endogenous ecdysone metabolites are produced in other tissues and accumulate in the carcass. The ovaries metabolized [^{3}H]ecdysone to essentially only ecdysone 22-fatty acyl esters. However, if a small amount of [^{3}H]ecdysone 22-acetate had also been formed, this would not have been readily detected. The fat-body extensively metabolized [^{3}H]ecdysone to produce both polar and apolar metabolites. The apolar metabolites were tentatively identified as ecdysone 22-fatty acyl esters and ecdysone 2,22-diacetate, which is supported by the findings in this report for the endogenous ecdysteroids in fat-body extracts. The [^{3}H]ecdysone metabolites formed by the fat-body *in vitro* remained at the origin on silica HPTLC (chloroform-methanol, 9:1 v/v, for development) and could therefore be polar conjugates or ecdysonoic acids. The fat-body was the major site for the production of polar metabolites. Yet, it is apparent from Table 3 that the methanol fraction derived from the fat-body extract did not contain hydrolysable conjugates. This indicates that the [^{3}H]polar metabolites are either non-hydrolysable, non-RIA-positive compounds (e.g., ecdysonoic acids) or that the polar metabolites are not retained in the fat-body, but released into the haemolymph and taken up by other tissues. Considering the evidence for the presence of polar hydrolysable conjugates in the haemolymph and ovaries (Table 3), the latter is a distinct possibility.

The gut tissues produced a wide variety of metabolites from [^{3}H]ecdysone *in vitro*. Among those tentatively identified were ecdysone 22-fatty acyl esters, ecdysone 2,22-diacetate and polar metabolites. In addition the crop produced a very apolar metabolite which migrated close to the solvent front on silica HPTLC (chloroform-methanol, 9:1 v/v, for development) and close to the origin on C_8-RP-HPTLC (with methanol-water, 9:1 v/v, for development). This metabolite is so apolar that it would probably elute in the chloroform fraction from the silica columns used in this study. Also, it is questionable whether such a metabolite would be RIA-positive or *Helix*-hydrolysable. Further studies with incubations of crop tissue with [^{3}H]ecdysone *in vitro* are needed to identify the nature of this metabolite prior to determining whether it occurs among the endogenous ecdysteroids.

No evidence is available either from the studies on the metabolism of [^{3}H]ecdysone *in vitro* or the separation of endogenous ecdysteroids for the presence of any 20-hydroxyecdysone in adult female insects. This would appear to be in contrast to adults of the Mediterranean field cricket *Gryllus bimaculatus* [39,40].

WORKING HYPOTHESIS AND CONCLUSIONS

By way of summary, our working hypothesis to rationalize the findings in this report and previously published results on the metabolism of [^{3}H]-ecdysone by organs of adult female house crickets is as follows: The ovaries appear to be the sole site of synthesis of ecdysteroids in adult female *A. domesticus* [22]. The ovaries synthesize ecdysone, ecdysone 22-fatty acyl esters and perhaps also a small amount of ecdysone 22-acetate. Large amounts of the ecdysone 22-fatty acyl esters are incorporated into the maturing oocytes and removed from the female as the eggs are laid, but significant amounts of ecdysteroids are also released into the haemolymph. The ratio of ecdysone to ecdysone 22-fatty acyl esters is considerably higher in the haemolymph and other non-ovarian tissues than in the ovary. This implies that either the ecdysone 22-fatty acyl esters are selectively taken up into the ovary or that the esterification occurs/continues in the oocyte. The ecdysteroids found in the haemolymph appear to be ecdysone together with smaller amounts of ecdysone 22-fatty acyl esters, ecdysone 2,22-diacetate and the unknown conjugate (peak 1b). We feel that the ecdysone and ecdysone 22-fatty acyl esters are released from the ovary, whereas the ecdysone 2,22-diacetate results from the metabolism of haemolymph ecdysone and/or ecdysone 22-acetate by non-ovarian tissues (gut and fat-body). The fat-body and gut are responsible for converting haemolymph ecdysone into polar metabolites (conjugates) which are, to a certain extent, accumulated by the ovary from the haemolymph. The carcass would appear to act as a sink for ecdysone and ecdysone 22-fatty acyl esters, although it does not metabolize ecdysteroids itself. It would be instructive to determine the excretion rate of ecdysteroids in mature adult female house crickets.

The identity of the putative ecdysone 2,22-diacetate and ecdysone 22-acetate must remain tentative at present. These identifications should be confirmed by spectral means (MS, NMR). This is one of our aims at present.

Although far lower amounts of ecdysteroid are detected in immature females and mature adult males, the identities of the ecdysteroids appear to be similar to those in mature females. For adult males, one could develop an analogous hypothesis to that developed above for adult females, where the testis is the source of ecdysteroids which are then distributed throughout the insect. This could be a further fruitful area of investigation.

The findings of this study raise a number of fundamental questions: why are ecdysteroids secreted by the ovaries and taken up by other tissues? Do ecdysteroids have a function in non-ovarian tissues? Why is the situation apparently so different in crickets from other orthopterans (locusts)? Why do locust ovaries produce polar ecdysteroid conjugates while cricket ovaries produce apolar conjugates? Why are concentrations of ecdysteroids 100-fold higher in newly-laid locust eggs than in cricket eggs? What function do ecdysteroids have during embryogenesis in crickets?

We hope that our future research will begin to answer these questions.

ACKNOWLEDGEMENTS

We thank Mr S. Sparks for the maintenance of the cricket colony. Profs. J. Koolman, L. I. Gilbert and J. D. O'Connor and Dr L. Reum generously provided the antisera. Anne Dinan and Del Bradbrook kindly read and commented on the manuscript.

REFERENCES

1. H. H. Hagedorn, The role of ecdysteroids in the adult insect, *in:* "Endocrinology of Insects", R. G. H. Downer and H. Laufer, eds., A. R. Liss Inc., pp 271-304 (1983).
2. H. H. Hagedorn, The role of ecdysteroids in reproduction, *in:* "Comprehensive Insect Physiology, Biochemistry and Pharmacology", G. A. Kerkut and L. I. Gilbert, eds., Pergamon Press, vol. 8:205-262 (1985).
3. M. Lagueux, Recherches sur les ecdysteroïdes dans les ovaires et les embryons d'insectes: revue des travaux chez l'orthoptère *Locusta migratoria, Rev. Canad. Biol. Experim.*, 42:215-221 (1983).
4. L. N. Dinan and H. H. Rees, The identification and titre of conjugated and free ecdysteroids in developing ovaries and newly-laid eggs of *Schistocerca gregaria, J. Insect Physiol.*, 27:51-58 (1981).
5. R. E. Isaac, M. E. Rose, H. H. Rees and T. W. Goodwin, Identification of the 22-phosphate esters of ecdysone, 2-deoxyecdysone, 20-hydroxyecdysone and 2-deoxy-

20-hydroxyecdysone from newly-laid eggs of the desert locust *Schistocerca gregaria*, *Biochem. J.*, 213:533-541 (1983).
6. T. Mizuno and E. Ohnishi, Conjugated ecdysone in the eggs of the silkworm, *Bombyx mori*, *Develop. Growth & Different.*, 17:219-225 (1975).
7. E. Ohnishi, M. Hiramoto, Y. Fujimoto, K. Kakinuma and N. Ikekawa, Isolation and identification of major ecdysteroid conjugates from the ovaries of *Bombyx mori*, *Insect Biochem.*, 19:95-101 (1989).
8. A. J. Slinger, L. N. Dinan and R. E. Isaac, Isolation of apolar ecdysteroid conjugates from newly-laid oothecae of *Periplaneta americana*, *Insect Biochem.*, 16:115-119 (1986).
9. A. J. Slinger and R. E. Isaac, Ecdysteroid titres during embryogenesis of the cockroach, *Periplaneta americana*, *J. Insect Physiol.*, 34:1119-1125 (1988).
10. B. Fournier and D. Radallah, Ecdysteroids in *Carausius* eggs during embryonic development, *Archives Insect Biochem. Physiol.*, 7:211-224 (1988).
11. M. Bownes, A. Shirras, M. Blair, J. Collins and A. Coulson, Evidence that insect embryogenesis is regulated by ecdysteroids released from yolk proteins, *Proc. Nat. Acad. Sci. USA*, 85:1554-1557 (1988).
12. J. A. Hoffmann and M. Lagueux, Endocrine aspects of embryonic development in insects, *in*: "Comprehensive Insect Physiology, Biochemistry and Pharmacology", G. A. Kerkut and L. J. Gilbert, eds., Pergamon Press, vol. 1:435-460 (1985).
13. D. Bulliere, F. Bullière and M. de Reggi, Ecdysteroid titres during ovarian development in *Blaberus craniifer*, *Roux's Archives Develop. Biol.*, 186:103-114 (1979).
14. H. Imboden and B. Lanzrein, Investigations on ecdysteroids and juvenile hormones and on morphological aspects during early embryogenesis in the ovoviviparous cockroach *Nauphoeta cinerea*, *J. Insect Physiol.*, 28:37-46 (1982).
15. R. Lanot, J. Thiebold, M. Lagueux, F. Goltzene and J. A. Hoffmann, Involvement of ecdysone in the control of meiotic reinitiation in oocytes of *Locusta migratoria* (insecta, Orthoptera), *Develop. Biol.*, 121:174-181 (1987).
16. R. Lanot, J. Thiebold, M.-F. Costet-Corio, P. Benveniste and J. A. Hoffmann, Further experimental evidence for the involvement of ecdysone in the control of meiotic reinitiation in oocytes of *Locusta migratoria* (Insecta, Orthoptera), *Develop. Biol.*, 126:212-214 (1988).
17. J. T. Warren, R. Steiner, A. Dorn, M. Pak and L. I. Gilbert, Metabolism of ecdysteroids during the embryogenesis of *Manduca sexta*, *J. Liquid Chromatog.*, 9:1759-1782 (1986).
18. M. J. Thompson, J. A. Svoboda, R. Lozano and K. R. Wilzer, Profile of free and conjugated ecdysteroids and ecdysteroid acids during embryonic development of *Manduca sexta* (L.) following maternal incorporation of [^{14}C]cholesterol, *Archives Insect Biochem Physiol.*, 7:157-172 (1988).
19. E. Ohnishi, T. Mizuno, N. Ikekawa and T. Ikeda, Accumulation of 2-deoxy-ecdysteroids in ovaries of the silkworm, *Bombyx mori*, *Insect Biochem.*, 11:155-159 (1981).
20. S. L. Smith and W. E. Bollenbacher, Ovarian ecdysteroids and their secretion in late-pharate adults of *Galleria mellonella*, *J. Insect Physiol.*, 31:419-424 (1985).
21. J. Koolman, K. Scheller and D. Bodenstein, Ecdysteroids in the adult male blowfly *Calliphora vicina*, *Experientia*, 35:134-135 (1979).
22. M. Renucci and A. Strambi, Evolution des ecdystéroides ovariens et hémolymph-atiques au cours de la maturation ovarienne chez *Acheta domesticus* L. (Orthoptère), *Comptes Rendus de L'Académie des Sciences, Paris, Série III*, 293:825-830 (1981).
23. Y. X. Zhu, H. Gfeller and B. Lanzrein, Ecdysteroids during oogenesis in the ovoviviparous cockroach *Nauphoeta cinerea*, *J. Insect Physiol.*, 29:225-235 (1983).
24. R. J. Weaver, A. Strambi and C. Strambi, The significance of free ecdysteroids in the haemolymph of adult cockroaches, *J. Insect Physiol.*, 30:705-711 (1984).
25. P. Whiting and L. Dinan, The occurrence of apolar ecdysteroid conjugates in newly-laid eggs of the house cricket, *Acheta domesticus*, *J. Insect Physiol.*, 34:625-631 (1988).
26. P. Whiting and L. Dinan, Identification of the endogenous apolar ecdysteroid conjugates present in newly laid eggs of the house cricket (*Acheta domesticus*) as 22-long-chain fatty acyl esters of ecdysone, *Insect Biochem.*, 19:759-767 (1989).
27. P. Whiting and L. Dinan, Formation of apolar ecdysteroid conjugates by ovaries of the house cricket, *Acheta domesticus in vitro*, *Biochem. J.*, 252:95-103 (1988).
28. L. Dinan, The chemical synthesis of ecdysone 22-long-chain fatty acyl esters in high yield, *J. Steroid Biochem.*, 31:237-245 (1988).
29. M. N. Galbraith and D. H. S. Horn, Insect moulting hormones: crustecdysone (20-hydroxyecdysone) from *Podocarpus elatus*, *Australian J. Chem.*, 22:1045-1057 (1969).

30. J. T. Warren and L. I. Gilbert, Ecdysone metabolism and distribution during the pupal-adult development of *Manduca sexta*, *Insect Biochem.*, 16:65-82 (1986).
31. L. Dinan, Separation of ecdysone acyl esters and their 2,3-acetonide derivatives by high-performance liquid chromatography, *J. Chromatog.*, 436:279-288 (1988).
32. R. E. Isaac, H. H. Rees and T. W. Goodwin, Isolation of ecdysone-3-acetate as a major ecdysteroid from the developing eggs of the desert locust, *Schistocerca gregaria*, *J. Chem. Soc., Chem. Commun.*, pp 594-595 (1981).
33. G. Tsoupras, Lau Bang, C. Hetru, J.-F. Muller and J. A. Hoffmann, Conversion *in vitro* de 20-hydroxyecdysone en metabolites phosphorylés et acetylés par des complexes tube digestif-tubes de Malpighi de larves de *Locusta migratoria*, *Comptes Rendus de l'Academie de Sciences de Paris, Serie III*, 296:77-80 (1983).
34. R. Lafont, C.Blais, P. Beydon, J.-F. Modde, U. Enderle and J. Koolman, Conversion of ecdysone and 20-hydroxyecdysone into 26-OIC derivatives is a major pathway in larvae and pupae of species from three insect orders, *Archives Insect Biochem Physiol*, 1:41-58 (1983).
35. J.-F. Modde, R. Lafont and J. A. Hoffmann, Ecdysone metabolism in *Locusta migratoria* larvae and adults, *Internat. J. Invertebrate Reprod. Develop.*, 7:161-183 (1984).
36. J. M. Gibson, R. E. Isaac, L. N. Dinan and H. H. Rees, Metabolism of [^{3}H]-ecdysone in *Schistocerca gregaria*: formation of ecdysteroid acids together with free and phosphorylated ecdysteroid acetates, *Archives Insect Biochem. Physiol.*, 1:385-407 (1984).
37. D. Bueckmann, G. Starnecker, K.-H. Tomaschko, E. Wilhelm, R. Lafont and J.-P. Girault, Isolation and identification of major ecdysteroids from the pycnogonid *Pycnogonum litorale* (Stroem) (Arthropoda, Pantopoda), *J. Comp. Physiol.*, B 156:759-765 (1986).
38. M. Garcia, J.-P. Girault and R. Lafont, Ecdysteroid metabolism in the terrestrial snail *Cepaea nemoralis* (L.), *Internat. J. Invertebrate Reprod. Develop.*, 9:43-58 (1986).
39. K. H. Hoffman and W. Behrens, Free ecdysteroids in male crickets, *Gryllus bimaculatus*, *Physiol. Entomol.*, 7:269-279 (1982).
40. D. Bulenda, A. Stecher, M. Freunek and K. H. Hoffmann, Ecdysone metabolism in adult crickets, *Gryllus bimaculatus*, *Insect Biochem.*, 16:83-90 (1986).

ECDYSTEROIDS IN EGGS OF *PERIPLANETA AMERICANA*: RAPID ANALYSIS USING 3μ C18 COLUMNS AND THE BINDING OF ECDYSTEROIDS TO VITELLIN

R. E. Isaac, F. El-Zoghby*, A. J. Slinger
and A.N. MacGregor

Department of Pure and Applied Biology
University of Leeds, Leeds LS2 9JT, UK, and
*Department of Plant Protection
University of Alexandria, Egypt

SUMMARY

The titres of free, polar and apolar conjugates of ecdysteroids during embryonic development of the cockroach, *Periplaneta americana*, were determined by radioimmunoassay (RIA). The different classes of ecdysteroid were resolved by both open column chromatography and reversed-phase high performance liquid chromatography (HPLC). HPLC using short 3μ C18 column and a gradient of acetonitrile in water permitted the rapid analysis of crude methanolic extracts without the need for prior clean-up procedures. Apolar ecdysteroid esters present in newly-laid oothecae were metabolized during early embryogenesis to release free ecdysteroids and to generate polar conjugates. Titres of free and polar compounds declined at around day 20 with the polar conjugates remaining at a low level for the remainder of embryogenesis. A large peak of free ecdysteroid (mainly 20-hydroxyecdysone) appeared at day 33 which, being 8 days before hatching, is likely to be involved in the formation of the first larval cuticle. At this late stage of development a new group of apolar ecdysteroids, probably inactivation metabolites, appeared and increased in importance as the free hormone titre declined. The maternal apolar esters present at day 0 were bound to vitellin which was degraded to low molecular weight polypeptides by day 20. This is coincident with the metabolism of the maternal esters to release free hormone.

INTRODUCTION

Recent developments in analytical techniques have resulted in the characterization of ecdysteroid metabolites and conjugates of diverse structure and polarity (for reviews see Refs. [1-2]. Such compounds are generally regarded as the products of inactivation pathways but in some situations they may, through the action of hydrolytic enzymes, also serve as a source of free active hormone [3]. It is therefore important to determine whenever possible not only the titres of free ecdysteroids (e.g., ecdysone and 20-hydroxyedysone) but also the quantities of polar and apolar conjugates. An established powerful tool in the separation of ecdysteroids that can range in polarity from ecdysteroid 22-phosphates to the apolar 22-fatty acyl esters is reversed-phase HPLC (see Lafont and Wilson, this volume). HPLC columns packed with 3μ material offer a number of advantages. Their increased efficiency allows the use of short (3-10 cm) columns with little loss in resolving power. The low void volumes of such columns can greatly reduce solvent consumption and run times, and the elution of compounds at higher concentrations can improve the detection limits of HPLC detectors. Collection of compounds in low volumes of eluent is also helpful when removal of solvent is required for the assay of particular compounds, e..g, radioimmunoassay (RIA). 3μ Reversed-phase columns have been routinely used in our laboratory to measure changes in ecdysteroid profiles during embryogenesis of the cockroach, *Periplaneta americana*, and for measuring the activities of ecdysteroid-metabolizing enzymes [4-6].

Chromatography and Isolation of Insect Hormones and Pheromones
Edited by A. R. McCaffery and I. D. Wilson
Plenum Press, New York, 1990

Newly-laid eggs of *P. americana* contain a mixture of apolar ecdysteroid esters that can be hydrolyzed enzymically *in vitro* to release mainly ecdysone, 2-deoxyecdysone and 20-hydroxyecdysone [7]. These compounds are metabolized during embryogenesis to free and polar ecdysteroids [5], many of which can be quantified by RIA either before or after enzymic hydrolysis. In our quest to quantify these different classes of ecdysteroids, a number of chromatographic methods were tested. Analysis by silica TLC was discounted in favour of open column chromatography and reversed-phase HPLC using a 7.5 cm, 3μ C18 column in conjunction with radioimmunoassay. The HPLC method was preferred as gradient elution quickly resolved polar, free, maternal apolar esters and a novel group of embryonic apolar ecdysteroids found in crude methanol extracts in a single run. These systems were used to monitor the fluctuations in ecdysteroid levels during embryogenesis. In addition, we have shown that in newly-laid oothecae ecdysteroid esters bind to vitellin and that the degradation of vitellin may determine the timing of ecdysteroid metabolism.

METHOD

General Methods

The sources of general chemicals and descriptions of routine methods (i.e., open column chromatography, silica TLC, radioimmunoassay (RIA) and the extraction of ecdysteroids from oothecae) have been reported previously by Slinger et al. [7] and Slinger and Isaac [4-6]. Enzymes were obtained from Sigma Chemical Co., Poole, Dorset, UK.

Chromatography

Ecdysteroid concentrations during embryogenesis were determined by RIA after resolution of the polar ecdysteroids, free ecdysteroids and apolar ecdysteroid esters by either open column chromatography or HPLC [5]. Ecdysteroids were resolved by HPLC with the following solvent systems.

System 1

System 1 fractionated ecdysteroids present in crude methanolic extracts into 5 groups of differing polarities using an Ultrasphere ODS 3μ column (75 x 4.6 mm, Beckmann, High Wycombe, Bucks., UK) and linear gradients of acetonitrile in water (15 to 45% (v/v), 0 to 3.5 min), followed by an increase to 100% acetonitrile (3.5 to 4 min) at a flow rate of 1 ml/min. After 7 min, the flow rate was increased to 2 ml/min and elution with 100% acetonitrile was continued for the duration of the run (12 min). Eluent was collected between 0 to 2 min (fraction A, polar ecdysteroids), 2.5 to 5.5 min (fraction B, free ecdysteroids), 5.5 to 7 min (fraction C, embryonic apolar ecdysteroids), 7 to 9.5 min (fraction D, maternal apolar ecdysteroids) and 9.5 to 12 min (fraction E). Each fraction was analyzed for ecdysteroids by RIA before and after enzymic hydrolysis.

System 2

System 2 was used to separate free ecdysteroids and consisted of the Ultrasphere column eluted, at a flow rate of 1 ml/min, with a linear gradient of acetonitrile in water changing from 19 to 50% (v/v) in 10 min. The column was then washed with 100% acetonitrile for a further 2 min.

System 3

System 3 resolved free ecdysteroids by normal phase chromatography on a Hypersil 5μ column (250 x 4.6 mm; Shandon Southern Products Ltd., Runcorn, UK) using dichloromethane-ethanol-water (83:15:2) and a flow rate of 1.5 ml/min.

System 4

System 4 was used to separate polar ecdysteroids and employed the Ultrasphere column eluted, at a flow rate of 0.8 ml/min, with a linear gradient of methanol in sodium citrate buffer (20 mM, pH 6.5) changing from 20 to 60% (v/v) in 10 min, followed by a change to 90% methanol in buffer (10 to 11 min). The column was eluted with 90% methanol until the end of the run (18 min).

For the chromatography of cockroach vitellin, a gel permeation column (Protein Pak 300, Waters Assoc.) was used with 0.2 M sodium chloride , 0.1 M TrisHCl, pH 7 as the eluting buffer.

Extraction and Purification of Vitellin

Vitellin was extracted from ovaries and newly-laid oothecae by homogenization in 50 mM TrisHCl, pH 8.1, 0.25 M sodium chloride, 1 mM phenyl methyl sulphonyl fluoride (PMSF) as described by Harnish and White [8]. Insoluble material was removed by centrifugation and the supernatant was kept for ecdysteroid and protein analysis. Repeated extraction of the insoluble material with Tris-buffered saline yielded extracts. Vitellin was purified either by ammonium sulphate precipitation [8] or by gel filtration chromatography using a TSK300 column.

Protein was assayed by the method of Markwell et al. [9].

Electrophoresis

SDS-PAGE was performed under reducing conditions using the discontinuous buffer system of Laemmli [10]. Samples were reduced by heating to 100°C for 2 min in 62.5 mM TrisHCl, pH 6.8, 2% (w/v) sodium dodecyl sulphate, 10 mM dithiothreitol, 5% (w/v) glycerol and 0.0025% bromophenol blue. Samples were applied to 0.75 mm slab gels containing 7.5% (w/v) acrylamide in the resolving gel and 4.5% (w/v) acrylamide in the stacking gel. Gels were stained with Coomassie blue and scanned using a Joyce-Loebl densitometer.

Analysis of Ecdysteroids in Protein Samples

Apolar ecdysteroids were extracted from soluble protein fractions by the addition of methanol (final composition, 70% w/v). Precipitated protein was removed by brief centrifugation and the aqueous methanol retained. The pellet was then washed with methanol (100%) and the combined extracts were taken to dryness before reconstituting in a known volume of methanol. Aliquots of the methanol extract were dried, reconstituted in 0.2 M sodium acetate buffer, pH 5.6, subjected to enzymic hydrolysis using *Helix pomatia* enzymes and analyzed for liberated ecdysteroids, as described by Slinger et al. [7]. Protein fractions collected from gel filtration chromatography were adjusted to pH 6 and incubated with *Helix* enzymes and analyzed for ecdysteroids as aforementioned.

Incubation with Protease Enzymes

Vitellin samples obtained by ammonium sulphate precipitation were passed down a PD10 column before incubation with a cocktail of proteinases (Sigma type XXI, 4 mg/ml; type IV, 1 mg/ml and trypsin type III, 1 mg/ml) and *Helix pomatia* sulphatase (80 units/ml) in 0.2 M sodium acetate buffer, pH 5.6. The pH of the vitellin solution was adjusted to 5.6 with HCl prior to adding the enzyme cocktail. Sometimes procine esterase was used to cleave ecdysteroid esters instead of the "sulphatase" preparation. Hydrolyses were carried out for 17 h at 30°C and the reactions terminated by the addition of ethanol to a final volume of 70% (v/v). After chilling, precipitated protein was removed by centrifugation and samples of the aqueous-ethanol were taken for RIA.

RESULTS

Chromatography of Ecdysteroids from Oothecae and Embryos

A chromatographic system was required that could resolve at least three classes of ecdysteroids (polar, free and apolar compounds) and could be used routinely in conjunction with RIA. Silica TLC was used initially but presented two problems. The crude methanolic extract was difficult to apply quantitatively to the silica gel and the elution of ecdysteroids from the silica for RIA was time consuming and gave variable recovery. Open column chromatography using a silica and a C2 bonded phase in sequence has been used successfully to isolate the various ecdysteroid fractions [7] but this method, although quantitative, was rather laborious for large numbers of samples but was used to analyze some samples up to day 13 of embryogenesis. It was decided to complete the study using HPLC with gradient elution (system 1) and a short 3μ C18 column (Ultrasphere ODS) to analyze the ecdysteroids present in extracts over the whole of embryogenesis. This method had a number of advantages. Firstly, crude methanolic extracts could be injected onto the column with no prior

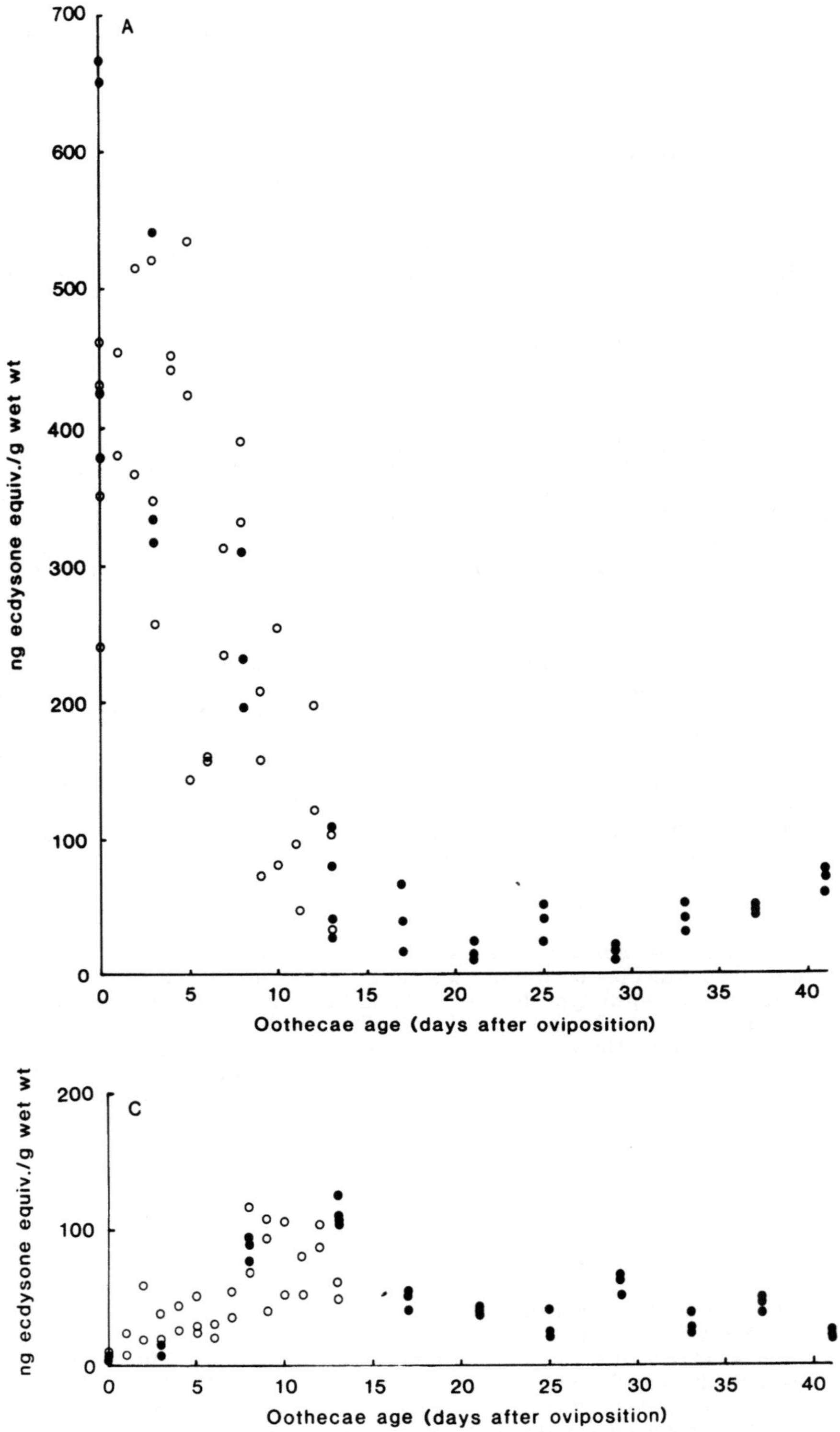

Fig. 1. Titres of (A) maternal apolar ecdysteroids (this page), (B) free ecdysteroids (facing page), (C) polar ecdysteroid conjugates (this page), and (D) embryonic apolar ecdysteroids (facing page) during embryogenesis in *P. americana*. Ecdysteroids present in methanolic extracts were resolved into the aforementioned groups either by silica and C2 open column chromatography (o) or by HPLC, system 1 (•). Free ecdysteroids were measured directly by RIA whereas conjugates (a,b,c) were first treated with *Helix* enzymes to release ecdysteroids which were then assayed by RIA.

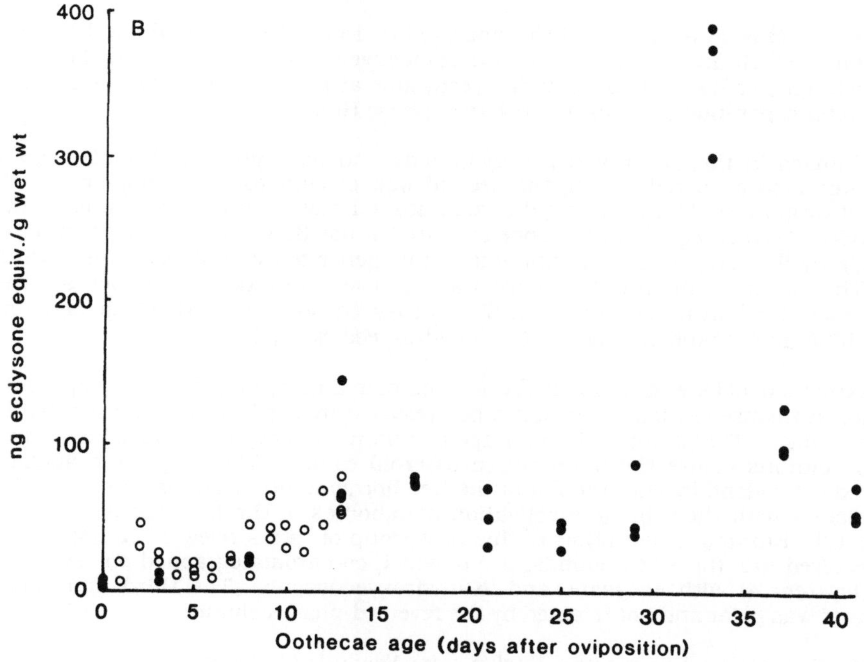

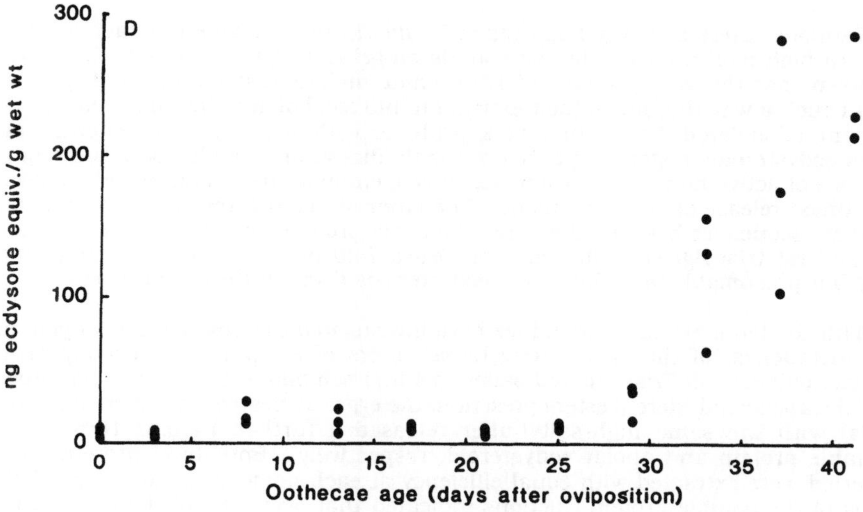

clean-up procedure. Secondly, the gradient chosen managed to resolve quickly not only the three classes mentioned but also a novel group of ecdysteroid metabolites that appeared in late embryos. Thirdly, the fractions collected were in a precise volume ready, after drying in a "Gyrovap", for analysis by RIA. Five fractions were collected vis. polar ecdysteroids, (0 to 2 min), free ecdysteroids (2.5 to 5.5 min), embryonic apolar ecdysteroids (5.5 to 7 min), maternal apolar ecdysteroids (7 to 9.5 min) and a further fraction eluting between 9.5 to 12 min. Each fraction was analyzed for ecdysteroids by RIA before and after enzymic hydrolysis.

Ecdysteroid Levels During Embryogenesis

Newly-laid oothecae contained mainly hydrolysable apolar esters with negligible amounts of free hormone being present (Figure 1a). The level of these esters dropped steadily from 440 to 65 ng ecd.eq./g over the first 13 days and this decline coincided with an increase in both free and polar hydrolysable ecdysteroids. The titre of free ecdysteroid appeared to rise from day 7 reaching a level of around 80 ng ecd.eq./g by day 13 (Figure 1b). The amount of polar ecdysteroid conjuates also increased over the same period reaching maximum levels of around 100 ng ecd.eq./g between days 8 and 13 (Figure 1c). Both the free ecdysteroid and polar ecdysteroid titres appeared to decline from around day 17, dropping to a minimum of 40 and 30 neg ecd.eq./g, respectively, by day 25. The molar ratio of 2-deoxyecdysone, ecdysone and 20-hydroxyecdysone at around day 13 was estimated to be 4:3:1.5 after separation on silica and reversed-phase HPLC.

Changes in the titres of free ecdysteroids and both polar and apolar ecdysteroid conjugates also occurred during the second half of embryonic development. After an apparent drop in the free ecdysteroid titre at around day 17, a relatively large peak (mean value, 350 ng ecd.eq./g) of free hormone appeared at day 33 which quickly dropped to 60 ng ecd.eq./g by day 41, just prior to the time of emergence of the first nymphal instar (Figure 1b). This peak of immunoreactivity was resolved into two components which co-chromatographed with ecdysone and 20-hydroxyecdysone on both silica and reversed-phase HPLC (molar ratio of ecdysone to 20-hydroxyecdysone, 1:2).

Extracts of older embryos were found to contain a new group of apolar ecysteroids that were not in themselves immunoreactive but released free ecdysteroids on incubation with *Helix* enzymes. This group, although apolar relative to ecdysone, eluted from reversed-phased columns before the maternal ecydsteroid esters. The embryonic apolar esters appeared coincidentally with the fall in the free hormone titre from day 33 onwards which is consistent with them being inactivation metabolites of the late free ecdysteroid peak (Figure 1d). Enzymatic hydrolysis of this new group of esters released ecdysteroids which were resolved into three compounds, two of which co-chromatographed on reversed-phase HPLC (system 4) with ecdysone and 20-hydroxyecdysone. The third immunoreactive compound was polar and not retained by the reversed-phase column.

Co-Extraction of Ecdysteroids and Vitellin from Newly-Laid Oothecae

Hormone esters in newly-laid eggs of *P. americana* can, when extracted with buffer, elute with high molecular weight compounds on gel filtration columns [11]. Recent work has shown that the yolk proteins of *Drosophila melanogaster* bind apolar ecdysteroid esters in such a way that limits their extraction into alcohol and their detection by RIA [12]. Treatment of isolated *Drosophila* yolk proteins with protease and esterase enzymes releases ecdysteroids suggesting that proteolytic digestion may also be a pre-requisite for the release of active hormone by esterases during embryogenesis and provide a mechanism for the timed release of active hormone. The latter study also revealed highly homologous amino acid sequences between the *Drosophila* yolk proteins and the substrate binding site of pig and rat triacylglycerol lipase. The *Drosophila* proteins do not display any lipase activity but presumably bind the apolar ecdysteroids through their lipid moiety.

With the latter results in mind we have investigated the possible role of yolk proteins in the metabolism of the apolar ecdysteroid esters of *Periplaneta*. Homogenization of newly-laid oothecae in Tris-buffered saline (pH 8.1) solubilized 86% of the egg protein and 83% of the apolar ecdysteroid esters present in the eggs. Repeated extraction of the insoluble material with the same high-salt buffer released a further 14 and 16% of the total extractable protein and apolar ecdysteroid, respectively. Both total protein and apolar ecdysteroid were extracted with equal efficiency at each homogenization step. SDS-PAGE analysis of the soluble protein fractions indicated that over 70% of the extracted protein migrated with protein bands (M_r of 105000, 100000, 98000, 94000 and 56000) present in adult female but not male haemolymph (Figure 2).

Co-Purification of Apolar Ecdysteroid Esters and Vitellin

Vitellin was isolated from the solubilized oothecal protein by ammonium sulphate precipitation. Various fractions were desalted and the ecdysteroid extracted from the protein by addition of ethanol. Ecdysteroid content was determined after enzymic hydrolysis of the apolar esters with *Helix* enzymes. Of the ecdysteroid present in various fractions (insoluble oothecal material, ammonium sulphate pellet and supernatant) over 90% was located with the precipitated vitellin.

The association of apolar ecdysteroid esters with vitellin was confirmed by chromatography of soluble protein extract on a gel filtration column. The bulk of the oothecal protein as measured by UV absorbance (280 nm) eluted with a retention volume corresponding to M_r of around 400,000 (Figure 3), a value similar to that already established for native vitellin from *P. americana* by other workers [8]. SDS-PAGE

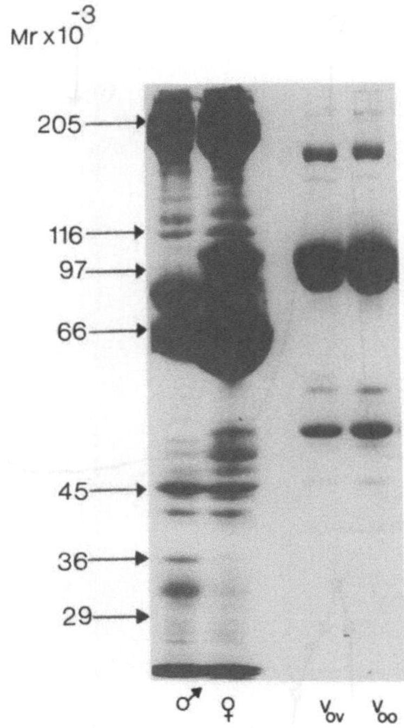

Fig. 2. SDS-PAGE of male haemolymph, female haemolymph, ovarian protein (V_{ov}) and oothecal protein (V_{oo}) under reducing conditions. Protein was detected by staining with Coomassie blue.

confirmed that this protein co-migrated with female specific polypeptides. RIA of the eluate after treatment with *Helix* enzymes showed that the apolar ecdysteroids co-chromatographed faithfully with the native vitellin and insignificant immunoreactivity was associated with lower molecular weight fractions (Figure 3). Similar results were obtained with vitellin extracted from terminal oocytes.

Vitellin Degradation During Embryogenesis

SDS-PAGE analysis of protein extracts from day 0, 6, 12 and 20 oothecae revealed that around 50% of the major yolk proteins (M_r, 94000 to 105000) were degraded by day 6 to smaller polypeptides with molecular weights below 66000 (Figure 4). Further proteolysis had occurred by day 12 with around 90% of the protein migrating as bands with M_r below 66000 and by day 20 only bands with M_r of 36000 or below 29000 were detected.

Effect of Incubation with Proteolytic Enzymes on Recovery of Apolar Ecdysteroid Esters from Vitellin

Vitellin extracted with high-salt buffer was precipitated with ammonium sulphate and desalted using a PD10 column before incubation with hydrolytic enzymes. Any ecdysteroid released was separated from protein by ethanol precipitation and the amount of immunoreactive steroid present in the aqueous ethanol was estimated by RIA (Table 1). As shown previously, *Helix* enzymes, but not porcine esterase, were able to efficiently release ecdysteroids. Unexpectedly, the protease cocktail in the absence of *Helix* enzymes also released strong immunoreactivity but did not appear to greatly enhance the release of hormone by *Helix* enzymes. SDS-PAGE of the vitellin after enzyme treatment confirmed that the protease mixture had extensively digested the protein and that incubation with *Helix* enzymes did not result in any significant protein breakdown.

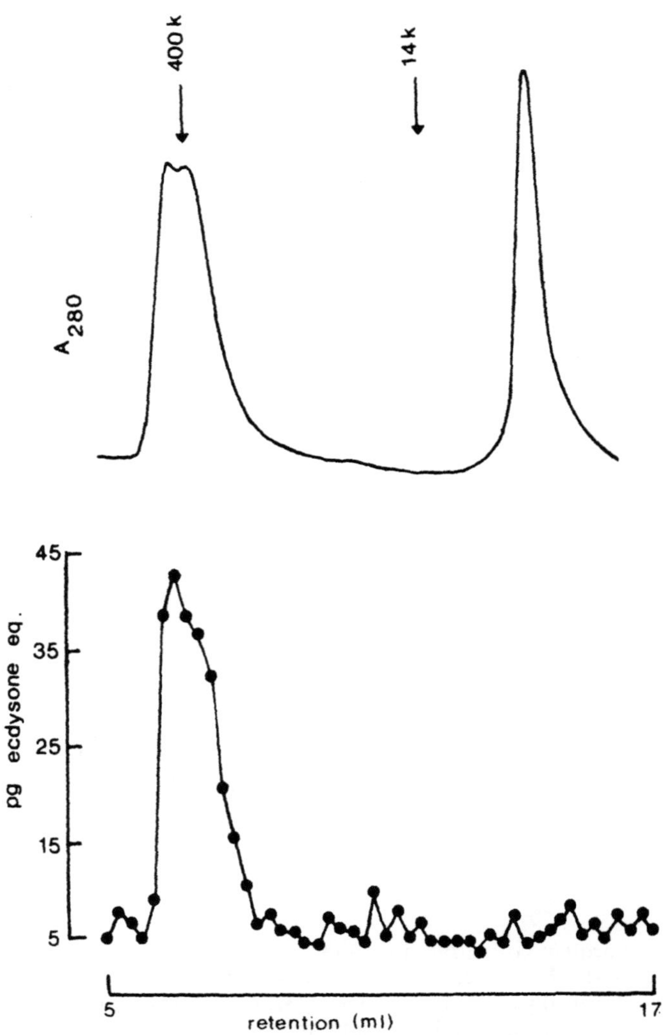

Fig. 3. Chromatography of cockroach vitellin on a gel filtration column (Protein Pak 300). The elution of protein was monitored using a UV detector set at 280 nm and fractions were treated with *Helix* enzymes to hydrolyze the apolar esters. Liberated ecdysteroids were extracted with ethanol before analysis by RIA.

Table 1. Effect of protease treatment on the hydrolysis and recovery of ecdysteroids bound to vitellin.

Treatment	Relative recovery of ecdysteroids (%)
None (buffer only)	10
Helix	67
Esterase	12
Helix + protease	84
Esterase + protease	100
Protease	97
† Methanol	60

Protein samples were incubated overnight with enzymes or mixtures of enzymes. Reactions were terminated by the addition of ethanol and the ecdysteroid released was measured directly by RIA. Maximum recovery was 1.5ng of ecdysteroid/mg of protein. Enzyme preparations did not contain ecdysteroids as measured by RIA.

†Apolar ecdysteroid esters were extracted from the vitellin with methanol being before hydrolysed with *Helix* enzymes and subjected to RIA.

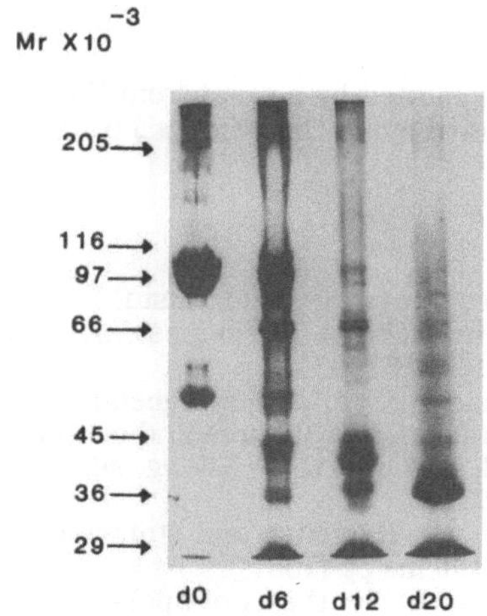

Fig. 4. SDS-PAGE of soluble protein from a) day 0, b) day 6, c) day 12 and d) day 20 oothecae run under reducing conditions. Proteins were stained with Coomassie-blue.

DISCUSSION

The metabolism of the maternal ecdysteroids leads to a rise in the free and polar conjugated ecdysteroids during early embryogenesis which may have an important hormonal role at this stage of development. It has been shown that peaks of free ecdysteroids occurring in young embryos of a number of insect species correlate with the synthesis of embryonic cuticle [13-20]. The appearance of a peak of free hormone (mainly 20-hydroxyecdysone) at day 33, 8 days before hatching, probably arises from *de novo* synthesis by the embryo and determines the synthesis and deposition of the first larval cuticle. This late rise in hormone levels is regulated by metabolism to a new family of apolar metabolites that are expected to be hormonally inactive and may be similar to the apolar inactivation metabolites of ecdysone found in nymphs and adults of *P. americana* [21]. The occurrence in eggs of ecdysteroid apolar esters that are structurally different from those apolar derivatives formed as inactive end-products of metabolism is consistent with a specific storage role for the maternal ecdysteroid esters.

The binding of ecdysteroid esters to vitellin is perhaps not unexpected but the nature of the binding is clearly different from that occurring in *Drosophila* eggs where the combined efforts of proteases and esterases are necessary for the release of hormone, at least *in vitro* [12]. These results may reflect the different structures of the vitellin and the hormone esters found in these two species. In *P. americana*, the early (day 0 to 20) fragmentation of the yolk proteins correlates well with the metabolism of the apolar esters to release free hormone but it is not known whether hydrolysis *in vivo* by egg esterases is dependent upon the prior digestion of the yolk protein or is just coincident.

CONCLUSION

We have demonstrated how reversed-phased HPLC and solvent gradients can be used to analyze crude insect extracts containing a range of ecdysteroids and their derivatives without resorting to a time consuming clean-up step. HPLC using 3μ columns not only allows the rapid through-put of samples but also saves on expensive HPLC-grade solvents. Similar chromatography has been used to develop an assay for the enzymic synthesis of apolar esters by ovarian membrane preparations [6]. A number of other enzymes involved in ecdysone metabolism (e.g., ecdysone 20-monoxygenase) can be quickly assayed using very small (3 cm) columns to rapidly separate and quantify reaction products whether by UV detection or radioactivity counting. The linking of such a HPLC column to a suitable radioactivity flow detector has obvious advantages as regards time, sensitivity and cost (see Lafont and Wilson, this volume).

ACKNOWLEDGEMENTS

Part of this work was supported by a grant from the Agricultural and Food Research Council to REI. FZ was supported by a British Council travel grant.

REFERENCES

1. R. Lafont and. Koolman, Ecdysone metabolism, *in*: "Biosynthesis, Metabolism and Mode of Action of Invertebrate Hormones", J. Hoffman and M. Porchet, eds., Springer, Berlin, pp 196-226 (1984).
2. H. H. Rees and R. E. Isaac, Biosynthesis and metabolism of ecdysteroids and methods of isolation and identification of the free and conjugated compounds, *in*: "Methods in Enzymology", J. H. Law and H. C. Rilling, eds., Academic Press, New York, vol. 111, pp 377-410 (1985).
3. R. E. Isaac and A. J. Slinger, Excretion and storage of ecdysteroids *in*: "Ecdysone", J. Koolman, ed., Verlag Thieme pp (1989).
4. A. J. Slinger and R. E. Isaac, Synthesis of apolar ecdysteroid conjugates by ovarian tissue from *Periplaneta americana, Gen. & Comp. Endocrinol.*, 70:74-82 (1988).
5. A. J. Slinger and R. E. Isaac, Ecdysteroid titres during embryogenesis of the cockroach, *Periplaneta americana, J. Insect Physiol.*, 34:1119-1125 (1988).
6. A. J. Slinger and R. E. Isaac, Acyl-CoA: ecdysone acyltransferase activity from the ovary of *P. americana, Insect Biochem.*, 18:779-784 (1988).
7. A. J. Slinger, L. N. Dinan and R. E. Isaac, Isolation of apolar ecdysteroid conjugates from newly-laid oothecae of *Periplaneta americana, Insect Biochem.*, 16:115-119 (1986).
8. D. G. Harnish and B. N. White, Insect vitellins: identification, purification and characterization from eight orders, *J. Experim. Zool.*, 220:1-10 (1982).
9. M. A. K.Markwell, S. M. Haas, L. L. Bieber and N. H. Tolbert, A modification of the Lowry procedure to simplify protein determination in membrane and lipoprotein samples, *Analyt. Biochem.*, 87:206-210 (1978).
10. U. K. Laemmli, Cleavage of structural proteins during assembly of the head of bacteriophage T4, *Nature*, 227:680-685 (1970).
11. A. J. Slinger, Ecdysteroids in ovaries and eggs of the American cockroach, *Periplaneta americana*, PhD Thesis, University of Leeds (1988).
12. M. Bownes, A. Shirras, M. Blair, J. Collins and A. Coulson, Evidence that insect embryogenesis is regulated by ecdysteroids released from yolk proteins, Proceedings of the National Academy of Science, USA, 85:1554-1557 (1988).
13. D. Bulliere, F. Bulliere and M. De Reggi, Ecdysteroid titres during ovarian and embryonic development in *Blaberus cranifer, Wilhelm Roux's Archives*, 186:103-114 (1979).
14. M. Lagueux, C. Hetru, F. Goltzene, C. Kappler and J. A. Hoffman, Ecdysone titre and metabolism in relation to cuticulogenesis in embryos of *Locusta migratoria, J. Insect Physiol.*, 25:709-723 (1979).
15. M. Cavallin and B. Fournier, Characteristics of development and variations in ecdysteroid levels in *Clitumnus* embryos deprived of their cephalic endocrine glands, *J. Insect Physiol.*, 27:527-534 (1981).
16. H. Imboden and B. Lanzrein, Investigations on ecdysteroids and juvenile hormones and on morphological aspects during early embryogenesis in the ovoviviparous cockroach *Nauphotea cinerea, J. Insect Physiol.*, 28:37-46 (1982).
17. B. Gharib and M. De Reggi, Changes in ecdysteroid and juvenile hormone levels in developing eggs of *Bombyx mori, J. Insect Physiol.*, 29:871-876 (1983).
18. N. Bordes-Alleaume and L. Sami, Ecdysteroid titres and cuticle depositions in embryos of the dipteran *Calliphora erythrocephala, Internat. J. Reprod. & Develop.*, 11:109-122 (1987).
19. P. Maroy, G. Kaufmann and A. Dubendorfer, Embryonic ecdysteroids of *Drosophila melanogaster, J. Insect Physiol.*, 34:633-637 (1988).
20. J. A. Hoffmann and M. Lagueux, Endocrine aspects of embryonic development in insects, *in*: "Comprehensive Insect Physiology Biochemistry and Pharmacology", G. A. Kerkut and L. I. Gilbert, eds., Pergamon Press, Oxford, vol. 1, pp 435-460 (1985).
21. J.-L. Connat and P. A. Diehl, Probable occurrence of ecdysteroid fatty acids in different classes of arthropods, *Insect Biochem.*, 16:91-97 (1986).

ADVANCES IN ECDYSTEROID HIGH PERFORMANCE LIQUID CHROMATOGRAPHY

R. Lafont and I. D. Wilson*

Department de Biologie, Ecole Normale Supérieure
CNRS-URA 686 and Université Paris
F-75230 Paris Cedex 05, France, and
*Department of Chemistry, University of Keele
Keele, Staffordshire ST5 5BG, UK

SUMMARY

High performance liquid chromatography (HPLC) is the most widely used method for the separation of ecdysteroids. Many efficient methods are presently available, which are adapted to the various classes of compounds (high, medium or low polarity ecdysteroids), for either analytical or preparative purposes.

This review includes a survey of classical chromatographic systems, including some recent developments of normal-phase HPLC (concerning especially the use of polar and non-polar bonded silica columns), a few general comments on rather sophisticated detection systems (diode-array detectors, radioactivity monitors) and some comments regarding the problem of sample preparation prior to HPLC analysis.

Several possible future developments of HPLC, including coupling with mass spectrometry, are discussed briefly.

INTRODUCTION

High performance liquid chromatography (HPLC) has become the most popular technique for ecdysteroid separations, both for analytical and preparative purposes. It offers a wide choice of techniques, that are adapted for polar or apolar metabolites, the number of which is continuously increasing [1, 2]. As a consequence, the identification of any ecdysteroid by comigration with a reference compound must rely on the simultaneous use of several (at least two) different HPLC systems, generally one normal-phase and one reversed-phase system. Of course with the increasing number of identified ecdysteroids, the criteria will become more and more demanding (let us remember that in "early times" one TLC system was considered adequate for assessing compound identity!).

Any Ecdysonist faced with a separation problem has the opportunity to make a logical choice, according to the results of a preliminary TLC step. He is thus able to solve easily almost all his separation problems after a limited number of experiments.

Current HPLC techniques have been designed about ten years ago and the HPLC of ecdysteroids may now be considered to be a mature technique (reviewed in references 3-5); as such what really new comments can be made in this field? Progress might *a priori* arise from (a) chromatographic systems, (b) column design, (c) detection (on-line or off-line) and/or (d) sample preparation procedures. In any case, such advances are usually limited, since efficient systems are already available and there is no real need for the design of new ones.

Chromatography and Isolation of Insect Hormones and Pheromones
Edited by A. R. McCaffery and I. D. Wilson
Plenum Press, New York, 1990

We shall review the current methods (see Table 1), and discuss some (minor) improvements which may be of interest for Ecdysonists.

CHROMATOGRAPHIC PROCEDURES (Table 1)

Normal-Phase Systems

Normal-phase systems generally use silica columns (sometimes APS- or DIOL-bonded columns), and a standard system of dichloromethane-isopropanol-water mixtures as initially designed by Lafont et al. [6] used in more than 50% of the literature. Some variations exist concerning the respective proportions of the three components, according to sample polarity. Specific ternary mixtures can be prepared for non-polar compounds, e.g., acetates or ecdysone precursors (e.g., 125:15:1 v/v/v) for medium polarity compounds (125:25:2 or 125:30:2 for ecdysone and 20-hydroxyecdyone) and even for more polar ecdysteroids (125:40:3 for 26-hydroxyecdysteroids) including glucoside conjugates (100:40:3). The latter has been used only for phytoecdysteroids (Figure 1 and Table 2).

When performing metabolic studies with insects or any biological material, normal-phase HPLC can be of interest to determine small modifications, e.g., oxidation, isomerization or hydroxylation: the influence of an additional hydroxyl group on the retention time greatly differs according to its localization on the ecdysone molecule (Figure 2).

With normal phase systems, retention time is particularly sensitive to temperature. Low temperatures result in greatly decreased retention times, possibly because they reduce ecdysteroid solubility in the stationary water phase adsorbed onto the silica particles. This effect is particularly striking between +10°C and +20°C. As a consequence, it is highly recommended that the HPLC system be placed in a room where the temperature is controlled.

Polar-bonded columns (-DIOL, -APS) can also be used instead of silica ones [3,7]. Diol-bonded columns have been used with the phasmid *Carausius morosus* [8], whereas APS ones proved particularly efficient for the separation of mixtures of 3α-OH, 3β-OH and 3-oxo ecdysteroids [7], indeed these separations were better than those achieved on silica columns [9]. In addition, they might allow the use of gradients without the problems linked with long re-equilibrium times encountered with silica columns.

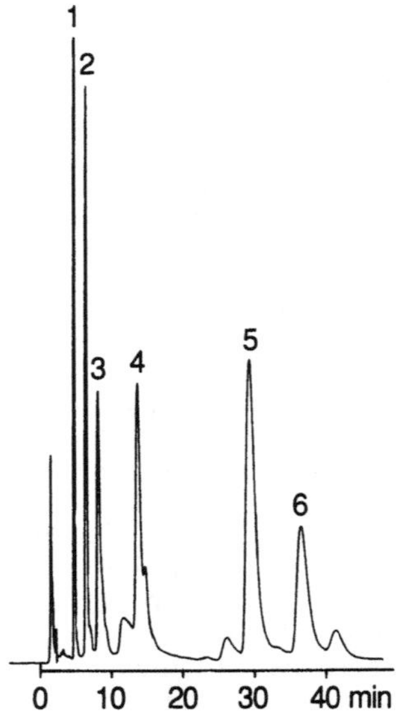

Fig. 1. Normal phase separation of a mixture of polar ecdysteroids. Operating conditions: column Zorbax -Sil, 250 x 4.6 mm, solvent dichloromethane-isopropanol-water, 100:40:3 v/v/v, flow-rate 1 ml/min: ecdysone; 2: 20-hydroxyecdysone; 3: 20,24-dihydroxyecdysone (= abutasterone); 4: 20,26-dihydroxyecdysone; 5: 20-hydroxyecdysone-25-glucoside; 6: 20-hydroxyecdysone-3-glucoside.

Table 1. Chromatographic systems commonly used for the HPLC analysis of ecdysteroids (see also Lafont et al., 1981 [3] for a more extensive review of data between 1968-1980).

Class of ecdysteroids	Mode of chromatography	Reference
POLAR	REVERSED-PHASE CHROMATOGRAPHY	
	Methanol-water	12
		13
	Methanol-sodium acetate	14
	Methanol-sodium phosphate	15
	Acetonitrile-Tris-$HClO_4$	16
	" "	17
	" "	18
	Acetonitrile-Tris-HCl	13
	Acetonitrile-sodium citrate	17
	Acetonitrile-1% trifluoroacetic acid	19
	ION-PAIR CHROMATOGRAPHY	
	Acetonitrile-Cetrimide-Phosphate	16
	Methanol-Tetrabutylammonium	14
	" "	20
	ION-EXCHANGE CHROMATOGRAPHY	
	Ammonium acetate	21
MEDIUM	REVERSED-PHASE CHROMATOGRAPHY	
	Ethanol-water	22
	Methanol-water	23
	" "	24, 25
	Acetonitrile-water	26
	" "	23
	Acetonitrile-methanol-water	27
	Acetonitrile-1% trifluoroacetic acid	28
	Acetonitrile-buffer	16
	" "	18
	Dioxan-water	24
	Tetrahydrofuran-water	24
	NORMAL-PHASE CHROMATOGRAPHY	
	SILICA:	
	Chloroform-95% ethanol	29
	Chloroform-methanol	30
	Dichloromethane-tetrahydrofuran-methanol	31
	Dichloromethane-methanol	32
	Dichloromethane-ethanol-water	6
	Dichloromethane-isopropanol-water	6, 16
	Dichloromethane-methanol-water-acetic acid	33
	Hexane-ethanol-methanol-acetonitrile	34
	Isooctane-isopropanol-water	11
	DIOL:	
	Dichloromethane-isopropanol-water	3
	" " "	8
	APS:	
	Dichloromethane-isopropanol-methanol	7
APOLAR	REVERSED-PHASE CHROMATOGRAPHY	
	Methanol-water	35, 36
	Methanol-Tris-$HClO_4$	37
	Methanol	38
	Acetonitrile-isopropanol	
	NORMAL-PHASE CHROMATOGRAPHY	
	SILICA:	
	Dichloromethane-isopropanol-water	37
	Dichloroethane-isopropanol-water	36

Abbreviations: SAX = strong anion exchanger; APS: aminopropyl silane

Table 2. Normal-phase analysis of ecdysteroids. [column Zorbax®-SIL, 250 x 4.6 mm, solvent dichloromethane-isopropanol-water 125:25:2 (A) or 125:40:3 (B) v/v/v; flow-rate 1 ml/min]

Compound	Retention time (min)	
	Solvent A	Solvent B
2-Deoxyecdysone 22-acetate	8.2	5.4
2-Deoxy-20-hydroxyecdysone 22-acetate	10.1	6.4
Poststerone	10.1	6.8
Rubrosterone	10.1	6.2
20-Hydroxyecdysone 25-acetate	10.4	5.9
2-Deoxyecdysone	11.6	7.2
20-Hydroxyecdysone 2-acetate	13.2	8.7
20-Hydroxyecdysone 3-acetate	13.8	8.7
2-Deoxy-20-hydroxyecdysone	17.2	9.4
20-Hydroxyecdysone 22-acetate	22.4	10.9
Ecdysone	22.8	12.0
Polypodine B (= 5,20-dihydroxyecdysone)	22.8	12.0
Makisterone A	26.8	12.7
22-Deoxy-20-hydroxyecdysone	28.2	13.4
20-Hydroxyecdysone	37.4	16.8
1,20-Dihydroxyecdysone (= integristerone A)	48.4	20.0
5,20,26-Trihydroxyecdysone	77.2	30.9
22-Deoxy-20,26-dihydroxyecdysone	87.6	32.9
20,26-Dihydroxyecdysone	-	43.9
20-Hydroxyecdysone 25-glucoside	-	101.5
20-Hydroxyecdysone 3-glucoside	-	115.2

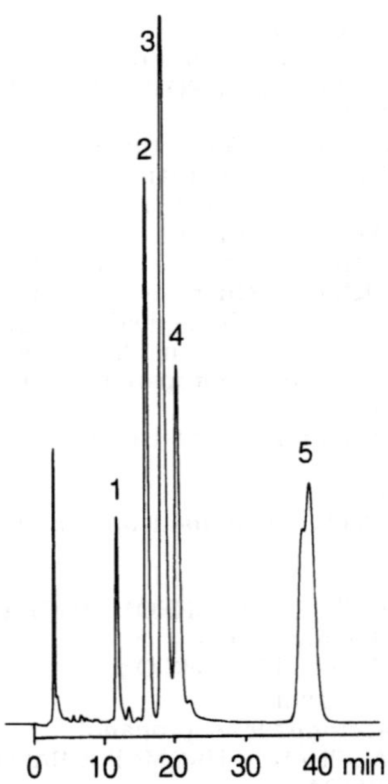

Fig. 2. Normal phase separation of various 20-hydroxyecdysone-derived compounds bearing one additional hydroxyl group at various positions. Operating conditions: column Zorbax -Sil, 250 x 4.6 mm, solvent dichloromethane-isopropanol-water, 125:40:3 v/v/v, flow-rate 1 ml/min: 5,20-dihydroxyecdysone (= polypodine B); 2: 20-hydroxyecdysone; 3: 1,20-dihydroxyecdysone (= integristerone A) + 11,20-dihydroxyecdysone (= turkesterone); 4: 20,24-dihydroxyecdysone (= abutasterone) + 16,20-dihydroxyecdysone (= malacosterone); 5: 20,26-dihydroxyecdysone.

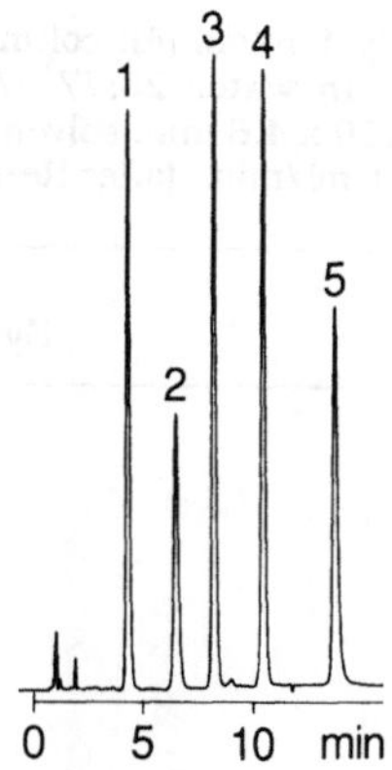

Fig. 3. Normal phase separation of ecdysteroids on a non-polar bonded silica. Operating conditions: column Zorbax® -TMS, 150 x 4.6 mm, solvent A dichloromethane-isopropanol-water, 125:25:2 v/v/v, solvent B: methanol, linear gradient O to 20% B in A in 30 minutes, flow-rate 2 ml/min. (1: 2-deoxyecdysone; 2: 2-deoxy-20-hydroxyecdysone; 3: ecdysone; 4: 20-hydroxyecdysone; 5: 20,26-dihydroxyecdysone).

We have recently observed that non-polar bonded-phase columns can also be used with the above solvents, and that they provide rather efficient separations with reduced retention times. In this respect, trimethylsilane (TMS) bonded phases proved particularly interesting (Figure 3): they gave very symmetrical peaks and a selectivity that differs from silica columns. We consider that the retention of ecdysteroids on non-polar bonded columns results from the presence of remaining free silanol groups on the bonded phases (this becomes especially evident when columns have previously been used over a long period) which would allow both NP and RP HPLC to be performed with a single column! In fact this finding is not really original: Ecdysonists engaged in SFC analyses also use C18-bonded columns instead of silica ones with a solvent system which is comparable with a normal phase one (CO_2 + methanol) - see Ref. [10]. Moreover, there exist some HPLC columns (CN: nitrile-bonded) that can be used both for NP and RP purposes and these might be used for normal phase chromatography of polar (but, of course, non-ionic) ecdysteroids.

Dichloromethane-based solvents, although very efficient for chromatographic separations, suffer from a high UV-cutoff and quenching properties of this compound, which preclude diode-array detection or in-line radioactivity monitoring (see below). This problem may be overcome with iso-octane-based mixtures [5,11] which in counterpart suffer from a lower efficiency (plate number) and poor ecdysteroid solubility, which may become inconvenient for (a) polar ecdysteroids and (b) preparative purposes.

Reversed-Phase Systems

Reversed-phase HPLC with C18-bonded columns is the most widely used system, and it provides efficient separations (Table 3). Although methanol-water is still being used, more and more ecdysonists are using acetonitrile/water or better buffer mixtures instead, especially when polar conjugates and/or ecdysonoic acids are present. Systems have been designed for polar or apolar metabolites, both of which may exist within the same animal. It has to be underlined that surprisingly apolar fatty acyl esters of ecdysteroids are not eluted with pure acetonitrile whereas they are with methanol (the same is true for cholesterol).

Selectivity linked with the nature of the organic phase is an important parameter for determining the most suitable solvent mixture for a given separation: according to the specific problem, ecdysonists may have better results with methanol, acetonitrile or tetrahydrofuran (THF) [23]. This is illustrated on Figure 4: pairs of 3β-OH and 3-oxo compounds are much better resolved with acetonitrile than with methanol [39]. On the other hand, methanol or THF can be more adequate for other specific cases (Figure 5 - from Ref. [24]).

Concerning polar ionisable metabolites, it may be of interest to use various pHs as mentioned earlier [3]. This gives access to the pK value (Figure 6), which is of major interest

Table 3. Reversed-phase analysis of ecdysteroids. (A): column: Spherisorb® 5-ODS2, 250 x 4.6 mm, solvent acetonitrile - 1% TFA in water 23:77 v/v, isocratic; flow-rate 1 ml/min. (B): column: Ultrasphere® - ODS, 150 x 4.6 mm, solvent linear gradient 40% to 80% of methanol in water over 20 min; flow-rate 1 ml/min. (after Rees and Isaac, 1985 [2]).

Compound	Retention time (min)	
	System A	System B
Ecdysone 22-phosphate		1.3
2-Deoxyecdysone 22-phosphate		2.0
20,26-Dihydroxyecdysone	4.1	4.9
1,20-Dihydroxyecdysone	4.6	
5,20,26-Trihydroxyecdysone	4.7	
20-Hydroxyecdysone	5.5	9.4
5,20-Dihydroxyecdysone	5.8	
26-Hydroxyecdysone		9.7
Inokosterone		9.9
3-Epi-20-hydroxyecdysone		9.9
3-Dehydro-20-hydroxyecdysone		9.9
Rubrosterone	6.0	
Makisterone A	7.3	
Poststerone	9.7	9.6
Ecdysone	10.1	13.1
3-Epiecdysone		13.6
3-Dehydroecdysone		13.5
20-Hydroxyecdysone 3-acetate	10.8	
20-Hydroxyecdysone 22-acetate	12.4	
2-Deoxy-20-hydroxyecdysone	12.9	14.8
3-Epi-2-deoxy-20-hydroxyecdysone		16.0
Muristerone A	13.3	
Ecdysone 3-acetate		16.0
Ecdysone 2-acetate		18.3
22-Deoxy-20-hydroxyecdysone	13.9	
20-Hydroxyecdysone 2-acetate	15.0	
Makisterone C	17.5	
20-Hydroxyecdysone 25-acetate	18.4	
2-Deoxyecdysone	34.1	18.6
3-Epi-2-deoxyecdysone		20.0
Ponasterone A	40.2	17.2

for the characterization of conjugates and can provide useful means of distinguishing between the different types of hydrolysable conjugates (Table 4).

Ion-Pair and Ion-Exchange Systems

The use of ion-pair and ion-exchange systems has remained limited. Ion-pairing with tetrabutylammonium or cetyltrimethylammonium salts can be used to increase the retention of anionic polar ecdysteroids on reversed-phase columns, but they are in fact of limited interest, as they allow these compounds to coelute with medium-polarity ecdysteroids [16]. Anion-exchange columns were used for the purification of polar conjugates (phosphate esters) (Table 5). They represent an efficient method, complementary to reversed-phase HPLC. However, the use of two different pHs (e.g., pH 7.5, then 2.5) with the latter provides an equivalent opportunity for obtaining pure ionizable conjugates, e.g., phosphate esters [19].

Quantitative Analyses

HPLC has been used extensively for the direct quantification of individual ecdysteroids in biological samples. This requires of course high sensitivity because of the low concentrations encountered and adequate sample clean-up (see below). Quantification is best obtained if an internal standard is added to the sample either before HPLC analysis or better before sample purification [3,25]. The early addition of the internal standard is preferable because it allows for the correction of losses resulting from variations in recovery efficiency during clean-up. Many phytoecdysteroids are available which can be used as internal standards. The choice has to be made after a preliminary run of a representative sample in order to select a compound that does not comigrate with major impurities or with ecdysteroids present in the biological extract. It is obviously also

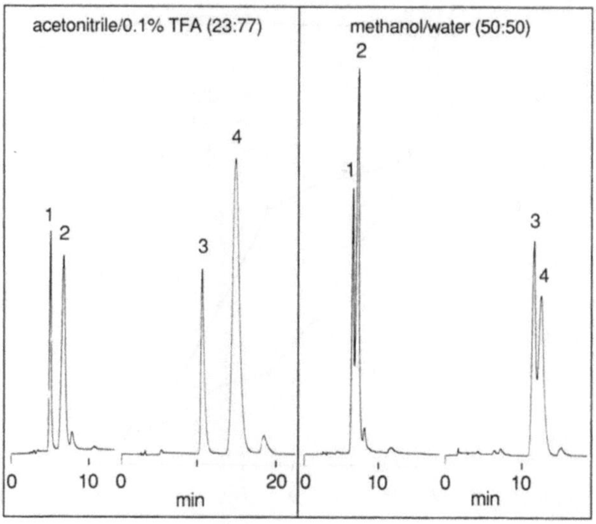

Fig. 4. Reversed-phase separation of 3β-OH/3-oxo pairs: solvent selective effects. Operating conditions: column Spherisorb® -5 ODS2, 250 x 4.6 mm, solvent as indicated, flow-rate 1 ml/min. 1: 20-hydroxyecdysone; 2: 3-dehydro-20-hydroxyecdysone; 3: ecdysone; 4: 3-dehydroecdysone. (From Girault et al., 1989 [39] reproduced with the permission from Alan R. Liss, Inc.)

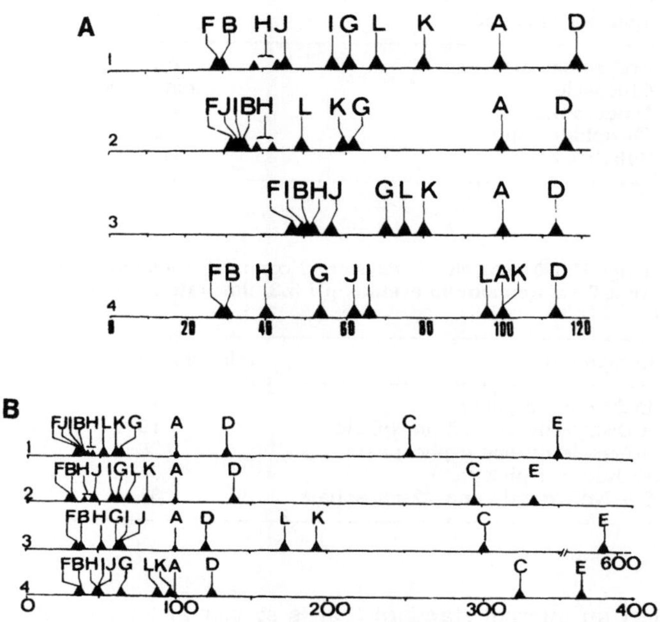

Fig. 5. Selectivity of reversed-phase systems is linked to the columns and/or the solvent systems. A: selectivity differences between Spherisorb®-ODS eluted with (1) acetonitrile-water (15:85) or (2) methanol-water (35:65) and Nucleosil®-ODS eluted with (3) methanol-water (50:50) or (4) acetonitrile-water(20:80). B: comparison of different solvents with a Spherisorb®-ODS column: (1) methanol-water (35:65), (2) acetonitrile-water (15:85), (3) tetrahydrofuran-water (10:90) and (4) dioxan-water(20:80). Retention is given relative to ecdysone [A] which is given a value of 100. [A] = ecdysone; [B] = 20-hydroxyecdysone; [C] = 2-deoxyecdysone; [D] = 2-deoxy-20-hydroxyecdysone; [E] = ponasterone A; [F] = polypodine B; [G] = makisterone A; [H] = inokosterone; [I] = cyasterone; [J] = poststerone; [K] = ajugasterone C; [L] = muristerone A. (From Wilson et al., 1982 [24], reproduced with the permission of Elsevier.)

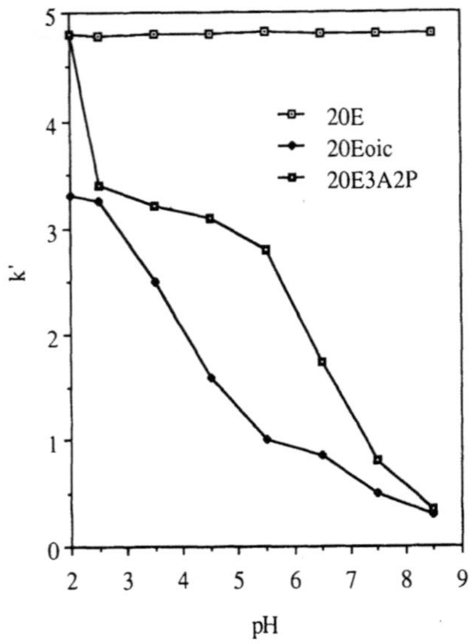

Fig. 6. Effects of pH on the capacity factor (k') of 20-hydroxyecdysone (20E), 20-hydroxyecdysonoic acid (20Eoic) and 3-acetyl-20-hydroxyecdysone 2-phosphate (20E3A2P). Operating conditions: column Zorbax® -C8, 250 x 4.6 mm, solvent acetonitrile-buffer 15:85 v/v, flow-rate 1 ml/min. (From Lafont et al., 1981 [3], modified.)

Table 4. Ionization characteristics of polar ecdysteroids (they will undergo a large change of k' between pK-1<pH<pK+1)

Type of compound	pK
Ecdysonoic acid	ca. 4.5
Glucoside	non-ionic
Glucuronide	ca. 4.5
Phosphate ester	ca. 7.2
Sulfate ester	ca. 1.9

Table 5. Ion-exchange HPLC analysis of ecdysteroid conjugates [column Partisil® -SAX, 250 x 4.6 mm, solvent 0.1M ammonium acetate, pH 5.3, flow-rate 2 ml/min] (after Rees and Isaac, 1985 [2]

Compound	retention volume (ml)
Ecdysone 22-phosphate	16
2-Deoxyecdysone 22-phosphate	19
3-Acetylecdysone 2-phosphate	28
Ecdysone 2-phosphate	31
20-Hydroxyecdysone 22-phosphate	31

necessary to select an internal standard that is similar in its extraction and chromatographic properties to those of the analyte(s) in question.

Preparative HPLC

Columns of different sizes exist with the same packings, and it is therefore very easy to scale-up any chromatographic method. Thus there exist analytical columns (i.d. 4.6 mm), semi-preparative ones (i.d. 9.6 mm) and preparative ones (i.d. ca. 20 mm or even more). By increasing the flow-rate in proportion to the cross-sectional-area, one obtains similar separations with increased maximal load.

In fact, amounts in the milligram range can be run on an analytical column without any loss of efficiency due to overloading. Larger amounts require of course wide-bore columns. It may happen that the separation of some pairs of compounds which are baseline resolved when loading is low elute as a single peak with a larger load. This was for instance observed with the pair 20-hydroxyecdysone-inokosterone run on a semi-preparative column (Zorbax®-SIL, 9.6 mm i.d.) even when less than 0.5 mg were injected. We have at present no explanation for such a result.

Microbore HPLC and Capillary Zone Electrophoresis (CZE)

Microbore HPLC has been designed essentially for coupling with mass spectrometry. It uses columns of small internal diameter (e.g., 1 mm) run with a reduced solvent flow-rate (ca. 50 µl/min) which is compatible with direct coupling with a MS. To our knowledge, this method has not received any serious application to ecdysteroids. It seems that in fact many problems remain with pumps and detection with such techniques.

CZE is also a microbore technique which provides very high efficiencies (ca. 1,000,000 theoretical plates!) and requires no pumps but detection is still a problem because only very small sample volumes may be applied to the columns (in the nl range). Although an electrophoretic technique CZE can nevertheless be used for neutral molecules which do not possess ionizable functional groups (for a recent review see Reference 40). Providing therefore that suitable detectors can be designed CZE and related techniques should prove useful for the resolution of complex mixtures of ecdysteroids.

DETECTION PROCEDURES

Diode-Array Detectors

Diode-array detectors represent valuable tools as, with no additional work, they give information about the absorbance spectrum of all eluted peaks. Thus in the case of ecdysteroids, it can be directly checked whether a compound that comigrates with a reference ecdysteroid has a UV absorbance with a maximum at 242 nm. This is an additional criterion for assessing the identity of peaks, and it works equally well with very small amounts of ecdysteroids (less than 100 ng - see Figure 7) provided that (a) the HPLC separation is efficient enough and (b) the solvent used does not absorb UV. The second condition would for instance exclude the use of dichloromethane with normal-phase systems. Several manufacturers supply very efficient, and sometimes very expensive, diode-array detectors.

Radioactivity Monitors

On-line radioactivity measurement is of considerable interest [41] for the following reasons: (a) it provides an immediate result, (b) it saves time and avoids the need for collecting many fractions, filling scintillation vials and waiting for their measurement, (c) it provides direct comparison with the signal of another detector, e.g., UV monitor (Figure 8), and this allows an easy check for the correspondence of radioactive peaks with reference compounds, (d) it does not significantly decrease the resolution when compared with UV signal: this means that at a flow-rate of 1 ml/min for column effluent and of 3 ml/min for the scintillation cocktail, with a cell size of 0.5 ml, the result is equivalent or even better than collecting 0.2 ml fractions (200 tubes for a 40 min analysis) and it might be improved even further by using a higher flow-rate for the scintillation cocktail.

There are nevertheless some limitations with the use of such detectors, when the amount of available radioactivity (especially tritium) is low. This is for instance the case when [^{3}H]-cholesterol labeling experiments are performed. In that case, it appears that fraction collection (and counting for 2 or 5 min/tube) provides more reliable results. This problem will however probably be overcome with a new generation of radioactivity monitors with improved background (< 5 cpm) that will conversely provide a higher sensitivity (e.g., the new Flo-One). There are presently several systems available (Berthold, Isomeas, Radiomatic, Beckman etc).

Fluorescent Derivatives

There have been several attempts to prepare fluorescent derivatives of ecdysteroids. The potential interest is great, as this would enhance the sensitivity of detection by at least

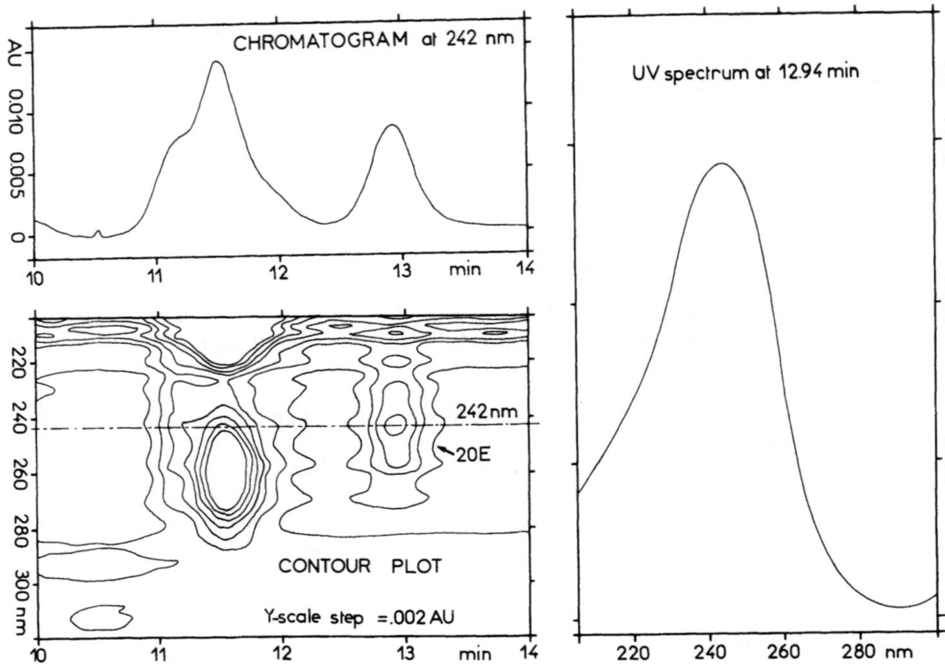

Fig. 7. Analysis of a haemolymph extract from *Pieris brassicae* pupae by normal phase chromatography (column: Zorbax -Sil, 250 x 4.6 mm, solvent iso-octane-isopropanol-water 100:40:3 v/v/v, flow-rate 2 ml/min), using a diode-array detector (Waters 990). The contour plot presented here represents the part of the chromatogram where 20-hydroxyecdysone (20E) is eluted (at 12.94 min), and gives evidence that the compound which elutes at that time shows a maximum of absorbance at 242 nm. Of interest is the high sensitivity reached, as the 20E peak gave a maximal absorbance of only 0.0085 AU, which was enough to obtain a characteristic UV spectrum after correcting for background and smoothing. (From Lafont, 1988 [5], reproduced with permission from Akademia Kiado, Budapest.)

100 times and possibly also the selectivity when compared with UV. Two possibilities exist in theory: pre-column or post-column derivatization.

There are of course several prerequisites for pre-column reactions: (a) they must be simple and if possible give rise to a single derivative for each ecdysteroid, (b) they must be quantitative, even at very low concentrations of ecdysteroids present in a crude sample, (c) they must be specific, and (d) the excess of reagent (if fluorescent itself) must be removed prior to HPLC analysis. Two approaches have been described using phenanthrene boronic acid [42] (this type of reagent is specific for diols - and would therefore react not only with ecdysteroids but also with sugars) and 1-anthroyl nitrile [43] (which reacts with alcohols, here the 2-OH of ecdysteroids). The interest in the second case would essentially be the increase in sensitivity, as this is not at all a specific reaction (the authors use the same reaction for the determination of prostaglandins). Some unsuccessful attempts have also been made using dansyl-hydrazine that reacts with the 6-keto group (Wilson, unpublished). These methods have not received any further application, no doubt in part because the reagents (except dansyl-hydrazine) are not commercially available. In addition, we can observe that the use of pre-column derivatization would suppress one major advantage of HPLC over GLC.

On the other hand, it has been known for many years that strongly acidic conditions induce a blue fluorescence in ecdysteroids, which can be used with silica TLC [44-46]. This reaction has also been used for developing an off-line procedure [47]. The reaction proceeds within a rather short time (20 to 30 min) and it would be conceivable to use it in a post-column reactor - the only practical problem might be that the reagent is pure sulfuric acid!

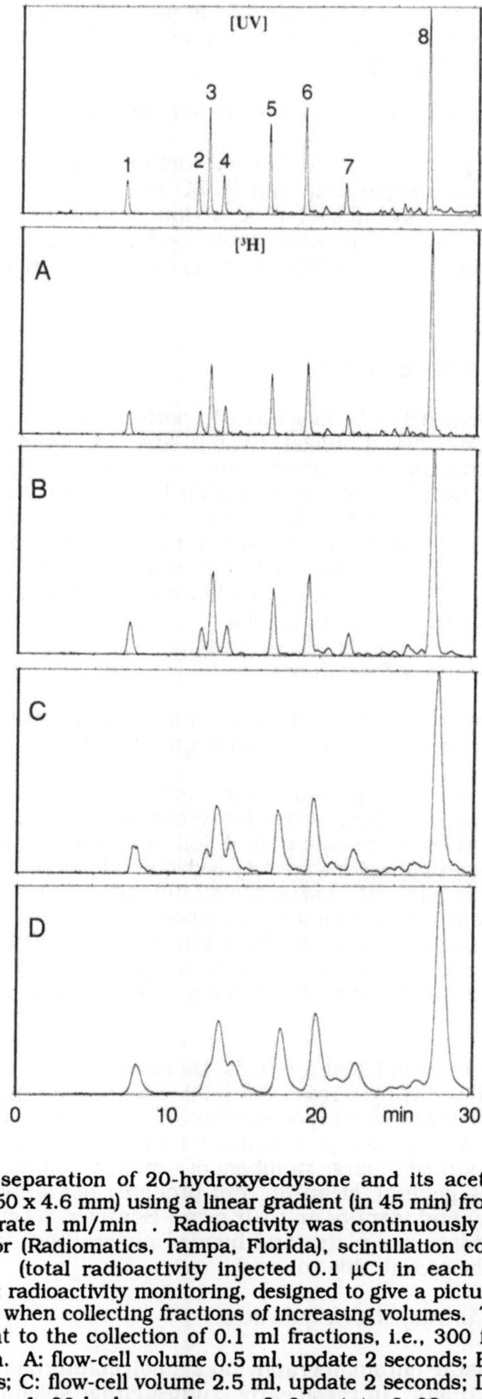

Fig. 8. Reversed-phase separation of 20-hydroxyecdysone and its acetates on a Spherisorb® - 5)DS2 column (250 x 4.6 mm) using a linear gradient (in 45 min) from 20% to 70% acetonitrile in 1% TFA, flow-rate 1 ml/min . Radioactivity was continuously monitored with a Flo-One model IC detector (Radiomatics, Tampa, Florida), scintillation cocktail Lumaflow III, flow-rate 3 ml/min (total radioactivity injected 0.1 μCi in each case). Upper trace: UV absorbance, A-D: radioactivity monitoring, designed to give a picture of the loss of resolution that would occur when collecting fractions of increasing volumes. The best resolution shown in A is equivalent to the collection of 0.1 ml fractions, i.e., 300 fractions over the 30 min analysis duration. A: flow-cell volume 0.5 ml, update 2 seconds; B: flow-cell volume 0.5 ml, update 6 seconds; C: flow-cell volume 2.5 ml, update 2 seconds; D: flow-cell volume 2.5 ml, update 6 seconds. 1: 20-hydroxyecdysone; 2: 3-acetate; 3: 22-acetate; 4: 2-acetate; 5: 3,22-diacetate; 6: 2,22-diacetate; 7: 2,3-diacetate; 8: 2,3,22-triacetate.

Mass Spectrometry

In-line HPLC-MS is a technique that would in theory provide direct and very useful information on the structure of ecdysteroids eluting from the column. Nevertheless this problem has attracted only very limited attention (see Evershed et al., this volume) and there is no report of its use for biological samples. From our own experience it proved rather disappointing and the limited access to the apparatus did not allow us to improve our preliminary results. Supercritical fluid chromatography (SFC) may prove to be a better candidate for coupling with MS [10].

HPLC-Radioimmunoassay (or Enzyme Linked Immunoassay)

HPLC-RIA (-EIA) is presently a widely used method that allows the quantification of individual ecdysteroids once separated with HPLC with a rather good accuracy and an excellent sensitivity. The drawback of this off-line technique is that it is very time-consuming, but in fact it remains the only viable method for HPLC when samples contain very low amounts of ecdysteroids. The alternative is GLC with either electron capture or MS for detection.

SAMPLE PREPARATION PROCEDURES

Analysis of biological extracts is generally performed after a few steps of sample clean-up. This is by no means an obligatory step, and it is quite possible to inject crude samples (in that case the use of a pre-column is recommended), especially for small samples. On the other hand, it seems advisable to eliminate most of the undesirable components of a crude extract before injecting it onto the HPLC column, and several methods are available for this purpose, which may differ depending on the type of compounds and the type of HPLC to be used. Generally, the aim is to eliminate very polar and/or apolar substances. We can distinguish between several major procedures, that are described below (see also Refs. [16,21,31,48,49]).

Conventional Sample Clean-Up Procedures

Conventional methods include: solvent partitions, preparative TLC, counter-current distribution and droplet counter-current chromatography (DCCC).

Solvent partitions are used in two ways: they can be used to eliminate non polar substances, e.g., by using chloroform-water (ecdysone and all the more polar ecdysteroids are present in the water phase - see Lafont et al. [6]) or to remove polar substances, e.g., by using a butanol-water system (medium-polarity ecdysteroids are found in the butanol phase [31]). None of these methods is able to separate all the ecdysteroids from other materials, as significant amounts of ecdysone precursors and apolar conjugates are found in chloroform in the first system, and polar conjugates are found in the water phase in the second case (the latter method can therefore be used to separate ecdysone and 20-hydroxyecdysone from their polar conjugates [21]). So the choice for a given method depends on the type of compounds of interest.

TLC and DCCC are more efficient methods, that can in some cases allow one to obtain rather pure compounds (see Figure 9 from Ref. [27]), but they are quite time-consuming and they are used rather for preparative purposes, where it is necessary to concentrate the active fractions prior to HPLC. Using such procedures for sample preparation for HPLC is rather time consuming and not suited to large numbers of samples. Indeed, such methods provide too much resolution for a general "clean-up" but probably not sufficient to unambiguously confirm compound purity and identity on their own. However, recent improvements of TLC with the use of high-performance plates and bonded phases have led to the design of very efficient systems (see Wilson et al. this volume).

Use of Reversed-Phase Cartridges (or Equivalent)

The use of low pressure chromatography on a small column has long been used for the fractionation of crude extracts. In these early studies either silica or alumina was used in such columns (normal phase systems) generally eluted with binary mixtures - a step-gradient of alcohol in chloroform or benzene. The availability of hydrophobic phases (resins like amberlite or hydrocarbon-bonded silica) has led to a complete renewal of these procedures [50]. Moreover, the design of small cartridges or syringes containing 0.2 to 1 g of HPLC phase has led to an ideally suited material for a rapid clean-up of small samples. Among them, Sep-pak cartridges from Waters are the most widely used [6,21,51-54]. In fact their use has been extended not only for biological extracts but also for desalting purposes, e.g., direct adsorption of ecdysteroids from culture media or from reverse-phase HPLC fractions containing an involatile buffer (Figure 10).

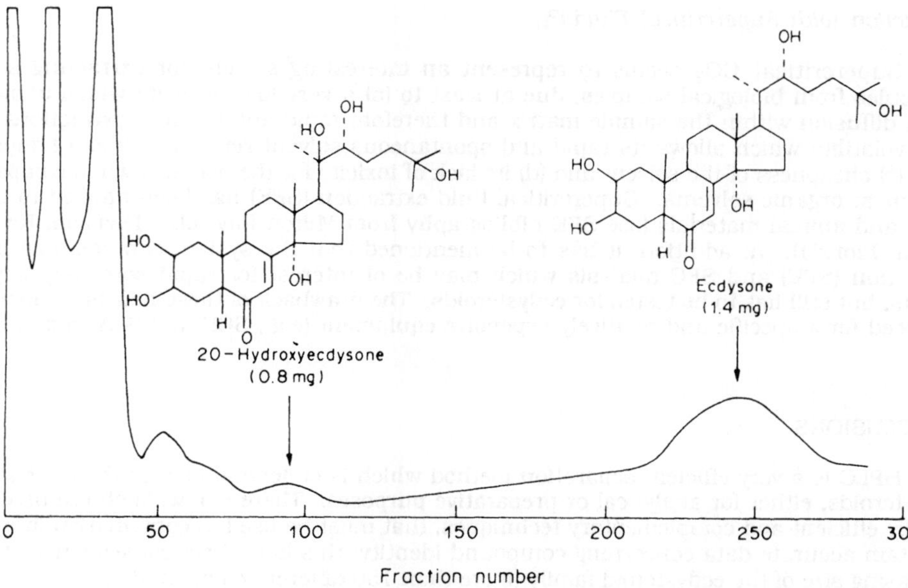

Fig. 9. Analysis by DCCC of a crude extract from *Bombyx mori* pupae. (From Kubo et al., 1985, [56] reproduced with permission from Pergamon Press.)

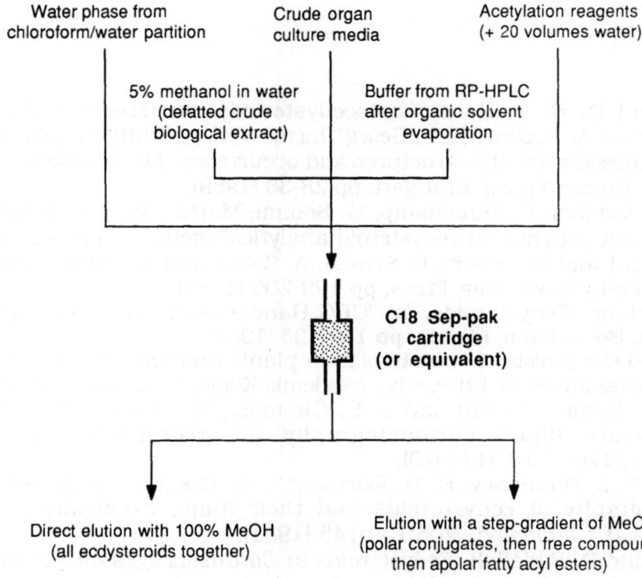

Fig. 10. Multiple uses of C18-bonded cartridges for sample processing before HPLC analysis.

Use of Other Types of Cartridges (or Low-Pressure Column Chromatography)

Normal phase cartridges or columns can be used to purify ecdysteroids using a step-gradient of methanol in chloroform (or benzene) which allows the easy separation of the three classes of ecdysteroids [15,21,31,49].

Recently, new silica-bonded phenyl boronic acid (PBA) cartridges have been designed, which appear of interest as they seem to selectively retain compounds bearing a free 20,22-diol (Murphy et al., this volume).

Extraction with Supercritical Fluids?

Supercritical CO_2 seems to represent an interesting solvent for extracting small molecules from biological samples, due at least to (a) a very low viscosity which allows a rapid diffusion within the sample matrix and therefore rapid and efficient extractions, (b) high volatility which allows its rapid and spontaneous solvent removal at room temperature, (c) cheapness of the solvent and (d) its lack of toxicity for the operator when compared with most organic solvents. Supercritical fluid extraction (SFE) has been applied to both plant and animal materials (see SFE bibliography from Milton Roy, LDC Division, Riviera Beach, Florida). In addition, it has to be mentioned that the system can combine both extraction (SFE) and SFC analysis which may be of interest for rapid screening experiments, but still has to be tested for ecdysteroids. The drawback of these new procedures is the need for a specific and relatively expensive equipment (e.g., SCE and SPA from Milton Roy).

CONCLUSIONS

HPLC is a very efficient separation method which is of general use for the analysis of ecdysteroids, either for analytical or preparative purposes. There is a wide choice between equally efficient and complementary techniques, that must be used in combination in order to obtain accurate data concerning compound identity: this is a direct consequence of the increasing size of the ecdysteroid family (more than 150 different compounds).

Although progress is still continuing, some areas appear underdeveloped until now. This concerns essentially the design of chemical reactions for forming fluorescent derivatives and of "routine" coupled HPLC-MS. Such developments would lead to significant improvements regarding identification and direct quantification of ecdysteroids from biological extracts.

REFERENCES

1. R. Lafont and D. H. S. Horn, Phytoecdysteroids: structures and occurrence, *in*: "Ecdysone", J. A. Koolman, ed., Georg Thieme-Verlag, Stuttgart, pp 39-64 (1989).
2. H. H. Rees, Zooecdysteroids: structures and occurrence, *in*: "Ecdysone", J. A. Koolman, ed., Georg Thieme-Verlag, Stuttgart, pp 28-38 (1989).
3. R. Lafont, P. Beydon, B. Mauchamp, G. Sommé-Martin, M. Andrianjafintrimo and P. Krien, Recent progress in ecdysteroid analytical methods, *in*: "Regulation of Insect Development and Behavior", F. Sehnal, A. Zabza and B. Cymborowski, eds., Technical University of Wroclaw Press, pp 199-206 (1981).
4. J. C. Touchstone, Ecdysteroids, *in*: "CRC Handbook of Chromatography. Steroids", CRC Press, Boca Raton, Florida, pp 119-135 (1986).
5. R. Lafont, HPLC analysis of ecdysteroids in plants and animals, *in*: "Chromatography '87", H. Kalasz and E. S. Ettre, eds., Akademia Kiado, Budapest, pp 1-15 (1988).
6. R. Lafont, G. Sommé-Martin and J. C. Chambet, Separation of ecdysteroids using high-pressure liquid chromatography on microparticulate supports, *J. Chromatog.*, 170:185-194 (1979).
7. L. N. Dinan, P. L. Donnahey, H. H. Rees and T. W. Goodwin, High performance liquid chromatography of ecdysteroids and their 3-epi, 3-dehydro and 26-hydroxy derivatives, *J. Chromatog.*, 205:139-145 (1981).
8. B. Fournier and D. Radallah, Ecdysteroids in *Carausius* eggs during embryonic development, *Archives Insect Biochem. & Physiol.*, 7:211-224 (1988).
9. C. Blais and R. Lafont, Ecdysteroid metabolism by soluble enzymes from an insect: metabolic relationship between 3β-hydroxy-, 3α-hydroxy- and 3-oxo-ecdysteroids, *Hoppe-Seyler's Zeitschrift für Physiologische Chemie*, 365:809-818 (1984).
10. M. W. Raynor, J. P. Kithinji, K. D. Bartle, D. E. Games, I. C. Mylchreest, R. Lafont, E. D. Morgan and I. D. Wilson, Packed column supercritical fluid chromatography and linked supercritical fluid chromatography - mass spectrometry for the analysis of phytoecdysteroids from *Silene nutans* and *Silene otites*, *J. Chromatog.*, 467:292-298 (1989).
11. P. Beydon, A. Fabre, J. Goichon and R. Lafont, Use of a diode-array detector for HPLC analysis of ecdysteroids in biological extracts, Communication presented at the VIIIth Ecdysone Workshop, Marburg, FRG (1987).
12. R. E. Isaac, N. P. Milner and H. H. Rees, High-performance liquid chromatography of ecdysteroids and ecdysteroid 22-phosphates, *J. Chromatog.*, 246:317-322 (1982).

13. C. Hetru, B. Luu and J. A. Hoffmann, Ecdysone conjugates: isolation and identification, *Methods in Enzymology*, 11:411-419 (1985).
14. R. E. Isaac, H. H. Rees and T. W. Goodwin, Isolation of 2-deoxy-20-hydroxyecdysone and 3-epi-2-deoxyecdysone from eggs of the desert locust, *Schistocerca gregaria*, *J. Chem. Soc., Chem. Communs.*, pp 418-420 (1981).
15. M. J. Thompson, G. F. Weirich, H. H. Rees, J. A. Svoboda, M. F. Feldlaufer and K. R. Wilzer, New ecdysteroid conjugate: isolation and identification of 26-hydroxyecdysone 26-phosphate from eggs of the tobacco hornworm, *Manduca sexta* (L.), *Archives of Insect Biochem. & Physiol.*, 2:227-236 (1985).
16. R. Lafont, G. Somme-Martin, B. Mauchamp, B. F. Maume and J. P. Delbecque, Analysis of ecdysteroids by high-performance liquid chromatography and coupled gas-liquid chromatography-mass spectrometry, *in*: "Progress in Ecdysone Research", J. A. Hoffmann, ed., Elsevier Biomedical Press, Amsterdam, pp 45-68 (1980).
17. J. F. Modde, R. Lafont and J. A. Hoffmann, Ecdysone metabolism in *Locusta migratoria* larvae and adults, *Internat. J. Invertebrate Reprod. & Develop.*, 7:161-183 (1984).
18. J. T. Warren, B. Steiner, A. Dorn, M. Pak and L. I. Gilbert, Metabolism of ecdysteroids during the embryogenesis of *Manduca sexta*, *J. Liquid Chromatog.*, 9:1759-1782 (1986).
19. P. Beydon, J. P. Girault and R. Lafont, Ecdysone metabolism in *Pieris brassicae* during the feeding last larval instar, *Archives of Insect Biochem. & Physiol.*, 4:139-149 (1987).
20. S. Scalia and E. D. Morgan, Simultaneous determination of free and conjugated ecdysteroids by liquid chromatography, *J. Chromatog.*, 346:301-308 (1985).
21. H. H. Rees and R. E. Isaac, Biosynthesis and metabolism of ecdysteroids and methods of isolation and identification of the free and conjugated ecdysteroids, *Methods in Enzymology*, 111:377-410 (1985).
22. M. Hori, Automatic column chromatographic method for insect moulting steroids, *Steroids*, 14:33-45 (1969).
23. D. A. Schooley and K. Nakanishi, Applications of high-pressure liquid chromatography to the separation of insect moulting hormones, *in*: "Modern Methods of Steroid Analysis", E. Heftmann, ed., Academic Press, New York, pp 37-54 (1973).
24. I. D. Wilson, C. R. Bielby and E. D. Morgan, Selective effects of mobile and stationary phases in reversed-phase high-performance liquid chromatography of ecdysteroids, *J. Chromatog.*, 238:97-102 (1982).
25. I. D. Wilson, C. R. Bielby and E. D. Morgan, Evaluation of some phytoecdysteroids as internal standards for the chromatographic analysis of ecdysone and 20-hydroxyecdysone from arthropods, *J. Chromatog.*, 236:224-229 (1982).
26. G. M. Holman and R. W. Meola, A high-performance liquid chromatography method for the purification and analysis of insect ecdysones: application to measurement of ecdysone titres during pupal-adult development of *Heliothis zea*, *Insect Biochem.*, 8:275-278 (1978).
27. I. Kubo and F. J. Hanke, Chemical methods for isolating and identifying phytochemicals biologically active in insects, *in*: "Insect-Plant Interactions - Springer Series in Experimental Entomology", J. R. Miller and T. A. Miller, eds. Springer-Verlag, New York, pp 225-250 (1986).
28. P. Nirde, G. Torpier, M. L. De Reggi and A. Capron, Ecdysone and 20-hydroxyecdysone: new hormones for the human parasite *Schistosoma mansoni*, *J. Insect Physiol.*, 23:317-319 (1977).
29. H. N. Nigg, M. J. Thompson, J. N. Kaplanis, J. A. Svoboda and W. E. Robbins, High-pressure liquid chromatography of the ecdysones - insect moulting hormones, *Steroids*, 23:507-516 (1974).
30. M. Valentin, W. E. Bollenbacher, L. I. Gilbert and H. Kroeger, Alterations in ecdysone content during the post-embryonic development of *Chironomus thumni*: correlations with chromosomal puffing, *Zeitschrift für Naturforschung*, 33c:557-560 (1978).
31. E. D. Morgan and C. F. Poole, The extraction and determination of ecdysones in Arthropods, *Advances in Insect Physiol.*, 12:17-62 (1976).
32. E. Ohnishi, T. Mizuno, N. Ikekawa, N. Awata and S. Sakurai, Occurrence of α-ecdysone in the developing embryos of the silkworm, *Bombyx mori*, *J. Insect Physiol.*, 23:317-319 (1977).
33. B. Stay, L. S. Ostedgaard, S. S. Tobe, A. Strambi and E. Spaziani, Ovarian and haemolymph titres of ecdysteroid during the gonadotrophic cycle in *Diploptera punctata*, *J. Insect Physiol.*, 30:643-651 (1984).

34. A. Moribayashi, H. Kurahashi and T. Ohtaki, Comparative studies on ecdysone metabolism between mature larvae and pharate pupae in the fleshfly, *Sarcophaga peregrina*, *Archives Insect Biochem. & Physiol.*, 2:237-250 (1985).
35. K. H. Hoffmann, D. Bulenda, E. Thiry and E. Schmid, Apolar ecdysteroid esters in adult female crickets, *Gryllus bimaculatus*, *Life Sciences*, 37:185-192 (1985).
36. T. Crosby, R. P. Evershed, D. Lewis, K. P. Wigglesworth and H. H. Rees, Identification of ecdysone 22-long-chain fatty acyl esters in newly laid eggs of the cattle tick *Boophilus microplus*, *Biochem. J.*, 240:131-138 (1986).
37. P. A. Diehl, J. L. Connat, J. P. Girault and R. Lafont, A new class of apolar ecdysteroid conjugates: esters of 20-hydroxyecdysone with long-chain fatty acids in ticks, *Internat. J. Invertebrate Reprod. & Develop.*, 8:1-13 (1985).
38. I. Kubo, S. Komatsu, I. Asaka and G. De Boer, Isolation and identification of apolar metabolites of ingested 20-hydroxyecdysone in frass of *Heliothis virescens* larvae, *J. Chem. Ecol.*, 13:785-794 (1987).
39. J. P. Girault, C. Blais, P. Beydon, C. Rolando and R. Lafont, Synthesis and NMR study of 3-dehydroecdysteroids, *Archives Insect Biochem. & Physiol.*, 10:199-213 (1989).
40. S. Terabe, Electrokinetic chromatography: an interface between electrophoresis and chromatography, *Trends Anal. Chem.*, 8:129-134 (1989).
41. I. D. Wilson, E. D. Morgan, P. Robinson, R. Lafont and C. Blais, A comparison of radio-thin-layer and radio-high-performance liquid chromatography for ecdysteroid metabolism studies, *J. Insect Physiol.*, 34:707-711 (1988).
42. C. F. Poole, S. Singhawangcha, A. Zlatkis and E. D. Morgan, Polynuclear aromatic boronic acids as selective fluorescent reagents for HPTLC and HPLC, *J. High Res. Chrom. & Chrom. Commun.*, 1:96-97 (1978).
43. I. Kubo and S. Komatsu, Micro analysis of prostaglandins and ecdysteroids in insects by high-performance liquid chromatography and fluorescence labeling, *J. Chromatog.*, 362:61-70 (1986).
44. T. Takemoto, S. Ogawa, N. Nishimoto, K. Y. Yen, K. Abe, T. Sato, K. Ogawa and M. Takahashi, The Isolation of ecdysterone from the radix of *Achyranthes obtusifolia* Lam, *Yakugaku Zasshi*, 87:1521-1524 (1967).
45. M. W. Gilgan and M. E. Zinck, Estimation of ecdysterone from sulphuric acid-induced fluorescence, *Steroids*, 20:95-104 (1972).
46. R. T.Mayer and J. A. Svoboda, Thin-layer chromatographic *in situ* analysis of insect ecdysones via fluorescence quenching, *Steroids*, 31:139-150 (1978).
47. J. Koolman, Analysis of ecdysteroids by fluorometry, *Insect Biochem.*, 10;381-386 (1980).
48. R. Lafont and P. Beydon, Methods for ecdysteroid analysis, *in*: "Recent Advances in Comparative Arthropod Morphology, Physiology and Development", A. P. Gupta, ed., Rutgers University Press, Vol. 1, pp 455-512 (in press) (1989).
49. E. D. Morgan and I. D. Wilson, Methods for separation and physico-chemical quantification of ecdysteroids, *in*: "Ecdysone", J. A. Koolman, ed., Georg Thieme-Verlag, Stuttgart, pp 114-130 (1989).
50. D. A. Schooley, G. Weiss and K. Nakanishi, A simple and general extraction procedure for phytoecdysones based on reversed-phase adsorption chromatography, *Steroids*, 19:377-383 (1973).
51. I. D. Wilson, C. R. Bielby, E. D. Morgan and A. E. M. McLean, Comparison of high-performance liquid chromatography and gas-chromatography for the analysis of ecdysteroids, *J. Chromatog.*, 194:343-352 (1980).
52. R. D. Watson and E. Spaziani, Rapid isolation of ecdysteroids from crustacean tissues and culture media using Sep-pak C18 cartridges, *J. Liquid Chromatog.*, 5:525-535 (1982).
53. G. D. Pimprikar, M. J. Coign, H. Sakurai and J. R. Heitz, High-performance liquid chromatographic determination of ecdysteroid titers in the house fly, *J. Chromatog.*, 317:413-419 (1984).
54. R. Lafont, J. L. Pennetier, A. Andrianjafintrimo, J. Claret, J. F. Modde and C. Blais, Sample processing for high-performance liquid chromatography of ecdysteroids, *J. Chromatog.*, 236:137-149 (1982).
55. I. Kubo, A. Matsumoto and S. Asano, Efficient isolation of ecdysteroids from the silkworm, *Bombyx mori* by droplet counter-current chromatography, *Insect Biochem.*, 15:45-47 (1985).

EXAMINATION OF CONDITIONS IN SUPERCRITICAL FLUID CHROMATOGRAPHY FOR ANALYSIS OF ECDYSTEROIDS

E. D. Morgan, H.-P. Huang and I. D. Wilson

Department of Chemistry
University of Keele
Keele, Staffordshire ST5 5BG, UK

SUMMARY

Ecdysteroids can be chromatographed with a supercritical fluid as the mobile phase in equipment adapted from high pressure liquid chromatography. A variety of columns (both normal and reversed-phase) have been applied under a range of conditions to learn something of the optimum conditions to be used. Temperature had little effect compared to the choice of stationary phase and proportion of methanol in the supercritical fluid. Chromatography was both fast and sensitive but as yet separation is not as efficient as that obtained in liquid chromatography.

INTRODUCTION

When a gas is compressed at a temperature exceeding its critical temperature, its viscosity and density increase smoothly, until, at sufficiently high pressure, they approach the values of a true liquid, without the substance ever going through the gas-liquid phase change that we normally observe at room temperature. In the region above the critical point, where the physical properties of a substance are somewhere between those of a gas and a liquid, that substance is said to be in the supercritical state. Lovelock [1] long ago suggested that a supercritical fluid might be used as the moving phase in chromatography, but it is only recently that the equipment necessary for the practical application of supercritical fluids in chromatography became commercially available.

Supercritical fluid chromatography (SFC) can use either the technology of gas chromatography (GC) or high pressure liquid chromatography (HPLC). Because the solubility of the analyte in the moving phase is closely related to density, and densities in SFC are much greater than in GC, substances which are too involatile, too polar, or too thermally unstable to be chromatographed by a moving gaseous phase will therefore be more readily chromatographed by a moving supercritical fluid. On the other hand the viscosity and diffusion coefficients of a supercritical fluid are closer to those of a gas than to those of a liquid, so that equilibrium will be reached more quickly, retention times should be shorter and there should be much less pressure drop across the column than in HPLC. The potential advantages of SFC were reviewed by Chester [2].

There is a range of gases with critical temperatures not far above ambient that might be used as mobile phase (e.g., Xe, CO_2, NH_3, SO_2, SF_6 or H_2S) but all can be eliminated on grounds of cost, toxicity or reactivity, except CO_2, which has a supercritical temperature of 32°C and requires a pressure of 73 atmospheres or Bars. Supercritical CO_2 however, is a relatively non-polar fluid, and for elution of polar compounds, a miscible polar substance must be added, in proportions such that the supercritical state is maintained. The "modifier" of the mobile phase can be methanol, acetonitrile or tetrahydrofuran. However, the use of the organic modifier in the mobile phase renders the flame ionization detector of GC unusable and so for highly polar substances, where a modified moving phase is required, it is necessary to use HPLC technology, and an ultraviolet or refractive index detector.

Chromatography and Isolation of Insect Hormones and Pheromones
Edited by A. R. McCaffery and I. D. Wilson
Plenum Press, New York, 1990

Because of the greater diffusivity and lower viscosity of a supercritical fluid compared with liquids, SFC promises to give equivalent resolution with much shorter retention times than HPLC. Faster elution gives narrower peak widths and this together with the greater UV transparency of CO_2 to short wavelength UV light, promises to give greater sensitivity of detection for SFC than for HPLC. Any increase in sensitivity of detection would be a great advantage in the chromatography of ecdysteroids, where the low concentration of the hormone in biological tissues and the limited amount of insect material available makes high sensitivity an important goal.

Ecdysteroids can be detected with great sensitivity by GC methods, using either electron capture detection [3,4] or by mass spectrometry (see Evershed, this volume), provided they are first converted into more stable and more volatile derivatives, such as their trimethylsilyl ethers. However, conversion of an ecdysteroid to a single derivative is required, and the four or more hydroxyl groups per molecule, with differing reactivities means that this is not a simple task [5].

Generally, HPLC determination of ecdysteroids is preferred, although it is not as sensitive (limit 10 to 50 ng injected), because no derivatization is required, and polar and non-polar derivatives can all be chromatographed in the same system. The strong UV chromophore of the α,β-unsaturated ketone, present in all ecdysteroids (λ_{max} 242 nm, ε ~10,000 to 14,000) makes them very suitable for UV detection.

It was therefore of interest to explore the SFC of ecdysteroids to see if it offers advantages of speed or sensitivity over HPLC for the detection of these hormones, their precursors and metabolites. Any increase in sensitivity or shortening of overall analysis time would be welcomed by insect physiologists.

Using SFC with GC methods, Raynor et al. [6] found that with CO_2 alone, ecdysteroid precursors with 2 or 3 hydroxyl groups could be chromatographed, but more polar compounds either required very long elution times (e.g., 2-deoxyecdysone at 32 min) or were not eluted from the column. From preliminary work done with the cooperation of Games and Mylchreest [7], we were able to show, that ecdysteroids could be chromatographed under supercritical conditions using CO_2-methanol mixtures on uncoated silica columns, though there was some evidence of peak tailing. To compare SFC with HPLC analysis, samples of crude phytoecdysteroids obtained from two flowering plants, *Silene nutans* and *Silene otites* were chromatographed under SFC conditions attempting to obtain resolution similar to that achieved with HPLC [8]. Separations that required 20 minutes for the elution of three ecdysteroids, from *S. otites* by HPLC were achieved in 2 min by SFC. A mixture of two major and three minor ecdysteroids in *S. nutans*, resolved in 45 min by HPLC were separated in 3 min on SFC, except that one of the minor components was lost. By further forcing the conditions, it was possible to get separation of three ecdysteroids from *S. otites* on a 2 cm column in just over one minute [8].

We therefore wished to study in more detail the effects of some of the variables encountered in SFC that may affect speed of separation and resolution of components. At the same time we have looked at the sensitivity of detection attainable and the linearity of response.

MATERIALS AND METHODS

The chromatographic system consisted of a cylinder of CO_2 fitted with a siphon tube to deliver liquid from the bottom of the cylinder. The CO_2 was passed through a column of molecular sieves to a Constametric 3000 pump (Milton Roy, Stone, Staffordshire) to compress the CO_2. The pump head was fitted with a clamp-on heat exchanger through which refrigerated water-methanol was circulated. Methanol (HPLC grade, degassed with helium) was pumped through a CP3000 pump (Milton Roy) to a magnetically stirred mixing chamber where it was mixed with the supercritical CO_2. The methanol flow rate was controlled by a GC-4000 Gradient Controller (Milton Roy). The modified supercritical fluid passed to a Rheodyne 7125 injection valve, with 20 µl injection loop (Anachem, Luton, UK) fitted inside a SpH99 column oven (Milton Roy), which also contained the guard column, column and a Tescon Model 26-1721-24-043 back pressure regulator. The fluid passed from the injection loop to the column assembly, then to a SM 3100X variable wavelength detector (Milton Roy) set to monitor at 254 nm, and then back to the back pressure regulator. From there the de-pressurized mixture of CO_2 and methanol was bubbled through water in a flask.

Five columns were used, four columns all of dimensions 25 cm x 4.6 cm from Milton Roy contained 5 μm silica, either uncoated, or coated with aminopropyl-, cyanopropyl- or octadecylsilyl groups. A 15 cm x 4.6 cm column containing 5 μm aminopropyl silica (HPLC Technology, Macclesfield, UK) was also used for comparison.

Solutions of ecdysone and 20-hydroxyecdysone (Simes, Milan), (1 μg/μl) were prepared in methanol and 10-fold dilutions made, down to 10 pg/μl. Samples of 10 μl of these solutions were injected into the system through the Rheodyne valve.

RESULTS AND DISCUSSION

The reproducibility of injection and response of the system were first tested by repeated injections of 10 μl of a solution of ecdysone and 20-hydroxyecdysone in methanol. In Figure 1 two such successive injections immediately after the equipment was set up are shown, demonstrating that good baseline and reproducibility was achieved immediately and that a reasonable separation of the compounds was obtained.

The linearity of response of the detector was tested by injecting samples in increasing dilution at different absorbance settings. The results are plotted in Figure 2, using absorbance range from 0.5 absorbance units to 0.001 units. Optimization of the system to use the most sensitive setting was not achieved in the time available. However, Figure 3 shows two injections of 100 pg ecdysone on two different aminopropyl columns with different integrator conditions, which show that peak integration of such small samples is possible and with further experience of the system and control of baseline noise, still greater sensitivity should be reached.

Raising the temperature at constant pressure may be expected to produce a slightly lower density and hence lower solubility of the solute in the mobile phase. In Figure 4(a) this is shown for two different proportions of methanol and on two different aminopropyl columns. It is seen that temperature had little effect on retention or on separation between ecdysone and 20-hydroxyecdysone, but the effect was a little stronger with lower methanol proportion, and hence longer retention times. On a different stationary phase (cyanopropylsilica), temperature increase had a much more pronounced effect, increasing retention and resolution (Figure 4b).

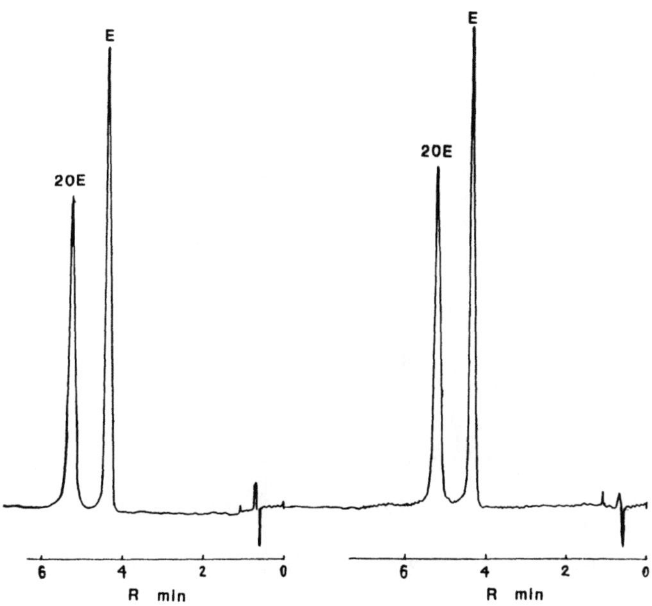

Fig. 1. Two successive injections of a mixture of ecdysone (E) and 20-hydroxyecdysone (20E) on an amino-propyl silica column (25 cm x 4.6 mm), pressure 3000 psi, mobile phase CO_2, containing 25% methanol, 4 ml/min and column temperature of 65°C.

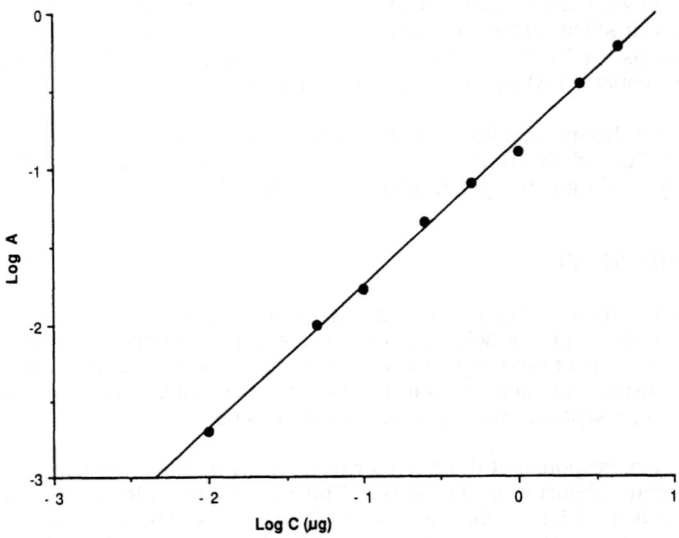

Fig. 2. Plot of log concentration of ecdysone in µg against log absorbance over a range from 10 µg to 10 ng. Conditions as in Figure 1.

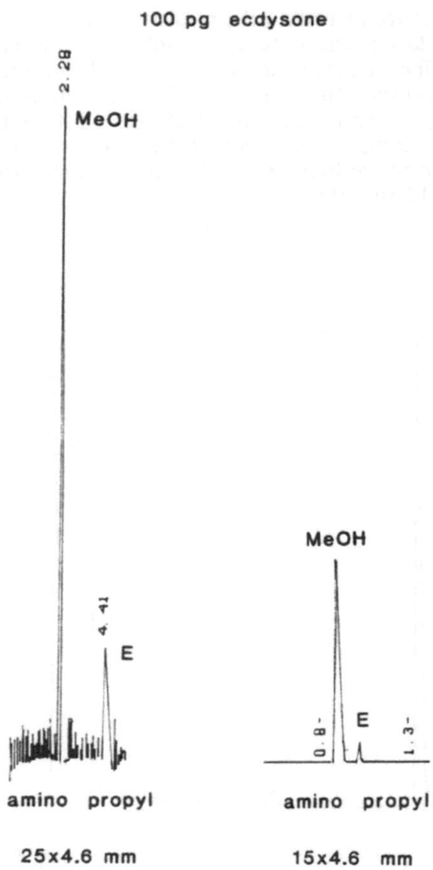

Fig. 3. Response obtained on injection of 100 pg ecdysone on two different aminopropyl columns. The noise level of the integrator was set differently for the two injections, but otherwise the conditions were the same as in Figure 1.

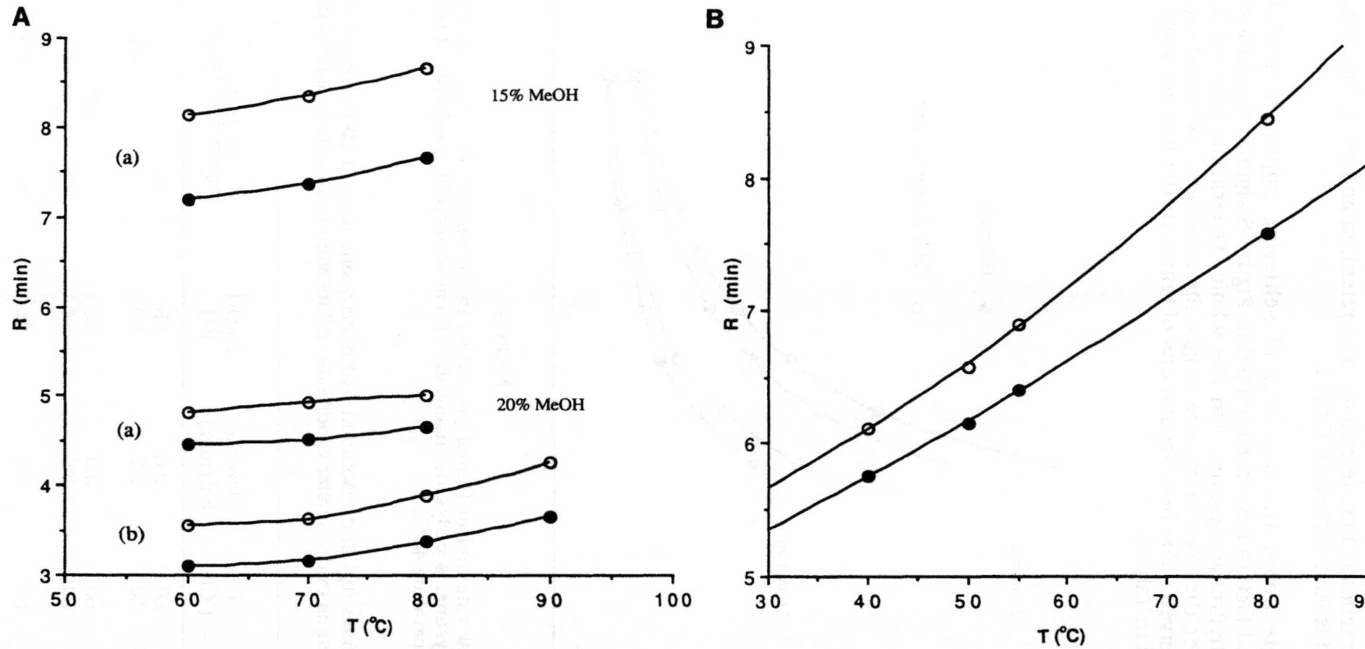

Fig. 4. Effect of temperature on retention time. ○, 20-Hydroxyecdysone, ●, ecdysone. In part A, (a) refers to a 15 cm aminopropyl column and (b) refers to a 25 cm aminopropyl column. Conditions otherwise as in Figure 1. In Part B is shown the effect of temperature for a cyanopropyl column, flow rate 4 ml/min, with 10% methanol and pressure 3000 psi.

The effect of flow rate on retention and separation are shown in Figure 5 against the amount of methanol in the mixture. For a given retention, separation was not affected by the flow rate. In other words, a retention time of 3 min for ecdysone gives the same separation from hydroxyecdysone whether this is achieved with a lower methanol % at 3.0 ml/min or with a higher methanol content at 2.0 ml/min. When all other conditions are constant, reducing the methanol proportion increased retention, and less than 10 to 12% methanol gave impractically long retention. The retention and resolution obtained on each of the columns used is summarized in Table 1.

To learn more about the best way of achieving improved resolution between ecdysteroids, several factors have been plotted in Figure 6 against separation (ΔR) between ecdysone and 20-hydroxyecdysone. On the whole, separation followed retention time closely. The greater the separation, the longer the retention. At constant pressure, no selectivity was observed. The best separations obtained in the limited range of conditions attempted are given in Table 2.

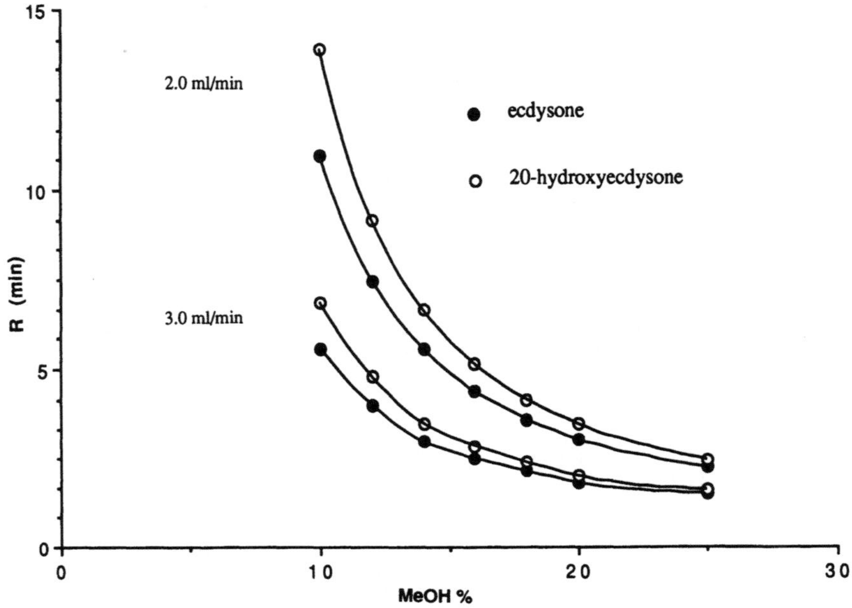

Fig. 5. Effect of flow rate and methanol proportion in the supercritical fluid on retention; O, 20-hydroxyecdysone, ● ecdysone, using the 15 cm aminopropyl column. Pressure 3000 psi, column temperature 40°C.

Table 1. Retention times (t_R) and resolution (Rs) of ecdysone (E) and 20-hydroxyecdysone (20E) on various columns under similar conditions. All measurements made using a pressure of 2000 psi

Column	T (°C)	Flow rate (ml/min.)	MeOH (%)	t_R (min.) E	t_R (min.) 20E	Rs
Silica, 25 cm	60	3.0	25	3.0	3.2	0.2
	60	2.0	20	7.1	8.1	5.0
ODS, 25 cm	60	3.0	25	2.3		0
	60	3.0	14	4.3	4.7	2.1
Aminopropyl 15 cm	60	3.0	25	1.5	1.6	0.3
Aminopropyl 25 cm	60	3.0	20	2.9		0
	60	2.0	20	4.5	4.8	0.8
Cyanopropyl 25 cm	60	3.0	20	2.8		0
	80	2.0	20	5.2	6.0	1.2

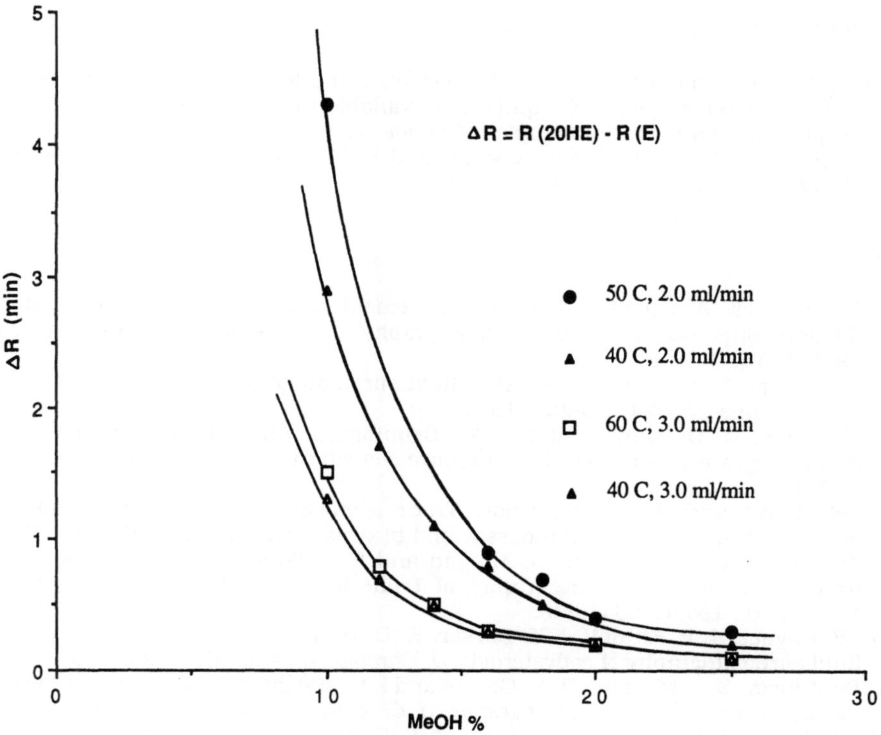

Fig. 6. Effect of various factors on separation (ΔR) between ecdysone and 20-hydroxyecdysone. Open symbols were at a flow rate of 3.0 ml/min, closed symbols at 2.0 ml/min.

Table 2. Maximum separation obtained for ecdysone (E) and 20-hydroxyecdysone (20E) in SFC at 2000 psi pressure

Column	T (°C)	Flow rate (ml/min.)	MeOH (%)	t_R (min.) E	20E	Δt_R	Rs	k'
Silica 25 cm	90	3.0	25	3.3	3.6	0.3	1.38	0.9
ODS 25 cm	40	3.0	12	4.9	5.5	0.6	3.15	2.0
Cyanopropyl 25 cm	80	3.0	11	6.7	7.3	0.6	3.65	2.1
Aminopropyl 25 cm	70	2.0	15	7.2	8.1	0.9	2.68	2.5
Aminopropyl 15 cm	40	2.0	10	11.0	13.9	2.9	9.1	6.9
	40	3.0	10	5.6	6.8	1.3	6.3	4.1

Supercritical fluid chromatography of ecdysteroids is clearly practical, rapid and, above all, sensitive. Resolution is not as great as that achievable in HPLC systems, but much more is yet to be learned about effect of pressure, nature of the stationary phase and the modifier. In the limited number of experiments so far performed, no pattern is evident which points towards the best column and conditions for their optimum separation.

ACKNOWLEDGEMENTS

We particularly thank G. Jones and P. Griffiths and Milton Roy Instruments, Stone, Staffordshire for making their SFC equipment available to us, Professor R. Lafont, Ecole Normale Superieure, Paris for the samples of *Silene* plant extracts, Professor D. E. Games, Swansea, for help with the preliminary work, and K. T. Alston for a great deal of willing help in setting up and operating the equipment.

REFERENCES

1. J. E. Lovelock, A signed and witness statement, unpublished, but illustrated, *in*: "Modern Supercritical Fluid Chromatography", C. M. White, ed., VCH, Heidelberg, 1988 (1958).
2. T. L. Chester, The role of supercritical-fluid chromatography in analytical chemistry, *J. Chromatog. Sci.*, 24:226-229 (1986).
3. C. F. Poole, E. D. Morgan and P. M. Bebbington, Analysis of ecdysones by gas chromatography using electron capture detection, *J. Chromatog.*, 104:172-175 (1975).
4. D. W. Borst and J. D. O'Connor, Trace analysis of ecdysones by gas-liquid chromatography, radioimmunoassay and bioassay, *Steroids*, 24:637-656 (1974).
5. C. R. Bielby, A. R. Gande, E. D. Morgan and I. D. Wilson, Practical aspects of the preparation and chromatography of trimethylsilyl ethers of ecdysteroids, *J. Chromatog.*, 194:43-53 (1980).
6. M. W. Raynor, J. P. Kithinji, I. K. Barker, K. D. Bartle and I. D. Wilson, Supercritical fluid chromatography of ecdysteroids, *J. Chromatog.*, 436:497-502 (1988).
7. E. D. Morgan, S. J. Murphy, D. E. Games and I. C. Mylchreest, Analysis of ecdysteroids by supercritical-fluid chromatography, *J. Chromatog.*, 441:165-169 (1988).
8. M. W. Raynor, J. P. Kithinji, K. D. Bartle, D. E. Games, I. C. Mylchreest, E. D. Morgan, R. Lafont and I. D. Wilson, Packed column supercritical fluid chromatography and linked supercritical-fluid chromatography-mass spectrometry for the analysis of phytoecdysteroids from *Silene nutans* and *Silene otites*, *J. Chromatog.*, 467:292-298 (1989).

CURRENT STATUS AND RECENT ADVANCES IN THE CHROMATOGRAPHY AND MASS SPECTROMETRY OF ECDYSTEROIDS

R. P. Evershed, M. Kabbouh, M. C. Prescott, J. L. Maggs* and H. H. Rees

Departments of Biochemistry, and *Pharmacology and Therapeutics, University of Liverpool
P.O. Box 147, Liverpool L69 3BX, UK

SUMMARY

The application of various mass spectrometric techniques to the structure elucidation, detection and quantification of free ecdysteroids and their polar and apolar conjugates is discussed. Three principal areas are considered: (i) gas chromatography-mass spectrometry and its complementarity to radioimmunoassay for the screening of ecdysteroids in invertebrates; (ii) the importance of fast atom bombardment mass spectrometry as an adjunct to nuclear magnetic resonance spectroscopy in the structure elucidation of polar and apolar ecdysteroid conjugates; and (iii) the potential of liquid chromatography-mass spectrometry for the separation and identification of ecdysteroids.

INTRODUCTION

Ecdysteroids occur widely in invertebrates in both free and conjugated forms (including polar and apolar conjugates; for reviews see Horn and Bergamasco [1], Rees [2]). Their occurrence at low concentration and in diverse forms imposes a considerable analytical demand. Until comparatively recently, analyses of conjugated ecdysteroids focused largely on the ecdysteroid moiety following its release by enzymic hydrolysis [3].

A wide variety of chromatographic methods have been employed for such studies. Thus, packed-column gas chromatography (GC) of trimethylsilyl (TMS) ether derivatives with electron capture detection provided a sensitive means of separation and analysis [4-6], while GC-mass spectrometry (GC-MS) also achieved selective and sensitive detection [7]. The development of high-performance liquid chromatographic (HPLC) techniques provided a versatile means by which intact ecdysteroid conjugates could be separated [8] without the need for derivatization. Additionally, highly polar ecdysone metabolites, not susceptible to enzymic hydrolysis such as ecdysonoic and 20-hydroxyecdysonoic acids [9,10], were also readily separated. Since recent developments in the HPLC of ecdysteroids are dealt with elsewhere in this volume (Lafont and Wilson), they will not be discussed further here, except to indicate that problems exist in the detection of components present at trace levels (nanogram to sub-nanogram) and characterization of eluting components. In trace analyses radioimmunoassay (RIA) provides a very sensitive and, in favorable circumstances, selective means of detection of ecdysteroids separated by HPLC [11]. In the characterization of eluting components, considerable reliance must be placed on prior knowledge of the HPLC behavior (elution orders) of known compounds. Where unknown compounds are encountered, or when unequivocal assignments are required, then recourse must be made to physicochemical techniques, such as MS or nuclear magnetic resonance spectroscopy (NMR).

Supercritical fluid chromatography/mass spectrometry (SFC/MS) with thermospray ionization of free ecdysteroids has been demonstrated [12]. However, the apparent incompatibility of polar compounds with SFC, particularly capillary SFC, currently limits its

Chromatography and Isolation of Insect Hormones and Pheromones
Edited by A. R. McCaffery and I. D. Wilson
Plenum Press, New York, 1990

general application [13]. The application of SFC to the analysis of ecdysteroids is discussed elsewhere in this volume (Morgan et al.).

The small amount of material often available for study, necessarily, places heavy emphasis on the use of MS techniques in investigations of ecdysteroids. Herein, we discuss the methodological aspects of the mass spectrometric and ancillary techniques currently used in ecdysteroid analyses in our laboratories. Additionally, we present preliminary data obtained on the analysis of a variety of ecdysteroids by combined liquid chromatography-MS (LC-MS).

MATERIALS AND METHODS

Gas Chromatography-Mass Spectrometry (GC-MS)

Free ecdysteroids, or ecdysteroids released from conjugates by enzyme treatment (*Helix pomatia*), were determined by GC-MS as their TMS ethers. Ecdysteroid TMS ethers were prepared according to previously published methods [7,14,15] by treatment with undiluted N-trimethylsilylimidazole (100°C, 15 hr). The silylated ecdysteroids were freed from the excess derivatising agent prior to GC-MS by adsorption chromatography (Al_2O_3, Brockman grade 1, 1.5 g, eluting with 4 ml 3:7 v/v ethyl acetate-hexane). After evaporation to dryness, the silylated ecdysteroids were dissolved in a minimum volume of hexane for analysis by GC-MS.

GC-MS was performed on a Pye-Unicam 204 GC coupled to a VG micromass 7070H double-focusing magnetic sector mass spectrometer. Data acquisition and processing were performed on a Finnigan INCOS 2300 data system. The interface oven had been modified for high temperature (>300°C) operation [14], the direct mode of interfacing being employed. Sample introduction was by on-column injection employing an SGE OCI-III injector. Both polyimide and aluminium-clad flexible fused-silica capillary columns (25 m x 0.2 mm i.d. BP-1 coated, OV-1 equivalent, 0.1 μm film thickness) have been used [16]. Helium was used as the carrier gas throughout. High sensitivity in trace analyses was achieved by selected ion monitoring (SIM) employing accelerating voltage switching. Quantitative analyses were performed as described previously [15].

Fast Atom Bombardment Mass Spectrometry (FAB-MS)

FAB-MS analyses were performed on the same mass spectrometer as the GC-MS work described above. The primary bombarding xenon beam was produced by an adapted saddle-field source atom gun (Ion Tech Ltd., Teddington, Middlesex, UK) operated at 8 to 9 kV and a tube current of 1 to 2 mA. The mass spectrometer was operated at a low resolution (ca. 1000), and at accelerating voltages of 4 kV and 2 kV to scan the m/z ranges 40 to 700 and 90 to 1400, respectively. Depending upon the nature of the compounds under investigation, both positive- and negative-ion scanning modes were employed, using either glycerol (polar conjugates) or tetraethylene glycol (apolar conjugates) liquid matrices.

A limited number of analyses were also performed on a VG Tritech TS-250 mass spectrometer employing fast ion bombardment MS (FIB-MS; synonymous with liquid secondary ion MS, LSIMS) in which Cs^+ ions, accelerated to 10 to 13 kV, provided the primary beam. Data were acquired by multiple channel analysis.

Liquid Chromatography-Mass Spectrometry (LC-MS)

Combined LC-MS was performed using an LKB 2150 solvent delivery system linked to the TS-250 double-focusing magnetic sector mass spectrometer. Ecdysteroids (low μg quantities) were dissolved in methanol and introduced via a Rheodyne 7125 loop injector. In preliminary investigations of ionization behavior, a 5 cm x 4.6 mm i.d. column containing CoPellicular ODS (Whatman) packing was used. The system was operated isocratically employing a methanol-water (7:3 v/v) mixture at a flow rate of 0.8 ml/min. Ionization was achieved using a VG Tritech Plasmaspray LC-MS interface-ion source at a vaporizer temperature of 200°C, source temperature of 275°C and discharge potential of 500 V. The mass spectra were recorded at low resolution (ca. 1000), at an accelerating voltage of 4 kV and by scanning the m/z range 100 to 800 at 1 scan/sec.

RESULTS AND DISCUSSION

During the course of investigations of ecdysteroids, the need often arises for techniques capable of rapid and routine detection of components at trace concentrations (sub-nanogram), and for structure elucidation where greater amounts of unknown compounds are encountered. The presence of ecdysteroids as mixtures requires the use of chromatographic techniques, while their occurrence at low concentration means that mass spectrometry is generally the only practically useful physiochemical technique for characterization purposes. In this contribution we discuss the uses of computerized GC-MS, FAB-MS and LC-MS in the analysis of free ecdysteroids and intact ecdysteroid conjugates (including polar and apolar conjugates).

Gas Chromatography-Mass Spectrometry

The aim in this work was to take advantage of the greater chromatographic efficiencies offered by modern capillary columns, compared to packed GC columns, in order to improve sensitivities and selectivities in GC-MS analyses. In doing this, it was hoped to provide a complementary technique to HPLC-RIA for the detection of ecdysteroids in trace amounts (typically sub-nanogram).

Even with the recent advent of capillary GC columns coated with highly immobilized GC phases, which allow programming of the GC oven temperatures to >400°C, derivatization of ecdysteroids is essential before GC analyses can be successfully performed. As shown previously [14,15] the behavior of fully trimethylsilylated ecdysteroids on flexible fused-silica capillary columns coated with immobilized apolar (OV-1 type) GC phases is excellent. One of the columns dedicated to this work (25 m x 0.22 mm i.d. BP-1 coated, 0.1 μm film thickness; SGE) has been in regular use for nearly four years, during which time approaching 800 ecdysteroid determinations have been performed. The exceptionally high quality GC performance that this column continues to display is shown in the chromatogram presented in Figure 1 obtained by GC-MS (SIM) of a mixture of authentic ecdysteroid TMS ethers. The relatively low volatility of ecdysteroid TMS ethers means that GC oven temperatures >300°C are required for elution. A shortcoming of extended use of polyimide-coated capillary columns at such high oven temperatures is the resulting brittlisation. The recently introduced aluminium-clad capillaries overcome this problem; however, care must be exersised when connecting these metal-clad columns directly into the sources of magnetic sector mass spectrometers where high voltages are present [16].

Losses of chromatographic performance have been observed to result from the accumulation of involatile impurities. Simply removing a short section (0.5 m) from the injection end of the column restored optimum performance. An alternative approach is to employ retention gaps [17] to serve as disposable pre-columns to protect the valuable analytical column.

A favorable signal-to-noise ratio is attainable in SIM mode (see Figure 1) such that limits of detection of 10 pg and 100 pg have been recorded for 20-hydroxyecdysone (**2**) and ecdysone (**1**) (structures in Figure 2), respectively [14,15]. With more modern mass spectrometers these detection limits would be improved still further. Thus, with respect to detection limits, capillary GC-MS (SIM) has been found to provide a complementary means of ecdysteroid analysis to HPLC with RIA monitoring of the eluant fractions.

Quantification of the ecdysteroids relies on an internal standard, most often makisterone A (**3**), a homologue of 20-hydroxyecdysone (**2**) possessing a methyl group at the C-24 position. For rigorous quantitative work, deuteriated internal standards would be the preferred option. However, preparation of deuteriated analogues of the wide range of ecdysteroids we routinely study would represent a major synthetic effort. While we are investigating possible approaches to deuteriation we regard the use of makisterone A (**3**) as an acceptable compromise, taking into account its occurrence in some insect [18] and crustacean [19] species.

Fast Atom Bombardment Mass Spectrometry

Ecdysteroid conjugates represent a particular challenge and although NMR spectroscopy (^{1}H and ^{13}C) provides valuable information on the site of conjugation, mass spectrometry is essential for the unequivocal determination of the conjugate.

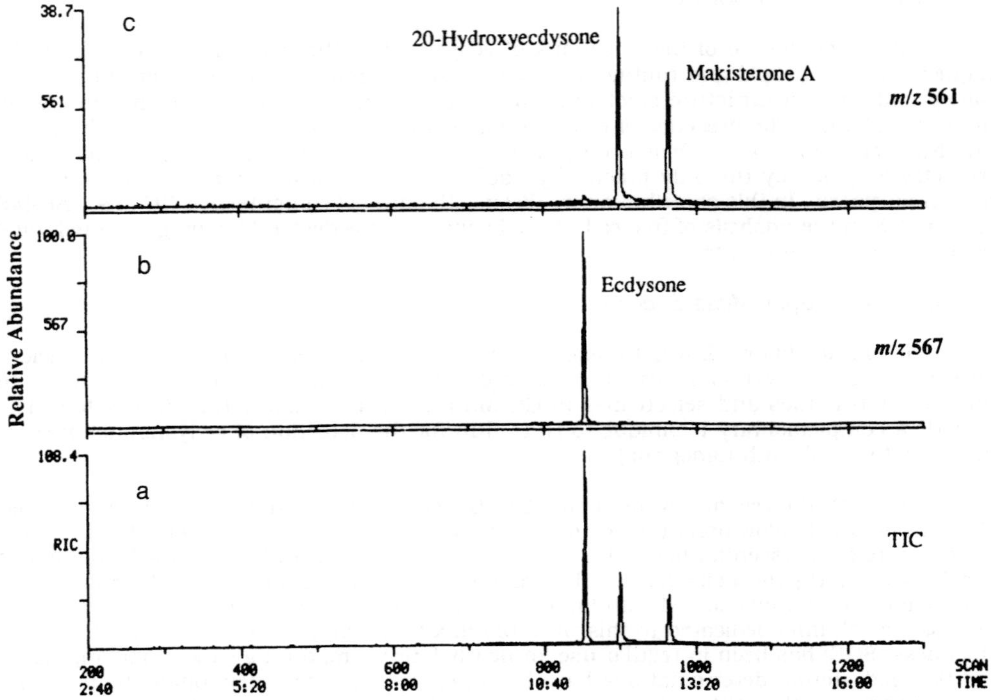

Fig. 1. Total ion current (TIC; a) and m/z 567 (b) and m/z 561 (c) mass chromatograms for the capillary GC (SIM)MS analysis of authentic fully trimethylsilylated ecdysteroids: ecdysone (**1**), 20-hydroxyecdysone (**2**) and makisterone A (**3**). The analysis was performed on a 25 m x 0.22 mm I.D. BP-1 coated (OV-1 equivalent, 0.1 μm film thickness) flexible fused-silica capillary column. Following on-column injection, the GC oven temperature was raised ballistically to 200°C, then programmed at 12°C/min to 310°C and held isothermally.

FAB-MS provides a very effective means of obtaining mass spectral information from the many free and conjugated forms of ecdysteroids observed in nature. FAB-MS is of special importance in the case of polar conjugates as neither electron impact (EI) nor desorption chemical ionization (DCI) [20] spectra display characteristic molecular or pseudo-molecular ions for this class of compounds.

It is perhaps not surprising that largely analogous FAB behavior is observed for ecdysteroids and bile salts [21] in view of their structural similarities. Negative-ion scanning FAB-MS is preferred for free ecdysteroids, ecdysteroid acetates, acids and glucosides, and ecdysteroid conjugates that contain an inorganic conjugate moiety. Ecdysone-22-phosphate (**4**) and 2-deoxyecdysone-22-phosphate (**5**) were not only the first ecdysteroid conjugates to be identified from insects [22] but were, to our knowledge, the first novel natural compounds to be determined by FAB-MS. The negative-ion spectra were characterized by $[M-H]^-$ ions. The ions appearing 22 amu higher than $[M-H]^-$ in FAB spectra correspond to deprotonated pseudo-molecular sodium adduct ions ($[M + Na - 2H]^-$). Lower mass ions are also seen which correspond to sequential losses of water ($[M-H-H_2O]^-$ and $[M-H-2xH_2O]^-$). These dehydrated species are much less pronounced in FAB owing to its markedly "softer" ionizing nature compared to electron impact or chemical ionization modes. Low abundance adducts of the intact pseudo-molecular species with matrix molecules are also observed (<10% of the $[M-H]^-$ anion). In the case of glycerol $[M+glycerol-H]^-$ ions occur 92 amu higher than the corresponding deprotonated molecular anion $[M-H]^-$. It is often argued that the appearance of adduct ions is a complicating factor in interpreting FAB spectra. However, in practice, alkali metal and matrix adduct ions are expected and predictable, and can be used corroboratively in confirming the presence of $[M-H]^-$ ions of ecdysteroids.

The ability to use FAB-MS to derive molecular weight information is vital to many structure investigations. However, beyond this, and in common with many other

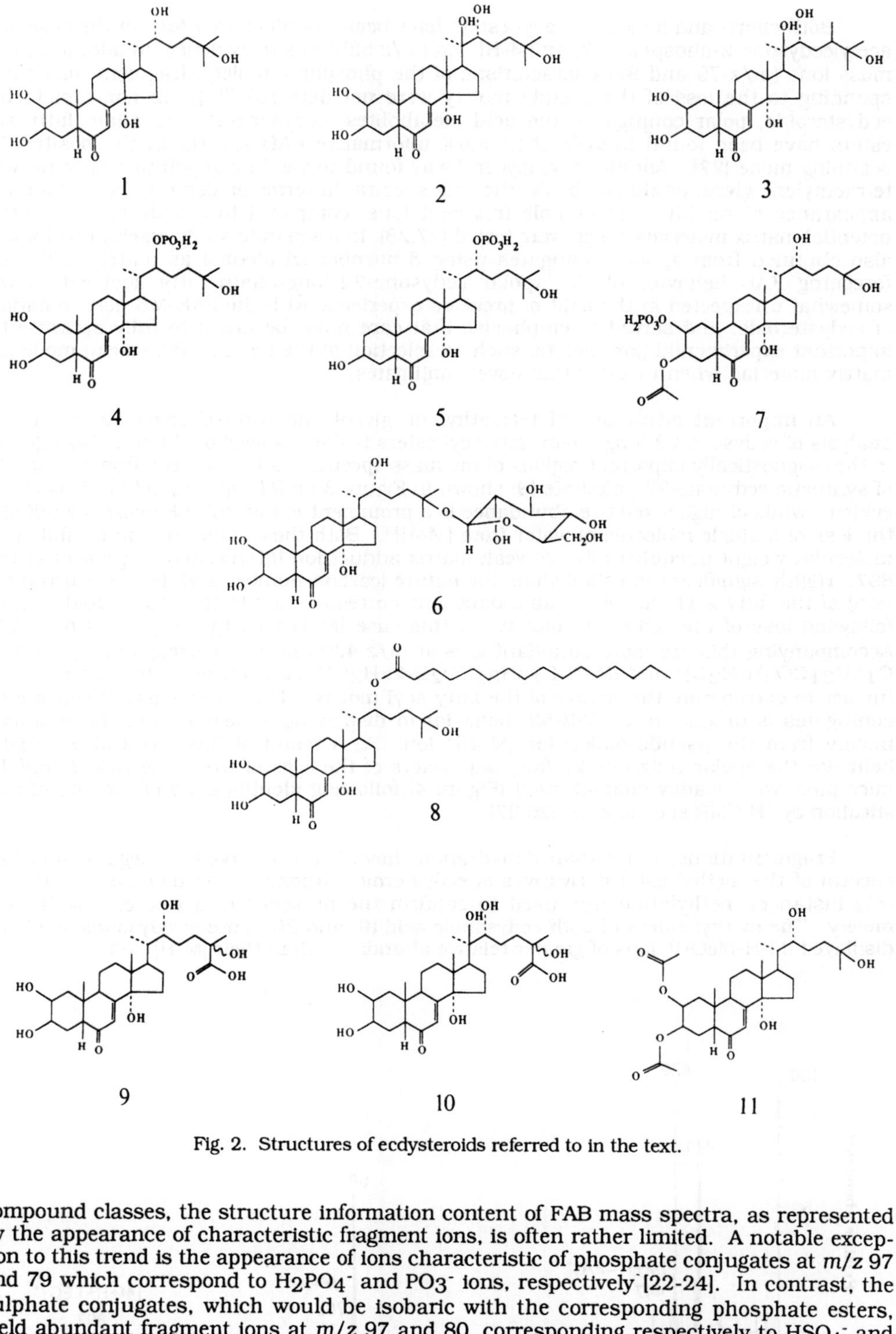

Fig. 2. Structures of ecdysteroids referred to in the text.

compound classes, the structure information content of FAB mass spectra, as represented by the appearance of characteristic fragment ions, is often rather limited. A notable exception to this trend is the appearance of ions characteristic of phosphate conjugates at *m/z* 97 and 79 which correspond to $H_2PO_4^-$ and PO_3^- ions, respectively [22-24]. In contrast, the sulphate conjugates, which would be isobaric with the corresponding phosphate esters, yield abundant fragment ions at *m/z* 97 and 80, corresponding respectively to HSO_4^- and $SO3^-$ [24,25]. The highly diagnostic nature of these abundant peaks in the FAB spectra is dictated by the stability of these oxyanions. More typically, the negative-ion FAB spectrum of 25-*O*-*β*-D-glucopyranoside (**6**), isolated from the nematode *Parascaris equorum*, displayed a prominent [M-H]⁻ ion at *m/z* 625. However, apart from the loss of water, fragmentations were otherwise lacking, and final confirmation of the precise nature of the conjugate's sugar moiety rested on GC-MS following release by enzymic hydrolysis [26].

Both short- and long-chain acyl esters have been examined by FAB. In the case of 3-acetylecdysone-2-phosphate (**7**) an [M-H]$^-$ ion (*m/z* 585) was in evidence, in addition to low mass ions (*m/z* 79 and 97) characteristic of the phosphate moiety; fragment ions corresponding to the loss of the acetate moiety were not detected [24]. In contrast to free ecdysteroids, polar conjugates and acid metabolites, ecdysone-22-long-chain fatty acyl esters have been found to yield their most informative FAB spectra in the positive-ion scanning mode [27]. Additionally, glycerol was found to yield disappointing spectra, with tetraethylene glycol producing by far the best spectra, in terms of signal-to-noise ratio and appearance of readily interpretable fragment ions, compared to a wide range of other potential matrix materials which were tested [27,28]. In a separate study, useful results were also obtained from apolar conjugates using 3-nitrobenzyl alcohol as matrix [29]. The foregoing FAB behavior of the apolar ecdysone-22-long-chain fatty acyl esters was somewhat unexpected in the light of previous experience with the FAB-MS determinations of ecdysteroids, and served to emphasise that care must be taken to fully optimise the important experimental parameters, such as selection of the most suitable scan mode and matrix material, when investigating novel conjugates.

An important advantage of tetraethylene glycol and 3-nitrobenzyl alcohol in the analysis of ecdysone-22-long-chain fatty acyl esters is the low level of chemical background in the diagnostically important regions of the mass spectra. As can be seen from the spectra of synthetic ecdysone-22-palmitate (**8**) shown in Figure 3 an [M+H]$^+$ ion, *m/z* 703, is clearly evident, while of higher relative abundance is a prominent ion at *m/z* 685 corresponding to the loss of a single molecule of water from [M+H]$^+$. Both these ions serve to establish the molecular weight unequivocally. A weak matrix adduct ion [M+H+TEG]$^+$ is present at *m/z* 897. Highly significant in establishing the nature (carbon number and degree of unsaturation) of the fatty acyl moiety are abundant ions corresponding to the ecdysteroid nucleus following loss of the fatty acid moiety, in this case [M+H-$C_{15}H_{31}CO_2H$]$^+$ at *m/z* 447. Accompanying this are more abundant ions at *m/z* 429 and 411 corresponding to [M+H-$C_{15}H_{31}CO_2H$-H_2O]$^+$ and [M+H-$C_{15}H_{31}CO_2H$-2xH_2O]$^+$, respectively, which can be used further to corroborate the nature of the fatty acyl moiety. Thus, this class of ecdysteroid conjugates is unique in its FAB-MS behavior in displaying a clear loss of the conjugate moiety from the pseudo-molecular [M+H]$^+$ ion. As a result of this favourable FAB-MS behavior the apolar ecdysone-22 fatty acyl esters of the eggs of the cattle tick, *Boophilus microplus*, were readily characterized (Figure 4) following identification of the site of substitution by ^{1}H NMR spectroscopy [26,27].

Fragmentations, other than dehydration, have been observed in negative-ion FAB spectra of the methyl ester derivatives of ecdysteroid carboxylic acid metabolites [10]. In this instance, methylation was used to confirm the presence of a free carboxylic acid moiety. The methyl esters of both ecdysonoic acid (**9**) and 20-hydroxyecdysonoic acid (**10**) displayed [M-H-MeOH]$^-$ ions of greater relative abundance than their [M-H]$^-$ ions.

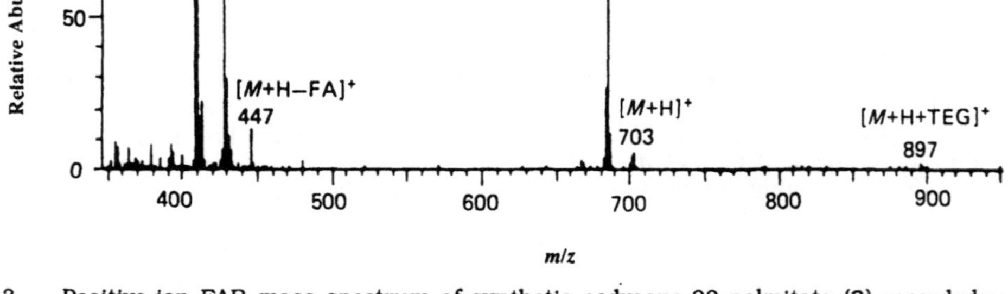

Fig. 3. Positive-ion FAB mass spectrum of synthetic ecdysone-22-palmitate (**8**) recorded using tetraethylene glycol as matrix and xenon as the primary atom beam.

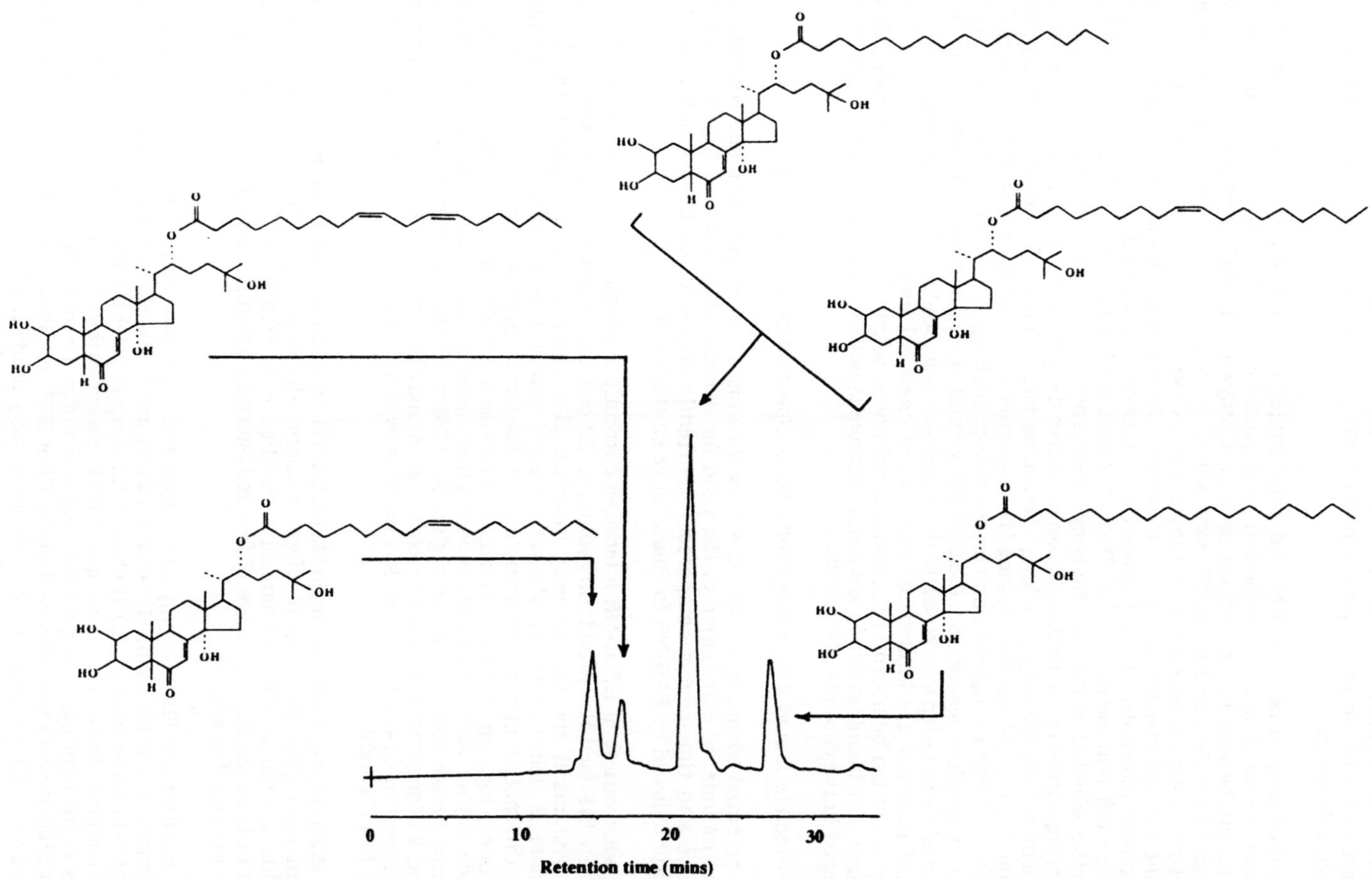

Fig. 4. Reversed-phase HPLC fractionation of the apolar ecdysone-22-long-chain fatty acyl esters from the eggs of the cattle tick, *Boophilus microplus*. Identification of the conjugate fatty acyl moiety relied largely on the positive-ion FAB-MS (see text for further details).

An alternative to employing high energy xenon atoms as the primary beam is to use fast ion bombardment (FIB, synonymous with liquid secondary ion mass spectrometry (LSIMS)) where Cs^+ ions constitute the bombarding species [30]. Cs^+ ion guns are finding increasingly wider applications owing to their more economical operation and enhanced yield of pseudo-molecular ionic species, particularly in the case of higher molecular weight (>1000 amu) substances.

Examination of a range of free and polar conjugated ecdysteroids by employing Cs^+ bombardment showed, much as expected, analogous behavior to that observed previously for FAB. As can be seen in Figure 5 the spectra of ecdysone (**1**) and 20-hydroxyecdysone (**2**) are dominated by the familiar $[M-H]^-$ ions which yield valuable molecular weight information. A notable difference between the Xe and Cs^+ spectra is the relatively higher abundance of the glycerol matrix adducts. In the case of 20-hydroxyecdysone (**2**) adduction of up to three matrix molecules $[M-H+3xglycerol]^-$ is apparent at m/z 735. As these data were recorded on a different instrument (VG TS-250) to that used in the FAB studies (VG 707H) it is as yet unclear whether or not the tendency for enhanced adduct ion formation is a function of the differing nature of the primary beam or source design. As in the FAB spectra, characteristic fragment ions are largely lacking. Low mass ions at m/z 143 and 159 in the spectra of ecdysone (**1**) and 20-hydroxyecdysone (**2**), respectively, are consistent with cleavage of the C-17/C-20 bond with charge retention on the side-chain fragment. A weak ion at m/z 361 is consistent with cleavage at the C-20/C-22 vicinal diol function. However, the low abundance of these latter ions restricts their usefulness in studies of natural isolates where amounts of material are limited or background chemical impurities are present. The possibility of Cs^+ ion bombardment providing enhanced sensitivity compared to FAB [30] is an important consideration where amount of sample is limited. However, this has yet to be fully assessed in the case of ecdysteroids.

High-Performance Liquid Chromatography-Mass Spectrometry (HPLC-MS)

The potential advantages of HPLC-MS for the analysis of ecdysteroids and ecdysteroid conjugates include: (i) elimination of the need for derivatization used in GC-MS; and (ii) convenience and time-saving compared to currently employed off-line sampling of the HPLC eluant followed by FAB-MS to characterize eluting peaks.

The wide variety of HPLC-MS interfaces currently available have been discussed in a recent review [31] (see also Mudd, this volume). Amongst the most popular of commercially available systems is the thermospray interface [32]. Briefly, in this system the HPLC mobile phase is delivered to the MS via a heated stainless steel capillary. The heating vaporizes the mobile phase and analytes. Ionization is promoted by the presence of a volatile buffer, e.g., ammonium acetate, to produce generally simple CI-like spectra. A prerequisite is that >20% v/v of the mobile phase must be aqueous, which in view of the high organic phase content (in the region of 100% methanol) necessary to elute apolar ecdysteroid-22-long-chain fatty acyl esters (e.g., **8**) may be limiting. In contrast, operation of thermospray in discharge ionization mode provides a means of operation compatible with most HPLC conditions.

The Plasmaspray interface available on the VG TS-250 mass spectrometer used in this study employed a glow discharge to effect ionization, the Plasmaspray tip acting as an electrode. This system provides enhanced versatility, ease of use, and absolute sensitivity compared to thermospray. A range of free ecdysteroids, conjugates and polar metabolites were examined on this system.

The positive-ion Plasmaspray mass spectrum and total ion current chromatogram obtained from 10 μg of ecdysone (**1**) is shown in Figure 6. Excellent ion current stability and a good signal-to-noise ratio (ca 20:1) shows that there is potential for using the technique at the sub-microgram level. The mass spectrum is characterized by a series of abundant high mass ions corresponding to sequential losses of water. In many respects this behavior is somewhat analogous to that observed under CI or DCI conditions [33]. However, under the Plasmaspray conditions employed in this preliminary study neither molecular nor pseudo-molecular ions were observed. The positive-ion Plasmaspray spectrum of 20-hydroxyecdysone shown in Figure 7a displays a predominance of ions at lower mass corresponding to sequential losses of water from the ecdysteroid nucleus following cleavage of the C-20/C-22 bond (see Figure 7). Cleavage at this position is peculiar to ecdysteroids possessing vicinal hydroxyl groups at C-20 and C-22 and has also been observed in EI, CI and DCI spectra. The 70 eV EI mass spectrum of 20-hydroxyecdysone (**2**) is shown in Figure 7b for comparative purposes. Apart from the higher abundance of fragment ions

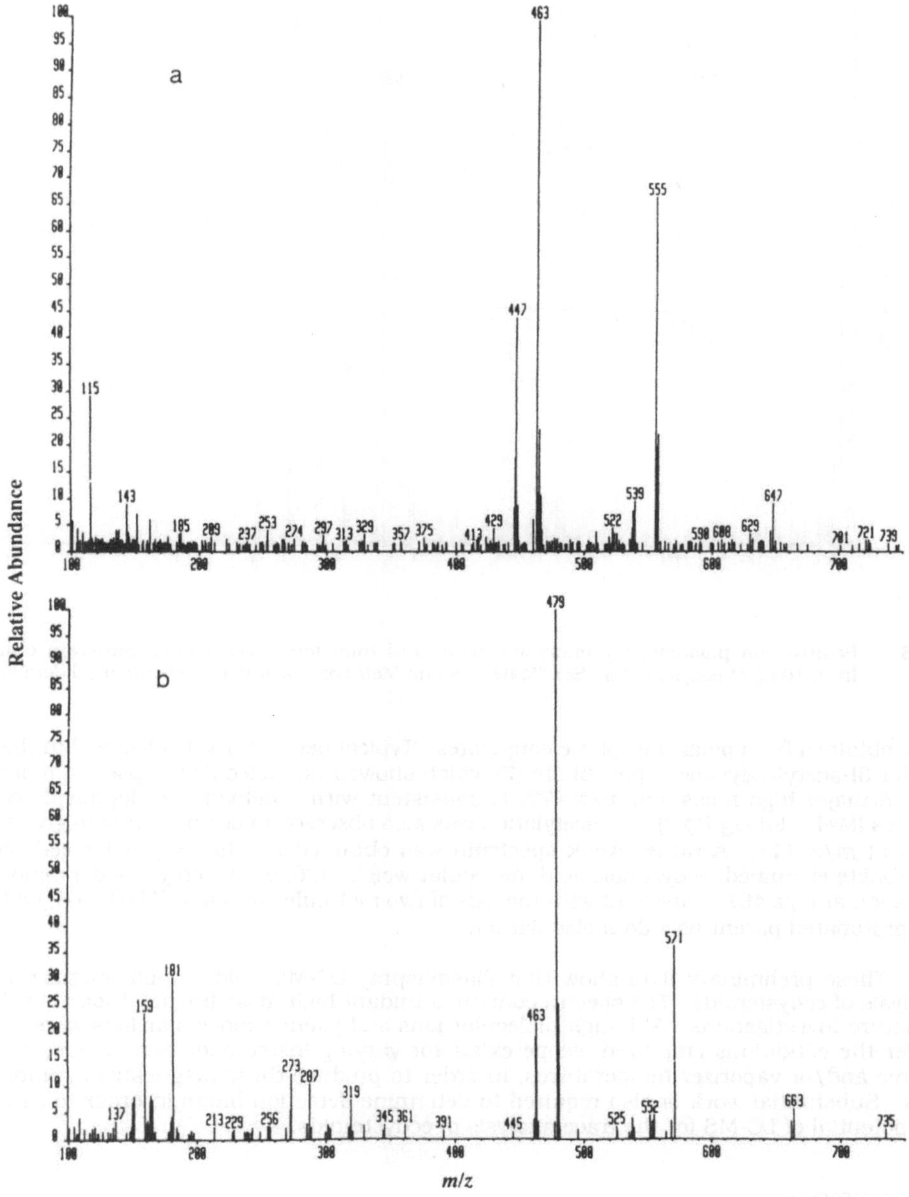

Fig. 5. Negative-ion FIB mass spectra of ecdysone (**1**; Figure 5a) and 20-hydroxyecdysone (**2**; Figure 5b) recorded using glycerol as matrix and Cs^+ ions as the primary bombarding beam.

at low mass (*m/z* <250) the spectra possess many similarities. The Plasmaspray mass spectrum of makisterone A (**3**) was largely congruent with that of 20-hydroxyecdysone (**2**) confirming the significance of the hydroxyl groups at C-20 and C-22 in directing fragmentation.

Of the conjugates that were examined, synthetic ecdysone-2,3-diacetate (molecular weight 548; **11**) yielded a major high mass ion at *m/z* 531. This would be consistent with the loss of water from the protonated pseudo-molecular ion, $[M+H-H_2O]^+$. Abundant fragment ions at lower mass *m/z* 513, 489, 471, 469, 455, 453 can be rationalized on the basis of further losses of water and fragmentations of the acetate moieties. Disappointing results

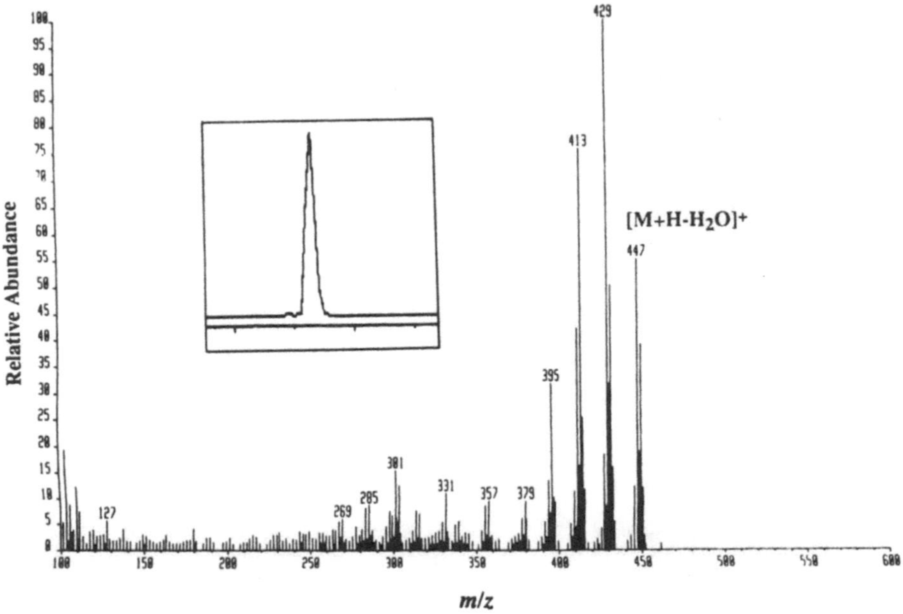

Fig. 6. Positive-ion plasmaspray mass spectrum and total ion current chramatogram obtained from 10 μg of ecdysone (**1**). See "Materials and Methods" for further experimental details.

were obtained from polar phosphate conjugates. Typical behaviour was observed in the case of 2(or 3)-acetylecdysone-2-phosphate (**7**) which showed no molecular or pseudo-molecular ion. A major high mass ion, *m/z* 472, is consistent with a dehydrated-dephosphorylated species $[M+H-H_2PO_3^{-}H_2O]^{-}$. Deacetylation was also observed to occur readily to give a base peak at *m/z* 411. A rather weak spectrum was obtained for the only polar ecdysteroid metabolite examined, ecdysonoic acid (molecular weight 496; **9**), which yielded a major high mass ion at *m/z* 461, consistent with the loss of two molecules of water ($[M+H-2xH_2O]^{-}$) from the protonated parent pseudo-molecular ion.

These preliminary data show that Plasmaspray LC-MS holds much promise for the analysis of ecdysteroids. The spectra contain abundant high mass fragment ions of value to structure investigations. Although molecular ions and pseudo-molecular ions were lacking under the conditions employed, scope exists for varying instrument parameters, such as source and/or vaporizer temperatures, in order to produce these diagnostically important ions. Substantial work is also required to determine detection limits in order to establish the potential of LC-MS for the trace analysis of ecdysteroids.

CONCLUSIONS

The chromatographic and mass spectrometric techniques discussed herein offer considerable scope for the separation, detection and characterization of ecdysteroids. Capillary GC-MS provides the most sensitive and selective physico-chemical technique currently available for the detection of free ecdysteroids as their TMS ether derivatives. With more modern instruments possessing improved source design, analyzer, pumping and detection systems, selectivity and detection limits should be improved still further. GC-MS SIM provides a means of trace analysis of ecdysteroids complementary to HPLC with RIA monitoring of the eluate fractions. FAB-MS continues to be an essential adjunct to NMR spectroscopy in the structure elucidation of novel ecdysteroid conjugates. In the case of polar conjugates and polar metabolites, FAB-MS currently provides the only effective means for obtaining the molecular weight information vital to establishing the nature of conjugate groups. Analogous mass spectral behavior has also been observed when employing Cs^+ ions (FIB or LSIMS) as the primary bombarding beam.

Plasmaspray HPLC-MS offers a potentially useful means of ecdysteroid analysis. Although only a limited range of ecdysteroids have been examined to date, indications are

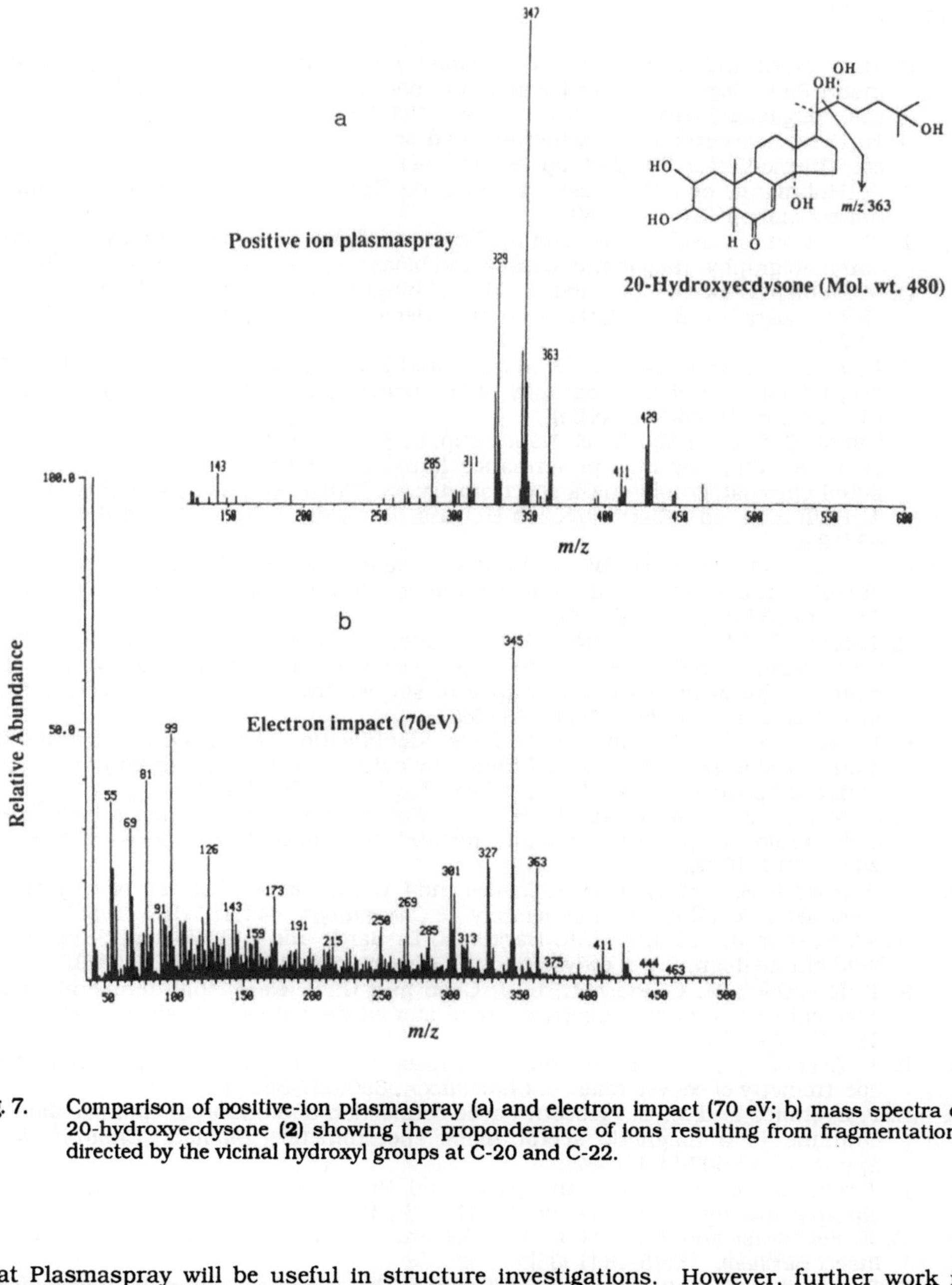

Fig. 7. Comparison of positive-ion plasmaspray (a) and electron impact (70 eV; b) mass spectra of 20-hydroxyecdysone (**2**) showing the proponderance of ions resulting from fragmentations directed by the vicinal hydroxyl groups at C-20 and C-22.

that Plasmaspray will be useful in structure investigations. However, further work is required to fully optimize conditions. It is unlikely that HPLC-MS will ever achieve the spectacular detection limits of GC-MS [34]. However, the detection of low picogram amounts of testosterone by SIM of the $[M+H]^+$ ion generated by thermospray [35] suggests that careful assessments of Plasmaspray HPLC-MS for the trace determination of ecdysteroids is required. The advantages of decreased sample handling and elimination of the need for derivatization makes routine HPLC-MS of ecdysteroids an exciting possibility.

ACKNOWLEDGEMENTS

We thank the SERC and The Leverhulme Trust for financial support and Dr L. N. Dinan for a generous gift of samples of apolar ecdysteroid conjugates. The VGTS-250 was purchased by means of grants from the Wellcome Trust and the University of Liverpool to the Department of Pharmacology and Therapeutics.

REFERENCES

1. D. H. S. Horn and R. Bergamasco, Chemistry of ecdysteroids, *in*: "Comprehensive Insect Physiology, Biochemistry and Pharmacology", G. A. Kerkut and L. I. Gilbert, eds., Pergamon Press, Oxford, Vol. 7:185-248 (1985).
2. H. H. Rees, Zooecdysteroids, structures and occurrence, *in*: "Ecdysone", J. Koolman, ed., Thieme-Verlag, Stuggart, pp 28-38 (1989).
3. J. A. Hoffmann, ed., "Progress in Ecdysone Research", Elservier/North-Holland, Amsterdam pp 431-465(1980).
4. J. D. O'Connor and D. W. Borst, Trace analysis of ecdysones by gas-liquid chromatography, radioimmunoassay and bioassay, *Steroids*, 24:637-656 (1974).
5. C. F. Poole, E. D. Morgan and P. M. Bebbington, Analysis of ecdysones by gas chromatography with electron capture detection, *J. Chromatogr.*, 104:172-175 (1975).
6. C. R. Bielby, A. R. Gande, E. D. Morgan and I. D. Wilson, Practical aspects of the preparation and chromatography of the trimethylsilyl ethers of ecdysteroids, *J. Chromatogr.*, 194:43-53 (1980).
7. R. Lafont, G. Somme-Martin, B. Mauchamp, B. F. Maune and J.-P. Delbeque, Analysis of ecdysteroids by high-performance liquid chromatography and coupled gas-liquid chromatography-mass spectrometry, *in*: "Progress in Ecdysone Research", J. A. Hoffmann, ed., Elsevier/North-Holland, Biomedical Press, Amsterdam, pp 45-68 (1980).
8. H. H. Rees and R. E. Isaac, Biosynthesis and metabolism of ecdysteroids and methods of isolation and identification of the free and conjugated compounds, *Methods in Enzymol.*, 111:377-410 (1985).
9. R. Lafont, C. Blais, P. Beydon, J.-F. Modde, U. Enderle and J. Koolman (1983) Conversion of ecdysone and 20-hydroxyecdysone into 26-OIC derivatives is a major pathway in larvae and pupae of species from three insect orders. *Arch. Insect. Biochem. Physiol.*, 1:41-58 (1983).
10. R. E. Isaac, N. P. Milner and M. M. Rees. Identification of ecdysonoic acid and 20-hydroxyecdysonoic acid isolated from developing eggs of *Schistocerca gregaria* and pupae of *Spodoptera littoralis*, *Biopchem. J.*, 213:261-265 (1983).
11. J. G. Mercer, A. E. Munn and H. H. Rees, *Echinococcus granulosus*: occurrence of ecdysteroids in protoscoleces and hydatid cyst fluid, *Mol. Biochem. Parasitol.*, 24:203-214 (1987).
12. E. D. Morgan, S. J. Murphy, D. E. Games and I. C. Mylchreest, Analysis of ecdysteroids by supercritical fluid chromatography, *J. Chromatogr.*, 441:165-169 (1988).
13. M. W. Raynor, J. P. Kithiji, I. K. Barker, K. D. Bartle and I. D. Wilson, Supercritical fluid chromatography of ecdysteroids, *J. Chromatogr.*, 436:497-502 (1988).
14. R. P. Evershed, M. C. Prescott, L. J. Goad and H. H. Rees, High-temperature gas chromatography mass spectrometry of steroid derivatives, *Biochem. Soc. Trans.*, 15:175-179 (1987).
15. R. P. Evershed, J. G. Mercer and H. H. Rees, Capillary gas chromatography-mass spectrometry of ecdysteroids, *J. Chromatogr.*, 390:357-369 (1987).
16. R. P. Evershed and M. C. Prescott, Facile connection of aluminium clad capillary GC columns to a magnetic sector mass spectrometer, *Biomed. Environ. Mass Spectrom.*, 18:503-506 (1989).
17. K. Grob, "Band broadening in space" and the "retention gap" in capillary gas chromatography, *J. Chromatogr.*, 237:15-23 (1982).
18. M. E. Feldlaufer and J. A. Svoboda, Makisterone A - a 28-carbon insect ecdysteroid, *Insect Biochem.*, 16:45-48 (1986).
19. K.-D. Spindler, R. Keller and J. D. O'Connor, The role of ecdysteroids in crustacean moulting cycle, *in*: "Progress in Ecdysone Research", J. A. Hoffman, ed., Elsevier/North-Holland, Biomedical Press, Amsterdam, pp 247-280 (1980).
20. M. J. Thompson, J. A. Svoboda and M. F. Feldlaufer, Analysis of free and conjugated ecdysteroids and polar metabolites of insects, *in*: "Analysis of Sterols and Other Biologically Significant Steroids", W. D. Nes and E. J. Parish, eds., Academic Press, San Diego, California, pp 81-105 (1989).
21. J. O. Whitney, Novel MS approaches to the analysis of free and conjugated bile salts, *in*: "Mass Sepctrometry in Biomedical Research", S. J. Gaskell, ed., John Wiley and Sons, Chichester, pp 61-74 (1986).
22. R. E. Isaac, M. E. Rose, H. H. Rees and T. W. Goodwin, Identification of ecdysone-22-phosphate and 2-deoxyecdysone-22-phosphate in eggs of the desert locust, *Schistocerca gregaria*, by fast atom bombardment mass spectrometry and NMR spectroscopy, *J. Chem. Soc. Chem. Commun.*, 249-251 (1982).

23 R. E. Isaac, M. E. Rose, H. H. Rees and T. W. Goodwin, Identification of the 22-phosphate esters of ecdysone, 2-deoxyecdysone, 20-hydroxyecdysone and 2-deoxy-20-hydroxyecdysone from newly laid eggs of the desert locust, *Schistocerca gregaria*, *Biochem. J.*, 213:533-541 (1983a).
24. R. E. Isaac and H. H. Rees, Isolation and identification of ecdysteroid phosphates and acetylated ecdysteroid phosphates from developing eggs of the locust, *Schistocerca gregaria*, *Biochem. J..*, 221:459-464 (1984).
25 M. E. Rose, M. P. Veares, I. A. S. Lewis and L. J. Goad, Analysis of steroid conjugates by fast atom bombardment mass spectrometry, *Biochem. Soc. Trans.*, 11:602-603 (1983).
26. G. M. O'Hanlon, O. W. Howarth and H. H. Rees, Identification of ecdysone 25-O-β-D-glucopyranoside a new metabolite of ecdysone in the nematode *Parascaris equorum*, *Biochem. J.*, 248:305-307 (1987).
27. T. Crosby, R. P. Evershed, D. Lewis, K. P. Wigglesworth and H. H. Rees, Identification of ecdysone 22-long-chain fatty acyl esters in newly laid eggs of the cattle tick *Boophilus microplus*, *Biochem. J.*, 240:131-148 (1986).
28. T. Crosby, Ph.D. Thesis, University of Liverpool, UK (1988).
29. L. Dinan, The chemical synthesis of ecdysone-22-long-chain fatty acyl esters in high yield, *J. Steroid Biochem.*, 31:237-245 (1988).
30. W. Aberth, K. Straub and A. L. Burlingame, Secondary ion mass spectrometry with cesium ion primary beam and liquid target matrix for analysis of bioorganic compounds, *Anal. Chem.*, 54:2029-2034 (1982).
31. T. R. Covey, E. D. Lee, A. P. Bruins and J. D. Henion, Liquid chromatography/mass spectrometry, *Anal. Chem.*, 58:1451A-1461A (1986).
32. C. R. Blakely and M. L. Vestal, Thermospray interface for liquid chromatography mass spectrometry, *Anal. Chem.*, 55:750-754 (1983).
33. W. R. Lusby, J.E. Oliver and M.J. Thompson, Application of disorption chemical ionization techniques for analysis of biologically active compounds isolated from insects, in: "Application of New Mass Spectromety Techniques in Pesticide Chemistry", J.D. Rosen, ed., John WIley and Sons, New York, pp 99-115 (1987).
34. S. J. Gaskell, K. Rollins, R. W. Smith and C. E. Parker, Determination of serum cortisol by thermospray with gas chromatography/mass spectrometry: comparison with gas chromatography/mass spectrometry, *Biomed. Environ. Mass Sepctrom.*, 14:717-722 (1987).
35. D. J. Liberato, A. L. Yergey, N. Esteban, C. E. Gomez-Sanchez and C. H. L. Schackleton, Theromospray HPLC/MS: a new mass spectrometric technique for the profiling of steroids, *J. Steroid Biochem.*, 27:61-70 (1987).

THIN-LAYER CHROMATOGRAPHY OF ECDYSTEROIDS: DETECTION AND IDENTIFICATION

I. D. Wilson, R. Lafont*, C. J. Porter**, R. G. Kingston**
K. Longden†, I. Fleming†† and P. Wall§

Department of Chemistry, University of Keele
Keele, Staffordshire ST5 5BG, UK
*Ecole Normale Superieure, Department de Biologie
CNRS-URA 686, 46 Rue D'Ulm, 75230 Paris Cedex 05, France
**V. G. Analytical Ltd., Floats Road
Wythenshawe, Manchester M23 9LE, UK
†V. A. Howe & Co. Ltd., 12-14 St Ann's Crescent
London SW18 2LS, UK
††Camlab Ltd., Nuffield Road, Cambridge CB4 1TH, UK, and
§BDH Chemicals Ltd., Broom Road, Poole BH12 4NN, UK

SUMMARY

Recent advances in the instrumentation available for thin-layer chromatography (TLC) which have greatly extended the potential of the technique for ecdysteroid analysis are described. In particular, improvements in scanning densitometers and radioactivity detectors have resulted in a versatile, rapid and quantitative technique which complements other chromatographic methods such as gas and high performance liquid chromatography. In addition, the viability of TLC-mass spectrometry (MS) for ecydsteroids, without prior isolation from the TLC plate, is shown, together with preliminary results for TLC combined with tandem mass-spectrometry (TLC-MS-MS) for studies on phytoecdysteroids.

INTRODUCTION

Thin-layer chromatography (TLC) has been an important and useful tool in studies on the isolation and identification of ecdysteroids for many years (e.g., see [1]). The reasons for this are not hard to find, TLC is simple, robust, technically undemanding, inexpensive and, at its simplest, does not require a knowledge of instrumental chromatography. However, the use of TLC in ecdysone research has, with only a few exceptions, been confined to the isolation of ecdysteroids or their qualitative analysis rather than as a means of quantification. Where the need for a quantitative result has been identified then recourse has usually been made to either gas or high performance liquid chromatography (GLC and HPLC) or, alternatively, to immunoassays. Partly this has been the result of the lower resolution and sensitivity afforded by TLC compared to, for example, HPLC, but in addition scanning densitometry has not been perceived to be an acceptable or viable alternative to these other techniques for quantitative work. However, TLC has, like all of the other major chromatographic techniques, undergone a process of continuous innovation. This has resulted in great improvements in both the quality of the TLC plates themselves (including the introduction of high performance (HP) TLC) and the types available such as the C_{18} bonded phases etc. (e.g., see [2]). These newer plates have provided the basis for better chromatographic performance, and this has been complimented by improvements in sample application (good sample application is essential in HP-TLC) and in methods for plate development. In the case of plate development automated multiple development [3] and overpressure TLC (OPTLC/OPLC) offer the possibility of significant improvements in resolution (see Read et al., this volume).

Chromatography and Isolation of Insect Hormones and Pheromones
Edited by A. R. McCaffery and I. D. Wilson
Plenum Press, New York, 1990

Here we present the results of recent studies in which the latest generation of scanning densitometers and radio-TLC detectors have been used to investigate the potential of TLC in metabolism studies with [^{3}H]-labeled compounds and in the study of ecdysteroid containing plant extracts. The problem of identification is also addressed using spectroscopic methods, without the need for the recovery of the sample from the TLC plate prior to analysis.

SAMPLE APPLICATION IN TLC

As alluded to in the Introduction good sample application is essential for good TLC and especially so for HP-TLC. Whilst careful application of the sample cannot of itself guarantee a good chromatographic result, poor sample application will certainly ensure that the system will not perform at its best. In HP-TLC, in particular, it is essential that the spot or band size is kept to a minimum and that the surface of the plate is not damaged during sample application. Obviously one method for ensuring that the spot size is kept to a minimum is to use volatile solvents of low eluotropic strength for application. In addition, a number of automated instruments are available which will apply small volumes of sample at a controlled rate to the plate. This ensures that the rate of application can be matched to that of solvent evaporation and enable a small spot to be obtained. Such instrumental methods of sample application can also help to compensate for the use of samples dissolved in strongly eluotropic solvents, which would otherwise give rise to large diffuse spots or bands.

This is well illustrated in Figure 1 where equal volumes of the same sample (a methanolic extract of phytoecdysteroids from *Silene otites*) have been applied to a silica gel HP-TLC plate either manually (Figure 1a) or instrumentally using the Camag Linomat IV (Figure 1b). The Linomat IV instrument is especially useful for bandwise sample application. The sample is delivered, via a syringe, in a stream of nitrogen (greatly aiding solvent evaporation). There is no contact between the syringe needle and the surface of the plate which ensures that the layer is not subjected to mechanical damage. Both sample volume (in this case 15 µl) and rate of application can be varied to suit the sample. As can be seen from the results manual and instrumental methods gave quite different chromatograms. In the case of manual application, even though this was performed as carefully as possible, the use of methanol resulted in very broad diffuse spots and this has led to peak splitting on chromatography (Figure 1a). The result obtained with instrumental application on the other hand was excellent (Figure 1b) as, despite the use of methanol, the band shape obtained was very narrow. Whilst an extreme case, these results do nevertheless re-emphasize the need for care at all stages of TLC if acceptable results are to be obtained.

SCANNING DENSITOMETRY

Having carefully applied the sample and then developed the plate with, hopefully, the desired result in terms of separation, there is the problem of obtaining quantitative information about the sample components.

The manufacture of reliable and sensitive scanning densitometers has proved to be a considerable technical challenge and it is really only with the ready availability of powerful personal computers that these devices have come of age. The capabilities of some of these instruments are illustrated with reference to studies performed on extracts of plants in the *Silene* family. Thus, in Figure 2 some of our earlier results obtained using a Camag scanner II, are shown for a separation of a crude extract of *Silene schafta* on silica gel with chloroform-ethanol as solvent [4]. As indicated, peaks were present with R_F values similar to those of 20-hydroxyecdysone (peak A) and polypodine B (peak B), together with an unknown component at higher R_F (peak C). Modern densitometers can be used to obtain further information about the nature of sample components by providing UV spectra. The spectra for the polypodine B and 20-hydroxyecdysone spots are shown in the inset and were identical to those of authentic samples. The UV spectrum of peak C on the other hand was quite different, suggesting that it was probably not an ecdysteroid, and therefore not worth isolating. In addition to simple UV spectra, this instrument can also be used to obtain spectra from various parts of the spot (i.e., tailing edge, middle and leading edge) to enable an estimate of peak purity to be obtained. Only if all three spectra were identical would the spot be considered to be a single component. This is rather like the information provided by diode array or fast scanning UV detection in HPLC.

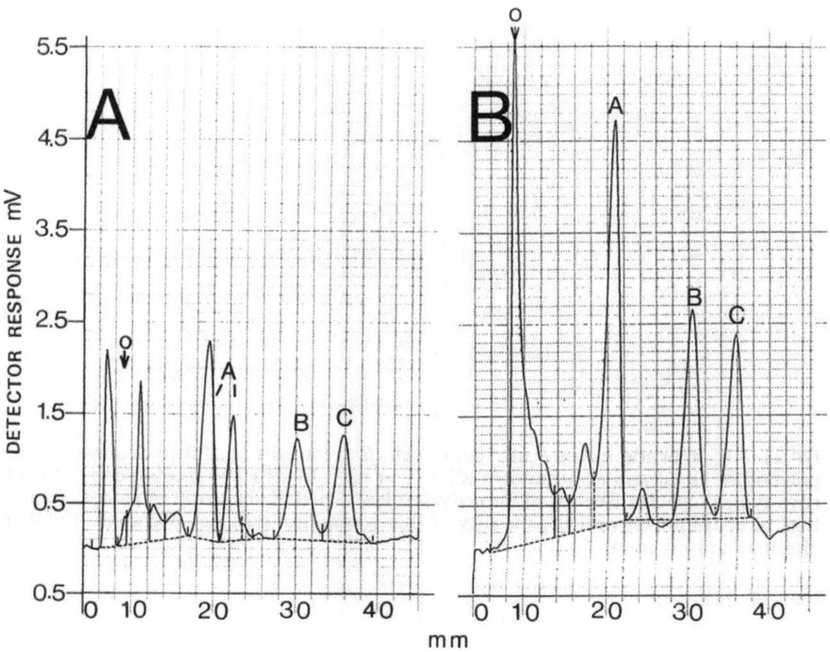

Fig. 1. Effect of sample application on the chromatography of an extract of *Silene otites*. A, manual application of 15 µl of methanolic extract; B, instrumental application (Linomat IV) of 15 µl of methanolic extract. Chromatography: HP-TLC on silica gel (E. Merck) with chloroform-ethanol 4:1 (v/v) (single development). Plate scanned at 254 nm with a Shimadzu CS 9000 scanning densitometer. Note the peak splitting of the origin (0) and peak A (20-hydroxyecydsone) and the distorted peak shape of B (2-deoxy-20-hydroxyecdysone) on the manually spotted plate (A) compared to B. Peak C (2-deoxyecdysone) gave good peak shape in both systems.

Similarly modern scanning densitometers can also provide chromatograms of the same track scanned at a number of wavelengths as shown in Figure 3. Here the sample has been scanned at nine different wavelengths between 200 and 335 nm in 15 nm increments once more providing output similar to that of a diode array detector. An important difference, however, between TLC and HPLC, is that in the latter all of this additional multi-wavelength and spectroscopic information has to be obtained in real time, on-line. In the case of TLC the plate acts as a "hard copy" of the chromatogram enabling this additional information to be obtained at a later date. Indeed, for most of the results shown here (for both plant extracts and [^{3}H]-metabolism studies), the plates were not scanned in the laboratory in which the separations were performed but were sent to one or more different laboratories for analysis.

Another area which we have been keen to explore is the use of scanning densitometers to obtain information from TLC plates following 2-dimensional (2D) TLC. 2D-TLC is of interest because it provides a method of greatly increasing the resolving power of TLC albeit at considerable cost to sample capacity (which is obviously reduced to one sample per plate) and an increase in analysis time. The increase in resolving power results from the ability of the chromatographer to alter the selectivity of the separation when going from the first to the second dimension. Indeed, in extreme circumstances the whole separation mechanism can be altered (e.g., normal to reversed-phase [5]). The problem is then that the components of interest may be scattered over a large proportion of the surface of the TLC plate.

In Figure 4a the distribution of components following 2D-TLC of a sample of the methanolic extract of *Silene otites* is shown as a "conventional" image of a TLC plate, i.e., spots (obtained with a Shimadzu scanning densitometer). In Figure 4b the same information has been converted to a conventional "linear" chromatogram. The 3D-image of the same plate shown in Figure 5 was obtained using the Desaga CD 60.

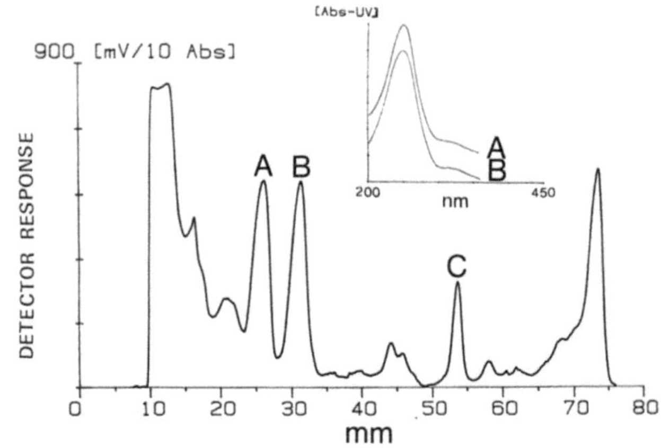

Fig. 2. HP-TLC of a crude extract of *Silene schafta* containing 20-hydroxyecdysone (A) and polypodine B (B) and an unknown (C) at 254 nm. In situ UV spectroscopy of peaks A and B gave typical spectra (see inset) for ecdysteroids. The UV spectrum of peak C was quite unlike that of ecdysteroids and the identity of this compound was not pursued. Further details of chromatography in [4].

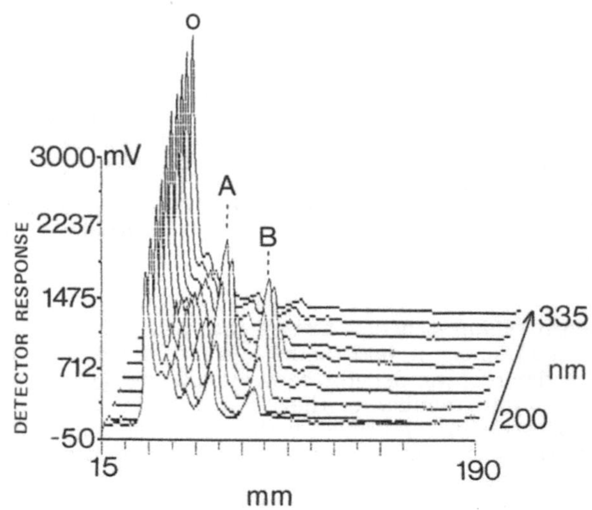

Fig. 3. HP-TLC of a partially purified extract of *Silene nutans* scanned at multiple wavelengths (200 to 235 nm in 15 nm increments). Chromatography: HP-TLC on silica gel (E. Merck) with chloroform ethanol 4:1, single development. Key: O, origin; A, 20-hydroxyecdysone; B, polypodine B.

Such images represent the current state of the art insofar as scanning the surface of TLC plates with commercial instruments is concerned. A considerable amount of work still remains to be done to enable the analyst to use 2D-TLC as flexibly as he might wish. Thus, the 3D-image from Figure 5 cannot yet be converted into a "linear" chromatogram such as that shown in Figure 4b, and had the chromatography been more complex than this it is unlikely that Figure 4b itself could have been produced. However, as will be demonstrated in the following section on radioactivity detection, these problems are not technical but merely require the right computer programs to be produced.

A further technical development which may be of interest for both one and two-dimensional TLC is the development of image analyzers for TLC [6]. Whilst beyond the scope of this review such devices have been developed and may soon become commercially available.

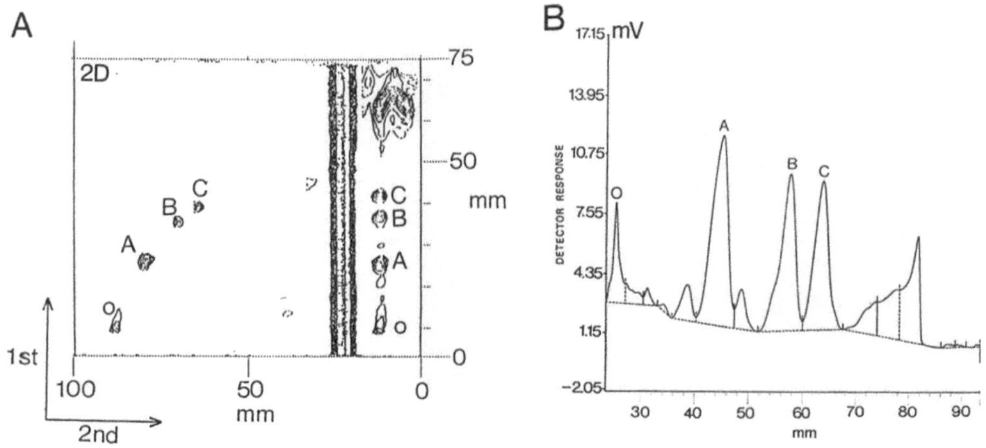

Fig. 4. 2D-TLC of an extract of *Silene otites*. A, "conventional" image showing the whole surface of the plate. The track on the righthand side of the plate shows a single development in 1-dimension for comparison with the 2D development which occupies the remainder of the plate. B, this chromatogram shows the conversion of the 2D chromatogram into a normal linear chromatogram. Chromatography: HP-TLC on silica gel (E. Merck), 1st dimension chloroform-ethanol 4:1 v/v (single development), 2nd dimension chloroform-ethanol 9:1 v/v (3 developments). Scanning densitometry was performed using a Shimadzu CS 9000 scanning densitometer at 254 nm with zig-zag scanning (16 mm swing). Key: O, origin; A, 20-hydroxyecdysone; B, 2-deoxy-20-hydroxyecdysone; C, 2-deoxyecdysone.

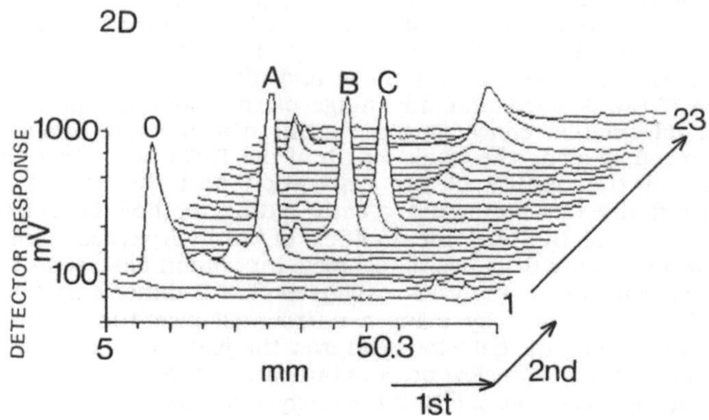

Fig. 5. The 2D chromatogram shown in Figure 4 shown as a 3D plot obtained on the Desaga CD 60 scanning densitometer.

RADIO-TLC DETECTORS

The ability to rapidly detect and quantify radioactivity on TLC plates can be used to great advantage in the study of the metabolic fate of [^{3}H]-ecdysteroids [7-10]. The introduction of the position sensitive proportional wire radio TLC detectors ("linear analyzers") has provided a detection system with a resolution of between 0.5 and 2 mm (depending upon the instrument and isotope) and the ability to detect as little as 100 dpm of [^{3}H].

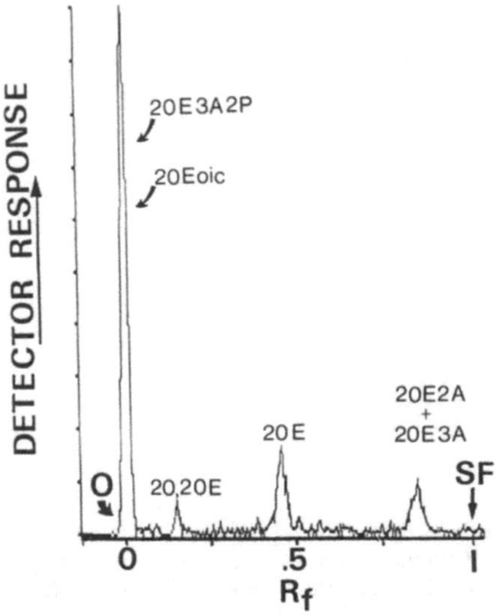

Fig. 6. HP-TLC of [^{3}H] metabolites of 20-hydroxyecdysone from *Locusta migratoria* [7]. Chromatography: HP-TLC on silica gel (E. Merck) using chloroform-ethanol 4:1 v/v (single development). [^{3}H]-detected using a Bethold LB 2842 linear analyzer. Key: O, origin; SF, solvent front; 20E2A, 20E3A, 20-hydroxyecdysone 2 and 3 acetates; 20E, 20-hydroxyecdysone; 20, 20E, 20,26-dihydroxyecdysone; 20-E3A2P, 20-hydroxyecdysone-3 acetate, 2 phosphate; 20-Eoic, 20-hydroxyecdysonoic acid.

We have found such instruments particularly useful as (a) a means of examining quite crude extracts that were unsuitable for HPLC; (b) for the rapid screening of a large number of samples in order to decide which ones should be further analyzed by HPLC; (c) to confirm results already obtained on HPLC concerning the nature and proportions of the radiolabeled metabolites present in a sample; and (d) to help guide the design of HPLC separations (see [7-9]). A particular advantage of the mode of operation of the linear analyzer is that information concerning the location of the radiolabeled metabolites is collected simultaneously along the whole length of the TLC track. This can enable a very rapid evaluation of the quality of the separation and the general distribution of radioactivity within the chromatogram greatly aiding method development. A typical result for the HP-TLC of the [^{3}H] metabolites of 20-hydroxyecdysone from *Locusta migratoria* is shown in Figure 6. A useful technical innovation has been the development of methods for using the linear analyzer to construct a 2-dimensional image of a plate following 2D-TLC. This is done by using a narrow slit over the length of the detector aperture and slowly moving the detector head over the plate in increments of 2.5 mm at a time [11-12]. This is obviously a slow process but can provide results of the type shown in Figure 7, for a mixture of acetates of [^{3}H]-20-hydroxyecdysone.

This 2D "digital autoradiograph" can, thanks to the software available, also be converted to the 3D image shown in Figure 7b. The data can be further processed to provide normal linear chromatograms of the type shown in Figure 8. Whilst potentially of great benefit for use with samples requiring the use of 2D separations for enhanced resolution, the length of time required to acquire the data has made its routine use rather impractical. A further minor problem is that the resolution of the instrument is different in the X and Y planes.

Recently however, a new position sensitive multiwire proportional chamber, the digital autoradiograph (DAR), has been introduced which can be used to image the whole surface of a 20 x 20 cm TLC plate simultaneously and which can be used for weak B-emitters such as [^{3}H] [13]. Such instruments are not entirely new, for example, the "AMBIS" multiwire proportional detector has been available for some years but it cannot be used with [^{3}H] [14]. The DAR consists of three layers of 100 wires, each layer being separated by a few mm. The chamber in which the three layers are placed contains a mixture of argon and methane (90:10) with the middle layer of wires set at a positive voltage of 1800 V. This middle layer of wires generates a charge signal from the ionizing radiation entering the

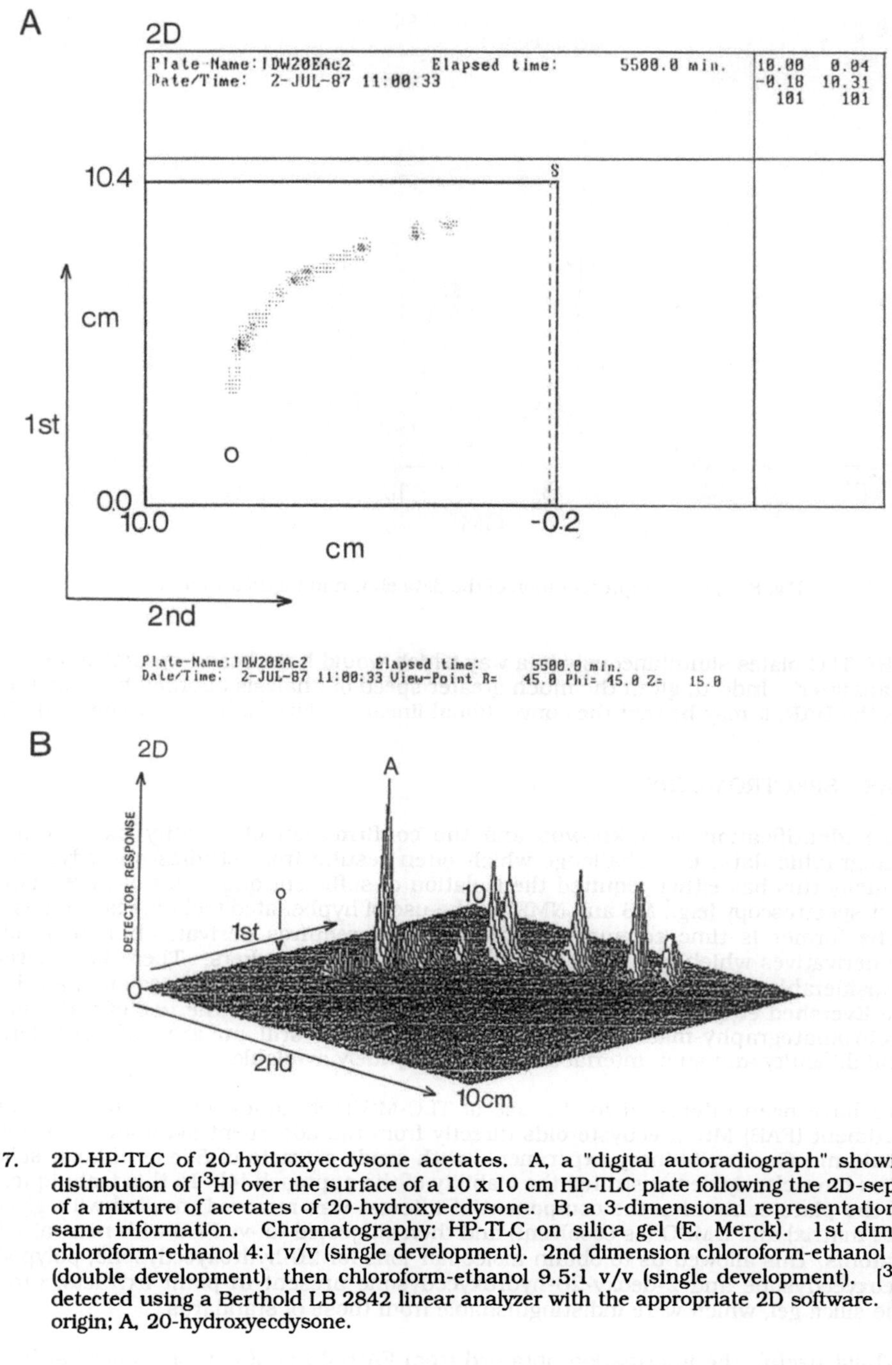

Fig. 7. 2D-HP-TLC of 20-hydroxyecdysone acetates. A, a "digital autoradiograph" showing the distribution of [^{3}H] over the surface of a 10 x 10 cm HP-TLC plate following the 2D-separation of a mixture of acetates of 20-hydroxyecdysone. B, a 3-dimensional representation of the same information. Chromatography: HP-TLC on silica gel (E. Merck). 1st dimension, chloroform-ethanol 4:1 v/v (single development). 2nd dimension chloroform-ethanol 9:1 v/v (double development), then chloroform-ethanol 9.5:1 v/v (single development). [^{3}H] was detected using a Berthold LB 2842 linear analyzer with the appropriate 2D software. Key: O, origin; A, 20-hydroxyecdysone.

chamber whilst the other two layers of wires pick up the charge and enable the position of the signal to be defined (see [11] for a full description of the apparatus).

Although the information provided by this device is very similar to that provided by the older linear analyzer used in the 2D mode and shown in Figure 7, the DAR is 100 times more sensitive. Even in situations where 2D-TLC has not been performed the DAR enables all the tracks on the plate to be analyzed simultaneously resulting in a considerable saving in time compared to determining each one individually. Sensitivity of 100 dpm of [^{3}H] per spot in 10 to 20 min has been claimed (for ^{14}C 1 to 2 min was sufficient). Using this device we have been able to image 2D separations of [^{3}H]-ecdysone metabolites on a number of 10 x

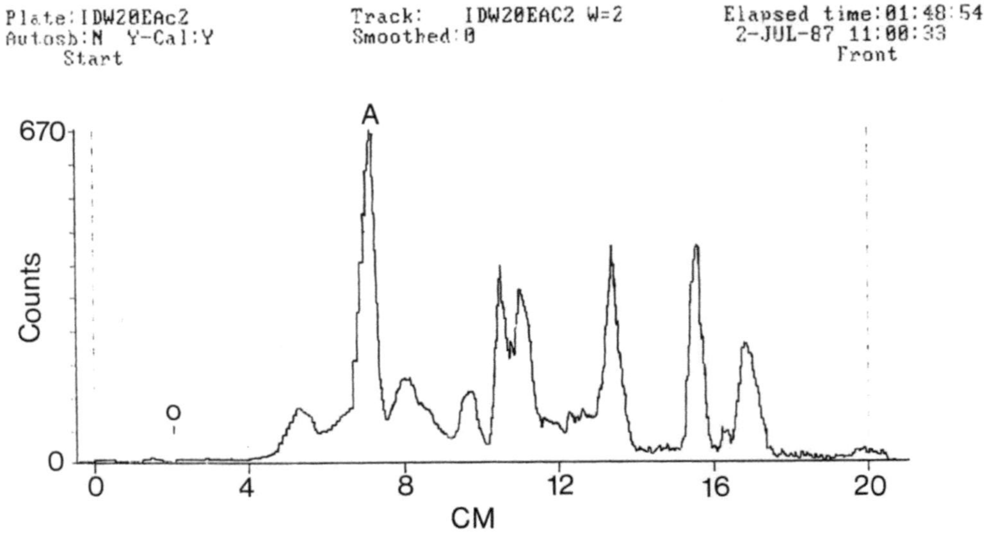

Fig. 8. A linear representation of the data shown in Figures 7a and b.

10 cm HP-TLC plates simultaneously in a way which would have been impractical using the "linear analyzer". Indeed, given the much greater speed of analysis afforded by instruments such as the DAR, it may be that the conventional linear analyzer will soon become obsolete.

TLC-MASS SPECTROMETRY

The identification of unknowns and the confirmation of identity based only on chromatographic data, is a challenge which often results from studies on ecdysteroids. Traditionally this has either required the isolation of sufficient quantities of pure material for direct spectroscopy (e.g., MS and NMR) or the use of hyphenated techniques such as GC-MS. The former is time consuming whilst GC-MS requires derivatization to produce volatile derivatives which has proved problematic for some workers. There has therefore been considerable interest in the use of alternative hyphenated techniques such as HPLC-MS (see Evershed et al., this volume) and SFC-MS [15]. However, the use of these linked liquid chromatography-mass spectrometry interfaces is still an area of considerable technical difficulty and such interfaces are not yet widely available.

We have been interested in the use of TLC-MS techniques with off-line fast atom bombardment (FAB) MS of ecdysteroids directly from the adsorbent (without the need for prior elution). Our preliminary experiments with crude extracts of *Silene nutans, schafta* and *otites* have clearly demonstrated the viability of the approach [4]. In this technique, the area of the plate containing the component of interest was removed, mulled with glycerol (the FAB matrix) and DMSO (as cosolvent) and then subjected to -ve FAB with neutral 7 KeV xenon atoms. This allowed us to obtain molecular ions for 20-hydroxyecdysone, polypodine B, 2-deoxyecdysone and 2-deoxy-20-hydroxyecdysone present in plant extracts, directly from the silica gel, which were indistinguishable from those of standards.

Whilst useful, the information obtained from FAB-MS is of course somewhat limited by the relative lack of fragmentation obtained. We have therefore been interested in extending the scope of the TLC-MS method by using tandem mass-spectrometry (or MS-MS) whereby ions are taken from one mass spectrometer, via a collision cell containing a reagent gas to induce fragmentation, into a second mass spectrometer from which a mass spectrum is obtained. This spectrum shows much greater fragmentation and therefore enables more structural information to be obtained. Indeed in our preliminary experiments we hoped to be able to eliminate the need for a chromatographic separation altogether and use the power of the MS-MS technique to enable the identification of the components directly from the crude plant extract. However, attempts to obtain MS-MS data on crude

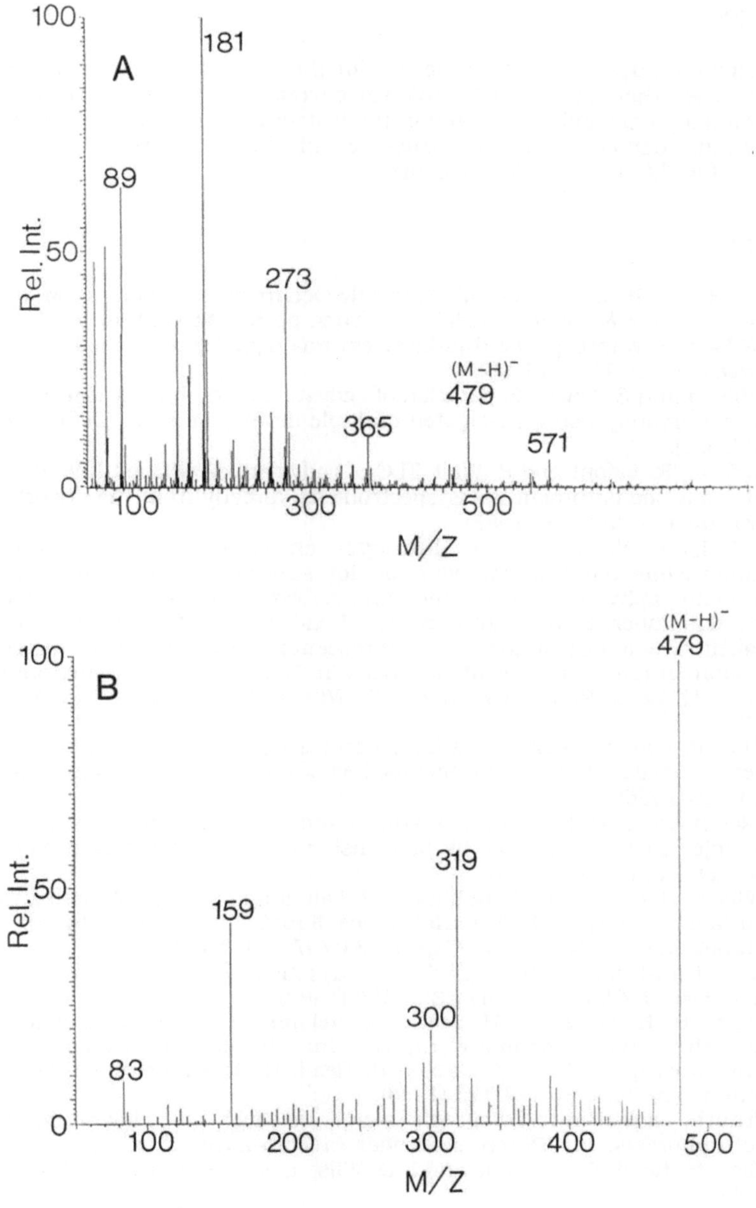

Fig. 9. FAB-MS and MS-MS of 20-hydroxyecdysone present in an extract of phytoecdysteroids from *Silene otites* following HP-TLC (all conditions as for Figure 1). A, -ve FAB spectrum of 20-hydroxyecdysone showing the $(M\text{-}H)^-$ ion together with ions resulting from the background and matrix. B, MS-MS of the 479 amu ion $(M\text{-}H)^-$ showing fragmentation due to 20-hydroxyecdysone.

extracts of *Silene nutans* and *otites* were only partially successful. When TLC-MS-MS of the same extracts was investigated, however, complete success was obtained and good spectra were produced from ecdysteroids. The TLC-FAB-MS and TLC-FAB-MS-MS results for 20-hydroxyecdysone, from *Silene otites*, following separation on an aluminium backed silica gel HP-TLC plate are shown in Figure 9.

Both TLC-MS and TLC-MS-MS are techniques of great potential which we have as yet hardly begun to examine in detail but which we will continue to investigate.

CONCLUSIONS

TLC remains a useful and versatile tool for the study of ecdysteroids which actively complements the other analytical techniques currently available. Continuing improvements in both plate type and quality and in the instrumentation available for modern TLC, such as scanning densitometers and sensitive radioactivity detectors, should ensure a continued role for TLC for some time to come.

REFERENCES

1. D. H. S. Horn, The ecdysones, *in*: "Naturally Occurring Insecticides", M. Jacobson and D. G. Crosby, eds., Marcel Dekker, New York, pp 333-459 (1971).
2. I. D. Wilson, Reversed-phase thin-layer chromatography of the ecdysteroids, *J. Plan. Chromatog.*, 1:116-122 (1988).
3. I. D. Wilson and S. Lewis, Separation of ecdysteroids by high-performance thin-layer chromatography using automated multiple development, *J. Chromatog.*, 408:445-448 (1987).
4. I. D. Wilson, R. Lafont and P. Wall, TLC of ecdysteroids with off-line identification by fast atom bombardment mass spectrometry directly from the adsorbent, *J. Plan. Chromatog.*, 2:357-359 (1988).
5. I. D. Wilson, Normal-phase thin-layer chromatography on silica gel with simultaneous paraffin impregnation for subsequent reversed-phase thin-layer chromatography in a second dimension, *J. Chromatog.*, 287:183-188 (1984).
6. R. M. Belchamber, S. J. Brinkworth, H. Read and J. M. Roberts, Qualitative and quantitative image analysis of fluorescents from high-performance thin-layer chromatography, *in*: "Recent Advances in Thin-Layer Chromatography", F. A. A. Dallas, H. Read, R. J. Ruane and I. D. Wilson, eds., Plenum, New York, pp 37-44 (1988).
7. I. D. Wilson and R. Lafont, Thin-layer chromatography and high-performance thin-layer chromatography of [^{3}H] metabolites of 20-hydroxyecdysone, *Insect Biochem.*, 16:33-46 (1986).
8. P. D. Robinson, E. D. Morgan, I. D. Wilson and R. Lafont, The metabolism of ingested and injected [^{3}H] ecdysone by final instar larvae of *Heliothis armigera*, *Physiol. Entomol.*, 12:321-330 (1987).
9. I. D. Wilson, E. D. Morgan, P. Robinson, R. Lafont and C. Blais, A comparison of radio-thin-layer and radio-high-performance liquid chromatography for ecdysteroid metabolism studies, *J. Insect Physiol.*, 34:707-711 (1988).
10. L. Dinan, Thin-layer chromatography of ecdysone acyl esters and their 2,3-acetonide derivatives, *J. Chromatog.*, 411:379-392 (1987).
11. G. Dietzel, H. Kubisak and H. Stelzer, Development of 2-dimensional detectors for radio-thin-layer chromatography, *in*: "Recent Advances in Thin-Layer Chromatography", F. A. A. Dallas, H. Read, R. J. Ruane and I. D. Wilson, eds., Plenum, New York, pp 117-123 (1988).
12. H. Filthuth, Digital autoradiograph: a new detector for radiochromatography and electrophoresis, *in*: "Recent Advances in Thin-Layer Chromatography", F. A. A. Dallas, H. Read, R. J. Ruane and I. D. Wilson, eds., Plenum, New York, pp 125-129 (1988).
13. H. Filthuth, A new detector for radiochromatography and radiolabeled multi sample distributions - the digital autoradiograph, *J. Plan. Chromatog.*, 3:198-202 (1989).
14. I. Smith, The AMBIS two-dimensional Beta scanner, *in*: "Recent Advances in Thin-Layer Chromatography", F. A. A. Dallas, H. Read, R. J. Ruane and I. D. Wilson, eds., Plenum, New York, pp 77-86 (1988).
15. M. W. Raynor, J. P. Kithinji, K. D. Bartle, D. E. Games, I. C. Mylchresst, R. Lafont, E. D. Morgan and I. D. Wilson, Packed column supercritical fluid chromatography and linked supercritical fluid chromatography-mass spectrometry for the analysis of phytoecdysteroids from *Silene nutans* and *Silene otites*, *J. Chromatog.*, 467:292-298 (1989).

A NOTE ON OVERPRESSURE THIN-LAYER CHROMATOGRAPHY OF ECDYSTEROIDS

H. Read, I. D. Wilson* and R. Lafont**

Department of Chemistry, Birkbeck College
29 Gordon Square, London WC1H 0PP, UK
*Department of Chemistry, University of Keele
Keele, Staffordshire ST5 5BG, UK, and
**Ecole Normale Supérieure, Department de Biologie
CNRS-URA 686, 46 Rue D'ulm
F-75230 Paris Cedex 05, France

INTRODUCTION

It is now clear that the ecdysteroid family of compounds is very large and comprises at least 150 structurally related compounds. This diversity imposes severe problems for the use of chromatographic methods in the isolation and identification of ecdysteroids. We have, therefore, been interested in the evaluation of new chromatographic systems for these compounds, such as supercritical fluid chromatography ([1-3] and Morgan et al., this volume), and new developments which promise increased resolution for more traditional methodologies such as thin-layer chromatography (TLC). In the case of TLC two methods have recently been introduced, namely automated multiple development (AMD) and overpressure TLC (OPTLC or OPLC), which offer the prospect of enhanced resolution compared to normal TLC. In AMD the plate is repeatedly redeveloped in a solvent which is allowed to migrate a little further up the plate with each development. With each development a reconcentration of the components occurs resulting in high resolution and very sharp bands. Preliminary results for the use of AMD with ecdysteroids have been described elsewhere [4]. In the OPLC technique the plate is held under an inert, impermeable membrane under hydrostatic pressure (typically ca. 25 bar) and solvent forced through the layer using a high pressure liquid chromatography (HPLC) pump. As the flow of solvent through the layer can be controlled and is no longer dependent on capillary action, plates can be developed very rapidly. In addition solvent can effectively be run off the end of the plate allowing the use of mobile phases of relatively low eluotropic strength to obtain increased resolution. Furthermore, such continuous development enables the collection of fractions or the use of HPLC detectors for quantification. The use of the technique has been recently reviewed [5].

Here we briefly describe experiments designed to compare OPLC with AMD for the resolution of a mixture of ecdysteroids and demonstrate the application of OPLC to the separation of the ecdysteroidal components in an extract obtained from the plant *Silene nutans*.

MATERIALS AND METHODS

Ecdysteroids were gifts from various sources. Sample preparation used to obtain the *Silene nutans* extracts has been described elsewhere [3]. Chromatography was performed on Merck silica gel TLC plates (20 x 20 cm) incorporating a fluorescent indicator (254 nm) supplied by BDH Ltd. (Poole, UK). Before carrying out OPLC a few mm of silica was removed from each edge of the plate and these edges sealed with "Impress", a water soluble polymer, followed by heating at 60°C for one hour. Samples were applied as solutions in methanol using a Camag Linomatt IV sample applicator (BDH, Poole, UK). OPLC was performed using the Chrompress 25 (Biotech, Luton, UK) with an overpressure of 25 bar and a solvent flow rate of either 2.5 or 5.5 ml/min. Solvents were composed of varying proportions of

Chromatography and Isolation of Insect Hormones and Pheromones
Edited by A. R. McCaffery and I. D. Wilson
Plenum Press, New York, 1990

chloroform-ethanol or dichloromethane-methanol ($CHCl_3$-EtOH, CH_2Cl_2-MeOH) (see text for details).

Following chromatography the plates were visualized under UV light at 254 nm and scanning densitometry performed using a Desaga CD60 scanning densitometer (Camlab, Cambridge, UK) at 254 nm in reflectance mode.

RESULTS

Initially, we studied the ability of the OPLC technique to separate a mixture of ecdysteroids composed of polypodine B, makisterone A, ecdysone, ponasterones A and C and polypodine B 2-cinnamate. This series of compounds covers a wide range of Rf values in normal linear TLC, and has been used previously by us to evluate AMD (see Table 1 and Ref. [4]) where we observed particular difficulty in resolving polypodine B and ecdysone. Our initial OPLC experiments were performed using the solvent system chloroform-ethanol 4:1 (v/v) and a solvent flow rate of 5.5 ml/min, with development for approximately 100 mm up the plate. At this flow rate the polar compounds tended to give rise to elongated streaks on the plate, probably due to the slow dissolution of the sample at the spot origin as the solvent was forced rapidly up the plate. Reducing the flow rate to 2.5 ml/min produced a marked improvement in spot profile. However, the results obtained with this solvent system gave no improvement in resolution over that obtained for standard linear TLC.

We therefore decided to apply a less eluotropic solvent system coupled with an increased development distance, initially to 150 mm, and then with continuous (or over) development whereby 10 ml solvent was collected from the exit capillary of the development chamber before solvent delivery was stopped. The best results were obtained using chloroform-ethanol 72.5:7.5 v/v which, coincidentally, was the system which gave the best results for AMD [4]. Maximum resolution was obtained using continuous development and these results are illustrated in Figure 1 and Table 1. We were able to achieve some separation between polypodine B and ecdysone, which had not proved to be possible with AMD, whilst the resolution between ponasterones A and C obtained with AMD was lost on OPLC.

We then proceeded to apply OPLC to the separation of polypodine B and 20-hydroxyecdysone present in an extract of the plant *Silene nutans*. Three solvent systems of increasing eluotropic strength were used composed of chloroform with 7.5, 10 or 15% (v/v) of methanol. Plates were eluted for 180 mm and the effects on Rf and resolution of these ecdysteroids in the three solvent systems noted. The results are shown in Figure 2a-c. Adequate resolution of polypdine B and 20-hydroxyecdysone was achieved with all three solvent systems although, not unexpectedly, with increasing eluotropic strength Rf increased and resolution decreased. The best resolution was obtained with chloroform-methanol 92.5:7.5 (Figure 2a). Had continuous development been employed the elution from

Table 1. Thin-layer Chromatographic Properties of Ecdysteroids Using Linear, AMD and OPLC Methods of Development: All Rf values have been normalised to ecdysone = 100

Compound	Linear TLC* single development CH_2Cl_2-MeOH 80:20	AMD* 25 steps CH_2Cl_2-MeOH 92.5:7.5	OPLC continuous development CH_2Cl_2-MeOH 92.5:7.5
Ecdysone	100	100	100
Makisterone A	88.5	87.5	90.1
Polypodine B	103.9	108.3	114.8
Ponasterone A	169.2	166.7	203.7
Ponasterone C	153.9	154.2	203.7
Polypodine B 2-cinnamate	208.8	200	388.8

* Data adapted from ref. 4

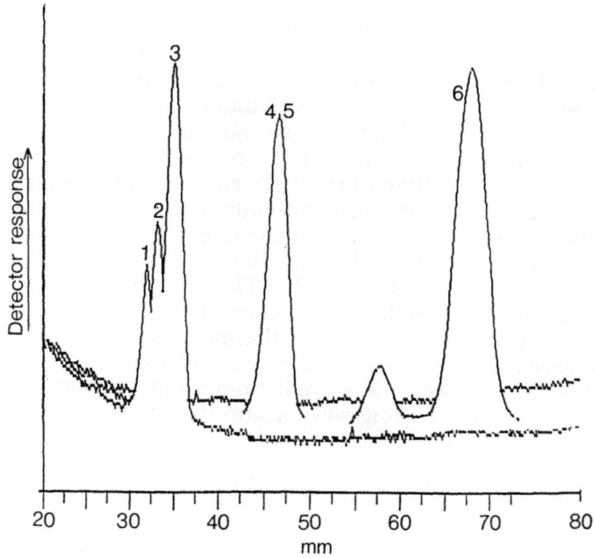

Fig. 1. OPLC of ecdysteroids (see text for details) on silica gel. Key - 1, makisterone A; 2, ecdysone; 3, polypodine B; 4, ponasterone A; 5, ponasterone C; 6, polypodine B 2-cinnamate.

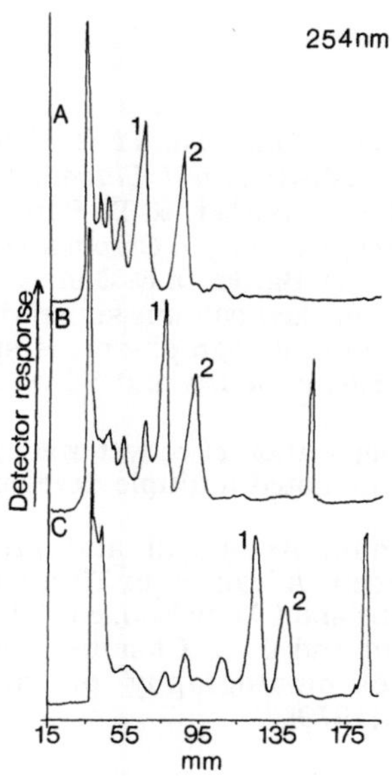

Fig. 2. Effect of solvent composition on R_f of ecdysteroids in an extract of *Silene nutans*. A, $CHCl_3$-EtOH 92.5:7.5; B, $CHCl_3$-EtOH 90:10; C, $CHCl_3$-EtOH 85:15. Peak 1, 20-hydroxyecdysone; peak 2, polypodine B.

the plate of both poypodine B and 20-hydroxyecdysone would have been easily achieved providing the basis for a simple and convenient preparative procedure.

DISCUSSION

It is clear from our results that OPLC of the ecdysteroids provided the best results when solvents of relatively low eluotropic strength, compared to those used for conventional linear development, were used. Continuous, or over, development would also seem to provide better resolution than merely running the solvent to the end of the plate. In this brief series of experiments our best results were achieved for the separation of the lower Rf components of our mixture, particularly ecdysone and polypodine B, where the separations, although not providing baseline resolution, were superior to those achieved using AMD. Compared to AMD the OPLC technique is very much faster with separations achieved within 0.5 min for a 150 mm development distance or approximately 5 min when the continuous development procedure was adopted. For AMD some 150 min were required to achieve a similar separation. Further work is clearly required to optimize both flow rate and solvent composition. However, these preliminary results clearly demonstrate the potential of the technique. A future area of investigation will be the use of OPLC connected in-line with UV detectors to enable the continuous mode of OPLC to be used for quantification, preparative and isolation work. This will require the use of HPLC type normal-phase eluents of the type developed by Lafont et al. [6] (and Lafont and Wilson, this volume) (because of the high UV absorbtion of chlorinated solvents) or alternatively the use of reversed-phase systems. In addition, we have recently adapted the OPLC system by the inclusion of a loop injector between the solvent pump and Chrompress chamber so that samples may be injected into the flowing solvent and this should allow multiple use of the same plate for preparative purposes.

CONCLUSIONS

The technique of OPLC is capable of separating complex mixtures of ecdysteroids and particularly good results were obtained for polar compounds such as ecdysone and polypodine B. The technique has also proved to be faster than AMD for the separation of the same compounds. The technique can also be applied to relatively crude extracts with good results.

REFERENCES

1. E. D. Morgan, S. J. Murphy, D. E. Games and I. C. Mylchreest, Analysis of ecdysteroids by supercritical-fluid chromatography, *J. Chromatog.*, 441:165-169 (1988).
2. M. W. Raynor, J. P. Kithinji, I. K. Barker, K. D. Bartle and I. D. Wilson, Supercritical fluid chromatography of ecdysteroids, *J. Chromatog.*, 436:497-502 (1988).
3. M. W. Raynor, J. P. Kithinji, K. D. Bartle, D. E. Games, I. C. Mylchreest, R. Lafont, E. D. Morgan and I. D. Wilson, Packed column supercritical fluid chromatography and linked supercritical-fluid chromatography-mass spectrometry for the analysis of phytoecdysteroids from *Silene nutans* and *Silene otites*, *J. Chromatog.*, 465:292-298 (1989).
4. I. D. Wilson and S. Lewis, Separation of ecdysteroids by high-performance thin-layer chromatography using automated multiple development, *J. Chromatog.*, 408:445-448 (1987).
5. E. Mincsovic and E. Tyihak, Analytical and preparative overpressured layer chromatography, *in*: "Recent Advances in Thin-Layer Chromatography", F. A. A. Dallas, H. Read, R. J. Ruane and I. D. Wilson, eds., Plenum, pp 57-66 (1988).
6. R. Lafont, G. Somme-Martin and J. C. Chambet, Separation of ecdysteroids using high-pressure liquid chromatography on microparticulate supports, *J. Chromatog.*, 170:185-194 (1979).

A NOTE ON SELECTIVE SEPARATION OF 20,22-DIHYDROXYECDYSTEROIDS FROM INSECT AND PLANT MATERIAL WITH IMMOBILIZED PHENYLBORONIC ACID

S. J. Murphy, E. D. Morgan and I. D. Wilson

Department of Chemistry
University of Keele, Keele
Staffordshire ST5 5BG, UK

INTRODUCTION

The isolation of ecdysteroids from biological material is a skillful and lengthy process. Short cuts in that isolation, taking advantage of any special properties of ecdysteroids, would be a considerable help in their isolation, identification and quantitative analysis. One possible strategy would be to exploit the presence of vicinal diol systems found in many, if not most ecdysteroids, either at C-2 and C-3, or at C-20 and C-22, or both.

Boronic acids react selectively with 1,2- and 1,3-diols, enediols, α-hydroxy acids and other such groups to form cyclic derivatives. Brooks [1,2] has thoroughly explored the use of boronic acids, including phenylboronic acid to give volatile cyclic boronate derivatives of steroids for gas chromatography. We have explored the use of cyclic phenylboronates as gas chromatographic derivatives [3] of ecdysteroids coupled with the use of the nitrogen-specific detector for selective detection of the boron, and Poole et al. [4] developed phenanthrene boronate derivatives for the fluorescence detection of ecdysteroids in high performance liquid chromatography (HPLC) and thin-layer chromatography (TLC).

Recently, immobilized phenylboronic acid has become available, either attached to silica, agarose or cellulose. We have therefore explored the use of solid phase extraction (SPE) with columns of immobilized phenylboronic acid (PBA) for the selective retention of ecdysteroids containing vicinal diol groups. Examination of molecular models shows that the side-chain hydroxyls at C-20 and C-22 can form an unstrained 5-membered ring in the boronate ester (Figure 1), but the axial C-2 and equatorial C-3 hydroxyls of ecdysteroids do not so readily form a cyclic boronate (Figure 1). We therefore considered the possibility that both ecdysteroids containing side-chain and A-ring diols might be selectively retained, but might then be separated by eluting first the A-ring diols followed by compounds with the side-chain diols.

The theory of extraction using immobilized PBA is straight-forward (Figure 2). The PBA and functional groups for which it has an affinity form a stable cyclic boronate in the presence of an alkaline buffer (Figure 2b). Non-bound contaminants are eluted, if necessary, using an organic modifier such as methanol. The complex is hydrolyzed in an acidic medium (Figure 2c) and the retained compounds eluted, again if necessary, with an organic modifier. If this form of extraction could be applied to ecdysteroids it would provide a more rapid and selective isolation method than those presently in use [5].

In this work the retention and elution characteristics of a number of ecdysteroids, on PBA, have been investigated in order to ascertain the possibilities of this new method of extraction. The use of PBA in the extraction of ecdysteroids from biological material has also been investigated, to see if the results obtained for standard ecdysteroids can be extended to biological samples.

Chromatography and Isolation of Insect Hormones and Pheromones
Edited by A. R. McCaffery and I. D. Wilson
Plenum Press, New York, 1990

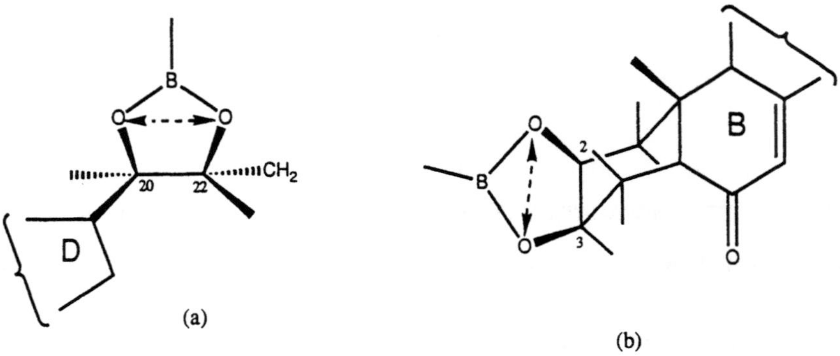

Fig. 1. Formation of cyclic boronate esters with the C-20,22 diol (a) and the C-2,3 diol of ecdysteroids (b). The O-O interatomic distance is 25.2 pm in (a) and 28.0 pm in (b), and the closest approach of the hydrogen nuclei of the parent OH groups is 12.0 pm in (a) and 16.8 pm in (b).

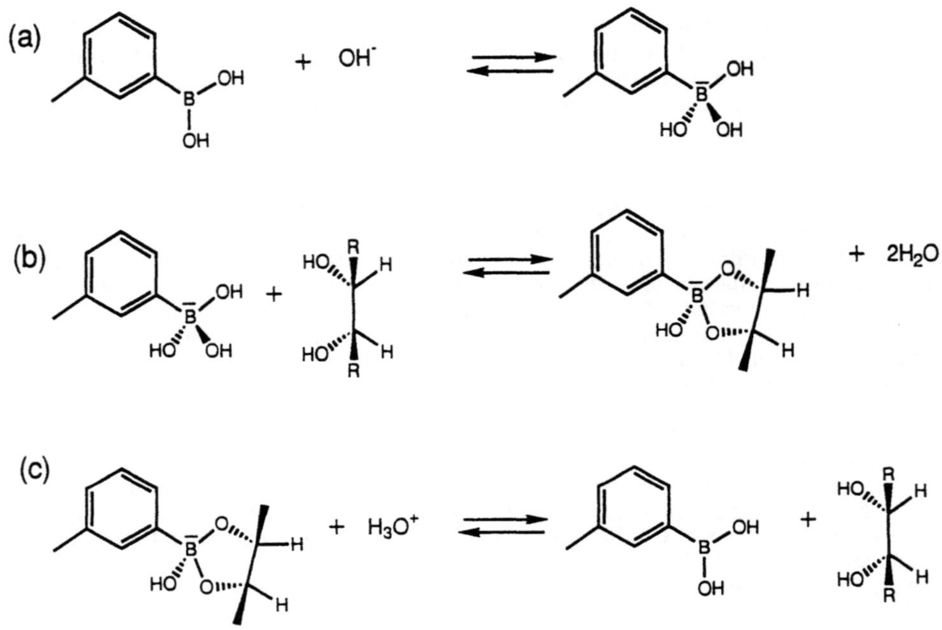

Fig. 2. Binding and elution of a diol-containing compound on PBA showing (a) equilibration, (b) retention, and (c) elution.

MATERIALS AND METHODS

Ecdysteroids (Figure 3) were obtained from the collection of I. D. Wilson or were purchased from Simes (Milan). The Sepralyte® PBA Bond Elut® cartridges were a gift from R. Calverly (Analytichem International).

For loading onto the PBA cartridges the ecdysteroid standards and extracts were dissolved in 2 ml of the appropriate alkaline buffer. The sample was then loaded on to the PBA cartridge, which had been conditioned with 1 ml methanol followed by 5 ml of the same alkaline buffer. The flow of liquid through the cartridge was achieved by applying a gentle vacuum from a water pump, with a flow rate maintained at ~1 ml/min. The PBA cartridge was then eluted with 2 ml fractions of the various buffers and solvent-buffer

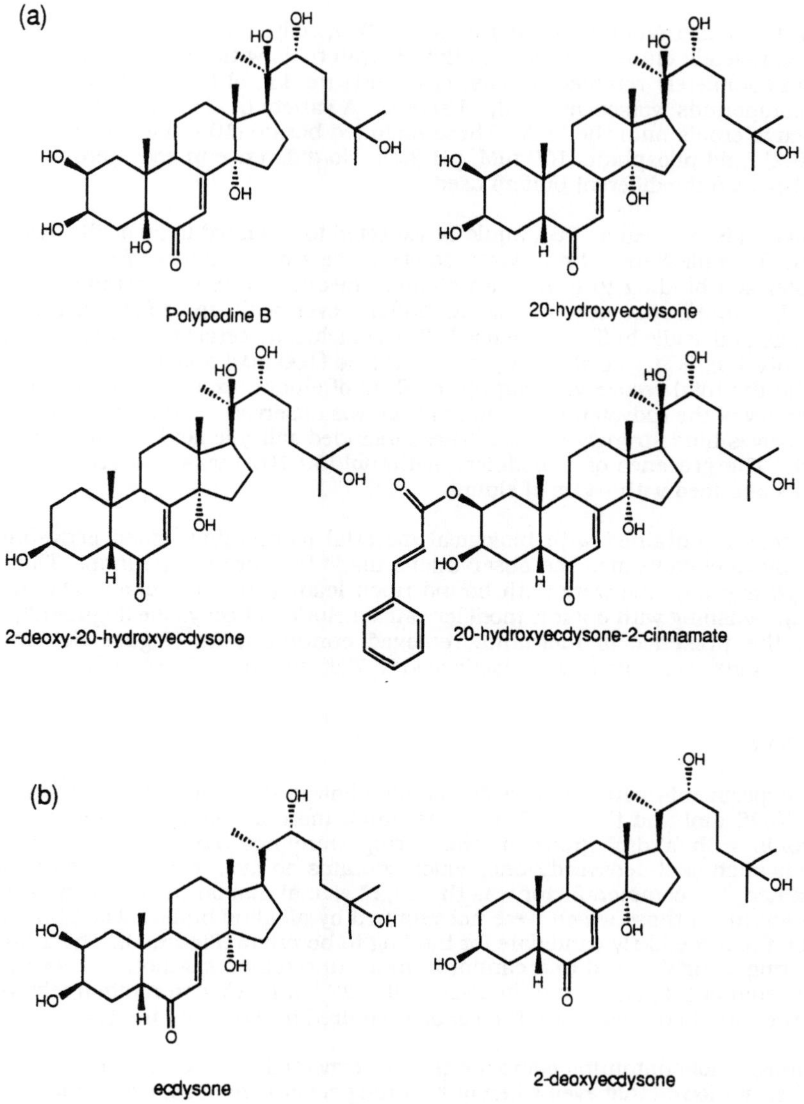

Fig. 3. **Ecdysteroids for which retention on PBA has been investigated (a) specifically retained compounds; (b) non-specifically retained compounds.**

mixtures (Figure 4). The eluant fractions collected were analyzed directly by reversed-phase HPLC on a Pye Unicam liquid chromatography system consisting of a LC3XP pump, a LC3-UV variable-wavelength UV detector set at 242 nm, a gradient elution control system and a Rheodyne 7125 injector. The insect extracts consisted of early 6th instar *Heliothis armigera* larvae (n = 10) with ecdysone and 20-hydroxyecdysone (1 µg each per insect) added. Extraction of the ecdysteroids was carried out by homogenization in water-chloroform (1:1, v/v; after Lafont et al. [6]) followed by centrifugation (3000 rpm for 10 min). The aqueous layer was evaporated. The residue was then extracted in buffer and fractionated on the PBA in the same method as that used for the standard ecdysteroids.

RESULTS

When loading ecdysteroids (structures in Figure 3a,b) on to PBA in alkaline buffer alone, complete retention of all the ecdysteroids studied was observed in most cases. Where some non-retention of ecdysteroids was observed no more than ~10% was eluted. However, upon addition of methanol to the eluant (50 to 70%), ecdysone and 2-deoxyecdysone (Figure 3b) were completely eluted. However, other ecdysteroids with hydroxyls at C-20 and C-22 were again completely retained in most cases (Figure 3a), although some slight elution of retained compounds was occasionally observed. A variety of alkaline buffers were used for loading ecdysteroids onto the PBA. These included borate (100 mM, pH 8.0), glycine (100 mM, pH 8.2), and phosphate (100 mM, pH 8.0). No difference in ecdysteroid retention was observed between the different buffers used.

Compounds retained on PBA would be expected to be eluted in an acidic medium, with a modifier such as methanol, if necessary, to overcome any secondary interactions. However, the ecdysteroids binding to PBA in an alkaline medium were very strongly retained, and could not be eluted by a variety of acidic buffers, even with up to 90% methanol present. Several acids and acidic buffers were used. These included acetate (100 mM, pH 5.0), formate (100 mM, pH 3.0), 50% glacial acetic acid, boric acid (100 mM) and 1% trifluoracetic acid. In no case did the total recovery, using up to 10 ml of eluant, exceed 20%. However, quantitative recovery of the ecdysteroids bound to PBA was observed upon elution with compounds that themselves bind strongly to PBA. These included salicylic acid (25 mM) and lactic acid (3%, w/v). The presence of a modifier, methanol (50-70%) was also required. Complete elution was obtained with ~4 ml of eluant.

The results obtained with biological material (containing added ecdysone and 20-hydroxyecdysone) confirmed the observations made for pure compounds. Thus ecdysone and 20-hydroxyecdysone were both bound upon loading in an aqueous alkaline medium (Figure 5a). Washing with organic modifier caused elution of ecdysone (Figure 5b). An acidic buffer in the presence of methanol removed contaminants (Figure 5c) and the 20-hydroxyecdysone was eluted by 3% lactic acid in 70% methanol (Figure 5d).

DISCUSSION

The experiments with model ecdysteroids showed that the different binding capacity of the C-20,22 diol and the C-2,3 diol was much more marked than had been expected. Ecdysteroids with a diol group in the A-ring were not more strongly retained than compounds such as 2-deoxyecdysone, which contains no diol. For those compounds selectively retained, the common factor was the 20,22 vicinal diol on the ecdysteroid side-chain, a feature absent in those which were not retained by alkaline buffer. The 20,22-diol would seem to be the more likely candidate for binding to boronate (Figure 1). The 2,3-diol on the steroid A ring is rigidly held and cannot form an unstrained *cis*-diol. However, the 20,22 diol on the steroid side-chain and the C-20 and C-22 hydroxyls can rotate freely with respect to each other and hence can form the *cis*-diol required for boronate binding.

Although diol-containing compounds were expected to be eluted from the PBA by an acidic buffer, we found that even when quite strong acids were used, e.g.,formate pH 3.0 or in high concentrations, they remained bound to the column. Indeed, 20-hydroxyecdysone was

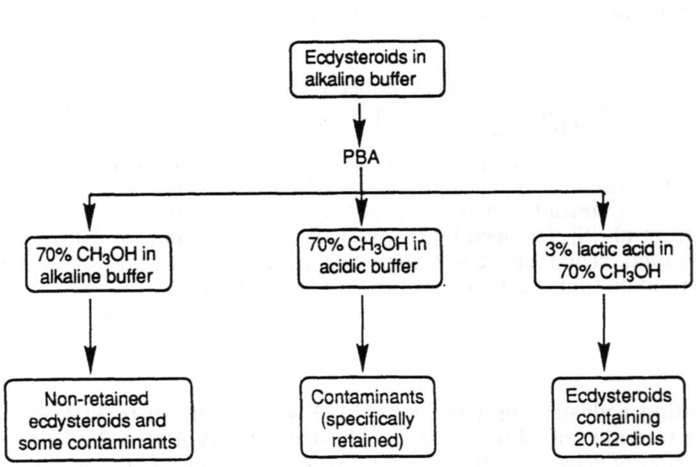

Fig. 4. Extraction procedure for ecdysteroids on PBA.

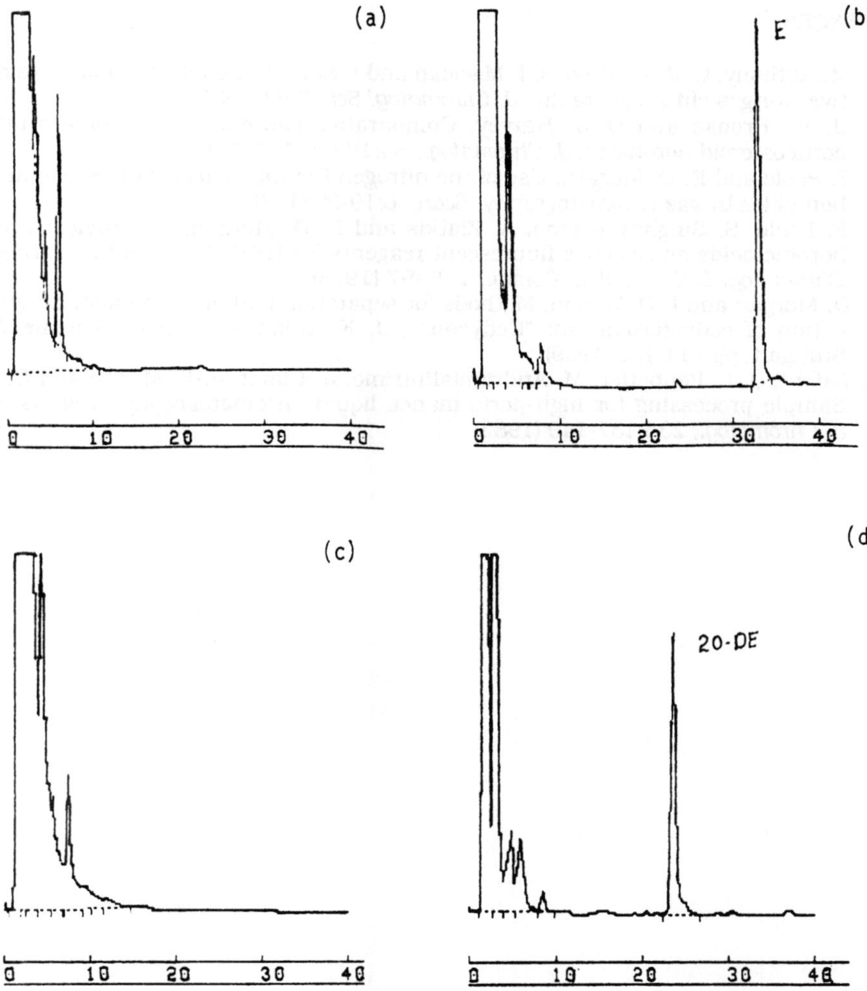

Fig. 5. RP-HPLC traces of the biological extract eluted from PBA (a) sample after loading; (b) after elution with 70% methanol in alkaline buffer; (c) after elution with 70% methanol in acidic buffer; and (d) after elution with 3% lactic acid in 70% methanol.

observed to be strongly bound after loading in an acetate buffer. The secondary interactions of the immobilized PBA must also be taken into account. Some reversed-phase interaction with the PBA is possible, hence the need for an organic modifier such as methanol. The interaction would not appear to be very strong however, as loading of ecdysteroids in a medium containing a small amount of modifier (~10% methanol) resulted in elution of a significant amount (~20%) of ecdysteroid. Another secondary interaction, not encountered here but which might affect the handling of biological material, is due to the amine linking the silica and the PBA. This protonates upon acidification and may cause retention of anionic species such as phosphates and carboxylic acids.

We have showed that, for a variety of neutral ecdysteroid species, separation on the basis of whether a molecule contains a 20,22 vicinal diol or not, is possible and that this can be extended to biological materials. The method could therefore be used as a simple, rapid, and selective means of concentration of 20-hydroxyecdysone, and similar substances.

Where a retained compound, such as 20-hydroxyecdysone, is found as a conjugate (e.g., phosphate or acyl ester), this method should quickly establish whether the conjugation is via the nucleus (usually C-2 or C-3) or the side-chain (usually C-22), depending upon whether or not the conjugate itself is retained.

REFERENCES

1. G. M. Anthony, C. J. W. Brooks, I. Maclean and I. Sangster, Cyclic boronates as derivatives for gas chromatography, *J. Chromatog. Sci.*, 7:623-631 (1969).
2. C. J. W. Brooks and D. J. Harvey, Comparative gas chromatographic studies of corticosteroid boronates, *J. Chromatog.*, 54:193-204 (1971).
3. C. F. Poole and E. D. Morgan, Use of the nitrogen thermal detector to determine sterol boronates by gas chromatography, *Scan*, 6:19-20 (1975).
4. C. F. Poole, S. Singhawangcha, A. Zlatkis and E. D. Morgan, Polynuclear aromatic boronic acids as selective fluorescent reagents for HPTLC and HPLC, *J. High Res. Chromatog. & Chromatog. Comm.*, 1:96-97 (1978).
5. E. D. Morgan and I. D. Wilson, Methods for separation and physico-chemical quantification of ecdysteroids, *in*: "Ecdysone", J. Koolman, ed., Georg Thieme-Verlag, Stuttgart, pp 114-130 (1989).
6. R. Lafont, J.-L. Pennetier, M. Andrianjafintrimo, J. Claret, J.-F. Modde and C. Blais, Sample processing for high-performance liquid chromatography of ecdysteroids, *J. Chromatog.*, 236:137-149 (1982).

PEPTIDES

ISOLATION AND CHARACTERIZATION OF A DIURETIC PEPTIDE, AP-I, FROM THE CORPORA CARDIACA OF THE HOUSE CRICKET, *ACHETA DOMESTICUS*

G. M. Coast, C. H. Wheeler, N. F. Totty*,
R. J. Philp* and G. J. Goldsworthy

Department of Biology, Birkbeck College
Malet Street, London WC1E 7HX, UK, and
*The Ludwig Institute for Cancer Research
The Courtauld Building, Riding House Lane
London WC1, UK

SUMMARY

Fractionation of extracts of the corpora cardiaca of the house cricket, *Acheta domesticus*, by reversed-phase high performance liquid chromatography (RP-HPLC) on octylsilica column packings, revealed a number of peptides which had diuretic activity. One of these peptides, AP-I, was associated with a major peak of UV absorbance. AP-I has been isolated using RP-HPLC, and the purity confirmed using a range of chromatographic procedures including: elution in different solvent systems; the use of octadecyl and phenyl columns; and separation on size-exclusion and ion-exchange columns. In all instances diuretic activity was associated with a peak of UV absorbance corresponding to AP-I. After treatment with pyroglutamate aminopeptidase, AP-I was sequenced by gas-phase Edman-degradation, and determined as <QRDIFHAQTDIFQVPK-OH. This sequence has been confirmed by peptide mapping, and the proposed molecular weight of 1924.98 conforms with that obtained by fast atom bombardment mass-spectrometry (FAB-MS). A peptide with this sequence has been synthesized using FMOC chemistry, and its molecular weight confirmed by FAB-MS. The synthetic peptide co-elutes with AP-I on a Lichrosorb RP8 column, but it does not stimulate fluid secretion in *Acheta*. Possible reasons for this lack of activity are discussed.

INTRODUCTION

Advances in the techniques of high-performance liquid chromatography (HPLC), automated Edman-sequencing, amino acid analysis, and fast atom bombardment mass spectrometry (FAB-MS), permit the isolation and sequencing of picomolar quantities of peptide material. These developments, and their growing usage, have been reflected by a rapid growth in the number of insect neuropeptides that have been fully characterized. In 1985 Mordue and Morgan referred to just two insect neuropeptides for which full sequence data was available [1]. That number is now over 40, and is still increasing rapidly. Although diuretic peptides have been described in many different species of insects [2,3], to date only one from *Locusta migratoria* [4] and one from *Manduca sexta* [5] have been sequenced fully.

Why is so little sequence data available for diuretic peptides? The lack of a quick and reliable assay has made the process of isolating diuretic peptides particularly slow and laborious [1]. Several methods, which do not rely upon the traditional Ramsay [6] technique for isolating Malpighian tubules *in vitro*, have been developed to circumvent this problem [7]. Significantly, the first diuretic peptide to be fully characterized, an arginine vasopressin-like diuretic peptide (AVP-LDP) from *Locusta*, was monitored during isolation using a radioimmunoassay (RIA) to vertebrate arginine vasopressin [4,8]. Not surprisingly, the primary structure of AVP-LDP bears a striking resemblance to that of arginine vasotocin (AVT) and arginine vasopressin [4].

Chromatography and Isolation of Insect Hormones and Pheromones
Edited by A. R. McCaffery and I. D. Wilson
Plenum Press, New York, 1990

Diuretic peptides have been described as being extremely labile, and as having a propensity for retention on chromatography columns and sticking to the walls of glass and plastic vessels [1,2]. Such problems are exacerbated by the fact that diuretic peptides appear to be present in small quantities. The amount of AVP-LDP stored in the suboesophageal ganglia of an individual locust is estimated to be less than 60 fmol [9] and some 13,000 suboesophageal and 38,000 thoracic ganglia were used as the starting material for the successful isolation and characterization of this peptide [8]. A diuretic peptide (DP-1) from the corpora cardiaca of *Locusta*, for which some partial sequence data is available [10], is estimated to be present at ca. 2 pmol/gland pair, and in *Manduca* 10,000 trimmed heads yielded only 5 nmol of pure diuretic peptide [5]. In marked contrast, peptides of the AKH family are present in relatively large amounts, typically several hundred pmol per individual. In the Lubber grasshopper, *Romalea microptera*, a single pair of corpora cardiaca contain several nanomoles of *Romalea* AKH-I (G. Gäde and J. H. Spring, personal communication), more than sufficient to yield a full sequence analysis! However, not all insect peptides that have been sequenced are present in such large amounts. For example, the leucokinins are present at < 1 pmol/head (Holman et al., this volume), but these peptides are quite small (< 1500 MW) and appear to be relatively stable compared with diuretic peptides, for which molecular weight estimates range up to 7000 [7].

In our laboratory, we have developed a rapid bioassay for diuretic activity [11] using a simple modification of the original Ramsay method [6]. With this assay we have identified a number of diuretic peptides in fractions from the separation of extracts of the corpora cardiaca of *Acheta domesticus* by HPLC, and some of these appear to be widespread in the central nervous system [12]. In some instances we were able to correlate peaks of UV absorbance with diuretic activity, and here we describe the isolation, sequencing and synthesis of one of these peptides, *Acheta* peptide-I (AP-I).

MATERIALS AND METHODS

Crickets were reared as previously described [11,13]. They were fed a diet of turkey starter crumbs, and water was provided *ad libitum*. Corpora cardiaca were removed from adults (> 5 days old) of either sex. Pairs of glands were dissected into 150 μl of ice-cold distilled water in batches of 100. Peptides were released by sonication at 0°C. Sonicated extracts were heated at 100°C for 5 min, to denature proteins, and then centrifuged at 12,000 g for 2 min. The supernatant was fractionated directly by injection onto HPLC columns, normally within 24 to 48 h [7,12].

Reversed-phase (RP)-HPLC was carried out using a variety of columns of differing selectivities. Initial separations of gland extracts were normally on a Lichrosorb RP8 (Merck) column (250 x 4 mm). Thereafter, additional purification was carried out using Lichrosorb RP18 (Merck; 250 x 4 mm), Phenyl (250 x 4 mm), or Aquapore RP300 (Brownlee Laboratories; 100 x 2.1 mm) columns. All reversed-phase separations were performed using linear gradients employing acetonitrile and water, together with ammonium acetate or trifluoroacetic acid (TFA) as mobile phase modifiers (see Table 1 for details). The eluant was monitored routinely by recording its absorbance at either 214 nm or 215 nm.

In addition, extracts of corpora cardiaca were separated by high performance size-exclusion (HP-SEC), and ion-exchange (HP-IEC) chromatography. HP-SEC was performed on a G2000SW (TSK) column (300 x 7.5 mm) eluted at 0.4 ml/min with 15% acetonitrile in 25 mM ammonium acetate buffer. The eluant was monitored at 215 nm. HP-IEC was performed using a DEAE-5PW (TSK) anion-exchange column (75 x 7.5 mm) eluted at 1ml/min with a linear gradient of 0 to 30% B (solvent A: 0.1 M ammonium carbonate, pH 9.0; solvent B: 1 M ammonium formate, pH 5.0). The eluant was monitored at 254 nm.

Table 1. Compositions of solvents used in reversed-phase separations

Solvent system 1:	Solvent A: 95% water, 5% acetonitrile, 0.11% TFA Solvent B: 40% water, 60% acetonitrile, 0.1% TFA
Solvent system 2:	Solvent A: 100% water, 10 mM TFA Solvent B: 95% acetonitrile, 5% water, 10 mM TFA
Solvent system 3:	Solvent A: 25 mM ammonium acetate, 100% water Solvent B: 25 mM ammonium acetate, 60% acetonitrile, 40% water

At all stages during the isolation of AP-I, diuretic activity was assessed by measuring the stimulation of fluid secretion by single isolated Malpighian tubules *in vitro*. This assay has been described in detail elsewhere [11]. Material to be assayed was taken to dryness at 50°C by centrifugal evaporation (Hetovac), and resuspended in an appropriate volume of *Acheta* Ringer. Fluid secretion rates were determined before and after the addition of assay material. Diuretic activity is expressed as the increase in rate of fluid secretion per unit length of tubule (pl/mm/min).

Reduction and alkylation of peptide material was performed as described by Tarr [14] using dithiothreitol and 4-vinyl pyridine. Products of this reaction were fractionated on an Aquapore RP300 column (solvent system 2). Absorbance was monitored at 214nm and 260 nm. The advantage of using 4-vinyl pyridine as an alkylating agent was that it reacted specifically with free cystyl residues to yield a product which absorbed strongly at 260 nm and was readily detected.

Peptide mapping by enzymic digestion was carried out using sequence grade enzymes (Boehringer-Mannheim; Miles). The ratio of enzyme to peptide was nominally 1:100 (w/w) and digestion was allowed to proceed overnight at room temperature unless otherwise stated. Incubation conditions were as recommended by the manufacturer. The reactions were terminated, and products separated, by injection directly onto an Aquapore RP300 column followed by elution in solvent system 2.

Sequence analysis, by gas phase Edman-degradation, was performed on an Applied Biosystems Peptide Sequencer (Model 477A) linked to an on-line PTH amino acid analyser (Applied Biosystems Model 120A) using the manufacturer's recommended protocols. For total amino acid analysis, purified AP-I was hydrolyzed using 6M hydrochloric acid and 1% phenol in the vapor phase at 165°C for 40 min after Dupont et al. [15]. Amino acid analysis was carried out using automated derivatization (Applied Biosystems 420A Derivatizer) followed by on-line PTH amino acid analysis (Applied Biosystems Model 130A Separation System).

Capillary zonal electrophoresis (CZE) of the purified peptide was performed on an Applied Biosystems Model 270A CZE. The separation was performed at 20 kV in citrate buffer pH 2.5. Absorbance was monitored at 200 nm and Dynorphin A 1-16 (Porcine) was included as an internal standard.

Peptide synthesis was performed on an Applied Biosystems Peptide Synthesizer (Model 430A) employing FMOC chemistry according to the manufacturer's recommended protocols.

RESULTS

Isolation and Purification of AP-I

Aqueous extracts of corpora cardiaca were injected onto a Lichrosorb RP8 column and eluted over 60 min at 1 ml/min in a linear gradient of 0 to 100% B (solvent system 1). Fractions were collected at 1 min intervals. The UV absorbance profile is shown in Figure 1, together with assay data relating to the diuretic activity in the fractions. As shown previously, diuretic activity was spread in many fractions [12]. Such activity was not resolved, but peaks of activity could be detected, superimposed upon this. One such peak was found consistently in fractions 34 and 35, and this was coincident with a major peak of UV absorbance, which was designated *Acheta* peptide-I (AP-I).

A separation of an aqueous extract from 50 pairs of corpora cardiaca on an Aquapore RP300 column eluted using solvent system 2 over 40 min with a linear gradient of 0 to 60% B is shown in Figure 2. The absorbance of the eluant was monitored simultaneously at 214 nm and 280 nm and fractions were collected at 0.5 min intervals and assayed for diuretic activity. Several peaks of diuretic activity were identified which were better resolved than on the Lichrosorb RP8 column. The diuretic activity present in fractions 38 to 40 was coincident with a major peak of UV absorbance at 214 nm which corresponded to AP-I identified on the Lichrosorb RP8 column. There was no corresponding absorbance peak at 280 nm, suggesting that neither tryptophan nor tyrosine were present in this material.

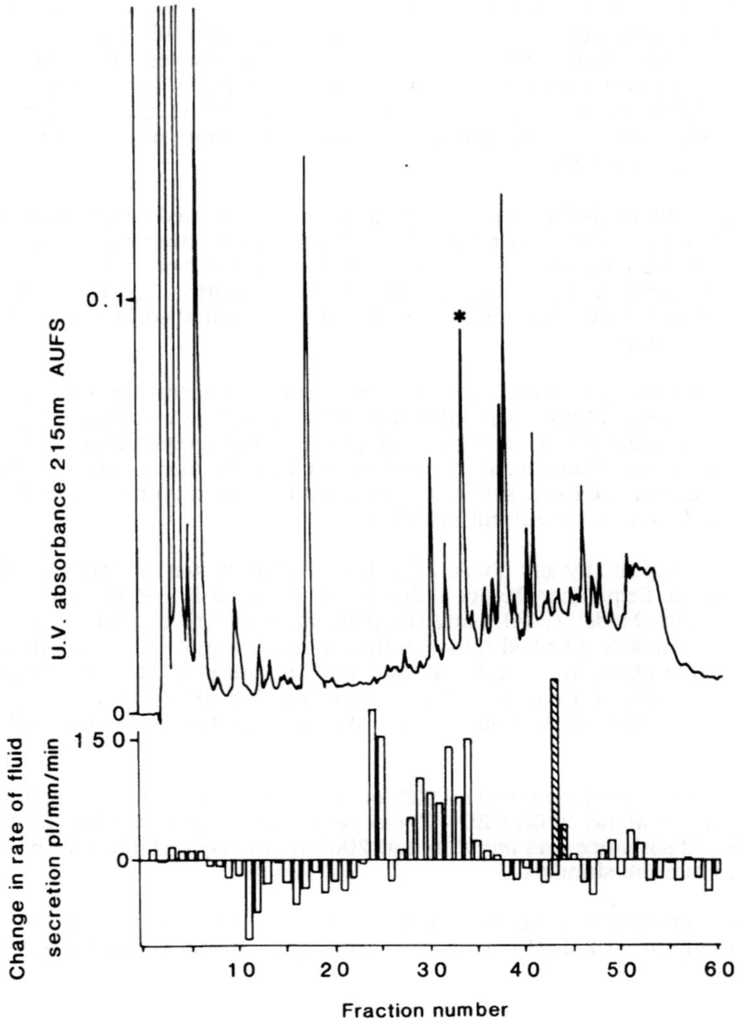

Fig. 1. Fractionation of an aqueous extract of 100 gland pairs of corpora cardiaca from *Acheta* on a Lichrosorb RP8 column (for chromatographic details see text). Histograms show the diuretic activity of individual fractions assayed on isolated Malpighian tubules. Hatched bars are data obtained from a different chromatographic separation. This latter activity appears to be quite labile and is not detected in all separations. The absorbance peak corresponding to AP-I is marked (*).

Fractions 34 and 35 from a separation of 900 pairs of corpora cardiaca on the Lichrosorb RP8 column were pooled, and rechromatographed on an Aquapore RP300 column eluted using a shallower gradient (0 to 60% B over 50 min; solvent system 2). A single major UV-absorbing peak was present (Figure 3) which corresponded to AP-I identified after fractionation of crude gland extracts on this column (see Figure 2). AP-I was quantified by amino acid analysis of an acid hydrolysate of an aliquot of the purified peptide (see below). A dose response curve for the diuretic activity of this purified material is shown in Figure 4.

For sequence analysis and peptide mapping, AP-I was isolated routinely on the Lichrosorb RP8 column, and further purified on the Aquapore RP300 column. To confirm the purity of AP-I, the material in the peak (Li8-AP-I) derived from fractionation on the Lichrosorb RP8 column was re-chromatographed using RP-HPLC systems with differing selectivities.

Li8-AP-I material was separated on the same column, but using acetonitrile with ammonium acetate as the mobile phase modifier at pH 6.75, with a linear gradient of 0 to

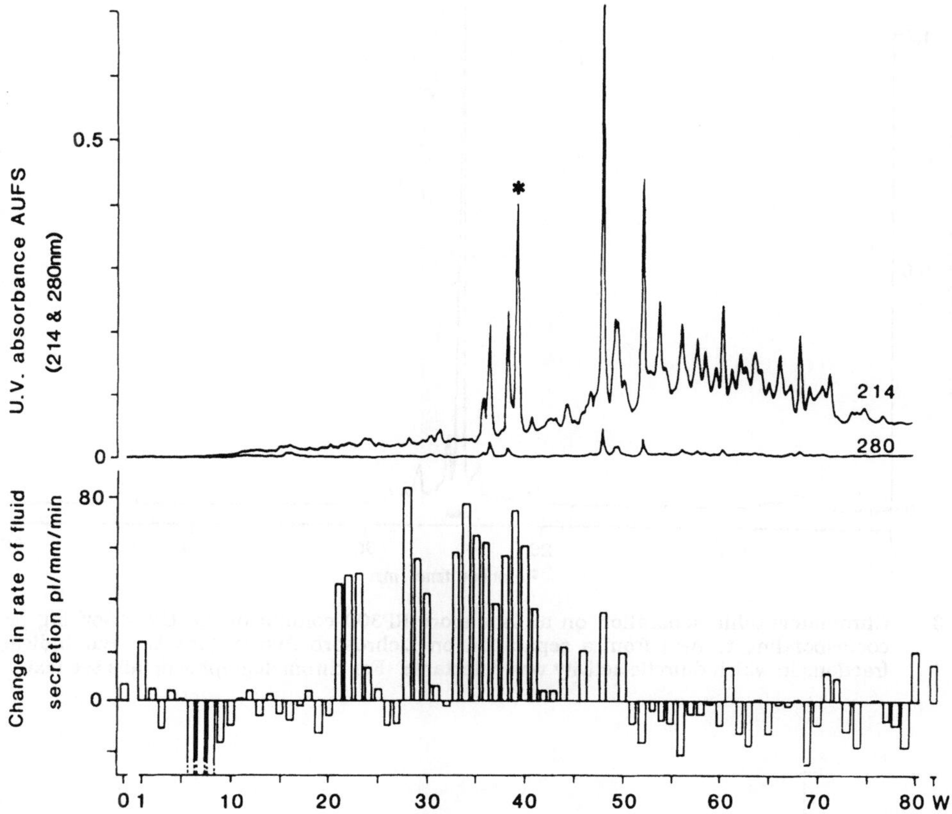

Fig. 2. Fractionation of an aqueous extract of 100 gland pairs of corpora cardiaca from *Acheta* on an Aquapore RP300 column. Elution was monitored simultaneously at 280 and 214 nm. Histograms slow the diuretic activity of individual fractions assayed on isolated Malpighian tubules. The absorbance peak corresponding to AP-I is marked (*). Note the lack of absorbance of the AP-I peak at 280 nm.

100% B (solvent system 3). Some minor peaks of UV-absorbing material were resolved, but diuretic activity remained associated with the main UV-absorbing peak corresponding to AP-I.

Chromatography on octyl columns at either pH 2.1 or 6.75 did not radically change the chemistry of the separation, which was dependent upon hydrophobic interactions between the peptide and the octylsilica packing. Separation on phenyl columns, although also dependent upon such hydrophobic interactions, is influenced considerably by the number and location of aromatic groups within peptides. Li8-AP-I was injected onto a phenyl column and eluted in a linear gradient of 0 to 100% B over 60 min in solvent system 1. This yielded a single major UV absorbing peak which contained diuretic activity (Figure 5).

The chromagraphy so far described employed RP-HPLC for both the initial isolation, and for the final purification step. An alternative approach was the use of HP-IEC or HP-SEC for the initial isolation, followed by purification by RP-HPLC. Our observations, from the separation of aqueous corpora cardiaca extracts, were that the recovery of diuretic activity and the resolution of peaks, were superior by HP-IEC (Figure 6) compared with HP-SEC (Figure 7). Assay of column fractions after HP-IEC revealed a number of peaks which contained diuretic activity, and these were re-chromatographed on the Lichrosorb RP8 column (eluted with solvent system 1; Figure 8). Two major UV-absorbing peaks were present in fractions 7 and 8 from the anion-exchange column. Fractions corresponding to these peaks had diuretic activity, and the more hydrophobic peak had the same column retention time as AP-I. The peak corresponding to AP-I (Figure 8) was then re-chromatographed on an octadecyl column (Lichrosorb RP18; solvent system 1). Previous experience had shown that the recovery of diuretic activity from this column was poor,

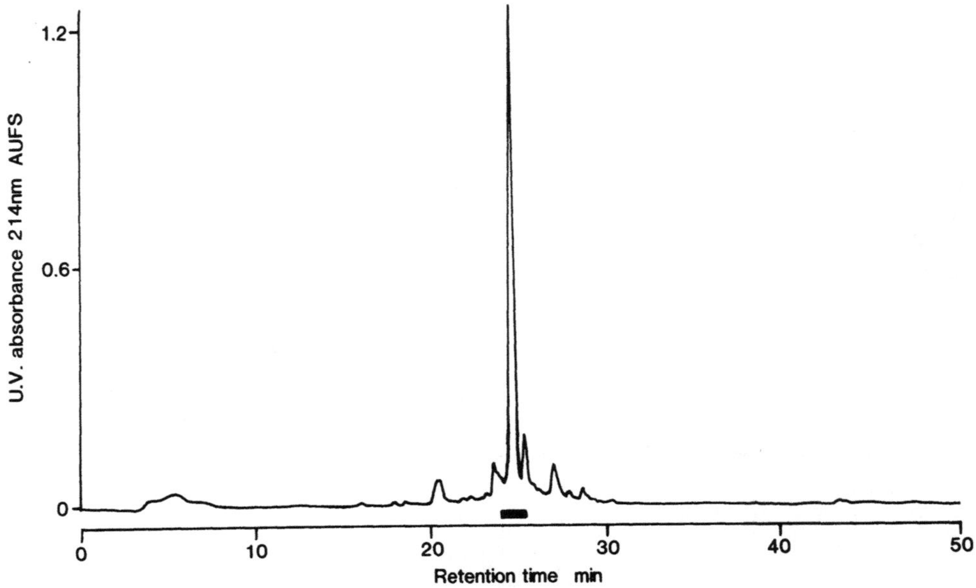

Fig. 3. Chromatographic separation on an Aquapore RP300 column of the UV-absorbing peak corresponding to AP-I from a separation on Lichrosorb RP8 (Figure 1). Bar indicates fractions in which diuretic activity was detectable. For chromatographic details see text.

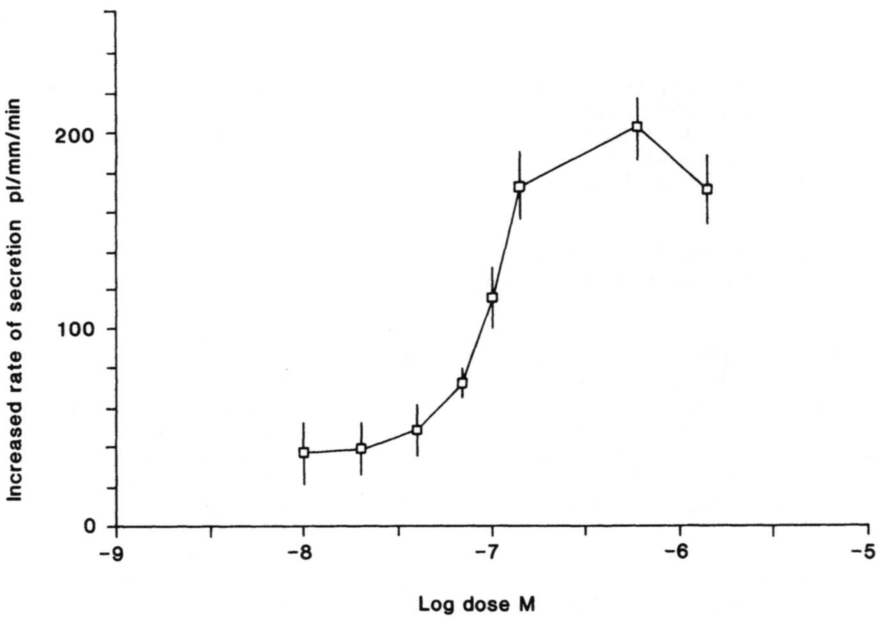

Fig. 4. Shows the dose response curve for natural AP-I purified as in Figure 3 when assayed for diuretic activity on isolated Malpighian tubules.

however, a single major UV-absorbing peak was obtained, and this contained diuretic activity.

A TSK G2000SW size-exclusion column was used to isolate diuretic peptides from methanolic extracts of cricket corpora cardiaca. Such extracts were prepared as described

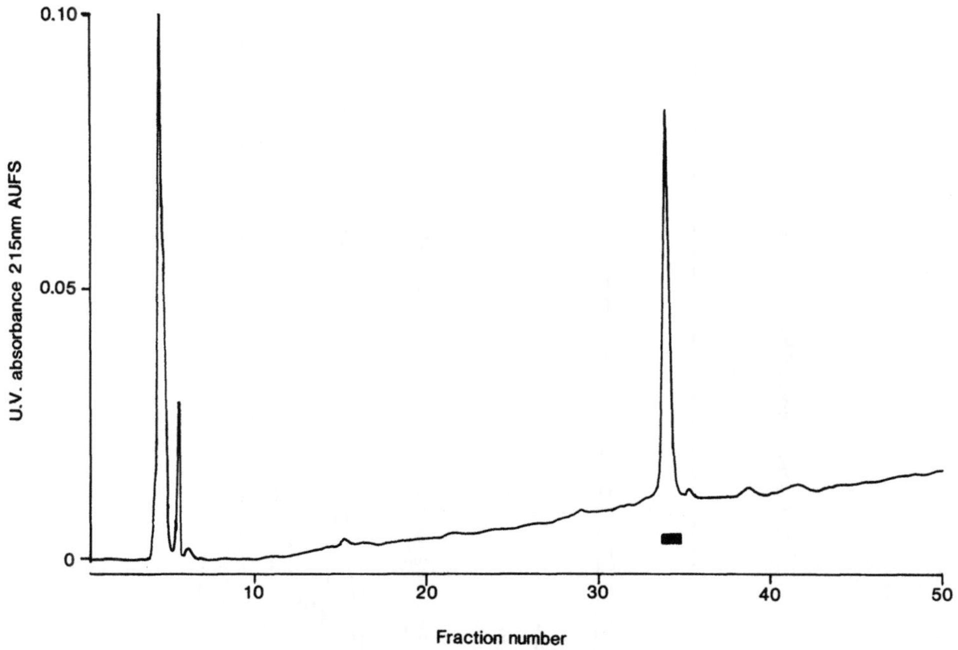

Fig. 5. Chromatographic separation of the UV-absorbing peak corresponding to AP-I from a separation on Lichrosorb RP8 (Figure 1) on a phenyl column. Bar indicates fractions in which diuretic activity was detected. For chromatographic details see text.

by Morgan and Mordue [16]. Peaks of UV-absorbing material were collected by hand, but were poorly resolved, and diuretic activity was widespread (Figure 7). Individual fractions from HP-SEC were re-chromatographed on the Lichrosorb RP8 column (solvent system 1). Fraction 3, from the HP-SEC separation, contained material having a large UV-absorbing peak with the same column retention time as AP-I and which had diuretic activity (Figure 9).

Sequence Analysis and Peptide Mapping

As a first step towards sequence analysis, a sample of AP-I (equivalent to 900 glands) after fractionation on Lichrosorb RP8 and Aquapore RP300 columns, was subjected to reduction and alkylation. After treatment with dithiothreitol and 4-vinyl pyridine, the reaction mixture was injected onto an Aquapore RP300 column which was eluted isocractically at 100% A for 10 to 15 min until non-peptide reactants had eluted and the baseline returned to zero. The gradient (0 to 100% B; solvent system 2) was then developed and the peptide eluted. Absorbance was monitored simultaneously at 214 and 260 nm. A single major peak of 214 nm-absorbing material was identified. There was no corresponding absorbance peak at 260 nm. Furthermore, the retention time of this material on the Aquapore RP300 column was identical to that of the untreated peptide and the material retained its diuretic activity. Reduction and alkylation of two further batches of natural AP-I confirmed these findings, and later samples of AP-I were not subjected to this treatment.

An aliquot of AP-I (equivalent to 100 gland pairs), that had been subjected to reduction and alkylation, was taken for N-terminal sequence analysis. No sequence data were obtained. Because the Edman degradation requires a free amino terminal, the presence of a blocking group was suspected. Diuretic activity of crude gland extracts was significantly reduced after incubation with pyroglutamate amino peptidase [17], indicating that some diuretic peptides may be blocked by a pyroglutamyl residue. A sample of the peptide was incubated (100 mM potassium phosphate; 5 mM EDTA; 10 mM DTT; 5% glycerol; 32°C) with pyroglutamate amino peptidase (Boehringer), and the reaction was stopped after 120 min by injection onto the Aquapore RP300 column and eluted in solvent system 2 (see Figure 11). Peaks of absorbance corresponding to the enzyme and impurities were readily identified by comparison with the UV-absorbance profile from a control incubation in the absence of

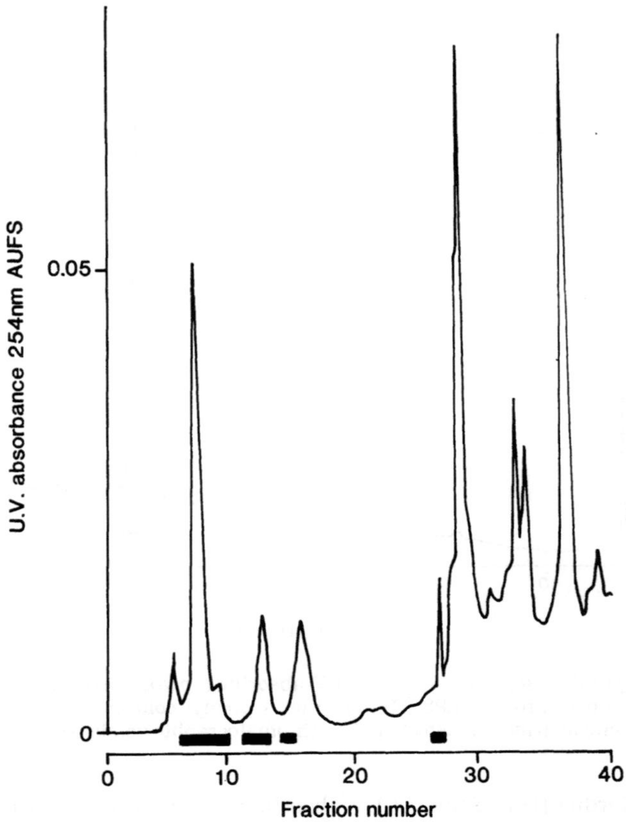

Fig. 6. Fractionation of an aqueous extract of 100 gland pairs of corpora cardiaca on a TSK DEAE-5PW anion exchange column (for chromatographic details see text). Bars indicate fractions containing diuretic activity as shown by assay on isolated Malpighian tubules. Fractions which contained diuretic activity were re-fractionated on a Lichrosorb RP8 column (see Figure 8).

AP-I. The peak corresponding to the native peptide disappeared, but two different UV-absorbing peaks were present, one more polar and the other less polar than the original material. Fractions corresponding to these peaks had no diuretic activity. Each peak was taken for sequence analysis. The more polar material (P1) gave the sequence RDIFHAQTDIFQVPK. The signal yield for the C-terminal lysine (K) residue of peak 1 was much reduced in comparison with the preceding signal for proline (P). The more hydrophobic material, P2, yielded the corresponding sequence RDIFHAQTDIFQVP, but no terminal lysine residue was detected.

To obtain further sequence information, particularly with reference to the C-terminal, peptide mapping of purified AP-I was undertaken using sequence-grade enzymes. The products of the reactions were isolated using the Aquapore RP300 column (solvent system 2).

Digestion of AP-I with TPCK-trypsin in the presence of 0.1 M ammonium acetate (pH 8.5) yielded a major UV-absorbing peak of material which was more polar than the native peptide. A distinct shoulder on the leading edge of this peak was not fully resolved, but was sequenced together with the main peak which yielded the sequence RDIFHAQTDIFQVP. Digestion of AP-I with TLCK-chymotrypsin at pH 7.2 produced three products having major peaks of UV-absorbance. Of these products only one peak could be sequenced which yielded HAQTDIF.

The native peptide was also treated with *Staphylococcus aureus* endoprotease (V8) in phosphate buffer (50 mM) at pH 7.8. Two major products, giving UV-absorbing peaks which were not well resolved, were obtained. The more hydrophobic peak yielded the sequence IFQVPK. Material from the more polar peak did not sequence initially, but after treatment with pyroglutamate amino peptidase gave the sequence DIFHAQT.

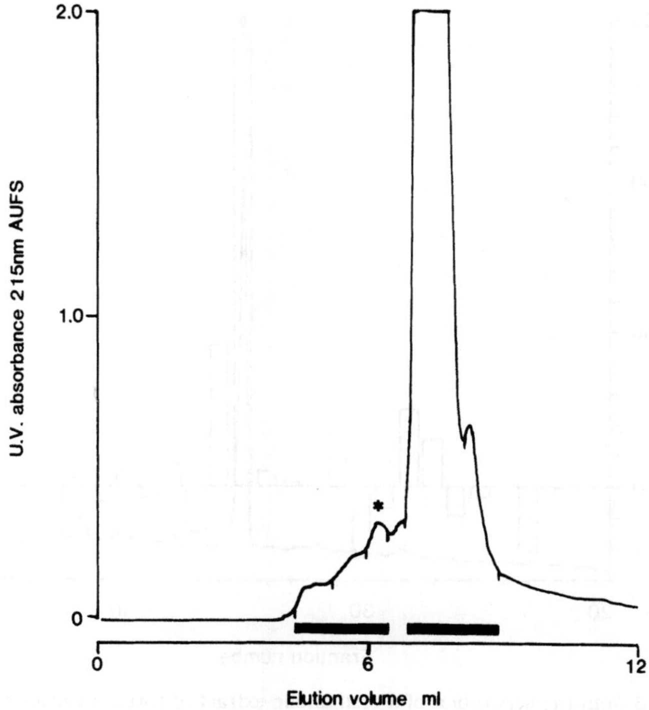

Fig. 7. Fractionation of a methanolic extract of 100 gland pairs of corpora cardiaca on a TSK 2000SW size exclusion column (for chromatographic details see text). Bars indicate fractions containing diuretic activity as shown by assay on isolated Malpighian tubules. Fractions containing diuretic activity were re-fractionated on a Lichrosorb RP8 column (see Figure 9). Fraction 3, which contained AP-I, is marked (*).

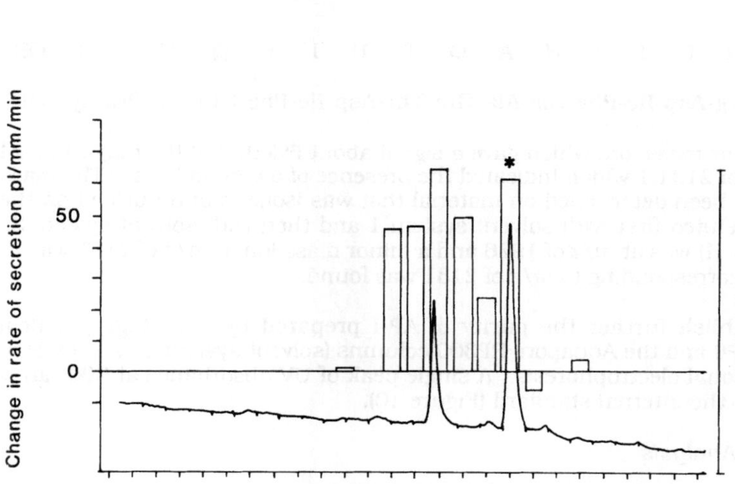

Fig. 8. Fractions 7 and 8 from the separation of an aqueous extract of corpora cardiaca by ion exchange chromatography (Figure 6), re-fractionated on a Lichrosorb RP8 column. Histograms show the diuretic activity of individual fractions assayed on isolated Malpighian tubules. The absorbance peak corresponding to AP-I is marked (*).

A sample of AP-I (200 pmol) was analyzed by positive ion FAB-MS. Two distinct mass ions (M + H) were obtained. The major peak gave an average m/z of 1926.8 which, after consideration of the isotopic ions present, corresponded to a mass of 1924.99 and conforms to the molecular weight of AP-I (1924.98) as calculated from the sequence:

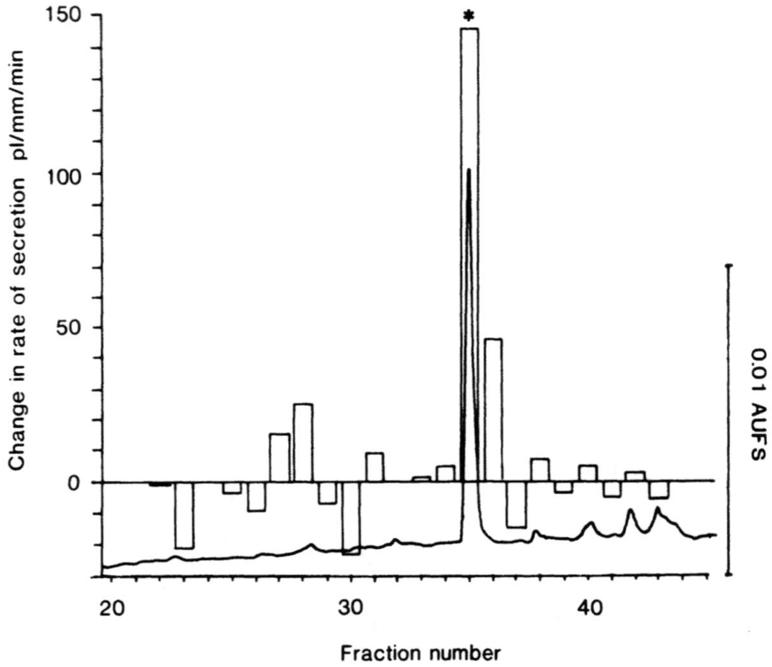

Fig. 9. Fraction 3 from the separation of a methanolic extract of corpora cardiaca by size exclusion chromatography (Figure 7), re-fractionated on a Lichrosorb RP8 column. Histograms show the diuretic activity of individual fractions assayed on isolated Malpighian tubules. The absorbance peak corresponding to AP-I is marked (*).

<Q R D I F H A Q T D I F Q V P K OH

pGlu-Arg-Asp-Ile-Phe-His-Ala-Gln-Thr-Asp-Ile-Phe-Gln-Val-Pro-Lys-OH

A second minor mass ion, which gave a signal about 5% that of the major mass ion, gave an average m/z of 2131.1 which indicated the presence of a contaminant. The mass of AP-I has subsequently been determined on material that was isolated and purified on the Lichrosorb RP8 column eluted first with solvent system 1 and then with solvent system 3. The major mass ion (M + H) was at m/z of 1926 and a minor mass ion at m/z of 2137 was also detected. No mass ion corresponding to m/z of 2131 was found.

To establish further the purity of AP-I prepared by a 2 stage purification on the Lichrosorb RP8 and the Aquapore RP300 columns (solvent system 1), a sample was analyzed by capillary zonal electrophoresis. A single peak of UV absorbance at 200 nm was detected in addition to the internal standard (Figure 10).

Amino Acid Analysis

The results of an amino acid analysis of a hydrolysate of AP-I (1 nmol) are presented in Table 2. Only those amino acids listed were detectable above background. The results agree well with the composition predicted from the sequenced peptide. Minute quantities of tyrosine and leucine were present as very minor contaminants of the purified peptide.

Synthesis of [pGlu]1--[Lys]16

Sequence and mass analysis of AP-I suggested that the entire sequence comprised pGlu-Arg-Asp-Ile-Phe-His-Ala-Gln-Thr-Asp-Ile-Phe-Gln-Val-Pro-Lys-OH ([pGlu]1..[Lys]16-OH). In an attempt to confirm this sequence, synthesis of both the free carboxyl, [pGlu]1..[Lys]16-OH, and the amidated form of the peptide, [pGlu]1..[Lys]16-NH_2, was performed using FMOC-chemistry. The deletion product [Arg]2..[Lys]16-OH from the carboyxl synthesis was obtained by removal and cleavage of the resin before coupling of the

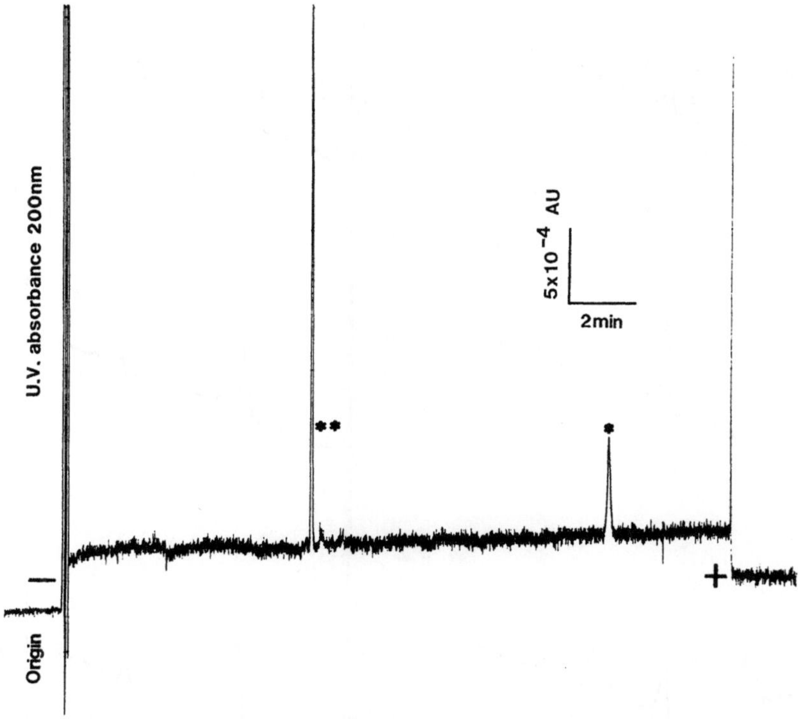

Fig. 10. Electrophoretogram from the separation of purified AP-I by capillary zonal electrophoresis (see text for details). Dynorphin 1-16 (**) was included as an internal standard. The single UV absorbing peak, assumed to correspond to AP-I, is marked (*).

N-terminal pyroglutamic acid. Synthetic peptides were purified by RP-HPLC on an Aquapore RP300 column (1 x 20 cm) eluted in solvent system 2 (10 to 50% B over 30 min at 5 ml/min). The primary structures of the synthetic peptides were validated by sequence analysis of the deletion product, [Arg]2..[Lys]16-OH, and by mass determination of the intact peptides by FAB-MS (School of Pharmacy, University of London).

The synthetic peptides [pGlu]1..[Lys]16-OH and [pGlu]1..[Lys]16-NH2 could readily be separated by RP-HPLC. Natural AP-I (nAP-I) and [pGlu]1..[Lys]16-OH co-eluted in a linear gradient of 20 to 40% B over 60 min (solvent system 1) on the Lichrosorb RP8 column. Both forms of the peptide were assayed at doses of 1×10^{-10} to 1×10^{-6} M on isolated Malpighian tubules, but a diuretic response could not be demonstrated for either peptide.

Table 2. Amino acid analysis of purified natural AP-I

Amino acid	Amount pmol	Ratio to valine*	Expected composition
Aspartic	2281.12	2.25	2
Glutamic	3390.16	3.34	3
Histidine	937.25	0.92	1
Arginine	1041.30	1.03	1
Threonine	909.94	0.9	1
Alanine	1106.09	1.09	1
Proline	1060.09	1.04	1
Tyrosine	66.02	0.06	0
Valine*	1015.09	1.00	1
Isoleucine	1897.82	1.89	2
Leucine	5.89	0.006	0
Phenylalanine	1865.62	1.84	2
Lysine	1079.00	1.06	1

Note the low levels of contaminating amino acids detected

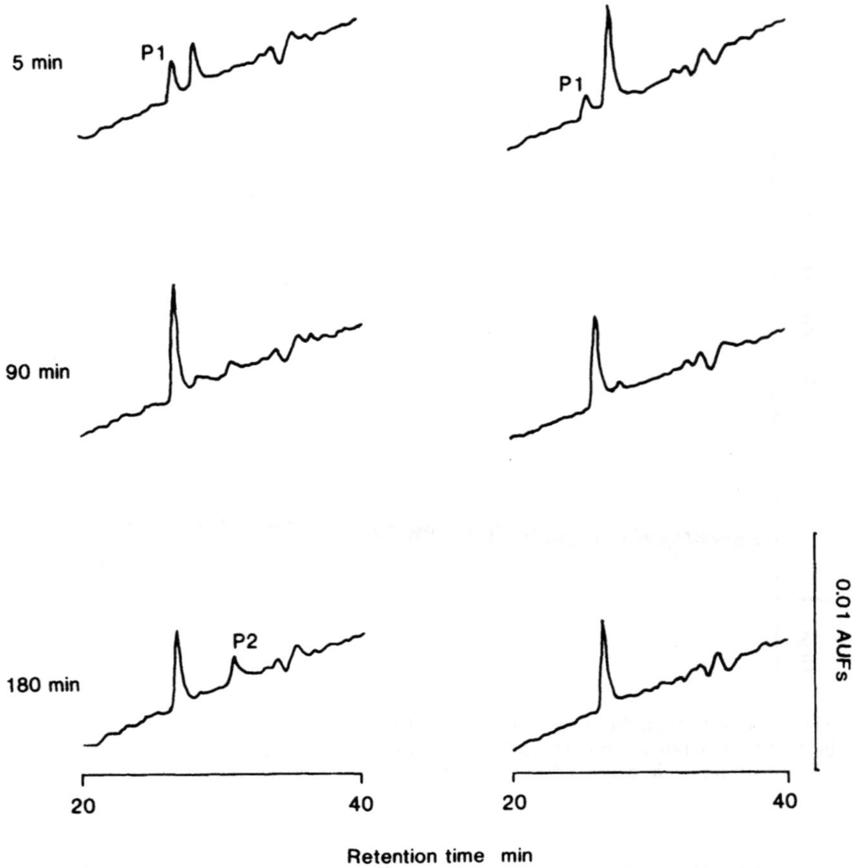

Fig. 11. Shows the time course for the action of pyroglutamate amino peptidase on natural AP-I and the synthetic peptide, [pGlu]1..[Lys]16-OH (see text for details).

Synthetic and natural peptides were characterized further by treatment with pyroglutamate amino peptidase followed by sequencing of the products. Treatment of [pGlu]1..[Lys]16-OH with pyroglutamate amino peptidase resulted in a single reaction product (Figure 11) which co-eluted with the deletion peptide [Arg]2..[Lys]16. This was in marked contrast to the effect of pyroglutamate amino peptidase on the natural peptide (see above). The behavior of [pGlu]1..[Lys]16-OH and nAP-I was investigated further by following the appearance of the reaction products (Figure 11) using RP-HPLC. When equimolar amounts of the two peptides were treated with pyroglutamate amino peptidase under identical conditions, the natural peptide yielded a more polar product very rapidly (P1; Figure 11), but this declined quickly to yield a second peak which was more hydrophobic than the original peptide (P2; Figure 11). On this evidence P2 appears to be a reaction product of P1. Co-elution of P1 with the deletion peptide, [Arg]2..[Lys]16-OH, suggested that they were identical, and this was supported by sequence analysis (see above). Synthetic [pGlu]1..[Lys]16-OH when treated with pyroglutamate amino peptidase, even after a 4 h incubation, yielded only one product which corresponded to peak 1.

DISCUSSION

Very little sequence information is available for insect diuretic peptides. A peptide with diuretic activity, AVP-LDP, has been isolated and sequenced from the suboesophageal and thoracic ganglia of *Locusta*. AVP-LDP is an antiparallel dimer of a 9 residue monomer which is linked by 2 disulphide bridges (Table 3). There is also some incomplete and partial sequence information for another diuretic peptide, DP-I, from the corpora cardiaca of *Locusta* (see Table 2). Clearly, although little information is available, DP-I and AVP-LDP are unrelated. A third peptide from *Locusta*, for which no sequence information is available, appears to be related to vertebrate adrenocorticotropic hormone [18]. Thus, there

Table 3. A comparison of primary structures of fully and partially characterised diuretic peptides from insects

Species	Peptide name	Structure
Locusta [4]	AVP-LDHa	C L I T N C P R G-NH_2 C L I T N C P R G-NH_2
Locusta [10]	DP-1b	? G ? G I Q A ? V/M Y K...
Manduca [5]	DHc	R M P S L S I D L P M S V L R Q K L S L E K E R K V H A L R A A A N R N F L N D I-NH_2
Acheta [this study]	AP-Id	<Q R D I F H A Q T D I F Q V P K-OH

appears to be considerable diversity even within a single species as to the nature of these peptides. A diuretic peptide has recently been isolated from the tobacco hornworm, *Manduca sexta* [5]. This peptide comprised 41 residues (see Table 3) and has considerable sequence identity with sauvagine, urotensin I, and corticotropin releasing factor. The hormone stimulates diuresis when injected into post-feeding, pre-wandering larvae, but may not have a direct effect upon the Malpighian tubules since it does not elevate cAMP levels or stimulate fluid secretion [5]. It has been suggested that the isolated peptide is a releasing factor for the true diuretic hormone.

In *Acheta*, a number of peptides with diuretic activity have been isolated from the neuroendocrine system by RP-HPLC [12]. AP-I, which is a major peptide from the corpora cardiaca of *Acheta*, has diuretic activity, and is coincident with a major peak of UV-absorbing material. This peptide was therefore selected for sequence study, to obtain structural information on diuretic peptides in *Acheta*. Routinely, AP-I was isolated using a two-stage purification on octyl columns eluted by acetonitrile gradients with TFA as a mobile phase modifier. A number of other chromatographic procedures were employed to assess the purity of this peak, and in some cases minor peaks of contaminating material were resolved from the major UV-absorbing peak. However, diuretic activity always remained associated with the main peak. On this basis material obtained after the two stage purification was used for sequence analysis and peptide mapping. The purity of this material is further supported by results from the amino acid analysis of a hydrolysate of AP-I which revealed only minor contamination, and capillary zonal electrophoresis which gave a single peak.

The biological activity of AP-I was investigated in some detail, and the availability of synthetic [pGlu]1..[Lys]16-OH (see below) enabled quantification of the amount of AP-I present in the corpora cardiaca. Aqueous extracts of 50 corpora cardiaca contain 500 to 600 pmoles of AP-I, i.e., about 10 to 12 pmole/gland pair. This was ca. 5 times the amount of DP-I estimated to be present in the corpora cardiaca of *Locusta*, and ca. 200 times greater than the amount of AVP-LDP in the suboesophageal ganglia of *Locusta*. A dose of 0.1 gland pair/5 µl elicited a maximum diuretic response in *Acheta* tubules [12,17], which corresponded to a concentration of AP-I of 2×10^{-7} M. Dose-response data for pure AP-I, which was quantified by amino acid analysis, indicated a similar potency. The potency of AP-I appears to be low in comparison with AKH and AVP-LDP which elicit maximum responses at 7×10^{-9} M [19] and ca. 1×10^{-9} M [20] respectively. Nevertheless, AP-I increased Malpighian tubule fluid secretion by about 200%. For *Manduca* DH an injection of 0.02 pmol of synthetic peptide [5] gave a half maximum diuretic response but, as noted above, this appears not to have a direct effect on tubule fluid secretion.

Preliminary chemical characterization of AP-I showed that it did not absorb strongly at 280 nm, and therefore did not contain either tryptophan or tyrosine residues. In addition, reduction and alkylation did not alter the elution time of the peptide on RP-HPLC, and there was no evidence of absorbance at 260 nm. This indicated that the peptide did not contain cystyl residues and thus, as expected, AP-I retained its biological activity after such treatment. In this respect, AP-I is different from both diuretic peptides in *Locusta* for which sequence information is available [4,10].

When subjected to gas-phase Edman-degradation, intact AP-I failed to sequence. This indicated the presence of a N-terminal blocking group. AP-I was therefore treated with pyroglutamate amino peptidase which surprisingly yielded two products, both of which were amenable to sequencing. Nevertheless, this confirmed pyroglutamic acid as the N-terminal residue. Sequence information at this stage indicated that AP-I was a 16 residue peptide with a C-terminal lysine (Table 3). Peptide mapping confirmed this sequence. In general, treatment of AP-I with sequence grade enzymes produced the expected number of fragments attributable to cleavage at specific peptide bonds. This suggested that the sequence obtained after pryoglutamate amino peptidase digestion encompassed the whole peptide. The sequences of the fragments conformed with that obtained previously (Table 3). Of the three fragments obtained after digestion with TLCK-chymotrypsin, only one yielded sequence information. One of the remaining fragments would correspond to the N-terminal fragment and therefore was blocked by a pyroglutamyl residue. The remaining C-terminal fragment, formed after cleavage at [Phe]12, would have an N-terminal glutamine residue. In the presence of TFA, this glutamine residue would very rapidly cyclize to yield a pyroglutamyl residue, which would explain why this fragment also did not sequence. Mass determination on a sample of AP-I was in agreement with the theoretical mass for the sixteen residue peptide, [pGlu]1..[Lys]16 and indicated that the C-terminus was a free carboxyl. Both the carboxyl ([pGlu]1..[Lys]16-OH) and amidated ([pGlu]1..[Lys]16-NH_2) forms of AP-I were synthesized and their elution on RP-HPLC compared with that of native AP-I. The two synthetic peptides were easily resolved from each other, but natural AP-I co-eluted with the carboxylated peptide, which suggested identity between nAP-I and [pGlu]1..[Lys]16-OH.

Unfortunately neither of these synthetic peptides possessed diuretic activity when assayed on single tubules *in vitro*. There are several possible reasons for this lack of activity. AP-I may not be a diuretic peptide; the sequence assigned to AP-I may be incorrect, or the sequence, although correct, may be misleading or incomplete; and lastly, the synthetic [pGlu]1..[Lys]16-OH may differ in some subtle fashion from nAP-I.

Isolation and purification of AP-I by a number of chromatographic procedures failed to separate diuretic activity from the major UV-absorbing peak which has been sequenced, and it appears unlikely that AP-I is not a diuretic peptide. The sequence data obtained was confirmed on a number of occasions and was corroborated by the enzyme mapping studies. The sequence assigned for AP-I (Table 3) is also supported by the molecular weight obtained by FAB-MS. Nevertheless, it is possible that the observed biological activity was due to a minor contaminant co-purified with AP-I. If a minor component accounted for all the observed diuretic activity, then it must share with AP-I the following properties: identical hydrophobicity on RP-HPLC at pH 2.1 or 6.75; a similar complement of aromatic groups; the same molecular size, as indicated by co-elution on HP-SEC; the same net charge, as indicated by co-elution on HP-IEC; pyroglutamic acid at the N-terminal; and contain no cystyl, tryptophanyl, or tyrosyl residues. Whilst this supposition cannot be rejected totally, it would appear unlikely.

Is it possible that the sequence data, whilst correct, are misleading? AP-I may have been modified during the isolation and purification procedures, so that the sequenced peptide is not identical to the native diuretic peptide. This seems unlikely since the material taken for sequence analysis, and that sent for FAB-MS analysis, were biologically active. Although the synthetic and natural peptides share the same molecular weights and column retention times, it is possible that the synthetic peptide does not adopt the same secondary structure as the natural peptide. Secondary structure prediction [21] suggests that AP-I should adopt an α-helix in solution, but preliminary studies using circular dichroism spectroscopy on synthetic AP-I do not support this prediction (Wheeler, unpublished observations). This may suggest that the synthetic peptide is in some way conformationally constrained.

Apart from the lack of diuretic activity, the strongest evidence that nAP-I and [pGlu]1..[Lys]16-OH are different from each other has been obtained after digestion of the peptides by pyroglutamate amino peptidase. The kinetics and end product formation observed with the two peptide substrates clearly, but inexplicably, show that there is a difference between them. Formation of two products after treatment of a peptide (AKH) with pyroglutamate amino peptidase has been described previously [22], and the presence of prolidase activity may account for this (T.K. Hayes, personal communication). There seems little doubt that product 1, from both the synthetic and the natural peptides, is

[Arg]2..[Lys]16-OH, but the identity of product 2 is equivocal. Our sequencing studies suggest that product 2 lacks a C-terminal lysine residue, but because the yield of this residue was poor in all sequencing runs, this cannot be taken as proof of its absence. Further characterization of the products of digestion of the natural peptide by FAB-MS will be necessary to clarify this situation.

In conclusion, we have characterized from *Acheta domesticus* a peptide which has diuretic activity when assayed on isolated Malpighian tubules. Although studies suggest that the sequence of AP-I is [pGlu]1..[Lys]16-OH, we are unable to confirm this by synthesis.

ACKNOWLEDGEMENTS

We would like to thank Professor M. Waterfield (Ludwig Institute for Cancer Research, London) for the use of facilities for peptide sequencing and synthesis, and Professor R. M. Caprioli (University of Texas, Houston) and Dr A. I. Mallet (St Thomas's Hospital, London) for providing us with mass analyses.

REFERENCES

1. W. Mordue and P. J. Morgan, Chemistry of peptide hormones, *in*: "Comprehensive Insect Physiology, Biochemistry and Pharmacology", G. A. Kerkut and L. I. Gilbert, eds., Pergamon Press, Oxford, Vol. 7, pp 153-183 (1985).
2. R. J. Aston and L. Hughes, Diuretic hormone-extraction and chemical properties, *in*: "Neurohormonal Techniques in Insects", T. A. Miller, ed., Springer, Berlin, pp 91-115 (1980).
3. J. E. Phillips, Endocrine control of salt and water balance: excretion, *in*: "Endocrinology of Insects", R. G. H. Downer and H. Laufer, Alan Liss, New York, pp 411-425 (1983).
4. J. P. Proux, C. Miller, J. P. Li, R. L. Carney, A. Girardie, M. Delaage and D. A. Schooley, Identification of an arginine vasopressin-like diuretic hormone from *Locusta migratoria*, *Biochem. Biophys. Res. Commun.*, 149:180-186 (1987).
5. H. Kataoka, R. G. Troetschler, J. P. Li, S. J. Kramer, R. L. Carney and D. A. Schooley, Isolation and identification of a diuretic hormone from the tobacco hornworm, *Manduca sexta*, *Proc. Nat. Acad. Sci., USA*, 86:2976-2980 (1989).
6. J. A. Ramsay, Active transport of water by the Malpighian tubules of the stick insect, *Dixippus morosus* (Orthoptera; Phasmidae), *J. Exp. Biol.*, 31:104-113 (1954).
7. C. H. Wheeler and G. M. Coast, Assay and characterization of diuretic factors in insects, *J. Insect Physiol.*, in press (1989).
8. D. A. Schooley, C. A. Miller and J. P. Proux, Isolation of two arginine-vasopressin-like factors from ganglia of *Locusta migratoria*, *Arch. Insect Biochem. Physiol.*, 5:157-166 (1987).
9. J. Proux, G. Rougon and A. Cuppo, Enhancement of excretion across locust Malpighian tubules by a diuretic vasopressin-like hormone, *Gen. Comp. Endocrinol.*, 47:449-457 (1982).
10. P. J. Morgan, K. J. Siegert and W. Mordue, Preliminary characterization of locust diuretic peptide (DP-I) and another corpus cardiacum peptide (LCCP), *Insect Biochem.*, 17:383-388 (1987).
11. G. M. Coast, Fluid secretion by single isolated Malpighian tubules of the house cricket, *Acheta domesticus*, and their response to diuretic hormone, *Physiol. Entomol.*, 13:381-391 (1988).
12. G. M. Coast and C. H. Wheeler, The distribution and relative potency of diuretic peptides in the house cricket, *Acheta domesticus*, *Physiol. Entomol.*, in press (1989).
13. C. W. Clifford, R. M. Roe and J. P. Woodring, Rearing methods for obtaining house crickets, *Acheta domesticus*, of known age, sex and instar, *Ann. Entomol. Soc. America*, 70:69-74 (1977).
14. G. E. Tarr, Manual Edman sequencing system, *in*: "Methods of Protein Microcharacterization", J. E. Shively, ed., Humana Press, New York, pp 155-194 (1986).
15. D. Dupont, P. Keim, A. Chui, I. Harding and K. J. Wilson, Gas-phase hydrolysis for PTC-amino acid analysis, Derivatizer-Analyzer Bulletin No. 2, Applied Biosystems Inc. (1987).
16. P. J. Morgan and W. Mordue, Separation and characteristics of diuretic hormone from the corpus cardiacum of *Locusta*, *Comp. Biochem. Physiol.*, 75B:75-80 (1983).

17. G. M. Coast, Stimulation of fluid secretion by single isolated Malpighian tubules of the house cricket, *Acheta domesticus, Physiol. Entomol.*, 14:21-30 (1989).
18. A. Rafaeli, S. W. Applebaum and P. Moshitzky, An ACTH-like diuretic hormone in locusts, *in*: "Insect Neurochemistry and Neurophysiology 1986", A. B. Borkovec and D. B. Gelman, eds., Humana Press, New York, pp 355-357 (1986).
19. G. L. Goldsworthy, K. Mallisson, C. H. Wheeler and G. Gade, Relative adipokinetic activities of members of the adipokinetic hormone/red pigment concentrating hormone family. *J. Insect. Physiol.*, 32:433-438 (1986).
20. J. P. Proux, M. Picquot, J.-P. Herault and B. Fournier, Diuretic activity of a newly identified neuropeptide - the arginine-vasopressin-like diuretic hormone: use of an improved bioassay, *J. Insect Physiol.*, 34:919-927 (1988).
21. P. Y. Chou and G. D. Fasman, Prediction of the secondary structure of proteins from their amino acid sequence, *Adv. Enzymol.*, 47:45-148 (1978).
22. K. J. Siegert and W. Mordue, Elucidation of the primary structures of the cockroach hyperglycaemic hormones I and II using enzymatic techniques and gas-phase sequencing, *Physiol. Entomol.*, 11:205-211 (1987).

MANDUCA ADIPOKINETIC HORMONE: QUANTIFICATION USING HIGH PERFORMANCE LIQUID CHROMATOGRAPHY AND RADIOIMMUNOASSAY

A. M. Fox and S. E. Reynolds

School of Biological Sciences
University of Bath
Claverton Down, Bath BA2 7AY, UK

SUMMARY

The nonapeptide hormone *Manduca AKH* (M-AKH) is involved in the regulation of lipid metabolism during flight in the tobacco hornworm, *Manduca sexta* (L.) and may also be involved in the regulation of carbohydrate metabolism in larval *Manduca*. We have used reversed-phase high performance liquid chromatography (RP-HPLC) and/or radioimmunoassay (RIA) procedures to quantify the M-AKH content of the corpora cardiaca (CC) and other neural tissues from *Manduca*. The M-AKH content of *Manduca* CC was determined using RP-HPLC on a C_{18} bonded column with detection by UV absorbance and by RIA. The M-AKH content of other tissues was determined by RIA alone. The identity of the M-AKH immunoreactive material from the CC and brain was investigated by RIA of RP-HPLC purified material. Both adult and larval CC contain immunoreactive material which coelutes with synthetic M-AKH on RP-HPLC. A small amount of M-AKH immunoreactive material was detectable in brain extracts, but this material did not coelute with M-AKH and is not thought to represent M-AKH.

The data from the quantification experiments were used to compare the bioactivities of synthetic M-AKH with those of CC extracts. Such comparisons suggest that M-AKH may not be the only factor involved in lipid regulation in adult moths, but they support the identification of M-AKH as the larval glycogen phosphorylase activating hormone (GPAH).

INTRODUCTION

Chromatographic and immunochemical techniques have been applied extensively in the isolation and localization of insect peptide hormones. The use of these techniques in the quantification of the peptide content of various tissues has been less widely reported. Siegert and Mordue [1] used reversed-phase high performance liquid chromatography (RP-HPLC) to determine the content of adipokinetic hormones I and II (AKH I and II) in the corpora cardiaca (CC) of larval and adult locusts. Their results confirmed preliminary estimates obtained from the quantitative amino acid analysis of purified hormone preparations and bioassay of CC extracts, respectively [2,3].

The tobacco hornworm, *Manduca sexta*, also possesses an AKH-family peptide known as *Manduca* adipokinetic hormone (M-AKH), which possesses lipid and carbohydrate mobilizing activities when injected into adult and larval *Manduca*, respectively. We have used RP-HPLC and radioimmunoassay (RIA) techniques to determine the M-AKH content of the CC and other neural tissues from larval and adult *Manduca*. In addition to the intrinsic value of this data, it has enabled us to speculate further on the physiological role of M-AKH in *Manduca*. Here we report the first combined use of HPLC and RIA to quantify a known peptide in an insect.

Chromatography and Isolation of Insect Hormones and Pheromones
Edited by A. R. McCaffery and I. D. Wilson
Plenum Press, New York, 1990

MATERIALS AND METHODS

Experimental Animals

The animals used in these experiments were from a laboratory culture of *Manduca sexta* maintained on artificial diet [4] on a long-day (LD 17:7) photoperiodic regime.

Sources of Peptides

M-AKH and [Tyr^1]-AKH were synthesized by us on a CRB Pepsynthesizer II using Fmoc polyamide chemistry and purified using RP-HPLC.

Preparation of Tissue Extracts

Brain, abdominal nerve cord and CC extracts were prepared from adult and larval tissues. Adults were day 2 moths of both sexes (which are of similar size); larvae were day 3 fifth instar (also unsexed). The extraction method was slightly modified from that reported by Jaffe et al. [5] for the extraction of the adipokinetic hormone from the corn earworm, *Heliothis zea*. The Millipore filtration step used by these workers was replaced by processing through a C_{18} Sep-Pak cartridge (Waters Associates, Milford, USA). The sample was loaded onto a Sep-Pak cartridge which had been primed with 100% acetonitrile followed by 0.1% trifluoroacetic acid (TFA) in water. M-AKH was eluted with 3 ml of 60% aqueous acetonitrile, and solvent removed in a rotary evaporator (Savant) prior to HPLC or RIA determination.

The efficiency of M-AKH extraction was assessed by adding [^{125}I-Tyr^1] M-AKH to 50 µl of extraction buffer containing 1 larval brain prior to homogenization. Subsequent loss of radioactivity was monitored using a LKB 1275 Minigamma counter. The recovery of the labeled peptide over the whole extraction process was 63.3 ± 2.2% (Mean ± S.E., n = 5). Of the total 37% lost, approximately 15% could be attributed to the homogenization step, 12% to the use of Sep-Paks, and 10% to the evaporation step.

HPLC Analyses

A Gilson HPLC system was used for the quantification of M-AKH. This consisted of two model 302 pumps, a model 702 gradient manager, a model 802 manometric module, a model HM Holochrome UV detector, a model 620 Data Master (all from Gilson International, Villiers-le-Bel, France), a Rheodyne 7125 injection valve, and a Rikadenki chart recorder, all controlled by an Apple IIe microcomputer.

A Spherisorb C_{18} 5 µm analytical column (0.46 x 25 cm, HPLC Technology, Macclesfield, UK) was used for quantification. Flow rate was 1 ml/min. Solvents used were: A, water containing 0.1% TFA; B, 0.1% TFA in 60% acetonitrile/water. Gradient conditions for quantification were: 33 to 77.5% B (0 to 20 min). The column was washed with 100% B for 5 min after each run, followed by a 15 min re-equilibration at 33% B before the next sample was introduced. The eluate was monitored by UV absorbance at 210 nm. Peak areas were calculated by the Data Master. Synthetic M-AKH standards were used to calibrate the system and to indicate the elution time for M-AKH under the above conditions.

To determine the elution profile of M-AKH immunoreactive material, a sample of CC or brain extract was chromatographed as above except that a linear gradient of 20 to 90% B was run over 40 min. Fractions were collected at 1 min intervals in 4 ml polypropylene tubes using a Gilson Model 201-202 fraction collector. An aliquot from each fraction was removed, freeze dried and resuspended in RIA buffer in order to assay for immunoreactivity.

Radioimmunoassay for M-AKH

A detailed description of the development of the radioimmunoassay (RIA) for M-AKH will be given elsewhere (Fox and Reynolds, in preparation). Briefly, the labeled ligand [^{125}I-Tyr^1] M-AKH was prepared by a chloramine-T method. The M-AKH antiserum was raised in rabbits. The antigen used was [Tyr]$^{-1}$ M-AKH conjugated to thyroglobulin using the bifunctional reagent BSOCOES [6]. Antibody titre was assessed by binding of the radiolabeled tracer. The RIA protocol was slightly modified from that described by Moshitzky et al. [7] for the determination of locust AKH-1. The antibody was used at a final dilution of 1:1800, which gave 30 to 35% binding of the labeled peptide in the absence of unlabeled peptide.

Bioassays

Synthetic M-AKH and CC extracts were assayed for adipokinetic activity in adult moths using the sulphophosphovanillin assay for total haemolymph lipids described by Goldsworthy et al. [8]. A sample of haemolymph (5 µl) was taken by puncturing the dorsal vessel in the abdomen immediately prior to the injection of 10 µl of distilled water containing either tissue extract or synthetic peptide. A second haemolymph sample of the same size was taken after 90 min. The adipokinetic response is given as the increase in lipid content of the second haemolymph sample over the first.

The bioactivity of M-AKH in larval *Manduca* was determined using an assay for the activation of fat body glycogen phosphorylase [9]. This measures glycogen phosphorylase activity in a fat body homogenate by the change in concentration of NADH. Because fat

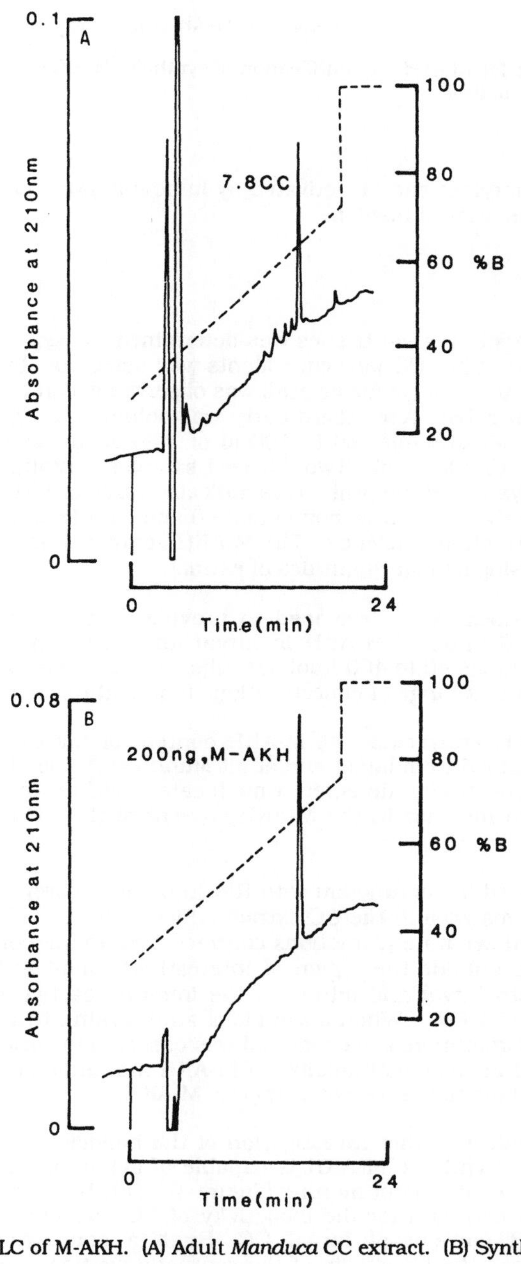

Fig. 1. RP-HPLC of M-AKH. (A) Adult *Manduca* CC extract. (B) Synthetic M-AKH.

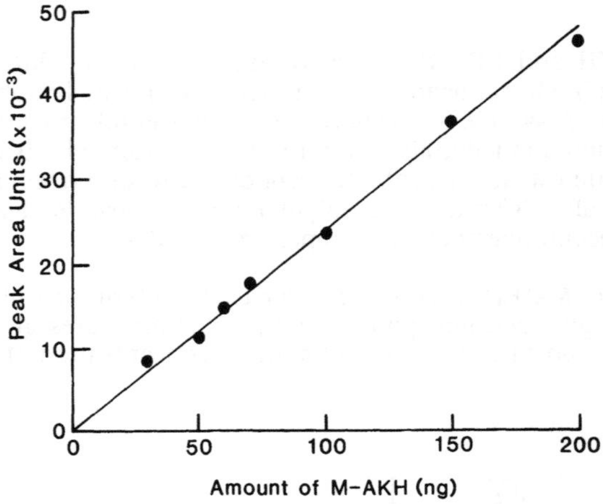

Fig. 2. Standard curve for RP-HPLC quantification of synthetic M-AKH. Standards were injected in 100 μl 10% acetonitrile.

body glycogen phosphorylase can be activated by handling, great care was taken to dissect larvae rapidly and in standard fashion.

RESULTS

The M-AKH content of larval tissues was determined by HPLC and RIA analysis. The content of less than 10 adult CC pair equivalents was detectable by UV absorbance of the HPLC eluate (Figure 1A). A single major peak was observed which coeluted with a sample of synthetic M-AKH (Figure 1B). A standard curve was obtained for the peak areas of varying amounts of synthetic M-AKH injected in 100 μl of 10% acetonitrile (Figure 2). This was linear in the range 25 to 200 pmol. Two different samples containing 3.9 and 7.8 CC pair equivalents were analyzed and the peak areas indicated titres of 19.0 and 19.9 (mean 19.45) pmol per CC respectively. This corresponds to 30.0 and 31.4 (mean 30.7) pmol per CC pair when adjusted for extraction efficiency. The M-AKH content of larval CC extracts was not detectable by HPLC using similar quantities of extract.

A radioimmunoassay was developed to provide a more sensitive method for the detection and quantification of M-AKH in larval and adult tissues. M-AKH could be quantified within the range 40 to 400 fmol per tube. The antiserum did not bind any other AKH family peptide in this range of concentrations (Fox and Reynolds, in preparation).

Using the M-AKH antiserum, the M-AKH content of the CC, brain and abdominal nerve cord was determined from larval and adult *Manduca* (Table 1). In both cases most of the M-AKH immunoreactivity detected was located within the CC. There was an approximately ten-fold increase in the M-AKH content of the CC between the fifth instar and the adult.

RP-HPLC was used in conjunction with RIA in order to confirm the identity of the M-AKH immunoreactive material in the CC extracts (Figure 3). Larval and adult CC extracts were chromatographed separately, fractions collected and an aliquot from each fraction (or from pooled fractions outside the region of interest) was analyzed by RIA. The results indicated that for both larval and adult CC the immunoreactive material coeluted with synthetic M-AKH on RP-HPLC. When a sample of adult brain extract was analyzed by this procedure, no peak of immunoreactive material was observed to coelute with M-AKH. It was therefore concluded that the small amount of M-AKH immunoreactive material quantified in these brain extracts did not represent authentic M-AKH.

These data permit a further investigation of the biological roles of M-AKH in larval and adult *Manduca*. Synthetic M-AKH is capable of activating larval fat body glycogen phosphorylase in a dose-dependent manner (Figure 4). This bioassay data can be compared with previously published data for the bioactivity of CC extracts by using the HPLC/RIA data for the M-AKH content of larval CC. Such a comparison indicates a close correspondence between the bioactivity of the synthetic peptide and the activity of the CC

Table 1. M-AKH Content of Various Nervous Tissues from *Manduca sexta*, determined by a specific M-AKH RIA

Tissue	M-AKH content (pmol per tissue)	n
Adult CC/CA	33.17 ± 1.39	5
Adult brain	0.71 ± 0.13	4
Adult nerve cord	0.13 ± 0.02	4
Larval CC	3.47 ± 0.26	4
Larval brain	0.09 ± 0.03	5
Larval nerve cord	0.17 ± 0.08	5

Values are corrected for extraction efficiency as described in the text, Means ± S.E.

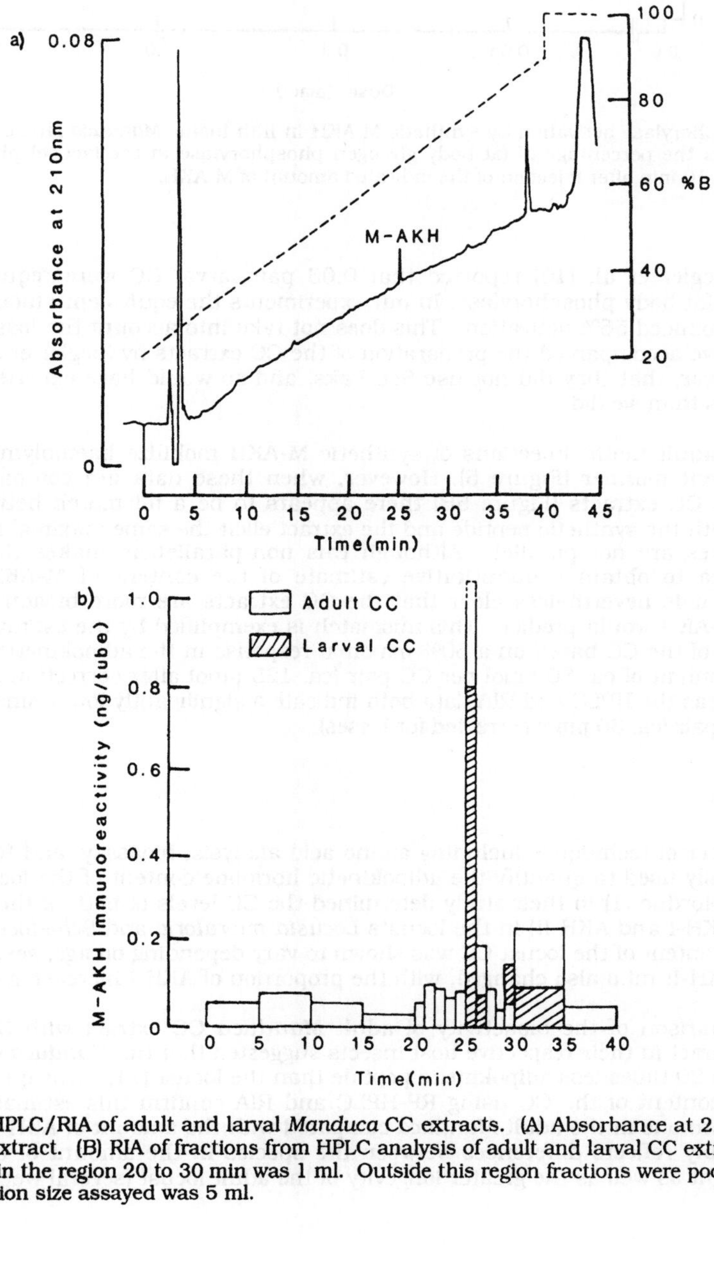

Fig. 3. RP-HPLC/RIA of adult and larval *Manduca* CC extracts. (A) Absorbance at 210 nm of adult CC extract. (B) RIA of fractions from HPLC analysis of adult and larval CC extracts. Fraction size in the region 20 to 30 min was 1 ml. Outside this region fractions were pooled so that the fraction size assayed was 5 ml.

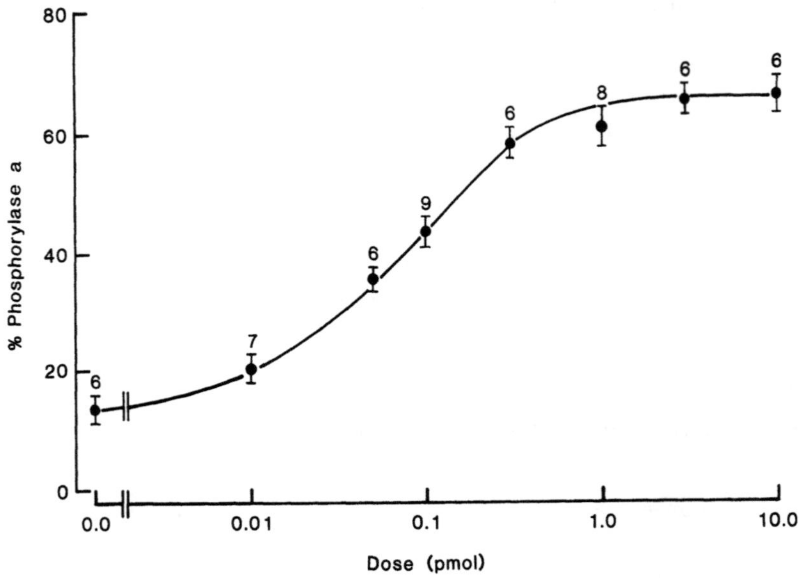

Fig. 4. Phosphorylase activation by synthetic M-AKH in fifth instar *Manduca* larvae. The ordinate shows the percentage of fat body glycogen phosphorylase in the (active) phosphorylase a form, 10 min after injection of the indicated amount of M-AKH.

extracts. Ziegler et al. [10] reported that 0.03 pair larval CC were required for 50% activation of fat body phosphorylase. In our experiments the equivalent amount of M-AKH (0.1 pmol) produced 56% activation. This does not take into account the losses of peptide that must have accompanied the preparation of the CC extracts by Ziegler et al. It is worth noting, however, that they did not use Sep-Paks, and so would have experienced slightly smaller losses than we did.

In the adult moth, injections of synthetic M-AKH mobilize haemolymph lipid in a dose-dependent manner (Figure 5). However, when these data are compared with the bioactivity of CC extracts (Figure 5B) there appears to be a mismatch between the two activities. Both the synthetic peptide and the extract elicit the same maximal response, but the two curves are not parallel. Although this non-parallelism makes the use of the bioassay data to obtain a quantitative estimate of the content of M-AKH in the CC problematic, it is nevertheless clear that the CC extracts are more bioactive than their content of M-AKH would predict. This mismatch is exemplified by the estimate for the M-AKH content of the CC based on a 50% maximal response in the adipokinetic assay which suggests a content of ca. 80 pmol per CC pair (ca. 125 pmol after correction for extraction losses), whereas the HPLC and RIA data both indicate a significantly lower amount of ca. 20 pmol per CC pair (ca. 30 pmol corrected for losses).

DISCUSSION

A number of techniques including amino acid analysis, bioassay, and RP-HPLC have been previously used to quantify the adipokinetic hormone content of the locust CC [1-3]. Siegert and Mordue [1] in their study determined the CC levels of both of the adipokinetic hormones (AKH-I and AKH-II) in the locusts *Locusta migratoria* and *Schistocerca gregaria*. The peptide content of the locust CC was shown to vary depending on age, sex, and species. The AKH-I:AKH-II ratio also changed, with the proportion of AKH-I increasing with age.

A comparison of the bioactivity of adult *Manduca* CC extract with the activity of locust CC extract in their respective host insects suggested that the *Manduca* CC contained roughly 10 to 20 times less adipokinetic peptide than the locust [11]. Our quantification of the peptide content of the CC using RP-HPLC and RIA confirm this estimate. The lower amount of hormone in the adult *Manduca* compared with the amount of AKH-I in the adult locust probably reflects differences between the species in the amount of time spent in sustained flight as well as the greater longevity of the adult locust (several weeks) compared

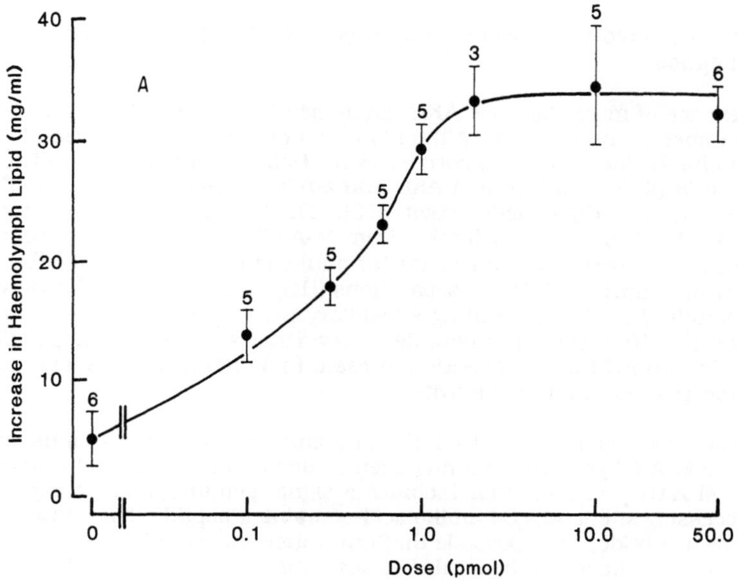

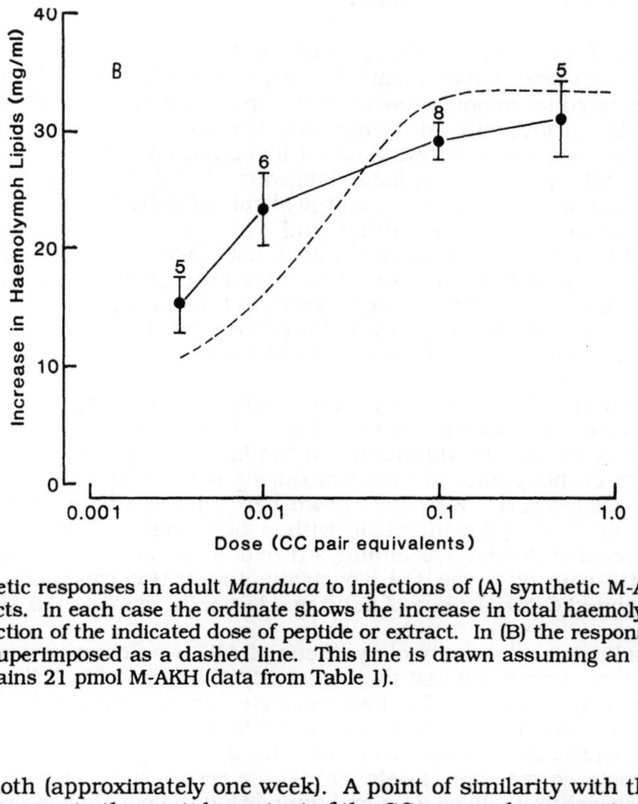

Fig. 5. Adipokinetic responses in adult *Manduca* to injections of (A) synthetic M-AKH and (B) adult CC extracts. In each case the ordinate shows the increase in total haemolymph lipid 90 min after injection of the indicated dose of peptide or extract. In (B) the response to synthetic M-AKH is superimposed as a dashed line. This line is drawn assuming an extract of one CC pair contains 21 pmol M-AKH (data from Table 1).

with the adult moth (approximately one week). A point of similarity with the locust is that a considerable increase in the peptide content of the CC occurs during metamorphosis.

It is worth noting that the procedure we used to extract M-AKH from nervous tissue involved appreciable losses of the peptide. These losses occurred in all stages of the procedure, so that it would be difficult to alter the protocol so as to reduce losses. Here we have presented both uncorrected and corrected M-AKH values so as to illustrate this point.

RP-HPLC analysis of the CC extracts indicated that almost all of the M-AKH immunoreactive material coeluted with synthetic M-AKH. The specificity of the RIA for M-AKH was thus confirmed. Both larval and adult CC extracts also contained immunoreactive material that eluted from the column in a much smaller peak later in the acetonitrile gradient. We considered the possibility that this subsidiary peak might represent a second AKH-family peptide.

The presence of more than one AKH-family peptide in the CC has now been noted to occur in a number of insects [12]. Although only one AKH-family peptide is known from *Manduca*, another lepidopteran, the corn earworm *Heliothis zea*, possesses two AKH-family peptides, a nonapeptide identical to M-AKH and a related decapeptide (Hez-HTH) with both hypertrehalosemic and adipokinetic activity [13]. The biological role of this second peptide is unclear. Hez-HTH differs considerably from M-AKH in its C-terminal sequence and it does not cross-react with our antiserum (unpublished). Additionally, Hez-HTH elutes earlier than M-AKH during RP-HPLC separations ([13]; our unpublished observations). We therefore conclude that the late-eluting subsidiary peak of immunoreactive material does not represent the *Heliothis* decapeptide. Nevertheless it remains possible that an additional, unknown AKH-family peptide is present in *Manduca* CC extracts, accounting for the late-eluting peak of immunoreactivity.

An alternative possibility is that the late-eluting peak corresponds to an M-AKH precursor. The M-AKH gene has recently been cloned and sequenced [14] and is known to encode a pro-M-AKH precursor that includes a signal peptide, a single copy of M-AKH, a tripeptide processing site, and a 34 amino acid-containing peptide that appears unrelated to any other known peptide. It is possible that our antiserum would recognize the intact pro-AKH peptide, but we have not been able to test this point directly. The behavior of the precursor on RP-HPLC columns is unknown.

The presence of AKH and AKH-like peptides in the insect brain has been the subject of some speculation. An immunocytochemical study using a "C-terminal specific" antiserum for AKH-I demonstrated immunoreactive neurones in the locust brain [6]. Subsequently the results of a similar study with an "N-terminal specific" antiserum suggested that the previously identified neurones contain an AKH-like material rather than AKH-I itself [15]. More recently, Moshitzky et al. [7,16] have claimed to identify AKH-I and to demonstrate its synthesis in the locust brain. The physiological role of AKH-I in the locust brain is not known. In the present study, RP-HPLC and RIA analysis indicated the absence of measurable quantities of M-AKH from the brain of both larval and adult Manduca (data not shown). A small amount of immunoreactive material corresponding to the late eluting peak observed in CC extracts was however present in the brain extracts. The structure and biological role of this material remain unknown. We conclude that the immunoreactive material quantified in the *Manduca* brain is not M-AKH.

When compared with bioassay data for synthetic M-AKH and for CC extracts, our quantitative data on the M-AKH content of the CC permit further speculation on the physiological role of M-AKH in *Manduca*. In the larva a peptide (GPAH) from the CC is known to regulate carbohydrate metabolism during starvation through the activation of glycogen phsophorylase [17]. We have shown here that synthetic M-AKH is capable of activating larval fat body phosphorylase with a 50% maximal response following the injection of 0.06 pmol of M-AKH. Assuming a titre of 33 pmol per CC pair for adult moths, or 3.5 pmol per CC pair for larvae, the bioactivity of M-AKH corresponds very closely with the previously reported GPAH activity of adult and larval CC extracts [10,18].

When the adipokinetic activity of synthetic M-AKH was compared with that of adult CC extracts, however, it appeared that the measured CC content of M-AKH could not account for all of the bioactivity observed. The dose-response curve for the synthetic peptide did not correspond well with that for the CC extracts, the latter being somewhat flatter, although the maximum response achieved was the same. Our dose-response data for the adipokinetic effect of CC extracts correspond closely with those reported by Ziegler and Schulz [19], indicating similar extraction efficiencies and supporting the conclusions reached here.

It seems possible therefore that the adult CC contain a second factor distinct from M-AKH which contributes to the adipokinetic activity of CC extracts. Such a factor might be a second AKH-family adipokinetic peptide (perhaps Hez-HTH or a related peptide), or possibly a non-peptidic factor with adipokinetic activity such as octopamine (known to be adipokinetic in locusts - [20]). Another possibility is that the adult CC contain a potentiating factor that enhances the response of the insect to M-AKH, but which does not have intrinsic adipokinetic activity. Further work is required to distinguish between these possibilities and to characterize fully any additional factors in the adult *Manduca* CC.

ACKNOWLEDGEMENTS

Thanks go to Dr Richard Kinsman (School of Chemistry, Bath University) for advice on peptide synthesis; to Mike Hinton and Maureen Norris (Animal House, Bath University) for help with raising antisera; and to many colleagues (but especially Jackie Rawlings) for help in maintaining the *Manduca* culture. AMF was supported by a Bath University Research Fund studentship.

REFERENCES

1. K. J. Siegert and W. Mordue, Quantification of adipokinetic hormones I and II in the corpora cardiaca of *Schistocerca gregaria* and *Locusta migratoria*, *Comp. Biochem. Physiol.*, 84A:279-284 (1986).
2. J. V. Stone, W. Mordue, K. E. Batley and H. R. Morris, Structure of locust adipokinetic hormone, a neurohormone that regulates lipid utilization during flight, *Nature*, 263:207-211 (1976).
3. W. J. A. Van Marrewijk, A. T. M. Van den Broek and A. M. T. Beenakkers, Regulation of glycogen phosphorylase activity in the fat body of *Locusta migratoria* and *Periplaneta americana*, *Gen. Comp. Endocrinol.*, 50:226-234 (1983).
4. R. A. Bell and F. G. Joachim, Techniques for rearing laboratory colonies of tobacco hornworms and pink bollworms, *Ann. Entomol. Soc. Am.*, 69:365-373 (1976).
5. H. Jaffe, A. K. Raina, C. T. Riley, B. A. Fraser, G. M. Holman, R. M. Wagner, R. L. Ridgeway and D. K. Hayes, Isolation and primary structure of a peptide from the corpora cardiaca of *Heliothis zea* with adipokinetic activity, *Biochm. Biophys. Res. Commun.*, 135:622-628 (1986).
6. H. Schooneveld, G. I. Tesser, J. A. Veenstra and H. M. Romberg-Privee, Adipokinetic hormone and AKH-like peptide demonstrated in the corpora cardiaca and nervous system of *Locusta migratoria* by immunocytochemistry, *Cell Tissue Res.*, 230:67-76 (1983).
7. P. Moshitzky, D. Yamashiro, L. Stuve, J. Ramachandran and S. W. Applebaum, Determination of locust AKH-1 by radioimmunoassay and the identification of an AKH-like factor in the locust brain, *Insect Biochem.*, 17:765-769 (1987).
8. G. J. Goldsworthy, W. Mordue and J. Guthkelch, Studies on insect adipokinetic hormones, *Gen. Comp. Endocrinol.*, 18:545-551 (1972).
9. R. Ziegler, M. Ashida, A. M. Fallon, L. T. Wimer, S. S. Wyatt and G. R. Wyatt, Regulation of glycogen phosphorylase in fat body of Cecropia silkmoth pupae, *J. Comp. Physiol.*, 131:321-332 (1979).
10. R. Ziegler, R. Hoff and M. Rohde, Storage site of glycogen phosphorylase activating hormone in larvae of *Manduca sexta*, *J. Insect Physiol.*, 34:143-150 (1988).
11. R. Ziegler, G. Kegel and R. Keller, Isolation and amino acid composition of the adipokinetic hormone of *Manduca sexta*, *Hoppe-Seyler's Zeitschrift fur Physiologische Chemie*, 365:1451-1456 (1984).
12. I. Orchard, Adipokinetic hormones - an update, *J. Insect Physiol.*, 33:451-463 (1987).
13. H. Jaffe, A. K. Raina, C. T. Riley, B. A. Fraser, T. G. Bird, C. M. Tseng, Y.-S. Zhang and D. K. Hayes, Isolation and primary structure of a neuropeptide from *Heliothis zea* with hypertrehalosemic and adipokinetic activities, *Biochem. Biophys. Res. Commun.*, 155:344-350 (1988).
14. J. Y. Bradfield and L. L. Keeley, Adipokinetic hormone gene sequence from *Manduca sexta*, *J. Biol. Chem.*, 264:12791-12793 (1989).
15. H. Schooneveld, H. M. Romberg-Privee and J. A. Veenstra, Phylogenetic differentiation of glandular cells in corpora cardiaca as studied immunocytochemically with region-speciific antisera to adipokinetic hormone, *J. Insect Physiol.*, 33:167-176 (1986).
16. P. Moshitzky, W. J. Henzel, A. Rafaeli, J. Ramachandran and S. W. Applebaum, Synthesis of adipokinetic hormone (AKH-1) in the locust brain, *Insect Biochem.*, 17:1133-1137 (1987).
17. K. J. Siegert and R. Ziegler, A hormone from the corpora cardiaca controls fat body glycogen phosphorylase during starvation in tobacco hornworm larvae, *Nature* 301:526-527 (1983).
18. R. Ziegler, Hyperglycaemic factor from the corpora cardiaca of *Manduca sexta* (L.) (Lepidoptera, Sphingidae), *Gen. Comp. Endocrinol.*, 39:350-357 (1979).
19. R. Ziegler and M. Schulz, Regulation of lipid metabolism during flight in *Manduca sexta*, *J. Insect Physiol.*, 32:903-908 (1986).
20. I. Orchard, B. G. Loughton and R. A. Webb, Octopamine and short-term hyperlipaemia in the locust, *Gen. Comp. Endocrinol.*, 45:175-180 (1981).

EXTRACTION, PURIFICATION AND SEQUENCING OF ADIPOKINETIC/RED PIGMENT-CONCENTRATING HORMONE-FAMILY PEPTIDES

Gerd Gäde

Institut für Zoologie, Lehrstuhl für Tierphysiologie
Heinrich-Heine-Universität, Geb. 26.12/00
Universitätsstr. 1, D-4000 Düsseldorf 1, FRG
Present address: Zoology Department, University of Cape Town
Private Bag, Rondebosch 7700, South Africa

SUMMARY

The adipokinetic and hypertrehalosaemic hormones of insects produced and stored in, and released from, the major neurohaemal organ of an insect, the corpora cardiaca, have been shown to be structurally related and belonging to what is now termed the adipokinetic hormone/red pigment-concentrating hormone peptide family (AKH/RPCH-family). Recently, we have elucidated the structures of new members of this family. Here a step-by-step account on the methods and techniques involved in the extraction, separation and sequencing of such peptides is given. The peptides have all been purified from corpus cardiacum material of different insect species. Various methods of extraction are compared. During isolation, bioassays in migratory locusts and American cockroaches were performed to monitor hyperlipaemic and hypertrehalosaemic effects. Separation from other compounds in the corpora cardiaca was achieved by a single-step method using reversed-phase high performance liquid chromatography (RP-HPLC). Examples are given for the use of HPLC systems in the isocratic or gradient mode. Separations on different support media are compared, and important features of detection are discussed. After separation, the amino acid composition of the purified peptides can be determined with pre- or post-column derivatization methods (phenylisothiocyanate or ninhydrin). Elucidation of the primary sequence by fast atom bombardment mass spectrometry or, after deblocking the pyroglutamate at the N-terminus, by gas-phase sequencing employing Edman degradation is outlined.

INTRODUCTION

It has been known for some time that insects employ a vast number of neuropeptides to regulate a multiplicity of physiological actions including secretion of the ecdysteroids and juvenile hormones as well as controlling water balance and intermediary metabolism. Until recently, our knowledge of the precise chemical composition of these peptides had been scarce due to the tiny quantities generally present in the nervous tissue and the lack of sensitive chromatographic and sequencing methods. The development of improved techniques and refined instruments for chromatographic separation using high-performance liquid chromatography (HPLC) in the reversed-phase (RP) mode and subsequent structure elucidation using fast atom bombardment mass spectrometry (FAB-MS) and automated gas-phase or pulsed-liquid phase sequencing have accelerated the discovery of new neuropeptides in insects tremendously. This is particularly true for peptides of the so-called adipokinetic hormone/red pigment-concentrating hormone (AKH/RPCH) family. The name of this family, which consists of structurally related but functionally diverse peptides, is based on the first members discovered - the octapeptide RPCH from the sinus gland of a prawn, which stimulates pigment concentration of erythrophores [1], and the decapeptide AKH (new nomenclature according to Raina & Gäde [2]: Lom-AKH-I) from the corpus cardiacum of locusts, which stimulates lipid mobilization from the fat body during

Chromatography and Isolation of Insect Hormones and Pheromones
Edited by A. R. McCaffery and I. D. Wilson
Plenum Press, New York, 1990

prolonged flight [3]. Here we have outlined how we have proceeded to screen various insect species for substances involved in the regulation of lipid and carbohydrate metabolism and which strategies we have used for extraction, purification and sequencing of the peptides.

BIOASSAYS

A prerequisite for the successful isolation of neuropeptides is the development of a sensitive and reliable bioassay. Such a bioasssay has to be quite rapid and not too cumbersome, since during purification many fractions derived from column chromatography have to be tested. Thus, bioassays should be available that allow the screening of a large number of fractions per day. The bioassay should ideally monitor a physiologically relevant function of the peptide that is to be isolated and it should be rather specific in order to distinguish the required activity from interferences. If possible, test animals for the bioassay should be commercially available or easy to rear, and the chemicals used for the test should be not too expensive.

In our research, we have routinely used three different bioassays for identifying substances that may be involved in the regulation of lipid and carbohydrate metabolism.

Hyperlipaemic and -Trehalosaemic Bioassays: Colorimetric Bioassays

In 1961 Steele [4] and later in 1969 Beenakkers [5], and Mayer and Candy [6], provided evidence for the hormonal control of carbohydrate and lipid mobilization. They discovered that injection of extracts from the corpora cardiaca of the American cockroach, *Periplaneta americana* L., and the migratory locust, *Locusta migratoria* L. and *Schistocerca gregaria* Forsk., increased the amounts of haemolymph sugars (trehalose) in the cockroach and of haemolymph lipids (diacylglycerol) in the locust. As a result the popular bioassays for the putative existence of hypertrehalosaemic or adipokinetic compounds have been based on measuring the increased concentrations of total carbohydrates or lipids in the haemolymph. We routinely take a 1 μl haemolymph sample from the insect under investigation at time zero, then inject the insect with the compound (or chromatographic fraction) to be analyzed (10 μl volume), and a second 1 μl sample is taken 120 min (cockroaches) or 90 min (locusts) later from the same animal. In some insects, e.g., adults of the Death's Head hawk moth and the Monarch butterfly, we require only 0.5 μl haemolymph samples due to the high lipid concentrations. Small insects (adult meal-worms, for example) receive the compound to be analyzed in a dose of 1 μl (see [7,8]). For analysis of the carbohydrates in the haemolymph we use the anthrone method [9] and for lipids the phosphovanillin method [10] as modified by Holwerda et al. [11]. The developed colors (green for the carbohydrates and pink for the lipids) can be easily determined in an inexpensive filter photometer at the appropriate wavelength. The amount of carbohydrates and lipids present in the sample is quantified by the use of standard curves and can be expressed as mg carbohydrates or lipids/ml of haemolymph. An increase in concentration compared to the pre-injection concentratirons is judged as a positive effect.

Phosphorylase-Activating Bioassay: An Enzymatic Bioassay

It has been demonstrated that the key enzyme for glycogen breakdown, the glycogen phosphorylase, can be activated in the cockroach fat body by an extract of the corpus cardiacum [12]. Later, the activation of locust fat body phosphorylase was reported upon injection of its own corpus cardiacum extract as well as upon injection of small quantities of synthetic Lom-AKH-I [13]. This enzymatic bioassay is easy to perform, and requires only a spectrophotometer (wavelengths 334, 340 or 365 nm for monitoring the absorbance of NADPH). Compared to the colorimetric bioassays described above, however, this bioassay is more time-consuming, and the insects can only be used once. For this bioassay, the desired material is injected into recipients (resting cockroaches or locusts) and the insects are then killed some 15 to 20 min later (e.g., see [13,14]). Individual fat bodies are removed as rapidly as possible and homogenized in 1 ml of ice-cold triethylamine acetate buffer (pH 7.0, 50 mM; containing 5 mM EDTA and 20 mM sodium fluoride). The homogenate is centrifuged for about 6 min at 12000 g, and the layer below the upper fat layer and above the pellet is used for the phosphorylase assay in the physiologically relevant direction of glycogen breakdown as outlined in detail by Ziegler et al. [15]. For calculations, the value for active phosphorylase (in the absence of AMP) is given as the percentage of total phosphorylase activity (in the presence of AMP). In our hands resting or control-injected insects have about 10 to 30% of the phosphorylase in the active state, whereas those insects injected with corpus cardiacum extract or active fractions from chromatographic steps have 60 to 80% of the enzyme in the active form. In general, this enzymatic bioassay is more sensitive than the colorimetric bioassay (e.g., [13]), but fewer assays can be performed per day.

TISSUE SOURCE AND EXTRACTION

Characteristically the nervous system of insects is much smaller than that of vertebrates. Thus, it is also anticipated that the amount of neuropeptide stored in the individual insect is minute. Successful purification of such neuropeptides requires the selection of a suitable insect species (easy to rear, no diapause, short lifespan or, alternatively, easy access to enable field collection to be made). Although rather small amounts of the neuropeptide may be stored in the nervous system (or in the animal) as a whole, peptide-rich regions or organs might be present. Regarding the AKH/RPCH-family peptides the major neurohaemal organ of an insect, the corpus cardiacum, is the best source of extraction. This can easily be shown by injecting extracts from various nervous tissues into the appropriate recipient and measuring the adipokinetic or hypertrehalosaemic effect quantitatively using the established bioassays. The ability to localize most of the biological activity in a specific tissue has many advantages: the bulk of irrelevant material without activity can be discarded and one can work with smaller volumes, and consequently smaller columns; most of the contaminating material (other peptides) is lost and fewer absorbance peaks occur during chromatography (see later); the material concerned is already enriched many times; fewer purification steps are necessary for a pure preparation. Clearly, these advantages outweigh the time spent with the dissection of the corpora cardiaca from individual insects.

Extraction procedures have to be carefully checked when attempting to purify neuropeptides. Although most of the chemical properties of the material to be isolated are not known exactly before the pure material is present and experiments have to be done by trial and error, some guide lines can be given. With the help of a reliable and quantitative bioassay the influence of temperature, pH, proteases inhibitors, solubility in aqueous and organic solvents etc. should be examined. Regarding the AKH/RPCH-family peptides heat-treatment of the extracts is possible, since the molecules are not thermolabile (see [16]). In a comparative study (unpublished results) we placed corpora cardiaca of *L. migratoria* into Eppendorf plastic vials containing various extraction media (c. 100% methanol, 80% methanol, distilled water containing 0.11% trifluoroacetic acid, 25% HPLC-solvent B, methanol:distilled water:acetic acid:thiodiglycol (0.9:9:1:0:1 v/v), 0.1 M acetic acid, 6% aqueous butanol, 50 mM ammonium acetate, pH 8.5) and disrupted the glands with the help of an ultrasonic homogenizer. The resulting homogenate was centrifuged for 3 min at about 10,000 g, the pellet discarded, and the supernatant dried by vacuum centrifugation. The material was taken up in distilled water for the adipokinetic bioassay or in 25% solvent B (0.1% trifluoroacetic acid in 60% acetonitrile) for use in reversed-phase high performance liquid chromatography (RP-HPLC, see later). It became evident that the two adipokinetic hormones from locust corpus cardiacum (Lom-AKH-I and Lom-AKH-II) were extracted identically in all media with the exception of the ammonium acetate (pH 8.5). This is depicted in Figure 1 for 3 examples. It can easily be seen that, although the peak height for the UV absorbance peak for Lom-AKH-I is almost identical for the three extraction media shown, the peak height for Lom-AKH-II is noticeably smaller when ammonium acetate is used as extractant indicating a different chemical behaviour of the latter compound compared to Lom-AKH-I.

Sometimes heavy losses of material do occur. Particularly the hydrophobic AKH/RPCH-family peptides absorb onto glass surfaces (a fact that has been successfully explored in the isolation of Lom-AKH-I by Stone et al. [3], see later), but also tend to stick to the walls of plastic tubes [17]. These authors used bovine serum albumin to counteract the absorbance, but, logically, such methods cannot be used during advanced or final stages of the purification protocol.

SEPARATION

Before the highly sophisticated instruments for HPLC became available, numerous chromatographic methods including paper chromatography and separation on a variety of gel filtration and ion-exchange columns were used to isolate compounds from the AKH/RPCH-family. In their pioneering work on the isolation of the locust adipokinetic hormone I, Stone et al. [3] fractionated methanolic corpora cardiaca extracts by gel

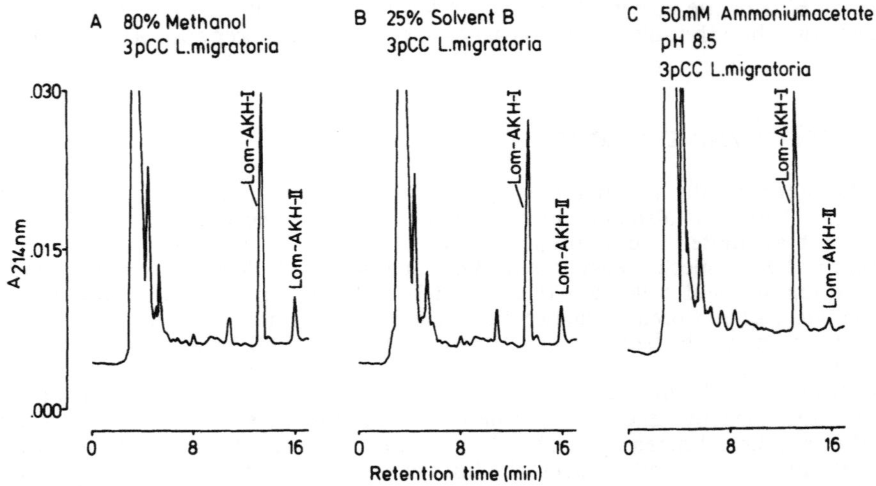

Fig. 1. Separation of an extract from 3 corpora cardiaca (CC) of *Locusta migratoria* on RP-HPLC (Chromatographic conditions: Nucleosil-100 C-18 column; 250 x 4.6 mm; plus 2 cm guard column; solvent A: H_2O + 0.11% TFA, solvent B: 60% acetonitrile + 0.1% TFA; gradient mode: 43 to 53% B in 20 min; flow rate: 1 ml/min). CC-material was extracted in different extraction media. A - 80% methanol extract; B - 25% HPLC solvent B; and C - 50 mM ammonium acetate.

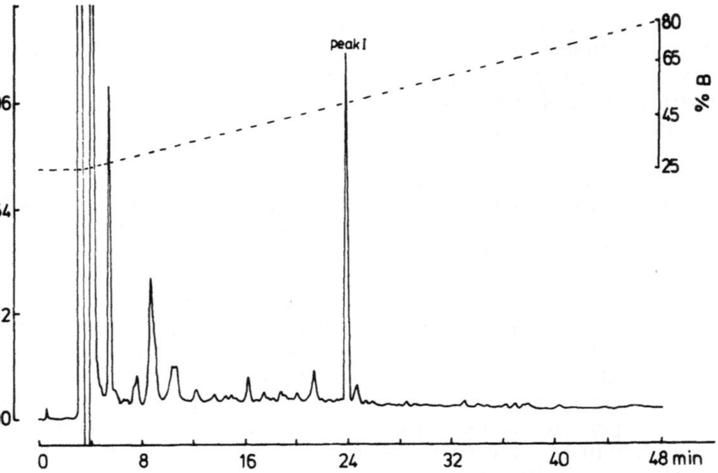

Fig. 2. Separation of an extract of thirty corpora cardiaca from *Nauphoeta cinerea* on RP-HPLC. (Chromatographic conditions: as in Fig. 1, but gradient ran from 25 to 80% B within 45 min; gradient started 3 min after injection). The absorbance peak with the highest absorbance ("peak 1") had the biological activity.

filtration on a controlled-pore glass bead column followed by subjection of the active fraction to thin-layer chromatography (TLC). Another quite successful isolation method was a combination of gel filtration on Sephadex G-25 followed by chromatography in aqueous butanol on Sephadex LH-20 that revealed a second AKH-compound in locust corpora cardiaca [18,19], two active fractions in the corpora cardiaca of the stick insect, *Carausius morosus* [19] and one or two fractions with glycogen phosphorylase activity and adipokinetic activity in the corpora cardiaca of the tobacco hornworm moth, *Manduca sexta* [20].

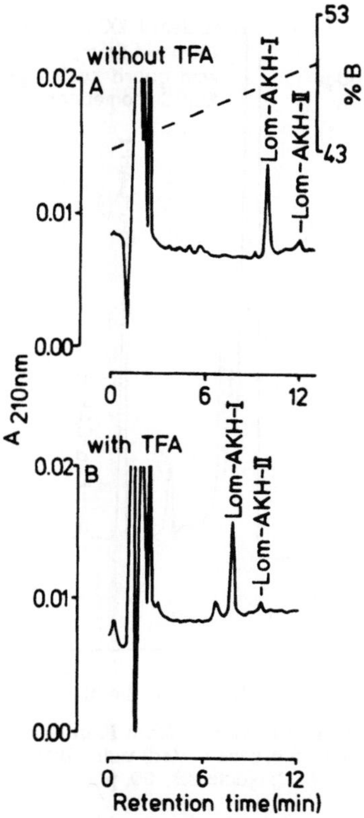

Fig. 3. Separation of a corpus cardiacum extract from *L. migratoria* on RP-HPLC. A: solvents contained no trifluoroacetic acid (TFA), B: solvents contained TFA. (Chromatographic conditions: Nucleosil 100 C-18 column, 125 x 4.6 mm, plus guard column; solvents and gradient as in Fig. 1 if not otherwise stated.)

Today, hydrophobic interactions are mainly used for the isolation of members from the AKH/RPCH-family with the RP-HPLC systems. There have been numerous reports on different columns and solvents used in the isolation. We have successfully introduced a single-step purification [21] by chromatography of the methanolic corpus cardiacum extract on a commercially available C18 bonded Nucleosil 100 column (5-7 μm particle size). Separation was achieved in a gradient mode with an aqueous solvent (A) and an aqueous solution mixed with acetonitrile (B); additionally, both solvents contained about 0.1% trifluoroacetic acid (TFA). Thus, the separation of the neuropeptides was based on partition between a relatively hydrophobic stationary phase and a relatively hydrophillic solvent.

An example of this method is depicted in Figure 2: a crude methanolic extract of corpora cardiaca from the cockroach *Nauphoeta cinerea* was applied to the Nucleosil column. It is evident that, by choosing the right tissue, extraction and separation method, the highest UV absorbance peak is the one containing the biological activity. Although the ion-pair reagent TFA is not necessary for a good separation of these peptides as can be seen for the separation of two adipokinetic hormones from the corpora cardiaca of *L. migratoria* (Figure 3), the incorporation of TFA into the solvents may be advantageous by "moving" other (contaminating) substances away from the peptides to be isolated. Other column support media can also be used for a successful separation. There is, for example, little overall effect in the efficiency of separation of the two hypertrehalosaemic peptides from the corpora cardiaca (Pea-CAH-I and II) of *P. americana* when using C8, C18 or Phenyl Nucelosil 100 columns, although characteristically different retention times occur (Figure 4).

In some cases it may be desirable to separate the corpora cardiaca peptides in the isocratic mode. This has the advantage of a stable baseline (important when low concen-

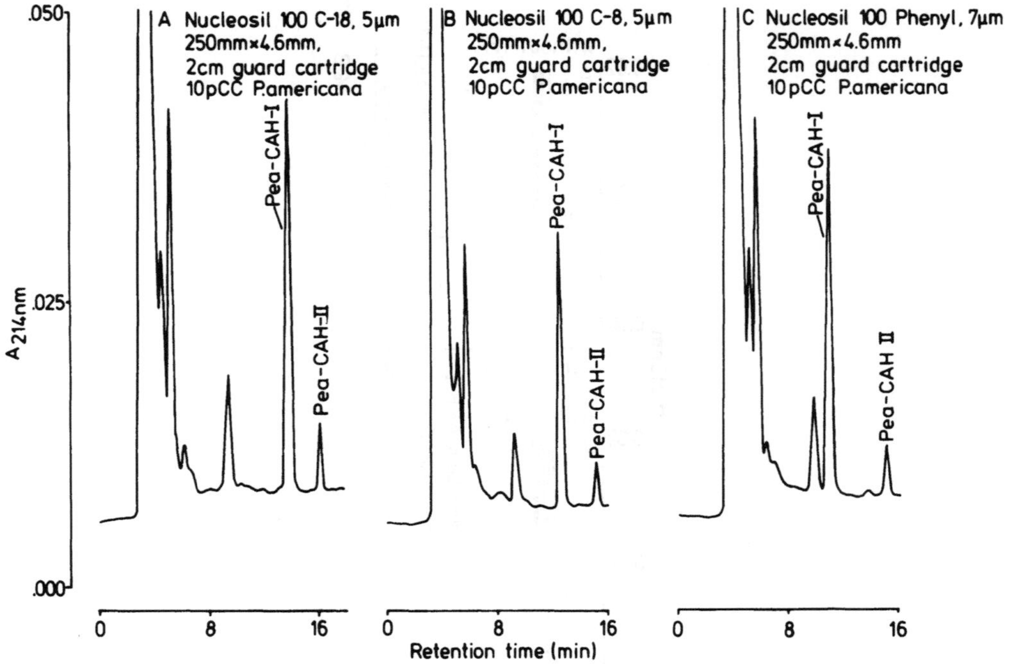

Fig. 4. Separation of a corpus cardiacum extract from *P. americana* on RP-HPLC (Chromatographic conditions as in Fig. 1 if not otherwise stated) using different stationary phases. A: Nucleosil 100, C8; B: Nucleosil 100, C18; C: Nucleosil 100, phenyl.

trations have to be detected at a high detector sensitivity). In addition, HPLC can be done without the need for re-equilibriation enabling more samples to be run. In the isocratic mode the question has to be answered what concentration of the organic modifier (acetonitrile) would provide optimal separation? An example for such isocratic HPLC-runs where the separation conditions differ primarily in the hydrophobicity of the mobile phase is given in Figure 5. We applied the synthetic cockroach hypertrehalosaemic peptide I, Pea-CAH-I, and the second adipokinetic peptide of *S. gregaria*, Scg-AKH-II, onto the column. Evidently, good separations were obtained under all the conditions employed; however, by decreasing the amount of acetonitrile three features become obvious: (1) the retention times were longer, (2) the two peptides moved further away from each other, and (3) the peak shapes become wider. If time of separation is of essence, in our example the higher amount of acetonitrile will separate the two peptides in less than 14 min (Figure 5A). However, if other contaminating compounds in an extract might interfere with complete separation, one might choose conditions as in Figure 5E even though the whole run will take up to 50 min.

Detection is very important during HPLC work, especially when only small amounts of corpora cardiaca extracts, and thus peptides, are available. First of all, for analytical work the detector should preferably be equipped with a small flow cell having a 10 mm path length. Some manufacturers only provide flow cells with a pathlength of 2.5 mm and, thus, a factor of 4 in sensitivity is lost. Detection in the low UV range, desirably at 206 or 210 nm, is optimal for these peptides (Figure 6), requiring acetonitrile with a low cut-off to be used. Filter photometers frequently offer filters of 214 or 227 nm but, as shown in Figure 6, absorbance of Lom-AKH-I is much reduced at these wavelengths.

The resolving power of the RP-HPLC separation technique reported above is best shown in two HPLC-traces from recent unpublished work. During studies on the isolation of adipokinetic peptides from the corpora cardiaca of the three locust species *L. migratoria*, *S. gregaria* and *S. nitans* [22] and the grasshopper *Romalea microptera* [23] we found that the major peak with adipokinetic activity in the three locust species, Lom-AKH-I, had the same retention time in all three species. A second (Scg-AKH-II) peptide eluted at the same time in the two *Schistocerca*-species, but was slightly more hydrophobic in *L. migratoria* (Lom-AKH-II), whereas both peptides in *R. microptera* (Rom-CC-I and II = Grb-AKH) eluted much earlier (thus were less hydrophobic) than the locust peptides. This is shown in Figure 7a for the synthetic peptides demonstrating that differences of only one amino acid residue

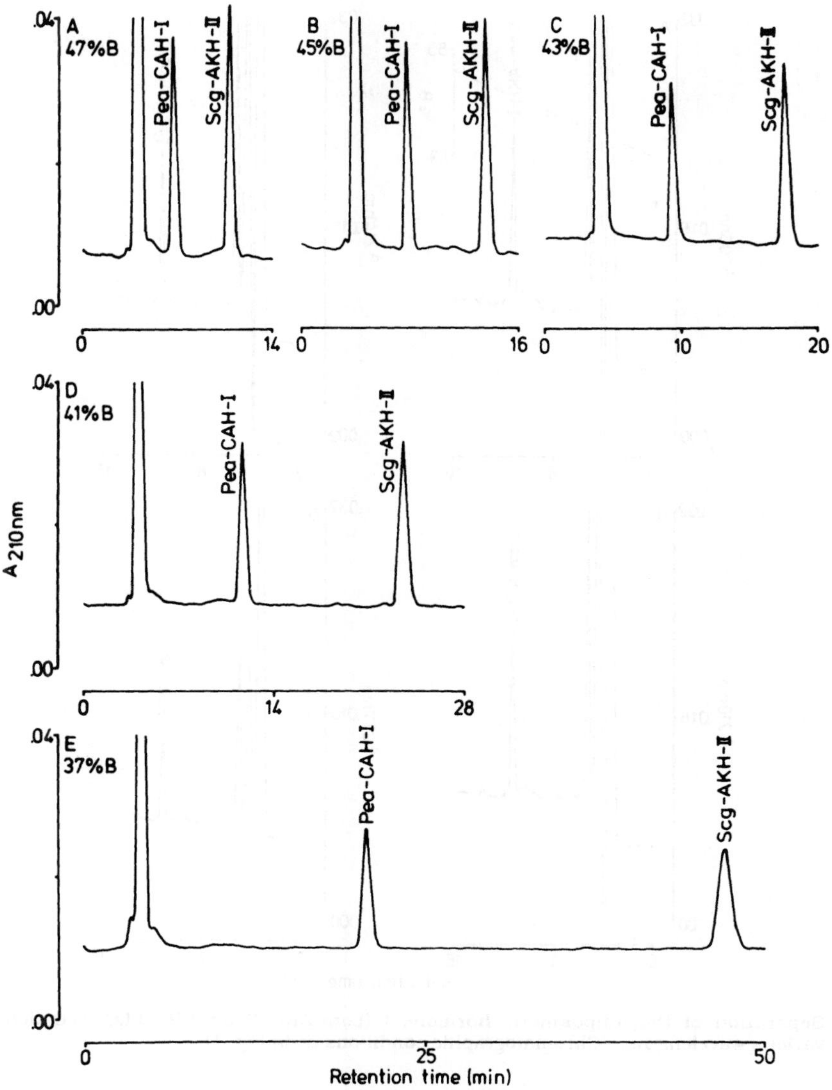

Fig. 5. Isocratic separation of two synthetic peptides (Chromatographic conditions as in Fig. 1) the hyper-trehalosaemic hormone I from *P. americana* (Pea-CAH-I) and the adipokinetic hormone II from *Schistocerca gregaria* (Scg-AKH-II) on RP-HPLC. Note the dependence of retention times upon the percentage of the organic modifier acetonitrile in solvent B.

in the structure in position two for the decapeptides (for sequences see Table 1) can result in different chromatographic properties. The other example (Figure 7b) shows HPLC traces of the synthetic hypertrehalosaemic peptide from *C. morosus* (Cam-HrTH-II, a decapeptide) and an N-terminal modified analogue that lacks the pyroglutamate residues (a nonapeptide) as well as a C-terminal modified analogue that has a Thr-acid instead of the Thr-amide (still a decapeptide). Both analogues are very well separated from the "parent" molecule.

With reliable and efficient bioassays, extraction and HPLC-methods at hand we developed a screening procedure to analyze extracts of corpora cardiaca obtained from different insect species for new members of the AKH/RPCH-family of peptides [21,24,25].

STRUCTURAL INFORMATION

Purified peptides can be subjected to amino acid analysis to obtain the amino acid composition data. Such information is needed to confirm the homogeneity of the material

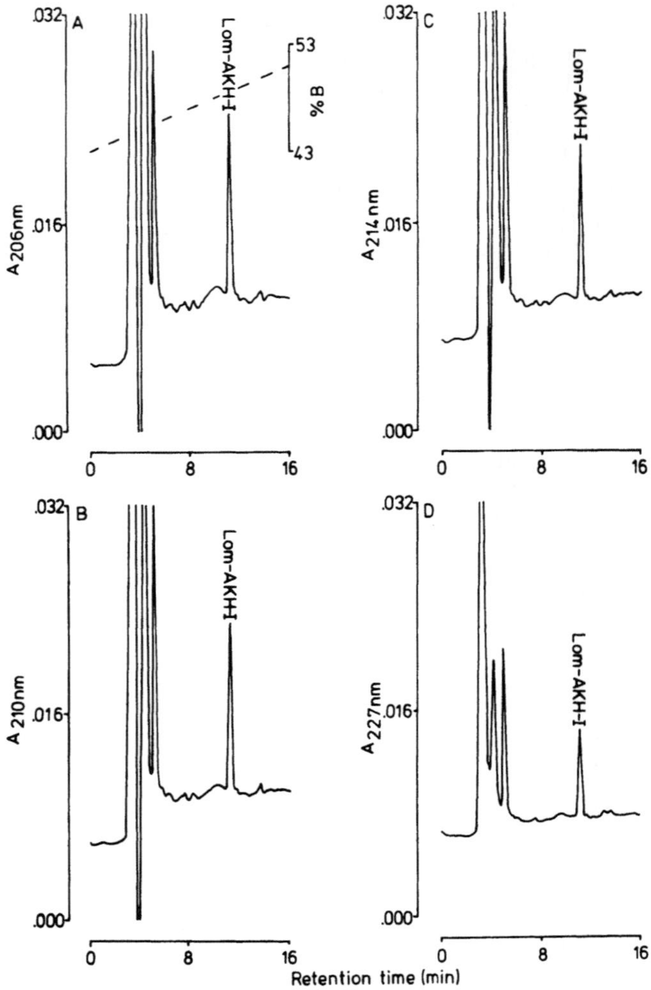

Fig. 6. Separation of the adipokinetic hormone I (Lom-AKH-I) on RP-HPLC and detection at various wavelengths. (Chromatographic conditions as in Fig. 1).

as well as helping in the assignment of the primary structure by gas-phase sequencing and fast atom bombardment mass spectrometry (FAB-MS).

Amino Acid Composition

The peptidic material at hand was hydrolyzed in constant boiling hydrocholoric acid (HCl) for various times. Although the amino acid tryptophan is destroyed by this procedure, it could be detected when a basic hydrolysis was performed in methanesulfonic acid. A second approach, when no tyrosine was found in the hydrolysate with HCl, is the detection of tryptophan in the peptide by HPLC-chromatography using a fluorimeter for detection. Hydrolysis in HCl also converts the amino acid residues asparagine and glutamine to aspartic and glutamic acid respectively.

After hydrolysis with HCl the amino acids can be separated and detected by classical ion-exchange chromatography and post-column ninhydrin derivatization. Alternatively, there are now a variety of pre- and post-column derivatization methods for HPLC available, which have successfully been used to detect amino acids in the low picomole and femtomole range. Most widely used nowadays is the derivatization with phenylisothiocyanate and detection of the phenylthiocarbomyl amino acids at 254 nm [26]. A number of studies have been conducted to analyze the amino acid content of members of the AKH/RPCH-family by the methods described above and have helped to decide whether new members of this family were present in the respective insect species (see, for example, [27-36]).

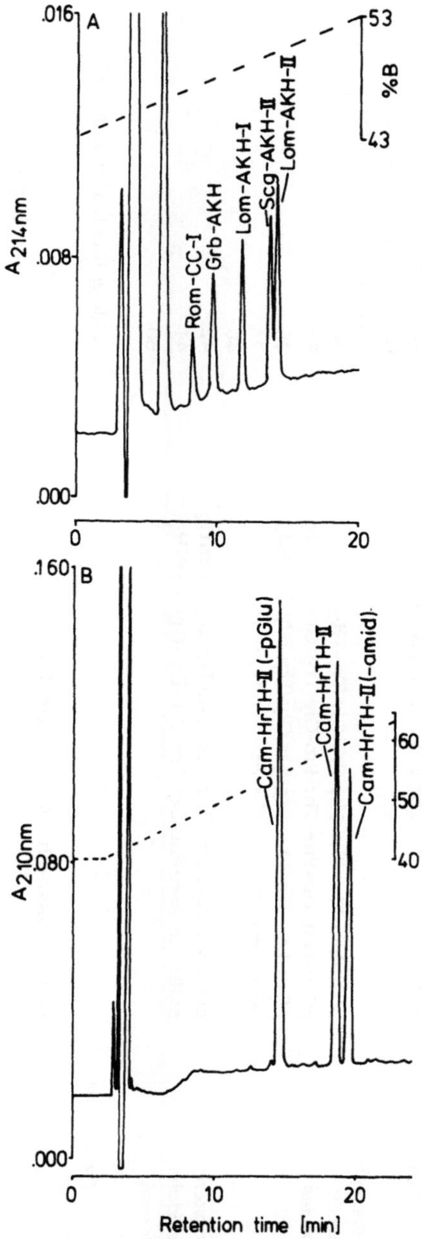

Fig. 7. Separation of A: synthetic naturally occurring peptides from corpora cardiaca of locusts and grasshoppers (see Table 1 for sequences). Separation was easily achieved despite the fact that the peptides differ only in one amino acid residue (conditions as in Fig. 2); B: synthetic hypertrehalosaemic peptide II from stick insect corpus cardiacum and its analogues, one lacking the N-terminal pyroglutamate residue and one without the C-terminal amide (Chromatographic conditions as in Fig. 1 - but the gradient ran from 43 to 65% B within 22 min.)

Sequence Determination

The classical Edman approach is still used successfully for the structural characterization of peptides; by selective cleavage from the amino terminal of the peptide one amino acid after the other is determined. However, the Edman degradation method is only applicable if the N-terminal residue of the peptide is not blocked. As shown by Stone et al. [3] the first member of the AKH/RPCH- family, Lom-AKH-I, was blocked by a pyroglutamate (pGlu) residue at the N-terminus and, additionally, by an amide at the C-terminus. Thus, a

Table 1. Primary structures of the AKH/RPCH-peptide family

Name	Source	Sequence	Reference
Lom-AKH-I	*Locusta migratoria*	pGlu-Leu-Asn-Phe-Thr-Pro-Asn-Trp-Gly-ThrNH_2	3
	Schistocerca gregaria		39
Cam-HrTH-II	*Carausius morosus*	pGlu-Leu-Thr-Phe-Thr-Pro-Asn-Trp-Gly-ThrNH_2	38
	Sipyloidea sipylus		47
	Extatosoma tiaratum		Gäde & Rinehart (submitted)
Pea-CAH-II	*Periplaneta americana*	pGlu-Leu-Thr-Phe-Thr-Pro-Asn-TrpNH_2	17, 43, 48
	Blatta orientalis		Gäde & Rinehart (submitted)
	Leptinotarsa decemlineata		Gäde & Kellner (in press)
Taa-AKH	*Tabanidae species*	pGlu-Leu-Thr-Phe-Thr-Pro-Gly-TrpNH_2	49
Taa-HoTH	*Tabanidae species*	pGlu-Leu-Thr-Phe-Thr-Pro-Gly-Trp-Gly-TyrNH_2	49
Hez-HrTH	*Heliothis zea*	pGlu-Leu-Thr-Phe-Ser-Ser-Gly-Trp-Gly-AsnNH_2	35
Mas-AKH	*Manduca sexta*	pGlu-Leu-Thr-Phe-Thr-Ser-Ser-Trp-GlyNH_2	37
	Heliothis zea		40
Rom-CC-I	*Romalea microptera*	pGlu-Val-Asn-Phe-Thr-Pro-Asn-Trp-Gly-ThrNH_2	42
Bld-HrTH	*Blaberus discoidalis*	pGlu-Val-Asn-Phe-Ser-Pro-Gly-Trp-Gly-ThrNH_2	41
	Nauphoeta cinerea		50
	Leucophaea maderae		
	Gromphadorhina portentosa		Gäde & Rinehart (submitted)
	Blattella germanica		
Pea-CAH-I	*Periplaneta americana*	pGlu-Val-Asn-Phe-Ser-Pro-Asn-TrpNH_2	17, 43, 48
	Blatta orientalis		Gäde & Rinehart (submitted)
	Leptinotarsa decemlineata		Gäde & Kellner (submitted)
RPCH	*Pandalus borealis*	pGlu-Leu-Asn-Phe-Ser-Pro-Gly-TrpNH_2	1
Lom-AKH-II	*Locusta migratoria*	pGlu-Leu-Asn-Phe-Ser-Ala-Gly-TrpNH_2	22, 39
Scg-AKH-II	*Schistocerca gregaria*	pGlu-Leu-Asn-Phe-Ser-Thr-Gly-TrpNH_2	22, 39
	Schistocerca nitans		22
Grb-AKH	*Gryllus bimaculatus*	pGlu-Val-Asn-Phe-Ser-Thr-Gly-TrpNH_2	51
	Romalea microptera		23

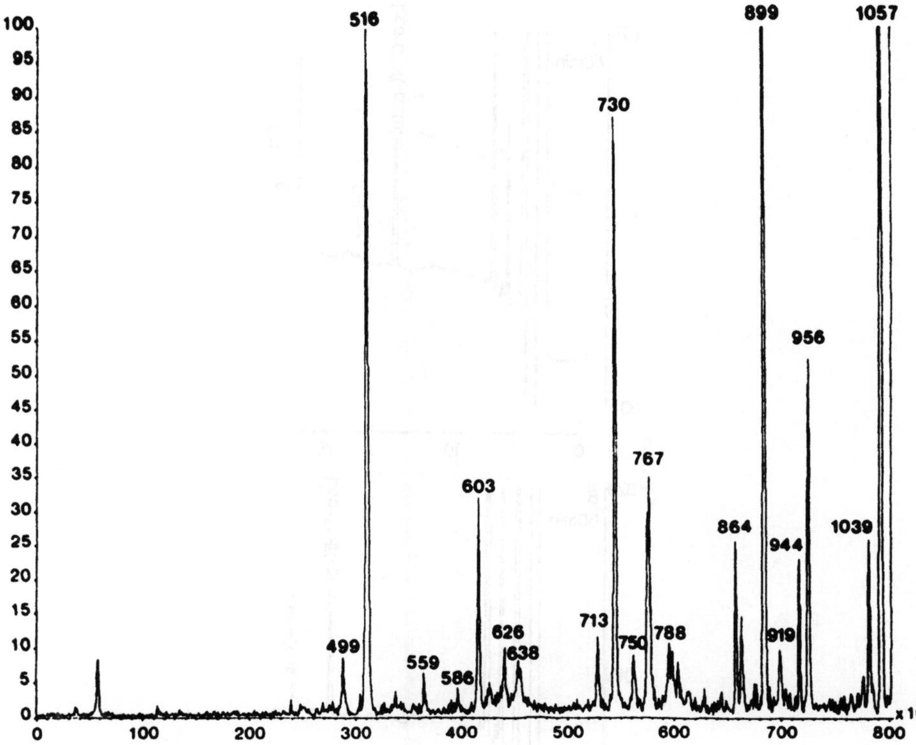

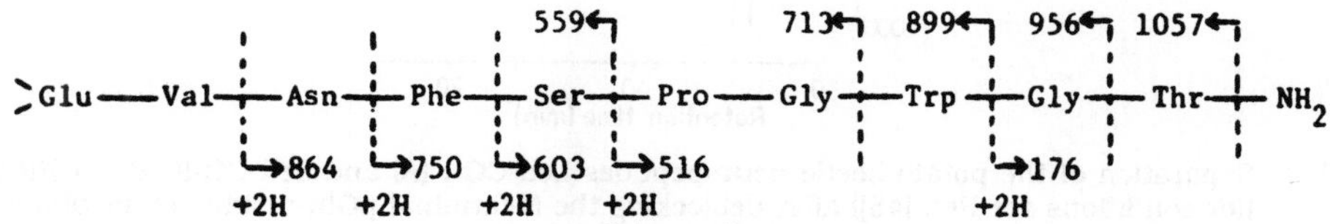

Scheme I

Fig. 8. B/E linked scan of the *m/z* 1074 ion in the FAB mass spectrum of the hypertrehalosaemic neuropeptide of *Nauphoeta cinerea* and structure of the peptide (after 50).

prerequisite for structure determination is a knowledge of whether the molecule has blocked termini or not. This can be demonstrated by lack of mobility in high-voltage electrophoresis, lack of derivatization using the dansyl-Edman procedure (e.g., see [3]) or by the failure to destroy the biological activity of the molecule when incubated with the exopeptidases leucin amino-peptidase (for cleavage at the NH_2-terminal) or carboxypeptidase A (for cleavage at the COOH-terminal) (e.g., [16,33]). The structure of Lom-AKH-I was sequenced by a combination of mass spectrometry and by dansyl-Edman and carboxypeptidase studies on fragments isolated following treatment with the protease thermolysin. Today, the blocked members of the AKH/RPCH-family are either sequenced using FAB-MS or, after deblocking enzymatically the pyroglutamate residue at the N-terminus, using the Edman degradation technique in the form of the most modern modification (gas-phase or pulse-liquid sequencing).

Fast Atom Bombardment Mass Spectrometry (FAB-MS)

In the FAB mode molecular ions are produced without derivatization of the peptide. The matrix used to obtain a FAB spectrum is critical to the success. In Professor Rinehart's laboratory, with whom we collaborate to assign sequences to peptides isolated from the corpora cardiaca of various insects, a liquid mixture of dithioerythritol and dithiothreitol is the favorite matrix used.

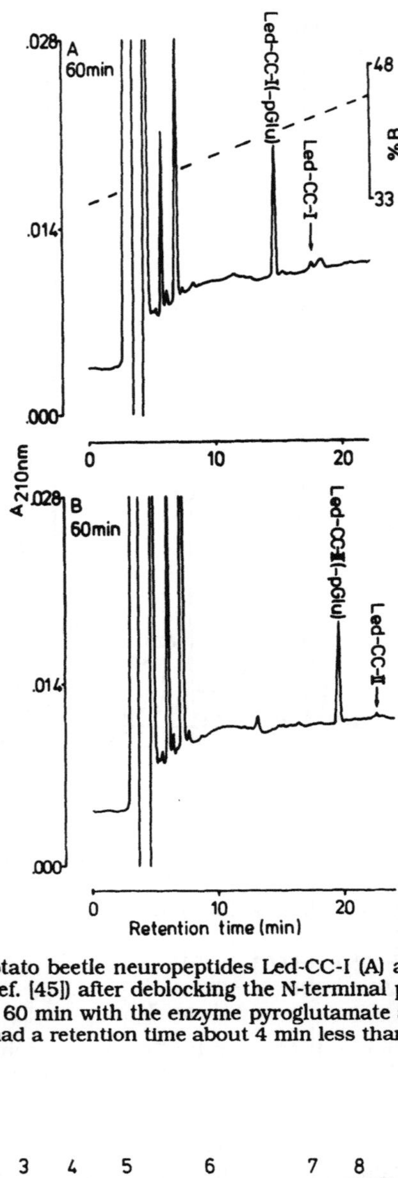

Fig. 9. Separation of the potato beetle neuropeptides Led-CC-I (A) and Led-CC-II (B) on RP-HPLC (for conditions see Ref. [45]) after deblocking the N-terminal pGlu-residue by incubating the purified peptides for 60 min with the enzyme pyroglutamate aminopeptidase. Note that the deblocked peptides had a retention time about 4 min less than those of the blocked peptides.

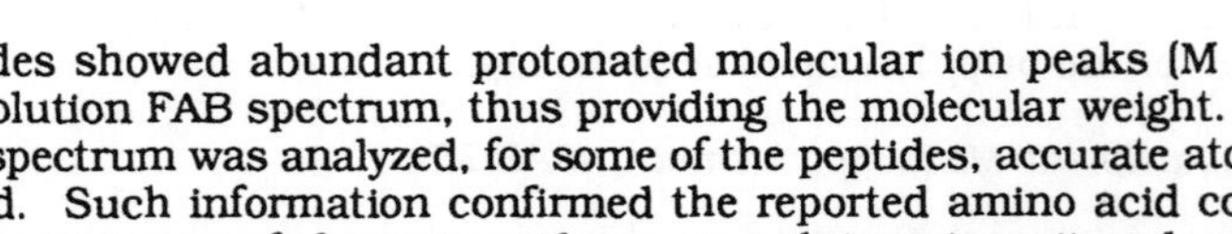

Fig. 10. Variant and invariant residues of the AKH-RPCH-family of peptides. For convenience, a parent molecule of Lom-AKH-I is assumed.

The peptides showed abundant protonated molecular ion peaks (M + H ion) in the normal, low resolution FAB spectrum, thus providing the molecular weight. When the high resolution FAB spectrum was analyzed, for some of the peptides, accurate atomic mass units were determined. Such information confirmed the reported amino acid composition data and indicated the presence of glutamic acid as a pyroglutamate unit and aspartic acid as a asparagine residue. It also showed that the C-terminus was amidated by the presence of an additional NH_2-group. However, some of the peptides, particularly those missing a proline residue in the molecule such as Lom-AKH-II, Scg-AKH-II and Grb-AKH, gave only limited or almost no fragmentation at amide bonds in the normal FAB spectrum. The sequences could, however, be assigned by consideration of the metastable fragmentation of the M + H

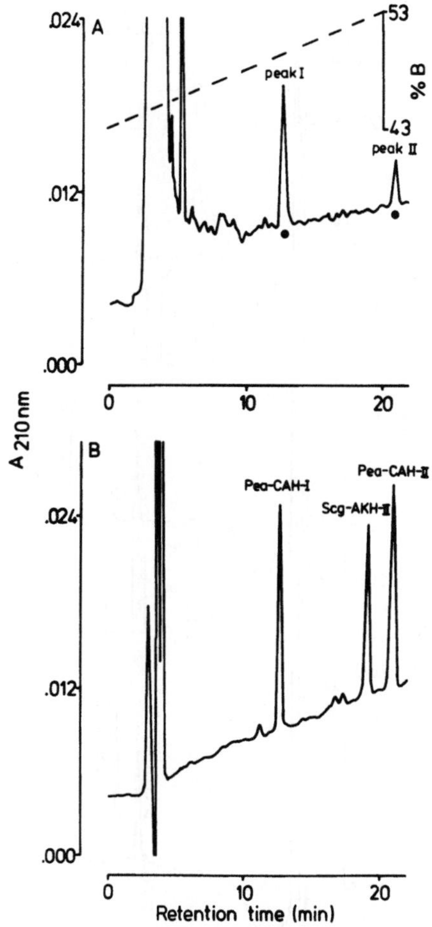

Fig. 11. Separation of A: an extract from corpora cardiaca of *Leptinotarsa decemlineata* (30 glands) on RP-HPLC (black dots indicate biological activity in cockroaches) and B: synthetic peptides (100 pmol). Note that the two neuropeptides from the Colorado beetle coelute with the cockroach peptides Pea-CAH-I and II. Chromatographic conditions as in Ref. [25].

residue in the molecule such as Lom-AKH-II, Scg-AKH-II and Grb-AKH, gave only limited or almost no fragmentation at amide bonds in the normal FAB spectrum. The sequences could, however, be assigned by consideration of the metastable fragmentation of the M + H ion, derived from a B/E scan (in this mode the flux density B/deflecting voltage E = constant; see Figure 8). In recent studies (peptides of *B. orientalis, L. maderae, B. germanica, G. protentosa* and *E. tiaratum*) a 4-sector ZAB-SE mass spectrometer (tandem mass spectrometry, MS/MS) was used and daughter ion peaks could be identified at much higher resolution and precision, thus giving accurate unit masses rather than 2 to 3 mass band width. Such a MS/MS technique has been previously successfully used to elucidate the sequence of Mas-AKH [37]. This technique was found to be superior to the B/E linked scan and MIKES techniques (mass-analyzed ion kinetic spectrometry) in a 2-sector mass spectrometer with a synthetic sample of Cam-HrTH [38].

Gas-Phase and Pulsed-Liquid Phase Sequencing

Using modifications of the Edman degradation technique very reliable and sensitive sequencing of small amounts of peptides can be performed using automated gas-phase and pulsed-liquid phase sequenators. After each cycle the cleaved amino acid is detected as the phenylthiohydantion derivatized amino acid by on-line HPLC chromatography.

However, before this method can be employed for peptides of the AKH/RPCH-family, the N-terminus has to be deblocked. As shown by Siegert et al. [39], for the two adipokinetic

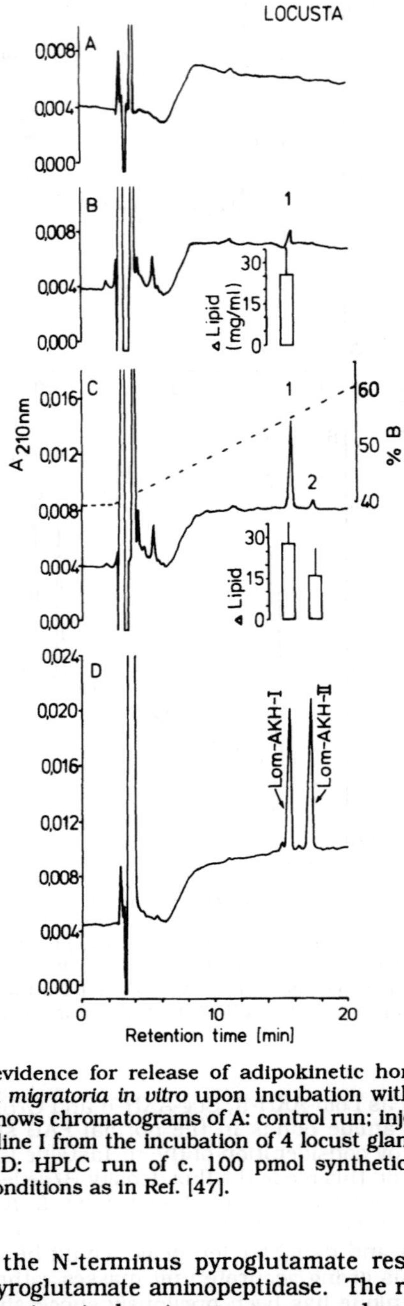

Fig. 12. Chromatographic evidence for release of adipokinetic hormones I and II from corpora cardiaca of *Locusta migratoria in vitro* upon incubation with high potassium and calcium saline. The figure shows chromatograms of A: control run; injection of 50 µl of 25% solvent B; B: HPLC runs of saline I from the incubation of 4 locust glands; C: HPLC run of saline III, 4 glands incubated; D: HPLC run of c. 100 pmol synthetic Lom-AKH-I and Lom-AKH-II. Chromatographic conditions as in Ref. [47].

hormones II of locusts, the N-terminus pyroglutamate residue can be enzymatically removed by the enzyme pyroglutamate aminopeptidase. The resulting HPLC-isolated fragment, which now had a free terminal amino group, was subsequently sequenced using the automated gas-phase mode. The primary sequences of other peptides of the family (e.g., Hez-HrTH [35], Mas-AKH [40], Bld-HrTH [41], Rom-CC-I [23] and Pea-CAH-I and II [43] have all been elucidated in part by this technique. A recent example of our work is given in Figure 9. Two peptides were isolated from the corpora cardiaca of the Colorado potato beetle, *Leptinotarsa decemlineata*, which had identical HPLC-retention times as the hypertrehalosaemic hormones Pea-CAH-I and II from *P. americana* [24] and also had the same amino acid composition [25]. After incubating the pure peptides, tentatively designated as Led-CC-I and II, separately with the protease pyroglutamate aminopeptidase for c. 60 min, the incubation mixtures were chromatographed by RP-HPLC (Figure 9) and new peaks of the deblocked peptides eluting about 4 min earlier than the blocked ones were found. Pulsed-liquid analyses of these fragments showed heptapeptides without the pyroglutamate residue and unequivocally confirmed that the complete peptides had indeed the same structure as the *P. americana* peptides Pea-CAH-I and II [45].

It is evident from Figure 10 that the peptides of this family share not only the blocked N- and C-termini, but also a pyroglutamate residue at position 1, phenylalanine at position 4, tryptophan at position 8, and, if longer than an octapeptide, glycine at position 9. The second residue may be leucine or valine, the third residue asparagine or threonine and the fifth residue serine or threonine. Position 6 shows repalcements of proline by serine, alanine or threonine, whereas the variations at position 7 can be asparagine, glycine or serine and at position 10 threonine, tyrosine or asparagine.

CONFIRMATORY STUDIES

Once a peptide has been isolated and sequenced, the ultimate proof of a correct structural assignment is the preparation of a synthetic peptide and confirmation of its chemical and biological properties by experimentation.

As an example, Figure 11 shows the chromatographic comparison of the natural (from a corpus cardiacum extract) peptides from the Colorado potato beetle and the synthetic peptides Pea-CAH-I and II on RP-HPLC. Such chromatographic evidence should be obtained for a number of different chromatographic systems (various gradients; different solvents). Further confirmation is acheived by comparing the fragmentation pattern of natural and synthetic peptides using FAB-MS (e.g., [38]) and by comparing the biological activity in the respective bioassay (e.g., [42]).

USE OF THE SYNTHETIC PEPTIDES

As an example of the use of synthetic peptides, studies on the release of peptides from the AKH/RPCH-family are discussed. Corpora cardiaca of *L. migratoria* were placed *in vitro* successively in saline containing different concentrations of potassium and calcium, and the respective incubation medium subsequently directly applied to RP-HPLC without any clean-up procedures. The resulting chromatograms showed that a massive release of material only occurred when the cells were depolarized by high potassium in the presence of calcium (Figure 12A-D). By comparison with synthetic Lom-AKH-I and II in the same chromatographic system it becomes evident that both compounds were released from the preparation. A standard curve (peak area versus concentration) can be constructed for the synthetic peptides (see [44]) allowing the released hormones to be quantified. If it is anticipated that very small amounts of such peptides will be released, detection by fluorimetric methods can be used instead of UV detection. Because of the tryptophan residue present in all members of the AKH-RPCH-peptides, excitation at 276 nm and emission at 350 nm allows as little as 5 pmol to be clearly detected (Figure 13). Even smaller quantities

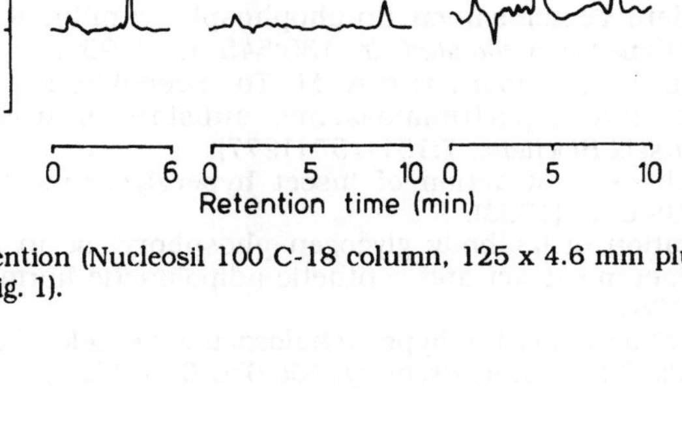

Fig. 13. RP-HPLC sequention (Nucleosil 100 C-18 column, 125 x 4.6 mm plus 2 cm guard column, solvents as in Fig. 1).

might be detectable at the highest instrument sensitivity. Undoubtedly, radioimmunoassay (RIA) would be an even better method for quantification. However, labeling by iodination of tyrosine residue using ^{125}I is not possible with the unmodified AKH/RPCH-peptides because they lack a tyrosine residue. Recently, Moshitzky et al. [46] published a RIA method using an antiserum raised to a Lom-AKH-I analogue having a glutamate residue at position one instead of the pyroglutamate; such a peptide, however, is biologically inactive.

CONCLUSIONS

Facilitated by advances in HPLC, gas-phase sequencing, FAB-MS and automated peptide synthesis many members of the AKH/RPCH-peptide family have been isolated, sequenced and are available commercially in synthetic form. Considering the vast numbers of insect species and the as yet discovered only slightly modified AKH/RPCH-peptides, it is highly probable that the number of new members of this family of insect peptides will grow in the near future.

ACKNOWLEDGEMENTS

The author is thankful for expert technical help with recent experiments by Miss Beate Schumacher, for secretarial help by Miss Heidi Sturm, and for financial support by the Deutsche Forschungsgemeinschaft (grants Ga 241/6-1 to 6-4). The author is a recipient of a Heisenberg Fellowship awarded by the Deutsche Forschungsgemeinschaft (Ga 241/5-2).

REFERENCES

1. P. Fernlund and L. Josefsson, Crustacean color-change hormone: amino acid sequence and chemical synthesis, *Science*, 177:173-175 (1972).
2. A. K. Raina and G. Gäde, Insect peptide nomenclature, *Insect Biochem.*, 18:785-787 (1988).
3. J. V. Stone, W. Mordue, K. E. Batley and H. R. Morris, Structure of locust adipokinetic hormone, a neurohormone that regulates lipid utilization during flight, *Nature*, 263:207-211 (1976).
4. J. E. Steele, Occurrence of a hyperglycaemic factor in the corpus cardiacum of an insect, *Nature*, 192:680-681 (1961).
5. A. M. Th. Beenakkers, The influence of corpus allatum and cardiacum on lipid metabolism in *Locusta*, *Gen. Comp. Endocrinol.*, 13:492 (1969).
6. R. J. Mayer and D. J. Candy, Control of haemolymph lipid concentration during locust flight: an adipokinetic hormone from the corpora cardiaca, *J. Insect Physiol.*, 15:611-620 (1969).
7. G. Gäde, Studies on the hypertrehalosaemic factor from the corpus cardiacum/corpus allatum complex of the beetle, *Tenebrio molitor*, *Comp. Biochem. Physiol.*, 91A:333-338 (1988).
8. G. Gäde, On the release and action of the hypertrehalosaemic hormone from the cockroach *Nauphoeta cinerea*, *Zeitschrift für Naturforschung*, 43c:108-116 (1988).
9. G. Spik and J. Montreuil, Deux causes d'erreur dans les dosages colorimetriques des oses neutres totaux, *Bulletin de la Societe Chimique et Biologique*, 46:739-749 (1964).
10. N. Zöllner and K. Kirsch, Uber die quantitative Bestimmung von Lipoiden (Mikromethode) mittels der vielen naturlichen Lipoiden (allen bekannten Plasmalipoiden) gemeinsamen Sulphophospho-vanillin Reaktion, *Zeitschrift für die Gesamte Experimentelle Medizin*, 135:545-561 (1962).
11. D. A. Holwerda, J. van Doorn and A. M. Th. Beenakkers, Characterization of the adipokinetic and hypertrehalosaemic substances from the locust corpus cardiacum, *Insect Biochem.*, 7:151-157 (1977).
12. J. E. Steele, The site of action of insect hyperglycaemic hormone, *Gen. Comp. Endocrinol.*, 3:46-52 (1963).
13. G. Gäde, Activation of fat body glycogen phosphorylase in *Locusta migratoria* by corpus cardiacum extract and synthetic adipokinetic hormone, *J. Insect Physiol.*, 27:155-161 (1981).
14. G. Gäde, Mode of action of the hypertrehalosaemic peptides from the American cockroach, *Zeitschrift für Naturforschung*, 40c:670-676 (1985).

15. R. Ziegler, M. Ashida, A. M. Fallon, L. T. Wimer, S. S. Wyatt and G. R. Wyatt, Regulation of glycogen phosphorylase in fat body of cecropia silkmoth pupae, *J. Comp. Physiol.*, 131:321-332 (1979).
16. G. Gäde, Further characteristics of adipokinetic and hyperglycaemic factor(s) of stick insects, *J. Insect Physiol.*, 26:351-360 (1980).
17. R. M. Scarborough, G. C. Jamieson, F. Kalish, S. J. Kramer, G. A. McEnroe, C. A. Miller and D. A. Schooley, Isolation and primary structure of two peptides with cardioacceleratory and hyperglycemic activity from the corpora cardiaca of *Periplaneta americana*, *Proc. Nat. Acad. Sci. USA*, 81:5575-5579 (1984).
18. J. Carlsen, W. S. Herman, M. Christensen and L. Josefsson, Characterization of a second peptide with adipokinetic and red pigment-concentrating activity from the locust corpora cardiaca, *Insect Biochem.*, 9:497-501 (1979).
19. G. Gäde, Studies on a phosphorylase-activating factor in the corpora cardiaca of stick insects: characterization and preliminary purification, *Zoologische Jahrbücher, Abt. Physiologie*, 85:266-277 (1981).
20. R. Ziegler and G. Gäde, Preliminary characterization of glycogen phosphorylase activating hormone and adipokinetic hormone from *Manduca sexta* corpora cardiaca, *Physiol. Entomol.*, 9:229-236 (1984).
21. G. Gäde, G. J. Goldsworthy, G. Kegel and R. Keller, Single step purification of locust adipokinetic hormones I and II by reversed-phase high performance liquid chromatography and amino acid composition of the hormone II, *Hoppe Seyler's Zeitschrift für Physiologische Chemie*, 365:391-398 (1984).
22. G. Gäde, G. J. Goldsworthy, M. H. Schaffer, J. C. Cook and K. L. Rinehart, Jr., Sequence analyses of adipokinetic hormones II from corpora cardiaca of *Schistocerca nitans*, *Schistocerca gregaria* and *Locusta migratoria* by fast atom bombardment mass spectrometry, *Biochem. Biophys. Res. Comm.*, 134:723-730 (1986).
23. G. Gäde, C. Hilbich, K. Beyreuther and K. L. Rinehart, Sequence analyses of two neuropeptides of the AKH/RPCH-family from the lubber grasshopper, *Romalea microptera*, *Peptides*, 9:681-688 (1988).
24. G. Gäde and M. Scheid, A comparative study on the isolation of adipokinetic and hypertrehalosaemic factors from insect corpora cardiaca, *Physiol. Entomol.*, 11:145-157 (1986).
25. G. Gäde, Characterization of neuropeptides of the AKH-RPCH-family from corpora cardiaca of Coleoptera, *J. Comp. Physiol.*, 89: 589 (1989).
26. R. L. Heinrikson and S. C. Meredith, Amino acid analysis by reverse-phase high performance liquid chromatography. Precolumn derivatization with phenylisothiocyanate, *Anal. Biochem.*, 136:65-74 (1984).
27. M. O'Shea, J. Witten and M. Schaffer, Isolation and characterization of two myoactive peptides: further evidence of an invertebrate peptide family, *Neurosci.*, 4:521-529 (1984).
28. G. Gäde, Adipokinetic and hyperglycaemic factors of different insect species: separation with high-performance liquid chromatography, *J. Insect Physiol.*, 30:729-736 (1984).
29. R. Ziegler, G. Kegel and R. Keller, Isolation and amino-acid composition of the adipokinetic hormone of *Manduca sexta*, *Hoppe Seyler's Zeitschrift für Physiologische Chemie*, 365:1451-1456 (1984).
30. G. Gäde, Amino acid composition of cockroach hypertrehalosaemic hormones, Zeitschrift für Naturforschung, 40c:42-46 (1985).
31. G. Gäde, Isolation of the hypertrehalosaemic factors I and II from the corpus cardiacum of the Indian stick insect, *Carausius morosus*, by reversed-phase high-performance liquid chromatography, and amino-acid composition of factor II, *Biol. Chem: Hoppe Seyler*, 366:175-179 (1985).
32. G. Gäde, Characterisation and amino acid composition of a hypertrehalosaemic neuropeptide from the corpora cardiaca of the cockroach, *Nauphoeta cinerea*, *Zeitschrift für Naturforschung*, 42c:225-230 (1987).
33. G. Gäde, Isolation and amino acid composition of the adipokinetic neuropeptide from the corpus cardiacum of the cricket, *Gryllus bimaculatus*, *Physiol. Entomol.*, 12:309-316 (1987).
34. G. Gäde, Isolation and amino acid composition of two new neuropeptides from the corpus cardiacum of the Eastern lubber grasshopper, *Romalea microptera*, belonging to the AKH/RPCH peptide family, *Zoologische Jahrbücher, Abt. Physiologie*, 91:307-313 (1987).
35. H. Jaffe, A. K. Raina, C. T. Riley, B. A. Fraser, T. G. Bird, C.-M. Tseng, Y.-S. Zhang and D. K. Hayes, Isolation and primary structure of a neuropeptide hormone from *Heliothis zea* with hypertrehalosaemic and adipokinetic activities, *Biochem. Biophys. Res. Comm.*, 155:344-350 (1988).

36. H. Jaffe, A. K. Raina, B. A. Fraser, P. Keim, K. R. Rao, Y.-S. Zhang, J. L. Lancaster and D. K. Hayes, Isolation of two neuropeptides in the AKH/RPCH-family from horseflies (Diptera), *Biochem. Biophys. Res. Comm.*, 151:656-663 (1988).
37. R. Ziegler, K. Eckart, H. Schwarz and R. Keller, Amino acid sequence of Manduca sexta adipokinetic hormone elucidated by combined fast atom bombardment (FAB)/tandem mass spectrometry, *Biochem. Biophys. Res. Comm.*, 133:337-342 (1985).
38. G. Gäde and K. L. Rinehart, Jr., Primary structure of the hypertrehalosaemic factor II from the corpus cardiacum of the Indian stick insect, *Carausius morosus*, determined by fast atom bombardment mass spectrometry, *Biol. Chem: Hoppe Seyler*, 368:67-75 (1987).
39. K. Siegert, P. Morgan and W. Mordue, Primary structures of locust adipokinetic hormones II, *Biol. Chem: Hoppe Seyler*, 366:723-727 (1985).
40. H. Jaffe, A. K. Raina, C. T. Riley, B. A. Fraser, G. M. Holman, R. M. Wagner, R. L. Ridgway and D. K. Hayes, Isolation and primary structure of a peptide from the corpora cardiaca of *Heliothis zea* with adipokinetic activity, *Biochem. Biophys. Res. Comm.*, 135:622-628 (1986).
41. T. K. Hayes, L. L.Keeley and D. W. Knight, Insect hypertrehalosaemic hormone: isolation and primary structure from *Blaberus discoidalis* cockroaches, *Biochem. Biophys. Res. Comm.*, 140:674-678 (1986).
42. J. H. Spring and G. Gäde, Factors regulating carbohydrate and lipid metabolism isolated from the corpus cardiacum of the Eastern lubber grasshopper, *Romalea microptera, J. Exp. Zool.*, 241:41-50 (1987).
43. K. J. Siegert and W. Mordue, Elucidation of the primary structures of the cockroach hyperglycaemic hormones I and II using enzyamtic techniques and gas-phase sequencing, *Physiol. Entomol.*, 11:205-211 (1986).
44. G. Gäde and R. Kellner, The metabolic neuropeptides of the corpus cardiacum from the potato beetle and the American cockroach are identical, *Peptides*, in press (1989).
45. K. J. Siegert and W. Mordue, Quantification of adipokinetic hormones I and II in the corpora cardiaca of *Schistocerca gregaria* and *Locusta migratoria*, *Comp. Biochem. Physiol.*, 84A:279-284 (1986).
46. P. Moshitzky, D. Yamashiro, L. Stuve, J. Ramachandran and S. W. Applebaum, Determination of locust AKH-I by radioimmunoassay and the identification of an AKH-like factor in the locust brain, *Insect Biochem.*, 17:765-769 (1986).
47. G. Gäde, Isolation, physiological characterization, release and sequence elucidation of a hypertrehalosaemic neuropeptide from the corpus cardiacum of the stick insect, *Sipyloidea sipylus*, *Physiol. Entomol.*, 14:405-418 (1989).
48. J. Witten, M. H. Schaffer, M. O'Shea, J. C. Cook, M. E. Hemling and K. L. Rinehart, Structures of two cockroach neuropeptides assigned by fast atom bombardment mass spectrometry, *Biochem. Biophys. Res. Comm.*, 124:350-358 (1984).
49. H. Jaffe, A. K. Raina, C. T. Riley, B. A. Fraser, R. J. Nachman, V. W. Vogel, Y.-S. Zhang and D. K. Hayes, *Proc. Nat. Acad. Sci. USA*, 86:8161-8164 .
50. G. Gäde and K. L. Rinehart, Jr., Amino acid sequence of a hypertrehalosaemic neuropeptide from the corpus cardiacum of the cockroach *Nauphoeta cinerea*, *Biochem. Biophys. Res. Comm.*, 141:774-781 (1986).
51. G. Gäde and K. L. Rinehart, Primary sequence analysis by fast atom bombardment mass spectrometry of a peptide with adipokinetic activity from the corpora cardiaca of the cricket *Gryllus bimaculatus*, *Biochem. Biophys. Res. Comm.*, 149:908-914 (1987).

IN VITRO BIOSYNTHESIS OF LOCUST ADIPOKINETIC HORMONES: ISOLATION AND IDENTIFICATION OF THE BIOACTIVE PEPTIDES AND THEIR PROHORMONES

R. C. H. M. Oudejans, F. P. Kooiman,
T. K. F. Schulz and A. M. Th. Beenakkers

Department of Experimental Zoology
University of Utrecht, 8 Padualaan
3508 TB Utrecht, The Netherlands

SUMMARY

The biosynthesis of adipokinetic hormones (AKH) I and II and their precursors was studied in the migratory locust *Locusta migratoria* using radiolabeled amino acids in an *in vitro* system. The peptides from the glandular part of the corpora cardiaca (CC) were extracted and separated by reversed-phase high performance liquid chromatography (HPLC). Incorporation of the [^{3}H] amino acids was measured by monitoring the radioactivity of the column eluate. Total time for biosynthesis and processing of the prohormones to the bioactive AKH I and II was 75 min. Only the glandular part of the CC synthesized these peptide hormones.

Protein fractions from CC were separated by gradient gel electrophoresis (PAGE or SDS-PAGE) and by two-dimensional gel electrophoresis (isoelectrofocusing followed by SDS-PAGE). After electrophoresis, proteins were transferred to nitrocellulose membranes by electroblotting and the incorporation of radiolabel was measured with an automatic linear analyzer. Time-course studies and pulse-labeling experiments revealed incorporation of radiolabel into most proteins.

By using monoclonal antibodies raised against AKH I, immunoreactive protein bands (prohormones and/or processing products) with a M_r of 72,000 and 33,000 were only detectable in the cytoplasm fraction of the glandular lobe of the CC.

INTRODUCTION

In locusts, adipokinetic hormones (AKHs) play a pivotal role in the mobilization of stored lipids as a fuel for sustained flight activity [1-3]. The hormones are synthesized in the glandular lobe of the corpora cardiaca (CC) and stored in secretory granula [4]; during flight small amounts are released into the haemolymph. From the migratory locust, *Locusta migratoria*, the amino acid sequence of AKH I (pGlu-Leu-Asn-Phe-Thr-Pro-Asn-Trp-Gly-Thr-NH_2) was determined by Stone et al. [5] and that of AKH II (pGlu-Leu-Asn-Phe-Ser-Ala-Gly-Trp-NH_2) by Siegert et al. [6]. Both hormones are blocked peptides and are members of the AKH/RPCH (red pigment concentrating hormone) family, all of which are characterized by a pyroglutamyl residue at the N-terminus and an amide at the C-terminus. All insect orders investigated to date contain neuropeptides belonging to this family and demonstrating several physiological actions such as hyperlipaemic, hypertrehalosaemic or myotropic effects [2,7].

Although the physiological aspects of the AKHs in locusts have been studied very extensively, only a few reports on their biosynthesis have been published [8-11].

We have studied the biosynthesis of AKH I and II in the migratory locust, *Locusta migratoria*, using [^{3}H] amino acids in an *in vitro* system. Apart from the incorporation of radiolabel into the bioactive peptides, we have also attempted to identify the precursor

Chromatography and Isolation of Insect Hormones and Pheromones
Edited by A. R. McCaffery and I. D. Wilson
Plenum Press, New York, 1990

proteins (prohormones) using monoclonal antibodies raised against AKH I. Preliminary reports have been published elsewhere [12,13].

MATERIALS AND METHODS

Insects

Locusta migratoria migratorioides were reared in the laboratory under crowded conditions as described previously [14]. Both adult males and females (12-16 days after the imaginal ecdysis) were used.

Materials

Synthetic AKH I, AKH II and tyr^1-AKH I were purchased from Peninsula Labs Inc. (Belmont, CA). L-phenyl-[2,3-^{3}H]-alanine (spec. act. 1.59 TBq/mmol) and L-proline-[5-^{3}H] (spec. act. 555 GBq/mmol) were obtained from Amersham Labs (Amersham, UK).

Grace's medium was obtained from Gibco (Paisley, UK). Streptomycin and penicillin were from Boehringer (Almere, The Netherlands), acetonitrile (HPLC-grade) was from Rathburn Chemicals (Walkerburn, UK) and trifluoroacetic acid (HPLC-grade) from Merck (Darmstadt, FRG). All other chemicals were of analytical grade and were purchased from local distributors.

Incubation Media

Several incubation media were tested for *in vitro* organ culture and AKH biosynthesis: locust haemolymph, deproteinized locust haemolymph, phosphate-buffered saline and media according to Grace [15], Pannabecker & Orchard [16] and Hekimi & O'Shea [8]. Haemolymph was collected and diluted 1:1 (by volume) with bidistilled water together with some crystals of phenylthiourea to prevent clotting [17]. After centrifugation for 6 min the lipid-free supernatant was applied onto a Sephadex G-25 column. The proteins in the void volume of the eluate were discarded and fractions containing the protein-free haemolymph were collected, freeze-dried and resolved in bidistilled water to a final concentration of 80% of the original haemolymph volume. Phosphate-buffered saline contained 8.0 g NaCl, 0.2 g KCl, 1.15 g Na_2HPO_4 and 0.21 g NaH_2PO_4/l. Grace's medium was as described by Grace [15], except changes made by the manufacturer for $CaCl_2$ which was lowered to 750 mg/l and KCl which was elevated to 4100 mg/l. Medium according to Hekimi & O'Shea [8] was modified by using TRIS- (pH 7.2) or HEPES- (pH 7.2) instead of TES-buffer.

In order to prevent growth of microorganisms, in the final incubation media a mixture of the antibiotics' streptomycin (0.1 mg/ml) and penicillin (100 units/ml) was always added.

In Vitro *Culture of Corpora Cardiaca and Biosynthesis of AKHs*

Complete corpora cardiaca or glandular parts of the organs were quickly excised and placed in 0.5 to 1 ml of incubation medium. Standard incubation experiments were performed at 30°C for up to 24 h in Eppendorff cups, using 10 to 15 CC in 50 µl of the appropriate incubation medium with addition of 185 to 370 KBq of [^{3}H] amino acid. Incubations were ended by thoroughly washing off the incubation medium and AKHs were extracted as described below.

In pulse-chase experiments, pulse-labeling was carried out for 120 min as described for standard incubations followed by an incubation in a chase medium in which the [^{3}H] amino acid was omitted.

Differential Pelleting

In some experiments, after the incubation time the CC were separated by differential pelleting according to Stone & Mordue [18] into nuclei, mitochondria, granula and cytoplasm.

Extraction of AKHs

CC were sonicated in 300 µl of methanol-water-acetic acid (100:10:1 v/v) and extracted three times. After centrifugation the supernatants were collected, pooled and evaporated at

40°C with a stream of nitrogen. The pellet containing the larger proteins was neutralized and further analyzed by electrophoresis.

Reversed-Phase High Performance Liquid Chromatography (HPLC) of AKHs

The AKHs were separated by reversed-phase (RP) gradient HPLC using an Alltech Lichrosorb RP-18 column (250 x 4.6 mm), eluted at 0.90 ml/min. Solvent A consisted of 15% acetonitrile containing 0.1% trifluoroacetic acid and solvent B of 60% acetonitrile containing 0.11% trifluoroacetic acid [19]. Elution was programmed as follows: a linear gradient (0 to 37% B in 18 min), an isocratic period (37% B for 8 min), a linear gradient (37 to 100% B in 5 min) and a linear gradient (100 to 0% B in 3 min). Total run time was 36 min. The eluate was monitored at 227 nm. Retention times of locust AKH I and II were identical to those of synthetic AKH I and II. Moreover, the elution profile was also checked by a bioassay for AKH [20].

Liquid Scintillation Counting

Radioactivity of the HPLC eluate was monitored by mixing 0.3 ml of the eluate with 4 ml Emulsifier-Safe (Packard-Becker, Groningen, The Netherlands). Analysis for radioactivity was performed with a Packard 4550 liquid scintillation spectrometer using settings for automatic dpm calculation with correction for quenching and background.

Gel Electrophoresis

Electrophoresis of CC proteins was carried out either on SDS polyacrylamide gradient gels (gradient 10 to 30%) (SDS-PAGE) in a buffer containing 0.1% SDS, essentially according to Laemmli & Favre [21] for 17 h at 25 mA using an LKB-2001 vertical electrophoresis unit, or on 12% homogeneous SDS-polyacrylamide for 2 h at 100 V using a BioRad electrophoresis unit and with the same buffer.

Prior to electrophoresis, proteins were denatured with 2.5% (w/v) SDS in the presence of 5% mercaptoethanol for 15 min at 70°C, followed by the addition of 20% of a 0.25% solution of bromophenol blue in glycerol (1:1, v/v). Protein markers (Pharmacia) were used for molecular weight calculation.

In some experiments native gradient PAGE (4 to 30%) gels were also run, without using a stacking gel.

Two-dimensional electrophoresis was carried out using isoelectrofocusing in the first dimension with Ampholines (pH range 3.5 to 9, Pharmacia) and gradient SDS-PAGE (4 to 30%) in the second dimension.

Transfer of Proteins onto Nitrocellulose Membranes

After electrophoresis proteins were electroblotted onto nitrocellulose membranes (Schleicher & Schüll, pore diameter 0.45 μm) using a vertical electroblot apparatus filled with a transfer buffer consisting of 25 mM TRIS base, 192 mM glycine and 20% methanol for 6 h at 60 V (400 mA), essentially according to Burnette [22]. After transfer, protein bands were stained with 0.05% (w/v) Amido Black 10B in 45% methanol and 10% acetic acid.

Radioscanning

Distribution of radiolabel over the protein bands on the nitrocellulose blots was measured with an automatic TLC linear analyzer LB 2842 (Berthold, Wildbad, FRG).

Monoclonal Antibodies

Monoclonal antibodies raised against AKH I were prepared as described in detail in an earlier paper [3]. Those used were 3F4 and D6; both raised against tyr^1 -AKH I-thyroglobulin as the antigen.

Immunostaining

Aspecific binding sites on the nitrocellulose membranes were blocked with 5% normal goat serum for 30 min at room temperature and washed three times for 10 min with 0.5% Tween-20 in phosphate buffered saline (PBS). Strips of 2 to 4 mm width were cut and incubated separately with the appropriate antibodies for 1 to 2 h at room temperature.

Unbound antibodies were washed off with Tween-PBS. Next the strips were incubated with peroxidase-conjugated goat-antimouse serum for 1 h at room temperature. After washing with Tween-PBS, bound antibodies were visualized with 3,3',5,5'-tetramethylbenzidine (TMB) reagent [23].

Gel Filtration HPLC of Proteins

In some experiments 25 to 100 CC or glandular lobes were sonicated in the column eluate used, centrifuged and the proteins in the supernatant were separated by gel filtration HPLC on a TSK-G 2000 SW (LKB, Bromma, Sweden) column (60 cm x 4.6 mm), using 0.05 M ammonium acetate (pH 7.0) - 15% acetonitrile as the eluent at a flow rate of 0.3 ml/min. The eluate was monitored at 227 nm. Protein fractions were pooled, freeze dried and subjected to electrophoresis on SDS-PAGE gradient gels.

RESULTS AND DISCUSSION

Incorporation of [^{3}H] Amino Acids into AKHs

After *in vitro* incubation of CC with [^{3}H] amino acid, the small peptides were extracted and separated by RP-HPLC. Both AKHs were well separated (Figure 1) and incorporation of radiolabel was measured by liquid scintillation spectrometry of the eluate (Figure 2). When [^{3}H] phenylalanine was used, incorporation of label was found in both AKHs (in the same ratio between AKH I and II as their molar ratio in the CC), whereas in the case of proline, only AKH I appeared to be radioactive (Figure 2 inset). These results are in accordance with the amino acid composition of both AKHs [5,6].

Apart from the radiolabeled AKHs, radioactivity was found in several other components. The high concentrations of radioactivity in the peak eluting after about 7 min were found to be due to non-incorporated [^{3}H] amino acid.

Additionally, it was observed that only the glandular part of the CC contained and synthesized AKHs, whereas the storage part of the CC and the hypocerebral ganglion (closely connected with the CC) did not.

Effect of Variation in the Composition of the Incubation Medium

Locust haemolymph contains high concentrations of amino acids (Oudejans & Kooiman, unpublished results). When using haemolymph or deproteinized haemolymph as the incubation medium, incorporation of radiolabel was detectable (data not shown), but due to dilution with non-labeled amino acids the radioactivity in the AKHs was less easily measured. Results of the Grace incubation medium were comparable with those obtained

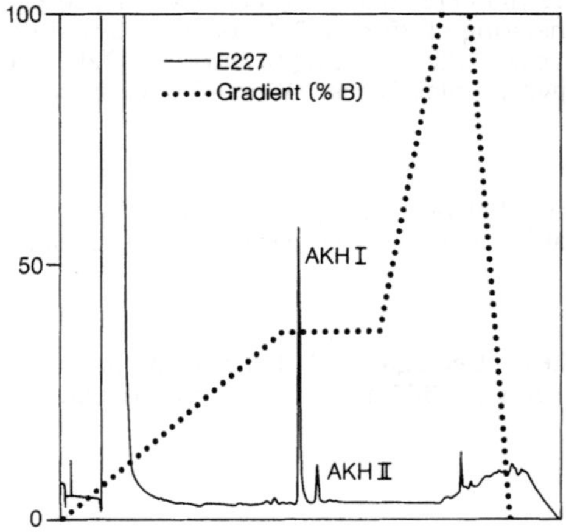

Fig. 1. Reversed-phase gradient HPLC profile of peptides extracted from the glandular lobes of corpora cardiaca of *Locusta migratoria*. Experimental parameters as described in Materials and Methods.

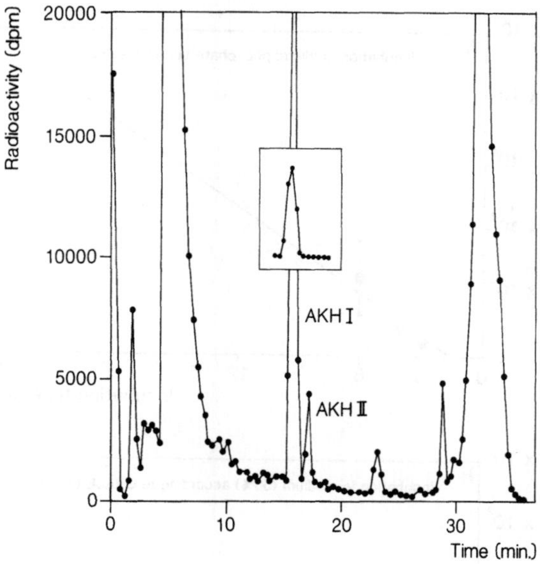

Fig. 2. Incorporation of [^{3}H] phenylalanine and proline (inset) into peptides of glandular lobes of corpora cardiaca of *Locusta migratoria*.

with haemolymph (Figure 3B). The highest incorporation into AKHs was found when no non-radiolabeled amino acids were added to the incubation medium. A medium containing trehalose (Figure 3C) was far superior to PBS (Figure 3A), which suggests that trehalose, the disaccharide of insect haemolymph, may be essential for energy supply during the biosynthesis of AKHs. The medium of Pannabecker & Orchard [16] gave results comparable with those of PBS (data not shown).

However, when all media were compared on a basis in which the biosynthesis of AKHs was calculated in pmol per unit time, the best results were obtained with haemolymph or the Grace medium, mimicking the composition of haemolymph.

Time-Course of Biosynthesis of AKHs

In time-course experiments all incubation media used resulted in a nearly rectilinear increase of label into AKHs with time (Figure 3A-C). Even an incubation time of up to 24 h showed no marked decrease, so we may conclude that the biosynthetic machinery is still active at the same level. It is, however, essential to add antibiotics to the incubation medium, particularly for longer incubation times. Otherwise, the CC are readily degraded and even lysed by microorganisms.

We have observed that the ultrastructure of the CC is far better maintained when using the trehalose containing medium than a saline based medium. In subsequent experiments the trehalose containing medium was used.

When extrapolating the results from the time-course studies, all incubation media used resulted in an identical starting point of first appearance of labeled AKHs at 75 min. This means that the time necessary for biosynthesis and processing of AKHs takes only 75 min. This is somewhat longer than the 20 to 30 min obtained for another locust species, *Schistocerca gregaria*, by Hekimi & O'Shea [8].

Search for the AKH-Prohormone(s)

The structure of the AKHs suggests that these peptide hormones are processing products from a larger protein: both the N-terminal pyroglutamyl residue and the C-terminal amide group indicate post-translational modification reactions of peptides, at the moment of or shortly after the cleaving of a prohormone into the end product peptides [24-26].

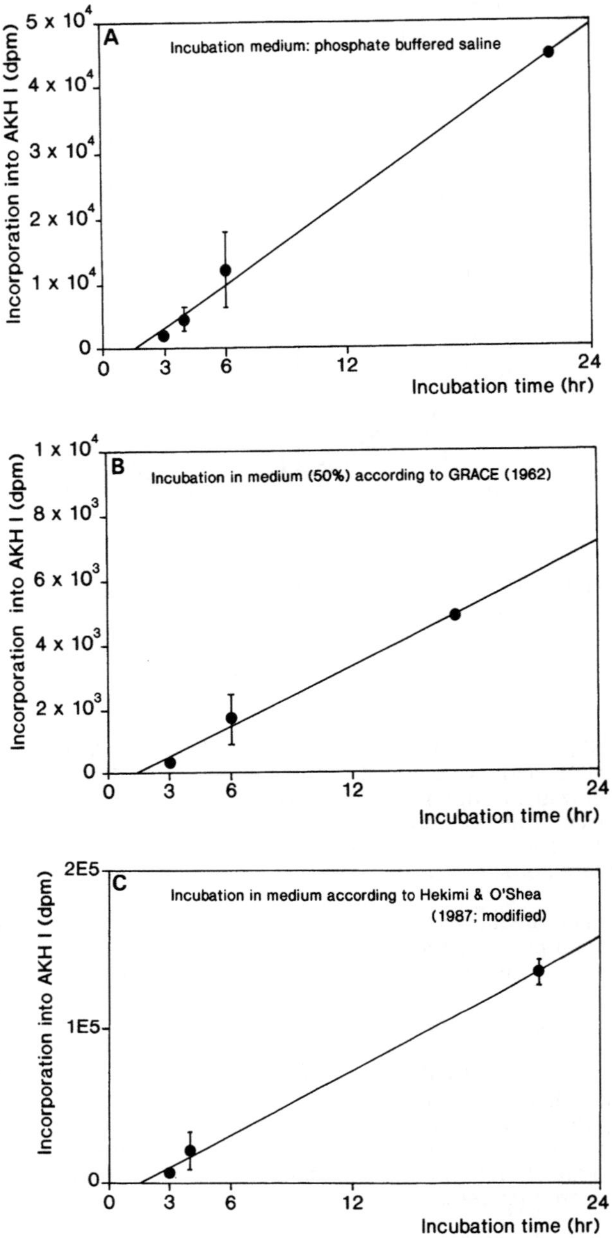

Fig. 3. Time-course of the incorporation of [^{3}H] phenylalanine into AKH I of *Locusta migratoria* in different incubation media (with SD). A: Phosphate-buffered saline; B: medium of Grace [15]; C: medium of Hekimi & O'Shea ([8]; modified).

Since in general a prohormone has no biological activity, it is nearly impossible to trace a prohormone in a neuroendocrine organ by any bioassay approach. We decided to identify the AKH-prohormone(s) in the complex protein composition of the glandular lobe of the CC by using monoclonal antibodies raised against AKH I [3,13], assuming that such an antibody may also recognize that particular epitope in the precursor molecule(s).

Separation of Proteins from the Glandular Lobe of CC by Gel Electrophoresis

SDS-PAGE of the proteins from glandular lobes resulted in a complex banding pattern (Figure 4A). After immunoreaction of the blots with monoclonal anti-AKH I (D6), one specific band was observed with a M_r of 33,000 (Figure 4B). When the glandular lobes were

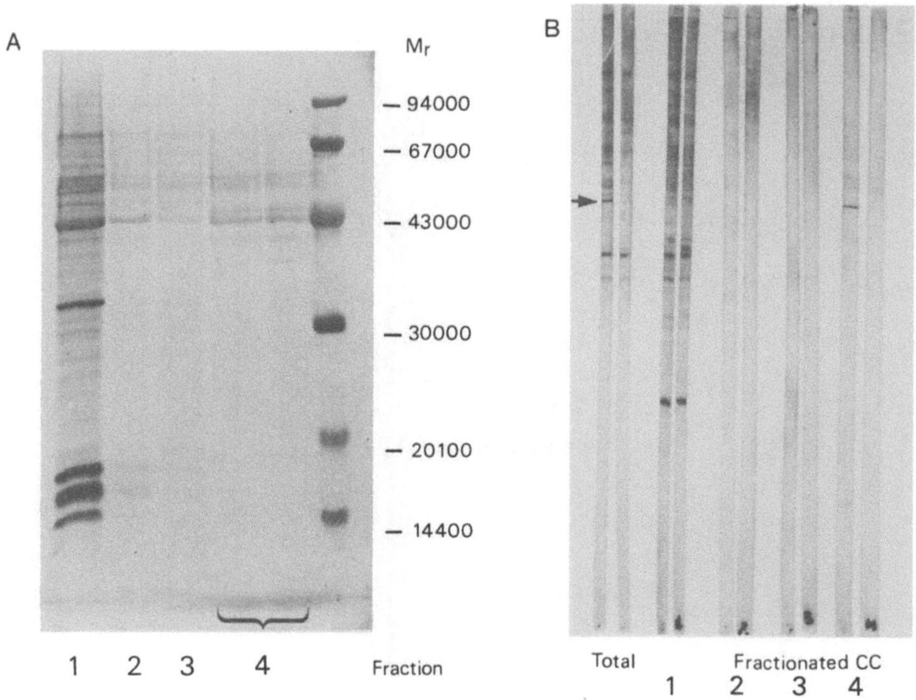

Fig. 4. Proteins from glandular lobes of corpora cardiaca of *Locusta migratoria*, after gradient electrophoresis (SDS-PAGE) and electroblotting onto a nitrocellulose membrane and staining with Amido Black 10B. A) Glandular lobes fractionated by differential pelleting into 1: nuclei (and coarse materials); 2: mitochondria; 3: granula and 4: cytoplasm. Marker proteins are indicated. B) Immunostaining of blots with anti-AKH I (D6) in the left lanes and controls (second antibody only) in the right ones.

fractionated by differential pelleting into several cell fractions [18], all fractions contained complex banding patterns with no obvious specific protein bands for a particular fraction (Figure 4A). However, the immunoreactive protein was only found in the proteins from the cytoplasm fraction (Figure 4B, lane 4) and, interestingly, not in the granula fraction. This may indicate that this protein, assumed to be an AKH precursor molecule, is only recognized by the antibody when newly synthesized. As soon as the molecule is incorporated into the granula the epitope is not exposed for recognition or, on the other hand, the precursor molecule might be directly processed by proteolytic enzymes to small end products, which are not recovered in the protein fraction. Similar results were obtained with antibody 3F4 which, however, recognized a protein in the cytoplasm fraction with a M_r of 72,000 (data not shown). As yet, we have no explanation for the recognition of different proteins by both antibodies.

Distribution of Radiolabel over the Proteins from the Glandular Lobe of CC

After *in vitro* incubation of glandular lobes of CC with [^{3}H] phenylalanine and fractionation of the lobes by differential pelleting, incorporation of radiolabel into the proteins of the four fractions after SDS-PAGE was measured (Figure 5). All fractions contained complex mixtures of proteins, nearly all of which appeared to be radiolabeled. The immunoreactive protein bands did not show more label than the non-reactive ones. Attempts to separate the proteins in their native state by electrophoresis without denaturing conditions were initially unsuccessful. However, when no stacking gel was used, excellent patterns were obtained on gradient gels. The immunoreaction with antibodies was in general very weak and not always consistent. It is possible that the immunoreactive regions were not particularly well exposed.

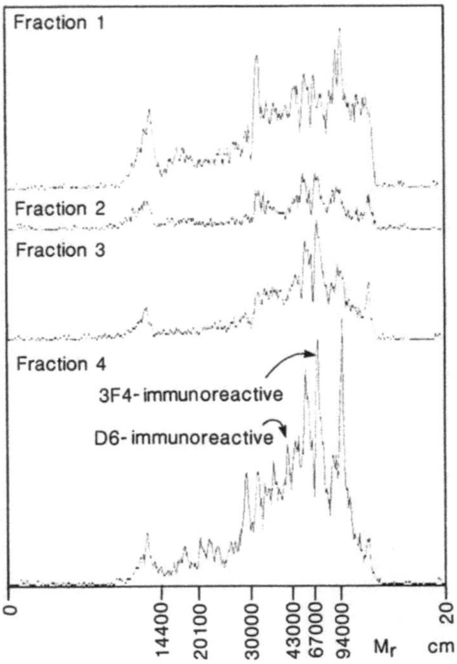

Fig. 5. Distribution of radiolabel over the proteins from fractionated glandular lobes of *Locusta migratoria* (as in Figure 4A), after incubation of the corpora cardiaca with $[^3H]$ phenylalanine for 160 min. Immunoreactive protein bands are only found, as indicated, in the proteins of the cytoplasm fraction.

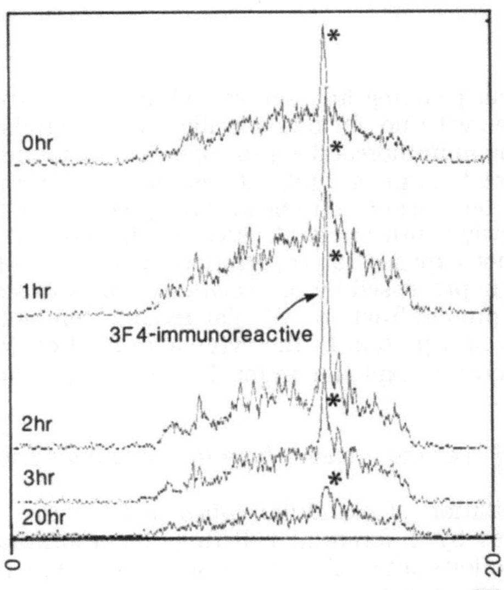

Fig. 6. Distribution of radiolabel over the proteins from corpora cardiaca of *Locusta migratoria* after pulse-labeling for 120 min with $[^3H]$ phenylalanine and chasing with non-labeled phenylalanine for 0, 1, 2, 3 and 20 h. The immunoreactive protein is indicated with an asterisk.

Distribution of Radiolabel over the Proteins after Pulse-Labeling of the CC

After pulse-labeling of the CC with [^{3}H] phenylalanine and chasing with non-labeled amino acid, the distribution of the radiolabel showed one prominent peak (Figure 6), which was coincident with the protein reacting with antibody 3F4. Shortly after pulse-labeling, the radioactivity in this protein still increased, probably because there was still a small pool of radioactive amino acid available in the cell. Obviously, after 20 h of chasing all radioactivity in the supposed precursor molecule will have been processed to small molecules, which are not recovered in the protein fraction. These may be, amongst others, AKH I and II, since the radioactivity found in both AKHs increased concurrently.

Two-Dimensional Electrophoresis of Proteins from Glandular Lobes of CC

In order to obtain more exact information about the M_r of the proteins which were reacting with anti-AKH I, we used two-dimensional electrophoresis. The proteins of the glandular lobes were separated according to their isoelectric point by isoelectrofocusing in the pH range 3.5 to 9 and subsequently separated according to their M_r by gradient SDS-PAGE (Figure 7A). After blotting and immunostaining, the protein band reacting with antibody D6 in the one-dimensional gradient SDS-PAGE (Figure 4B, lane 4) appeared to consist of two proteins with an identical isoelectric point of 6.2 and a M_r estimated as 33,000 and 32,000, respectively (Figure 7B). It is interesting to note that, in Figure 7A, the immunoreactive proteins were hardly visible (arrow), which is in good agreement with the expected low concentration of the AKH-prohormone(s) based on their high turn-over rate. Experiments in order to obtain this type of information for the other polyclonal [8,27,28] and monoclonal antibodies available are currently in progress.

Separation of Proteins from Glandular Lobes of CC by Gel Filtration HPLC

The proteins from the glandular lobes of the CC were separated by gel filtration HPLC on a TSK-G 2000 SW column (Figure 8). The simple elution profile for such a complex mixture of proteins in the lobe, pointed to the poor separation obtained with this technique. This was confirmed after electrophoresis of the pooled fractions (Figure 9), showing little difference in the composition of consecutive fractions with only a gradual change. Unfortunately, immunoreaction procedures on the blots of these fractions revealed no positive protein bands. Separation of native proteins on other columns might be a better approach, a wide-pore C8 reversed-phase column being the most promising one.

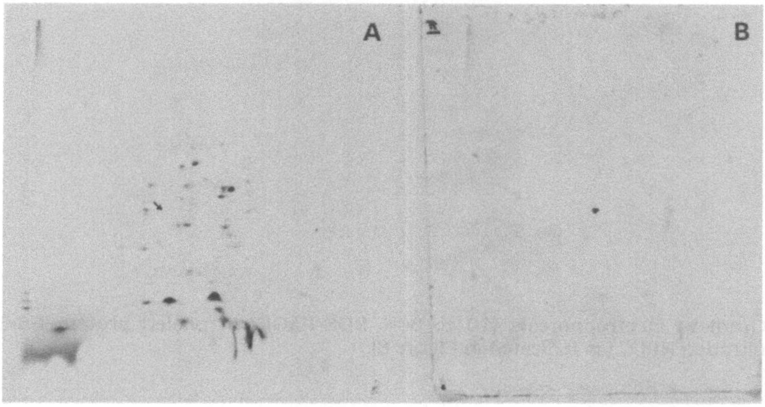

Fig. 7. Two-dimensional electrophoresis (first dimension: isoelectrofocusing pH 3.5 to 9 and second dimension: gradient SDS-PAGE) of proteins from glandular lobes of corpora cardiaca of *Locusta migratoria*. A) After electroblotting onto a nitrocellulose membrane and staining with Amido Black 10B. The position of the immunoreactive proteins are indicated with an arrow. B) After immunostaining the same blot with monoclonal anti-AKH I (D6). Two immunoreactive proteins are visible both with an isoelectric point of 6.2, and a M_r estimated as 33,000 and 32,000, respectively.

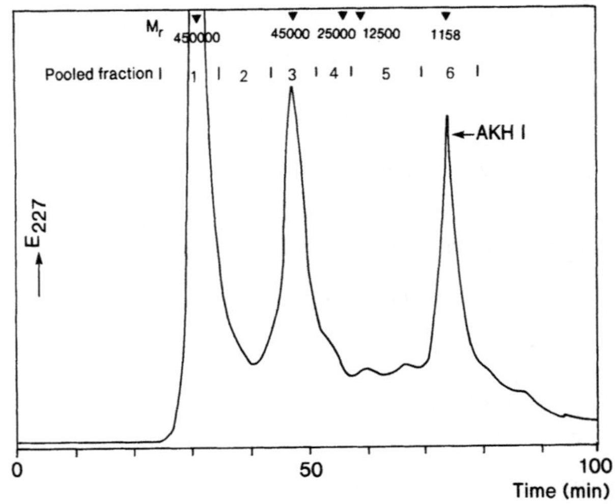

Fig. 8. HPLC elution profile of total proteins from glandular lobes of corpora cardiaca of *Locusta migratoria* on a TSK-G 2000 SW column, using 0.05 M ammonium acetate (pH 7.0) - 15% acetonitrile as the eluent at a flow rate of 0.3 ml/min. Marker proteins and pooled protein fractions as indicated.

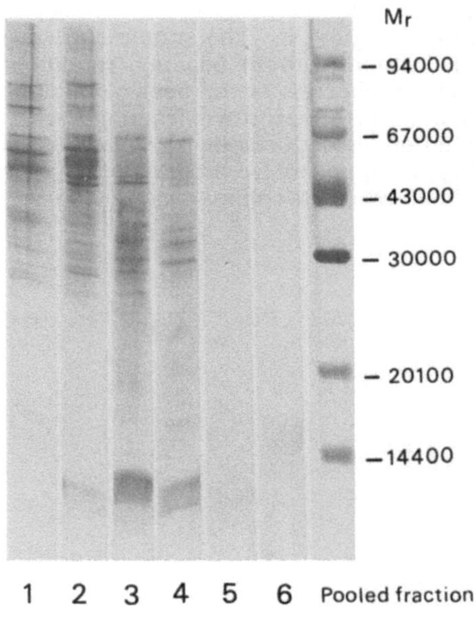

Fig. 9. Gradient electrophoresis (10 to 30% SDS-PAGE) of pooled protein fractions from gel filtration HPLC (as indicated in Figure 8).

CONCLUSIONS

The biosynthesis of AKHs *in vitro* can be studied relatively easily at the level of end products. To identify the precursor molecule(s) is a task less easily attained. We have found precursor molecules (prohormones or AKH-related proteins) with a M_r of 72,000 and 33,000. These differ from the polypeptides identified by Hekimi & O'Shea [8], who calculated a M_r of about 8,000 for two different prohormones in *Schistocerca gregaria*, although their calculations have changed considerably as judged from their earlier work [29,30].

A very intriguing, although discouraging, observation is that, as yet, all monoclonal and polyclonal antibodies raised against AKH available to us ([8,27,28], this paper) seem to react with several different proteins from blots made with the proteins of glandular lobes of the CC: obviously there is no protein in common reacting with all antibodies, which should be expected when one or more AKH-prohormones are present. Further studies are in progress to explain these observations.

REFERENCES

1. A. M. Th. Beenakkers, R. E. B. Bloemen, T. A. De Vlieger, D. J. Van der Horst and W. J. A. Van Marrewijk, Insect adipokinetic hormones, *Peptides*, 6 (Suppl. 3):437-444 (1985).
2. I. Orchard, Adipokinetic hormones - an update, *J. Insect Physiol.*, 33:451-463 (1987).
3. R. C. H. M. Oudejans, H. Voshol, T. K. F. Schulz and A. M. Th. Beenakkers, Time-resolved fluorescence immunoassay, *in*: "Immunological Techniques in Insect Biology", L. I. Gilbert and T. A. Miller, eds., Springer-Verlag, New York, pp 75-92 (1988).
4. J. H. B. Diederen, H. A. Maas, H. J. Pel, H. Schooneveld, W. F. Jansen and H. G. B. Vullings, Co-localization of the adipokinetic hormones I and II in the same glandular cells and in the same secretory granules of corpus cardiacum of *Locusta migratoria* and *Schistocerca gregaria*.An immunoelectron-microscopic study, *Cell Tissue Res.*, 249:379-389 (1987).
5. J. V. Stone, W. Mordue, K. E. Batley and H. R. Morris, Structure of locust adipokinetic hormone, a neurohormone that regulates lipid utilization during flight, *Nature*, 263:207-211 (1976).
6. K. Siegert, P. Morgan and W. Mordue, Primary structures of locust adipokinetic hormones II, *Biol. Chem. Kappe Seyler*, 366:723-727 (1979).
7. G. Gäde, New structures of insect neuropeptides, *in*: "Endocrinological Frontiers in Physiological Insect Ecology", F. Sehnal, A. Zabza and D. L. Denlinger, eds., Technical University Press, Wroclaw, pp 625-650 (1988).
8. S. Hekimi and M. O'Shea, Identification and purification of two precursors of the insect neuropeptide adipokinetic hormone, *J. Neurosci.*, 7:2773-2784 (1987).
9. P. Moshitzky, W. J. Henzel, A. Rafaeli, J. Ramachandran and S. W. Applebaum, Synthesis of adipokinetic hormone (AKH-I) in the locust brain, *Insect Biochem.*, 17:1133-1137 (1987).
10. M. H. Schaffer and B. E. Noyes, The potential of adipokinetiic hormone to teach us about neuropeptides, *Bio-Essays*, 7:67-71 (1987).
11. M. H. Schaffer and B. E. Noyes, The peptides, mRNA, and genes of the AKH family, *in*: "Molecular Entomology", J. H. Law, ed., Alan R. Liss, New York, pp 141-153 (1987).
12. R. C. H. M. Oudejans, F. P. Kooiman and A. M. Th. Beenakkers, Biosynthesis of locust adipokinetic hormones: incorporation of radiolabelled amino acids, *Gen. Comp. Endocrinol.*, 74:306 (1989).
13. T. K. F. Schulz, F. P. Kooiman, R. C. H. M. Oudejans and A. M. Th. Beenakkers, Biosynthesis of locust adipokinetic hormones: an immunological approach, *Gen. Comp. Endocrinol.*, 74:306 (1989).
14. D. J. Van der Horst, A. M. C. Baljet, A. M. Th. Beenakkers and E. Van Handel, Turnover of locust haemolymph diglycerides during flight and rest, *Insect Biochem.*, 8:369-373 (1978).
15. T. C. C. Grace, Establishment of four strains of cells from insect tissues grown *in vitro*, *Nature*, 195:788-789 (1962).
16. T. Pannabecker and I. Orchard, Regulation of adipokinetic hormone release from locust neuroendocrine tissue: participation of calcium and cyclic AMP, *Brain Res.*, 423:13-22 (1987).
17. C. M. Williams, Physiology of insect diapause. IV. The brain and prothoracic glands as an endocrine system in the cecropia silkworm, *Biol. Bull.*, 103:120-138 (1952).
18. J. V. Stone and W. Mordue, Isolation of granules containing adipokinetic hormone from locust corpora cardiaca by differential centrifugation, *Gen. Comp. Endocrinol.*, 39:543-547 (1979).
19. G. Gäde, G. J. Goldsworthy, G. Kegel and R. Keller, Single step purification of locust adipoinetic hormones I and II by reversed-phase high-performance liquid chromatography, and amino-acid composition of the hormone II, *Hoppe-Seyler's Zeitschrift für Physiologische Chemie*, 365:393-398 (1984).
20. G. J. Goldsworthy, W. Mordue and J. Guthkelch, Studies on insect adipokinetic hormones, *Gen. Comp. Endocrinol.*, 18:545-551 (1972).
21. U. K. Laemmli and M. Favre, Nutrition of the head of bacteriophage T4, *J. Mol. Biol.*, 80:575-599 (1973).

22. W. N. Burnette, "Western bloting": electrophoretic transfer of proteins from sodium dodecyl sulfate-polyacrylamide gels to unmodified nitrocellulose and radiographic detection with antibody and radioiodinated protein A, *Anal. Biochem.*, 112:195-203 (1981).
23. T. K. F. Schulz, D. J. Van der Horst, H. Amesz, H. O. Voorma and A. M. Th. Beenakkers, Monoclonal antibodies specific for apoproteins of lipophorins from the migratory locust, *Arch. Insect Biochem. Physiol.*, 6:97-107 (1987).
24. F. Wold, In vivo chemical modification of proteins (Post-translational modifications), *Ann. Rev. Biochem.*, 50:783-814 (1981).
25. A. J. Turner, Processing and metabolism of neuropeptides, *Essays in Biochem.*, 22:69-119 (1986).
26. J. M. Fisher and R.H. Scheller, Prohormone processing and the secretory pathway, *J. Biol. Chem.*, 263:16515-16518 (1988).
27. H. Schooneveld, G. I. Tesser, J. A. Veenstra and H. M. Romberg-Privee, Adipokinetic hormone and AKH-like peptide demonstrated in the corpora cardiaca and nervous system of *Locusta migratoria* by immunocytochemistry, *Cell Tissue Res.*, 230:67-76 (1983).
28. H. Schooneveld, H. M. Romberg-Privee and J. A. Veenstra, Immunocytochemical differentiation between adipokinetic hormone (AKH)-like peptides in neurons and glandular cells in the corpus cardiacum of *Locusta migratoria* and *Periplaneta americana* with C-terminal and N-terminal specific antisera to AKH, *Cell Tissue Res.*, 243:9-14 (1986).
29. S. Hekimi and M. O'Shea, Adipokinetic hormones in locusts: synthesis and developmental regulation, Society for Neuroscience, Abstracts of 15th Annual Meeting, Abstract nr. 282.12 (1985).
30. S. Hekimi and M. O'Shea, Neuropeptide synthesis and its regulation in an insect model system, Society for Neuroscience, Abstracts of 16th Annual Meeting, Abstract nr. 260.12 (1986).

A STRATEGY FOR THE ISOLATION AND STRUCTURAL CHARACTERIZATION OF CERTAIN INSECT MYOTROPIC PEPTIDES THAT MODIFY THE SPONTANEOUS CONTRACTIONS OF THE ISOLATED COCKROACH HINDGUT

G. M. Holman, R. J. Nachman and M. S. Wright*

Veterinary Toxicology and Entomology Research Laboratory
Agricultural Research Service, US Department of Agriculture, Route 5, Box 810
College Station, TX 77840, USA and
*Department of Entomology, Texas A & M University
College Station, TX 77843, USA

SUMMARY

A four-step purification system using high performance liquid chromatography (HPLC) was developed and used in conjunction with a rapid and sensitive bioassay to isolate and purify a number of myotropic peptides present in methanolic head extracts of the cockroach, *Leucophaea maderae*, and the house cricket, *Aceta domesticus*. The extracts were sequentially fractionated on a series of four HPLC columns exhibiting different separation characteristics. The procedure yielded peptides of sufficient purity for primary sequence determination. A total of 17 unique peptides, including the first gastrin/cholecystokinin-like invertebrate peptides, were structurally characterized and synthesized.

INTRODUCTION

Although Cameron [1] initially observed that the corpus cardiacum (CC) of insects contained a factor that stimulated the muscles of the heart and alimentary tract, Davey [2] was the first to describe the effect of CC homogenates on the spontaneous contractile activity of the isolated hindgut of the cockroach, *Periplaneta americana*. Exposure of the isolated hindgut to solutions containing 0.025 CC equivalents/ml produced increases in tonus, frequency and amplitude of contraction, and coordination of contractions. Three years later paper chromatography was used to separate *P. americana* CC extracts into several biologically active fractions [3]. Two components, designated Factors P_1 and P_2, were responsible for most, if not all, of the hindgut-stimulating effects of the crude homogenate. Factor P_1 also stimulated contractions of the heart while Factor P_2 stimulated only the hindgut. The myotropic activity of both factors was destroyed by incubation with chymotrypsin, which differentiated them from proctolin, the myotropic neuropeptide first described by Brown [4]. Proctolin (RYLPT-OH) was subsequently isolated from whole body extracts of *P. americana* and structurally characterized [5]. This success was the result of a multi-year effort that can only be described as heroic [6]. They processed 125 kg of *P. americana* with the contemporary purification techniques (open column ion-exchange, adsorption thin-layer chromatography) to obtain 180 μg of pure peptide which was sufficient for structural characterization with hand sequencing methods.

In the early 1980s, dramatic improvements occurred in both peptide isolation techniques and the instrumentation for structural characterization. The discovery of ion-pairing reagents allowed the powerful separation characteristics of reversed-phase high-performance liquid chromatography (RP-HPLC) to be applied to peptide isolation and purification. Fast-atom bombardment mass spectrometry (FAB-MS) required substantially less peptide for sequence analysis than did older MS systems and in addition detected post-translational modifications. However, the technology of gas-phase sequencing had the greatest impact upon the successful determination of the primary sequences of peptides.

Chromatography and Isolation of Insect Hormones and Pheromones
Edited by A. R. McCaffery and I. D. Wilson
Plenum Press, New York, 1990

The gas-phase sequencer requires about 1000-fold less peptide than the manual techniques it replaced.

In 1983, we HPLC fractionated a *Leucophaea maderae* cockroach head extract on a Waters μBondapak phenyl column [7]. Five of the twenty fractions collected contained myotropic activity when assayed on the isolated *L. maderae* hindgut. However, the UV-214 chromatogram of the run indicated a complex sample matrix requiring additional purification steps to obtain pure peptides suitable for sequencing procedures. During 1984, the additional purification steps were developed and integrated into a four-step HPLC purification system. This system, which is described below, yielded 17 novel myotropic peptides (12 from the cockroach, 5 from the house cricket) of sufficient purity for successful structural characterization.

MATERIALS AND METHODS

Leucophaea maderae cockroaches were taken from stock colonies maintained at 27°C and fed dry dog food and water *ad lib*. Preparation of the isolated cockroach hindgut bioassay including dissection procedures and composition of the saline have been described in detail elsewhere [8,9]. House crickets (*Acheta domesticus*) were purchased from Flukers Cricket Farm, Baton Rouge, Louisiana, USA.

Groups of 250 cockroach heads or 500 house cricket heads were dissected into 250 ml centrifuge bottles containing 200 ml of a methanol-water-acetic acid solution (90:9:1 v/v) and homogenized for 3 min with a Polytron® homogenizer. Following centrifugation (30 min, 9820 g, 5°C), the supernatant from two bottles was decanted into a 1 l round-bottom flask and the solvent removed by rotary vacuum evaporation until only the water and a portion of the acetic acid remained. This viscous solution was transferred into a 250 ml centrifuge bottle. The evaporation flask was rinsed 3 times with 40 ml aliquots of water which were added to the centrifuge bottle and additional water was added to bring the volume to 200 ml. One-half of this solution was transferred to a second centrifuge bottle. The aqueous solution in each bottle was extracted sequentially with 100 ml of ethyl acetate then 100 ml of hexane. The upper (organic) layer was removed in each case by aspiration and discarded. The aqueous portion from four such bottles was transferred to a 1 l round-bottom flask. After removal of residual hexane by rotary evaporation, the remaining aqueous solution was shell-frozen with liquid nitrogen and taken to dryness by lyophilization. Each flask prepared in this manner contained 1000 cockroach head equivalents or 2000 cricket head equivalents. Three 3000 cockroach head extracts and a single 12,800 cricket head extract were prepared during late 1984 and early 1985.

Prior to HPLC fractionation, the lyophilized head extracts were fractionated on Waters C-18 Sep-paks®. The Sep-paks were activated with 4:1 acetonitrile-water (made to 0.01% trifluoroacetic acid, TFA) and equilibrated with 0.1% TFA as previously described [7]. The lyophilizate was dissolved in 0.1% TFA (4 cockroach head equivalents/ml, 8 cricket head equivalents/ml) and 5 ml was eluted onto a Sep-pak cartridge followed by a rinse with 10 ml 0.1% TFA (both solutions discarded). Elution with 5 ml of a 7:3 water-acetonitrile solution (made to 0.1% TFA) removed the myotropic peptides from the Sep-pak. All of the water-acetonitrile aliquots from an extract were pooled and the acetonitrile was removed by rotary evaporation, shell-frozen and lyophilized. This dry residue was dissolved in sufficient 0.1% TFA to give concentrations of 20 cockroach head equivalents/ml or 40 cricket head equivalents/ml. Five ml of this TFA solution was eluted through a fresh Sep-pak followed by a 5 ml rinse of 0.1% TFA (both eluates discarded). Myotropic peptides were eluted with 5 ml of a 1:4 water-acetonitrile solution (made to 0.01% TFA) which was collected and pooled in centrifuge bottles and held in the refrigerator prior to HPLC fractionation. This final Sep-pak step was included to concentrate the samples, stabilize the myotropic peptides during storage, and facilitate the rapid concentration of sample prior to HPLC fractionation.

All HPLC fractionations and analyses were performed on a Waters ALC-100 liquid chromatograph equipped with two pumps, a solvent programmer, a septumless injector, and a variable wavelength UV detector. The extract was fractionated through a series of four HPLC columns having different separation characteristics. The columns and operating conditions (in sequence) were as follows:

A. Waters μbondapak phenyl, 300 x 4.6 mm (i.d.) (Waters Associates, Milford, MA). Conditions: Solvent A - 0.1% TFA in water. Solvent B - 25% acetonitrile in water, made to 0.1% TFA. Initial conditions: 100% solvent A for 8 min then linear program to 100% solvent B over 1 hr. Flow rate: 1.5 ml/min. Temperature: ambient. Detector range: 2.0 aufs at 214 nm.

B. Rainin Microsorb C-1, 250 x 4.6 mm (i.d.) (Rainin Instrument Co., Woburn, MA). Solvents identical with system A above. Initial conditions: 100% Solvent A for 8 min then linear program to 50% solvent B over 1 hr. Flow rate: 1.5 ml/min. Temperature: ambient. Detector range: 2.0 aufs at 214 nm.

C. Techsphere 3 C18, 150 x 4.6 mm (i.d.) (Phenomenex, Rancho Palos Verdes, CA). Solvents identical with systems A and B above. Initial conditions: 50% Solvent A for 8 min then linear program to 100% solvent B over 40 min. Flow rate: 1 ml/min. Temperature: ambient. Detector range: 0.5 aufs at 214 nm.

D. Waters I-125 Protein-Pak, 300 x 7.8 mm (i.d.) (Waters Associates, Milford, MA). Conditions: Solvent A - 95% acetonitrile-5% water, made to 0.01% TFA. Solvent B - 75% acetonitrile-25% water, made to 0.01% TFA. Initial conditions: 100% solvent A for 8 min then linear program to 100% solvent B over 40 min. Flow rate: 1.5 ml/min. Temperature: ambient. Detector range: 0.2 aufs at 214 nm.

The initial fractionation of the extracts was performed on the µbondapak phenyl column. An aliquot of the appropriate Sep-pak purified extract, containing 120 cockroach head equivalents or 400 cricket head equivalents, was prepared and injected onto the µbondapak phenyl column as previously described [7]. Beginning at 36 min post-injection, twenty consecutive 2 min (3 ml) fractions were collected. An aliquot containing 8 cockroach head equivalents (200 µl) or 10 cricket head equivalents (75 µl) was removed from each of the twenty fractions, evaporated, and submitted to bioassay on the isolated cockroach hindgut. Myotropin-containing fractions were capped and stored in the refrigerator. Inactive fractions were discarded. All 25 of the cockroach-head runs and the 32 cricket-head runs were evaluated in this fashion to detect changes in retention times of the myotropins. None was found. At this point, the appropriate active fractions were pooled and concentrated. For example, all 25 of the 54 to 56 min fractions of the cockroach head extract were subjected to vacuum centrifugation to remove acetonitrile then transferred onto a single Sep-pak cartridge. Following a 5 ml rinse with 0.1% TFA, the myotropins were eluted from the Sep-pak with 5 ml of 80% acetonitrile (0.01% TFA) into a polypropylene test tube which was capped and stored in the refrigerator. Six active fractions (52 to 54, 54 to 56, 56 to 58, 58 to 60, 62 to 64 and 68 to 70 min) from the cockroach head extract and three active fractions (40 to 42, 64 to 66 and 66 to 68 min) from the cricket head extract were processed in this manner and retained for additional purification on the C-1 column. The fractionation of the extracts on the µbondapak phenyl column had reduced the amount of residue in the retained fractions to a level that allowed each entire fraction (both cockroach and cricket) to be fractionated on the C-1 column in a single pass.

Fractions were collected from the C-1 column every 2 min and, following bioassay of an aliquot of each, the active fractions were concentrated, stabilized and stored in 80% acetonitrile (0.01% TFA) as described above. Active fractions collected from the C-1 column were further purified on the Techsphere 3 C18 column. In this case, samples were collected as peaks. The eluent containing active peaks was concentrated and stabilized as described above. The final HPLC purification step utilized the Waters I-125 Protein Pak column run in the normal phase mode. The active peaks isolated from the Techsphere 3 C18 were evaporated to near dryness, taken up in 95% acetonitrile (0.01% TFA) and injected onto the Waters Protein Pak column. Fractions were collected as peaks into polypropylene tubes and an aliquot was withdrawn for bioassay. The tube containing the active peak was capped and held in the refrigerator for amino acid analysis, evaluation with proteases, and sequence analysis. A total of 12 cockroach peptides and 5 cricket peptides were isolated in high purity from the extracts.

Amino acid content of the 17 peptides was determined with the Waters Picotag® amino acid analysis system as previously described [10]. The values obtained for phenylalanine were used to calculate total peptide recovered, titre of peptide/insect head, and potency (threshold of activity) on the isolated cockroach hindgut [10].

Approximately 20 head equivalents of each peptide were incubated with immobilized aminopeptidase M (Pierce Chemical Co., Rockford, IL) in cockroach saline for 30 min at 37°C. After a 1 min centrifugation to precipitate the enzyme gel, the supernatant was applied to the isolated cockroach hindgut. The myotropic activity of those peptides with a free (unblocked) N-terminus was destroyed while the myotropic activity of peptides with a blocked N-terminus was retained. Blocked peptides (a total of 4) were deblocked by enzymatic removal of the blocking group (pyroglutamic acid) with pyroglutamylamino-peptidase (Sigma Chemical Co., St Louis, MO). The resulting fragment with a free N-terminus was isolated by HPLC and verified by amino acid analysis [11]. The enzymatic deblocking step was necessary since the Edman chemistry used with automated sequencing techniques required a free amino group at each cycle for derivatization.

Pure peptides and deblocked fragments were submitted to the Model 470A protein sequencer (Applied Biosystems, Foster City, CA) for sequential degradation. PTH-amino acids generated by the sequencer were identified by HPLC analysis [10].

Synthetic peptides, based upon the sequences obtained from the protein sequencer were synthesized as previously described [12]. Crude peptides were purified with the same HPLC systems used for isolation of the natural products. The two sulfated peptides (LSK, LSK II) were prepared as previously described [13]. Following HPLC purification, the structures of these two synthetic sulfated peptides were verified by mass spectrometry [12].

RESULTS

A total of 17 myotropic peptides (16 stimulators and 1 inhibitor) were isolated and structurally characterized from head extracts of the cockroach (12) and house cricket (5). These structures are shown in Table 1.

The retention times of the cockroach and cricket peptides in the four step HPLC purification system are shown in Table 2. The 12 cockroach peptides were collected in six of the 2 min μbondapak phenyl fractions. The fraction eluting from 52 to 54 min contained Leucokinin (LK)-III and LK-IV. These two peptides were easily resolved on the C-1 column. Fraction 54 to 56 min contained LKs V, VI and Leucosulfakinin (LSK)-II. LK-V was resolved on C-1 from LK-VI and LSK-II which eluted together in two consecutive fractions. LSK-II and LK-VI were well separated on the C-18 column (Figure 1). However, LK-VI was collected in association with a large, off-scale peak. Chromatography of that peak in the I-125 system cleanly separated LK-VI from this interference (Figure 2). Fraction 56 to 58 min contained only one myotropin, LK-VII, which was easily isolated. Fraction 58 to 60 min contained three myotropins, LKs I and II, and Leucopyrokinin (LPK). When this fraction was run in the C-1 system, LPK was eluted 20 min after the two leucokinins and required only one more step (C-18) to achieve a purity suitable for sequence analysis. LSK and LK-VIII were isolated from Fraction 62 to 64 min. LSK and LK-VIII were easily separated on the C-1 column (15

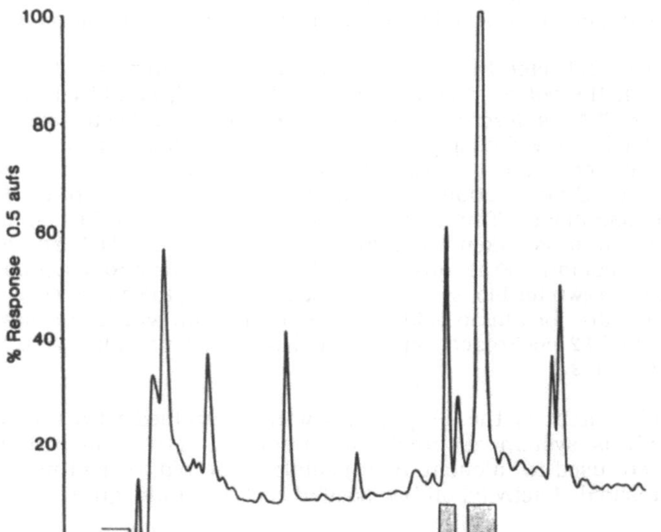

Fig. 1. Fractionation on Techsphere 3 C18 of an area eluting 32 to 36 min from Microsorb C-1. Fractions were collected manually as peaks. Peaks over stippled areas contained myotropic activity. HPLC conditions given in text.

Table 1. **Structures of the myotropic peptides isolated from head extracts of the cockroach *Leucophea maderae* and the house cricket *Acheta domesticus*.**

Leucokinin I	Asp-Pro-Ala-Phe-Asn-Ser-Trp-Gly-NH_2
Leucokinin II	Asp-Pro-Gly-Phe-Ser-Ser-Trp-Gly-NH_2
Leucokinin III	Asp-Gln-Gly-Phe-Asn-Ser-Trp-Gly-NH_2
Leucokinin IV	Asp-Ala-Ser-Phe-His-Ser-Trp-Gly-NH_2
Leucokinin V	Gly-Ser-Gly-Phe-Ser-Ser-Trp-Gly-NH_2
Leucokinin VI	pGlu-Ser-Ser-Phe-His-Ser-Trp-Gly-NH_2
Leucokinin VII	Asp-Pro-Ala-Phe-Ser-Ser-Trp-Gly-NH_2
Leucokinin VIII	Gly-Ala-Asp-Phe-Tyr-Ser-Trp-Gly-NH_2
Leucopyrokinin	pGlu-Thr-Ser-Phe-Thr-Pro-Arg-Leu-NH_2
Leucosulfakinin	Glu-Gln-Phe-Glu-Asp-Tyr(SO_3)-Gly-His-Met-Arg-Phe-NH_2
Leucosulfakinin II	pGlu-Ser-Asp-Asp-Tyr(SO_3)-Gly-His-Met-Arg-Phe-NH_2
Leucomyosuppressin	pGlu-Asp-Val-Asp-His-Val-Phe-Leu-Arg-Phe-NH_2
Achetakinin I	Ser-Gly-Ala-Asp-Phe-Tyr-Pro-Trp-Gly-NH_2
Achetakinin II	Ala-Tyr-Phe-Ser-Pro-Trp-Gly-NH_2
Achetakinin III	Ala-Leu-Pro-Phe-Ser-Ser-Trp-Gly-NH_2
Achetakinin IV	Asn-Phe-Lys-Phe-Asn-Pro-Trp-Gly-NH_2
Achetakinin V	Ala-Phe-His-Ser-Trp-Gly-NH_2

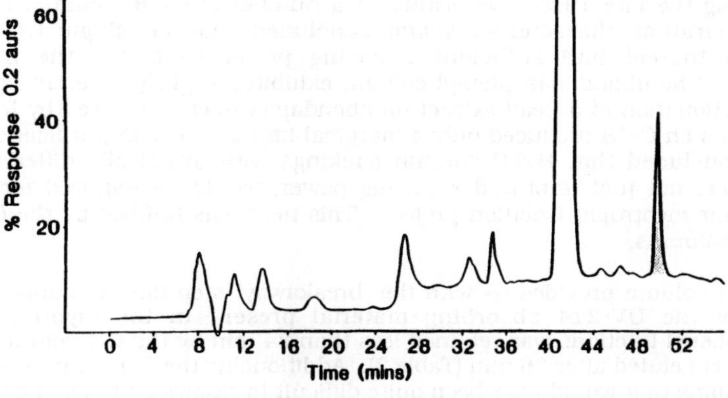

Fig. 2. **Fractionation on Waters I-125 Protein Pak column of peak eluting 35 to 37 min from Techsphere 3 C18. Stippled peak at 49.5 min contained all of the myotropic activity in the sample, and was designated leucokinin VI. HPLC conditions given in text.**

min difference). The inhibitory myotropin, LMS, was the sole activity present in Fraction 68 to 70 min. Eluting as a broad peak from C-1, it was collected as a single sharp peak from C-18. Amino acid analysis of an aliquot suggested an impurity which was easily separated from the LMS on the I-125 column. Of the cricket myotropins, only Achetakinins (AKs) I and II were not resolved after C-1 but were collected as separate peaks from C-18. However, had they not separated, they would have been easily resolved on the I-125 column as their retention times differed by almost 7 min.

Head titres for each of the seventeen peptides were calculated based upon the value obtained for phenylalanine from quantitative amino acid analysis of the natural products. These results are shown in Table 3. In cockroach heads, LPK and LSK-II were by far the most abundant of the peptides (1.36 and 1.30 pmol/head, respectively) while LK-VIII was present in the smallest quantity (0.06 pmol/head equivalent). The Achetakinin titres varied from a maximum of 0.15 pmol/cricket head (AK-III) to a minimum of 0.02 pmol/head (AK-II).

Potencies (threshold of activity concentrations) were determined for all seventeen peptides on the isolated cockroach hindgut (Table 3). Among the cockroach peptides, LK-VIII was the most potent (0.029 nM) and LPK the least potent (0.60 nM). Four of the peptides (LK-V, LK-VI, LK-VIII, LSK-II) exhibited potencies similar to proctolin, which has ranged from 0.03 nM to 0.05 nM. Potencies of the Achetakinins ranged from 0.27 nM (AK-II) to 0.043 nM (AK-V).

DISCUSSION

The single most important requirement for a successful isolation of a biologically active natural product is a reliable bioassay system for following the compounds of interest through the purification procedure. In our case, we were truly blessed. The isolated hindgut preparation of *L. maderae* requires minimal dissection skills and the associated instruments are simple to operate. The preparation is physically rugged and if care is taken to assure constant aeration, a single gut is viable for an entire day. After dissection and suspension of the gut in the chamber, a 30 min to 1 hr period of equilibration was allowed during which the saline in the chamber was replaced 3 to 5 times with fresh saline. At that point, a relatively constant pattern of spontaneous contractions was established and the assay of fractions for myotropic activity begun. Alteration of the pattern of spontaneous contractile activity (either stimulatory or inhibitory) indicated the presence of a myotropic activity. Reaction of the preparation to an active fraction was usually immediate and a yes/no decision was always possible within 1 min. When the myotropin-containing saline was replaced with fresh saline, effects of the myotropins ceased immediately and within 2 to 5 min the spontaneous contractile activity had returned to the pre-exposure pattern and the next fraction was assayed. During the isolation of the cockroach and cricket peptides, up to 80 samples/day were evaluated for myotropic activity on a single isolated cockroach hindgut. Finally, the hindgut receptors were very sensitive to the cockroach and cricket myotropins (see Table 3) so only small quantities of material were expended during the isolation steps resulting in yields of pure peptides in sufficient quantities for structural characterization.

The four-step HPLC purification procedure was developed over a period of several years. During the late 1970s, we evaluated a number of C-18 columns having slightly different separation characteristics and concluded that no single C-18 column or combination thereof had sufficient resolving power to purify the myotropins to homogeneity. The μbondapak phenyl column exhibited slightly different characteristics; however, fractionation of a head extract on μbondapak phenyl followed by fractionation of active fractions on C-18 produced only a marginal improvement in purification. From this result, we concluded that HPLC column packings with drastically different separation characteristics, not just improved resolving power, would be required for a successful outcome of our myotropin isolation project. This need was fulfilled by the C-1 and I-125 Protein Pak columns.

The C-1 column provided us with the "breakaway" step that we badly needed. More than 95% of the UV-214 absorbing material present in the myotropin-containing μbondapak phenyl fractions was retained less than 24 min by the C-1 column yet all of the myotropins were eluted after 26 min (Table 2). Additionally, the C-1 column separated some of the myotropins that would have been quite difficult to resolve on C-18. For example, LK-I and LPK were isolated in the same 2 min fraction (58 to 60) from μbondapak phenyl. On the C-18 column, their retention times differed by 36 sec (0.6 min) and the two peaks were only partially resolved (Holman, unpublished observation). However, the problem of separating

Table 2. Retention times (min) of the cockroach and cricket myotropic peptides in the four HPLC systems used for purification*

	Column			
Peptide	Phenyl	C-1	C-18	I-125
LK-III	52-54	26-28	37.3	49.8
LK-IV	52-54	32-34	42.8	50.3
LK-V	54-56	26-28	38.4	48.2
LK-VI	54-56	32-36	35-37	49.5
LSK II	54-56	32-36	32.6	54.3
LK-VII	56-58	36-40	40.2	45.7
LK-I	58-60	38-40	41.2	43.7
LK-II	58-60	38-40	42.4	43.3
LPK	58-60	58-60	40.6	Not run
LK-VIII	62-64	26-28	51.4	42.3
LSK	62-64	43-45	51.0	55.3
LMS	68-70	70-74	59.8	42.7
AK-V	40-42	28-32	42-45	46.0
AK-I	64-66	40-44	55.8	43.2
AK-II	64-66	40-44	57.7	36.4
AK-III	66-68	52-56	63.5	37.8
AK-IV	66-68	60-64	83.0	47.5

* Column conditions described in Methods. Abbreviations: LK, Leucokinin; LSK, Leucosulfakinin; LPK, Leucopyrokinin; LMS, Leucomyosuppressin; AK, Achetakinin. Peptides are listed in ascending order of retention on the μbondapak phenyl system. Two-minute fractions were collected from phenyl and C-1 systems. Single numbers represent apex of specific peak. LK-VI was collected as a large, off-scale peak from C-18. AK-V was collected in association with a complex of at least three ill-defined peaks.

LK-I and LPK on C-18 never materialized as LPK was retained 20 min longer than LK-I on the C-1 column (Table 2). Similarly, LK-VIII and LSK were collected in the same μbondapak phenyl fraction and exhibited very similar retention times (51.4 min and 51.0 min, respectively) on C-18. Again the problem of separation never materialized, since LK-VIII and LSK were collected from C-1 in fractions that were 15 min apart (Table 2).

The Waters I-125 Protein Pak column (operated in a normal phase mode) was chosen as the final purification step for two reasons: 1) the tiny (sometimes invisible) amount of residue remaining after the previous purification steps was easily dissolved in Solvent A (95% acetonitrile, 0.01 TFA); and 2) the peaks collected from this column were already in a solution of 75% to 85% acetonitrile (0.01 TFA) and required no further manipulation (other than capping) prior to storage. In our experience, the manipulation of samples containing small amounts of pure peptides should be minimized to reduce chances for contamination. Of the 17 peptides, only LPK was judged sufficiently pure after the C-18 purification to proceed directly to sequencing [11]. LK-I, LK-III and AK-II were probably sufficiently pure for sequence analysis after C-18 as single peaks were obtained from the Protein Pak column. However, Protein Pak purification was required for the remaining 13 peptides as none were pure following C-18 fractionation. In several cases, the myotropic peptides were minor components of the C-18 purified peak [14]. The power of the Protein Pak purification is illustrated in Figures 1 and 2. The chromatogram in Figure 1 was obtained from the fractionation of a 32 to 36 min C-1 fraction on C-18. Two active peaks were collected, the first at 32.6 min (LSK-II) and a second large, off-scale peak eluting from 35 to 36 min. When this 35 to 37 min peak was fractionated on Protein Pak (Figure 2), the myotropic peptide (LK-VI) was cleanly separated from the off-scale peak which contained a large majority of the UV-214 absorbing material from the C-18 fraction.

Table 3. Head titres, recovery from extract and potency on the isolated cockroach hindgut of the *Leucophaea maderae* and *Acheta domesticas* myotropic peptides.

Peptide	pmol/head*	nmol recovered*	Potency (nM)† $\bar{x} \pm SD$
LK-I	0.48	1.34	0.20 ± 0.03
LK-II	0.35	0.97	0.16 ± 0.07
LK-III	0.22	0.61	0.07 ± 0.004
LK-IV	0.23	0.64	0.14 ± 0.05
LK-V	0.17	0.47	0.052 ± 0.013
LK-VI	0.16	0.44	0.056 ± 0.014
LK-VII	0.09	0.24	0.12 ± 0.09
LK-VIII	0.06	0.17	0.029 ± 0.005
LPK	1.36	3.79	0.60 ± 0.26
LSK	0.34	0.95	0.22 ± 0.11
LSK II	1.30	3.63	0.045 ± 0.014
LMS	0.23	1.28	0.078 ± 0.003
AK-I	0.04	0.52	0.082 ± 0.022
AK-II	0.02	0.27	0.048 ± 0.005
AK-III	0.15	1.86	0.27 ± 0.028
AK-IV	0.07	0.87	0.18 ± 0.086
AK-V	0.09	1.12	0.043 ± 0.005

* Based upon 2785 cockroach head equivalents except for LMS which was based upon 5550 cockroach head equivalents. The value for the Achetakinins was calculated on the basis of 12,400 cricket head equivalents.
† Purified synthetic peptides were used throughout.

Three separate 3000-head extracts were required for the isolation and characterization of the *Leucophaea* peptides. Nine of the peptides obtained from the first 3000-head extract were submitted to analysis by mass spectrometry but no useful data were obtained from that effort. An additional 1500-head equivalents of one peptide (later named LSK-II) from the second 3000-head extract were submitted to mass spectrometry with the same result. Leucokinins I-VIII [10,14-16] and the two LSKs [12-13] were structurally characterized from this second extract. Unfortunately (for us) we had neglected to collect LPK during the C-1 fractionations of the second extract. Therefore, a third 3000-head extract was performed. LPK was successfully purified, enzymatically deblocked and structurally characterized from the material obtained in this extract. Mass spectrometry was not attempted [11]. The LMS-containing μbondapak phenyl fractions from the second and third 3000-head extracts were pooled and LMS was isolated and structurally characterized as described elsewhere [17].

Although we felt secure with our structural characterizations of the sulfated LSKs, our evidence was based primarily upon the synthesis of sulfated peptide sequences that matched the HPLC retention times and biological activities of the natural products. The LSKs isolated from the third head extract provided us with the material to obtain additional proof of the sulfated nature of these peptides. About 1 nmol of natural LSK-II and an equivalent amount of synthetic LSK-II were submitted to fast-atom bombardment mass spectrometry at the USDA Western Regional Research Center at Berkeley, CA. The spectra for both compounds showed molecular ions at 1317.2 (MH^+) and 1237.3 (MH^+ - SO_3H) consistent with a single sulfate group and a pyroglutamyl residue in the structure. Basic hydrolysis of the remaining LSK-II natural product and all of the LSK natural product followed by amino acid analysis demonstrated the presence of a tyrosine sulfate residue in both peptides [12,13]. LSK and LSK-II were the first gastrin/cholecystokinin-like peptides to be isolated from an invertebrate source.

A modified version of the four-step HPLC peptide procedure was recently used, in conjunction with the isolated *Leucophaea* hindgut preparation, to isolate over 40 myotropic neuropeptides from an extract of 9000 *Locusta migratoria* brain-corpora cardiaca/allata-suboesophageal ganglion complexes. The μbondapak phenyl, C-1, and Protein Pak systems were altered to ensure that all of the myotrpoic neuropeptides present were eluted from the columns during the programmed run. Additionally, a Supelco C-8 Deactivated Base column was substituted for the Techsphere 3 C18 column used in the present system. The C-8 column produced superior separations along with increased recoveries of biologically active peptides (for details and peptide structures see De Loof et al. and Schoofs et al., this volume).

It does not seem unreasonable to predict that the accelerated rate of insect neuropeptide discovery will continue for several years since the necessary instrumentation for isolation and structural characterization is becoming universally available. Isolation systems will be based upon traditional physiological and biochemical assays (both homologous and heterologous), immunological similarity of some insect neuropeptides to known vertebrate and invertebrate peptide structures, and molecular biology techniques. Some initial successes with the latter two methods have already been reported. Antibodies raised against the molluscan peptide FMRFamide were used to isolate and structurally characterize a nonapeptide containing C-terminal FMRFamide from head extracts of *Drosophila melanogaster* [18]. In addition, the gene that coded for the nonapeptide and eight other FMRFamide-related neuropeptide structures was isolated and sequenced [18]. Independently, and virtually simultaneously, molecular biological techniques were used to isolate and sequence the same *Drosophila* gene [19]. Finally, oligonucleotide probes based upon the structures of LSKs and gastrin/cholecystokinin were used to elucidate a gene coding for two LSK-like structures from *Drosophila* genomic DNA and head cDNA libraries [20].

As new insect neuropeptide structures are revealed, pure synthetic replicas will become available commercially or through custom synthesis and immediately stimulate advances in our understanding of these chemical messengers and the functions they control. New physiological and biochemical functions influenced by these peptides will be discovered. Evolutionary relationships among insect species and between insects and other animals, both invertebrate and vertebrate, will be clarified. Immunocytochemical studies will reveal specific identifiable neurons that synthesize these peptides and biochemical studies will demonstrate the precise mechanism of peptide biosynthesis, processing, post-translational modification, secretion, transport and metabolism. Analogue, structure-activity, and computer modeling studies will demonstrate how biological information is encoded upon the structure of these peptides and will increase our knowledge of messenger-receptor interactions. Insect neuropeptides are the master regulators of insect development, metabolism, homeostasis, behaviour and reproduction. They are the molecular messengers that provide coordination between the tissues that make up the organisms we call insects.

REFERENCES

1. M. L. Cameron, Secretion of an orthodiphenol in the corpus cardiacum of the insect, *Nature, London*, 172:349-350 (1953).
2. K. G. Davey, The mode of action of the corpus cardiacum on the hindgut in *Periplaneta americana, J. Exp. Biol.*, 39:319-324 (1962).
3. B. E. Brown, Pharmacologically active constituents of the cockroach corpus cardiacum: resolution and some characteristics, *Gen. Comp. Endocrinol.*, 5:387-401 (1965).
4. B. E. Brown, Neuromuscular transmitter substance in insect visceral muscle, *Science*, 155:595-597 (1967).
5. B. E. Brown and A. N. Starratt, Isolation of proctolin, a myotropic peptide from *Periplaneta americana, J. Insect Physiol.*, 21:1879-1881 (1975).
6. A. N. Starratt and B. E. Brown, Structure of the pentapeptide proctolin, a proposed neurotransmitter in insects, *Life Science*, 17:1253-1256 (1975).
7. G. M. Holman, B. J. Cook and R. M. Wagner, Isolation and partial characterization of five myotropic peptides present in head extracts of the cockroach, *Leucophaea maderae, Comp. Biochem. Physiol.*, 77C:1-5 (1984).
8. G. M. Holman and B. J. Cook, Pharmacological properties of excitatory neuromuscular transmission in the hindgut of the cockroach, *Leucophaea maderae, J. Insect Physiol.*, 16:1891-1907 (1970).

9. B. J. Cook and G. M. Holman, Comparative pharmalcological properties of muscle function in the foregut and the hindgut of the cockroach *Leucophaea maderae*, *Comp. Biochem. Physiol.*, 61C:291-295 (1978).
10. G. M. Holman, B. J. Cook and R. J. Nachman, Isolation, primary structure and synthesis of two neuropeptides from *Leucophaea maderae*: members of a new family of cephalomyotropins, *Comp. Biochem. Physiol.*, 84C:205-211 (1986).
11. G. M. Holman, B. J. Cook and R. J. Nachman, Primary structure and synthesis of a blocked myotropic neuropeptide isolated from the cockroach *Leucophaea maderae*, *Comp. Biochem. Physiol.*, 85C:219-224 (1986).
12. R. J. Nachman, G. M. Holman, B. J. Cook, W. F. Haddon and N. Ling, Leucosulfakinin II, a blocked sulfated insect neuropeptide with homology to cholecystokinin and gastrin, *Biochem. Biophys. Res. Comm.*, 140: 357-364 (1986).
13. R. J. Nachman, G. M. Holman, W. F. Haddon and N. Ling, Leucosulfakinin, a sulfated insect neuropeptide with homology to gastrin and cholecystokinin, *Science*, 234:71-73 (1986).
14. G. M. Holman, B. J. Cook and R. J. Nachman, Isolation, primary structure and synthesis of leucokinins VII and VIII: the final members of this new family of cephalomyotropic peptides isolated from head extracts of *Leucophaea maderae*, *Comp. Biochem. Physiol*, 88C:31-34 (1987).
15. G. M. Holman, B. J. Cook and R. J. Nachman, Primary structure and synthesis of two additional neuropeptides from *Leuophaea maderae*: members of a new family of cephalomyotropins, *Comp. Biochem. Physiol.*, 84C:271-276 (1986).
16. G. M. Holman, B. J. Cook and R. J. Nachman, Isolation, primary structure and synthesis of leucokinins V and VI: myotropic peptides of *Leucophaea maderae*, *Comp. Biochem. Physiol.*, 88C:27-30 (1987).
17. G. M. Holman, B. J. Cook and R. J. Nachman, Isolation, primary structure and synthesis of leucomyosuppressin, an insect neuropeptide that inhibits the spontaneous contractions of the cockroach hindgut, *Comp. Biochem. Physiol.*, 85C:329-333 (1986).
18. J. R. Nambu, C. Murphy-Erdosh, P. C. Andrews, G. J. Feistner and R. H. Scheller, Isolation and characterization of a *Drosophila* neuropeptide gene, *Neuron*, 1:55-61 (1988).
19. L. E. Schneider and P. H. Taghert, Isolation and characterization of a *Drosphila* gene that encodes multiple neuropeptides related to Phe-Met-Arg-Phe-NH_2 (FMRFamide), *Proc. Nat. Acad. Sci. USA*, 85:1993-1997 (1988).
20. R. Nichols, S. A. Schneuwly and J. E. Dixon, Identification and characterization of a *Drosophila* homologue to the vertebrate neuropeptide cholecystokinin, *J. Biol. Chem.*, 263:12167-12170 (1988).

ISOLATION AND IDENTIFICATION OF NEUROPEPTIDES IN *LOCUSTA MIGRATORIA*

A. De Loof, L. Schoofs, G. M. Holman*
T. Hayes**, R. Nachman* and J. Vanden Broeck

Zoological Institute of the University
Naamsestraat 59, B-3000 Leuven, Belgium
*Veterinary Toxicology and Entomology Research Laboratory
Agricultural Research Service, US Department of
Agriculture, College Station, TX 77841, USA, and
**Laboratories for Invertebrate Neuroendocrine Research
Department of Entomology, Texas A and M University
College Station, TX 77843, USA

SUMMARY

By means of high performance liquid chromatography (HPLC) using 4 columns with different separation characteristics, 44 neuropeptides were isolated from the migratory locust, *Locusta migratoria*, starting from an extract of 9000 brain-corpora cardiaca-corpora allata-suboesophageal ganglion complexes. A heterologous bioassay, based on the motility of the isolated hindgut of *Leucophaea maderae* was used to monitor active fractions during the purification steps. 32 hindgut stimulating and 12 hindgut inhibitory peptide fractions were isolated in a pure form. The amino acid sequences of 10 of the isolated locust peptides were established, belonging to at least three distinct families. The primary structures of some of the isolated peptides resemble vertebrate peptide hormones like gastrin, cholecystokinin and tachykinins which are known to stimulate contractions of vertebrate smooth muscle. The sequence homologies and analogous myotropic activities in insects and vertebrates substantiate evidence for a widespread and early evolution of these peptides. In an alternative approach the possibilities of recombinant DNA technology for unravelling the structures of insect neuropeptides are being investigated. A cDNA expression library (lambda gt11) was prepared starting with mRNAs from brains of *Locusta*. Screening was done by means of antisera directed against peptide hormones. Transformed bacterial clones synthesized molecules immunologically related to bovine growth hormone, amphibian melanophore stimulating hormone, hypertrehalosaemic factor II (*Carausius*) and caudo-dorsal cell hormone (*Lymnea*). So far, no spectacular base sequence homologies with the "authentic materials" have been observed.

INTRODUCTION

The tremendous improvement of chromatographic, immunological and recombinant DNA methods achieved during the last decade and the insight that heterologous bioassays can be very useful, is now resulting in rapid progress in insect endocrinology. The small size of insect brains and the scarcity of available homologous bioassays are no longer unsolvable obstacles for the isolation and determination of the amino acid sequences of a large number of peptides. In our laboratories we have used the *Leucophaea* hindgut motility bioassay to monitor the purification of peptides with myotropic activity from brains of *Locusta migratoria*. In a second approach we have used recombinant DNA methods to try to unravel the sequence of brain peptides for which suitable bioassays are not available at this time. These peptides are, however, recognized by antisera directed against peptides which have a hormonal function in vertebrates.

Chromatography and Isolation of Insect Hormones and Pheromones
Edited by A. R. McCaffery and I. D. Wilson
Plenum Press, New York, 1990

MATERIALS AND METHODS

Locusta migratoria was raised under laboratory conditions [1]. Brain-corpora cardiaca-corpora allata-suboesophageal ganglion complexes were dissected from 9000 adult animals of both sexes, quickly frozen in liquid nitrogen and stored at -70°C until use.

Methods for the culture of *Leucophaea maderae* and the preparation of the *Leucophaea* hindgut bioassay have been described previously [2].

Extraction Methods

The extraction of the brain complexes and subsequent purification steps are described by Schoofs et al. (this volume). The brain complexes were extracted and homogenized in a methanol-water-acetic acid (90:9:1 v/v/v) solution. Following centrifugation (30 min, 9820 g, 5°C) supernatants were decanted and pellets were resuspended by vigorous shaking. The tubes were recentrifuged and both supernatants were removed by rotary evaporation at 30°C until only the water and a portion of the acetic acid remained. Ethylacetate was added and after centrifugation and removal of the upper layer, n-hexane was added to each tube and centrifuged. The lower aqueous portions of each centrifuge tube were pooled, frozen and freeze dried. All water and acetonitrile-water solutions used for Sep-pak purification and HPLC-analysis contained 0.1% trifluoracetic acid (TFA) as an ion pairing agent. Procedures for Sep-pak purification have been previously described [3].

The dry residue of the extracted material from 250 brain equivalents was dissolved in 50 ml 0.1% TFA. Aliquots of 5 ml were applied to a Sep-pak and eluted each time with 5 ml of 30% acetonitrile. The Sep-pak was rinsed with 10 ml of 75% acetonitrile solution. All of the 30% (10) and 75% (10) fractions were pooled, evaporated and freeze dried.

High Performance Liquid Chromatography (HPLC) Analysis

HPLC analysis was performed on a Waters ALC 100 Liquid Chromatograph equipped with two pumps, a solvent programmer, a Rheodyne septumless injector and a variable wavelength UV detector, set at 214 nm (Waters Assoc. Inc., Milford, MA, USA). Initially, separation of the biologically active compounds was carried out on a Waters μ-Bondapak phenyl column (4.6 mm x 30 cm) at ambient temperature and with a solvent flow rate of 1.5 ml/min. A lyophilized sample of 250 brain equivalents was dissolved in 0.5 ml 0.1% TFA in water and after injection of this sample a linear gradient of 0.1% TFA in water (Solvent A) against 50% acetonitrile made to 0.1% TFA (Solvent B) was started. Final concentration was achieved after 2 hours; fractions containing 3 ml were arbitrarily collected every 2 min starting at 36 min until 100 min. Thirty-six HPLC runs, each containing 250 brain equivalents were required to process the entire extract through μ-Bondapak fractionation. The various fractions containing myotropic compounds were identified by bioassay and retained. All 36 of the appropriate fractions were pooled and concentrated on a Sep-pak. The 80% acetonitrile/water eluate was taken to dryness and further fractionated through a series of three HPLC columns, having different separation characteristics, in the following sequence:

(1) Rainin Microsorb C1 (4.6 mm x 25 cm). Conditions: Solvent A: 0.1% TFA in water. Solvent B: 25% acetonitrile in water, made to 0.1% TFA. Initial conditions: 100% Solvent A for 8 min, then a linear program to 100% Solvent B over 2 hours. Flow rate 1.5 ml/min. Detector range: 2.0 absorbance units full scale (AUFS) at 214 nm. Fractions were manually collected every 4 min starting at 0 min.

(2) Supelco-C8 Dibasic (4.6 mm x 15 cm). Solvent A: 0.1% TFA in water. Solvent B: 50% acetonitrile in water, made to 0.1% TFA. Initial conditions: 25% solvent B for 8 min, then a linear program to 100% solvent B over 80 min. Flow rate: 1.5 ml/min. Detector range: 1.0 AUFS at 214 nm.

(3) Waters I-125 Protein-Pac column (7.8 mm x 30 cm). Solvent A: 95% acetonitrile made to 0.01% TFA. Solvent B: 50% acetonitrile made to 0.01% TFA. Initial conditions: 100% solvent A for 8 min then a linear program to 100% solvent B over 80 min. Flow rate: 1.5 ml/min. Detector range: 0.2 AUFS at 214 nm.

The amino acid composition of peptides isolated from the HPLC was determined with the Waters Picotag amino acid analysis system. For amino acid analysis and enzymatic analyses procedures, see Schoofs et al. (this volume). Residues of HPLC purified fractions of an aliquot of the purified peptides (about 5%) were incubated with immobilized aminopeptidase M (Pierce Chemical Company) and evaluated for biological activity. When biological activity was unaffected, the N-terminal of the pure peptide might be blocked by pyroglutamic acid (pGlu). Using pyroglutamate aminopeptidase (PGAP) digestion to remove the pGlu, followed by HPLC separation, a fragment was obtained with a free terminal amino group which could be easily sequenced with the automated gas phase sequencer.

Recombinant DNA Technological Methods

A cDNA library starting from RNAs of 500 *Locusta* brains was prepared in the expression vector lambda gt11. Following induction by IPT6, screening was done with several antisera directed against peptide hormones from other species. The method of Sanger et al. [4] was used for sequencing.

RESULTS

Identification of Locust Peptides by Means of Chromatographic Isolation and Subsequent Amino Acid Sequencing

Thirteen hindgut stimulating and five inhibiting fractions were isolated by HPLC fractionation of Sep-pak purified brain-CC-CA-SOG extracts on the μ-Bondapak phenyl column (Figure 1). The first fraction containing a hindgut stimulating substance eluted between 42 and 44 min (fraction 4). All other active substances had longer retention times, with the last active material eluting between 88 and 90 min. The μBondapak phenyl column was purged with 75% acetonitrile-water, following each run and eluates were collected and tested for myotropic activity. In all cases, bioassay of 5 brain-CC-CA-SOG equivalents demonstrated that no biological activity on the roach hindgut was retained by the column. All myotropic peptides were eluted from Sep-paks by the 30% acetonitrile fractions. The insect neuropeptide proctolin (Arg-Tyr-Leu-Pro-Thr) was also analyzed with the HPLC-system. The retention time for proctolin was 42.8 min, which corresponded to fraction 4. Based on retention times, the myotropic peptide in fraction 4 appears to be proctolin. Since this peptide has already been sequenced, this fraction was discarded and not further purified.

All pooled biologically active fractions obtained from the μ-Bondapak phenyl column were concentrated and further fractionated on Microsorb C1. Again, biological activity was located by bioassaying fractions of each of the eighteen runs. In several cases, separation on Microsorb C1 revealed two or more biologically active fractions. All 39 active fractions obtained after μBondapak phenyl and Microsorb C1 fractionation are shown in Figure 2. These fractions were further purified on Supelco C8 and final purification was achieved on the Waters Protein-pac column. In this way, a total of 44 locust peptides, that influenced the motility of the cockroach hindgut, were obtained in pure form. The pure peptides were further enzymatically analyzed by incubation with aminopeptidase M and subjected to PGAP-HPLC analyses when N-terminally blocked. After isolation, the highly purified peptides which were present in sufficient amounts (at least 500 pmol) were subjected to amino acid analysis. About 25 of the isolated peptides were present in amounts high enough to do gas phase sequence amino acid determination.

We report here the amino acid analysis and gas phase sequence amino acid determination of 3 of these peptides. The retention times of these peptides on the four columns, Bondapak phenyl, Microsorb C1, Supelco C8 and Waters Protein Pac I-125, used in this sequence, are shown in Table 1.

Amino acid analysis showed that the molar ratio of fraction 10.11B was as follows: Asp (1), Glx (1), Gly (1), Arg (1), Thr (1), Pro (A), Leu (1), Phe (1) and a total of 7.2 nmoles of pure peptide was recovered. Sequence analysis revealed the following sequence of the peptide which was termed locustamyotropin or Lom-MT:

Glu-Gly-Asp-Phe-Thr-Pro-Arg-Leu

The molar ratio of fraction 14.10 was Asp (1 or 2), Glu (1 or 2), Ser (1), Gly (1 or 2), His (1), Arg (1), Tyr(1), Met (1), Phe (1), Leu (1) and Ala (1). 4.8 nmoles of pure peptide, designated as locustasulfakinin or Lom-SK were recovered from the extract of 9000 brain equivalents. After removal of the blocking pGlu by means of PGAP-HPLC analysis, the peptide was subjected to amino acid sequence determination. Interpretation of the results yielded the following sequence:

pGlu-Leu-Ala-Ser-Asp-Asp-Tyr-Gly-His-Met-Arg-Phe

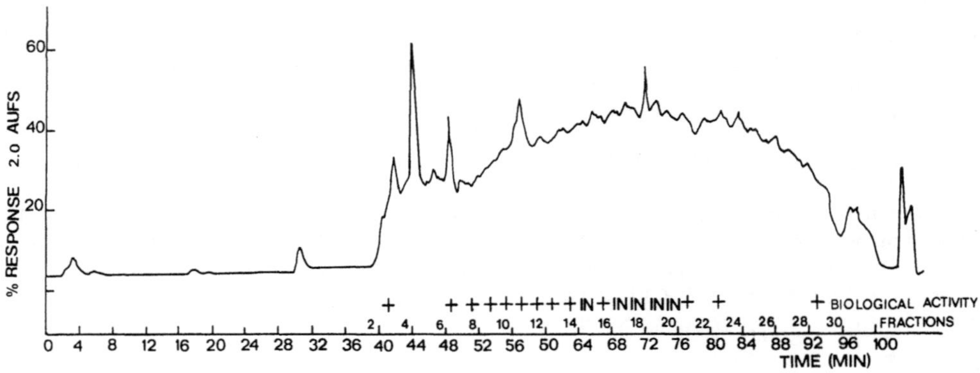

Fig. 1. HPLC fractionation of a locust extract containing 250 brain-corpora cardiaca-corpora allata-suboesophageal ganglion equivalents. Fractions containing inhibitory (In) or stimulating (+) activity are indicated. HPLC and bioassay procedures are described in the text.

A \ B	6	7	8	9	10	11	12	13	14	15	16	17	18	19	20	21
8								+	+							
9	++		++					++								
10			+			++	+++	+++			+++	+++				
11											+++					
12				+												
13			++					++						++		
14					+++			+++								
15									In	In				++		
16						++									++	
17			In		++											
18																
19										In						
20											In	In				+
21												In			In	+
22																+
23								In		In	+					+
24																+
25																+

Fig. 2. Representation of biologically active fractions eluting after μ-Bondapak phenyl (A: horizontal columns) and Microsorb C1 (B: vertical columns) fractionations. (+: hindgut-stimulating fractions, In: hindgut inhibitory fractions).

Fraction 21FB had the following molar ratio of amino acids: Asx (2), Glx (3), Ser (1), Gly (1), Arg (1), Pro (3), Val (1), Leu (1) and Phe (1). The pGlu was removed and interpretation of the results after sequencing yielded the following primary structure:

pGlu-Asp-Ser-Gly-Asp-Glu-Trp-Pro-Gln-Gln-Pro-Phe-Val-Pro-Arg-Leu

Table 1. Retention times (min) of hindgut stimulating and hindgut inhibiting peptides of *Locusta migratoria* on A) Waters μ-Bondapak phenyl column; B) Supelco Cl-column; C) Supelco C8-column; D) Waters I-125-Protein-Pac.

1) Hindgut stimulating fractions

Fraction	A	B	C	D
8.14	50-52	48-52	24	55
9.6	52-54	20-24	26.8	39.8
9.8	52-54	28-32	25.2	51.8
9.12	52-54	44-48	25	48.3
9.13	52-54	48-52	26	50.2
10.11A	54-56	40-44	25.5	55
10.11B	54-56	40-44	29.5	46.9
10-12	54-56	44-48	25-26	48.7
10.16	54-56	60-64	30.0	50.8
10.17	54-56	64-68	31-32	51.6
11.16	56-58	60-64	30	51.6
12.9	58-60	32-36	34-36	41
13.8	60-62	28-32	35	50.9
13.13A	60-62	48-52	37.2	51.9
13.13B	60-62	48-52	37.2	63.2
13.19	60-62	72-76	37	54.2
14.10	62-64	36-40	37.8	49.9
14.13	62-64	48-52	38-39	44.1
15.19A1	64-66	72-76	41	42.1
15.19A2	64-66	72-76	41	44.6
15.19B1	64-66	72-76	41-42	51.44
15.19B2	64-66	72-76	41-42	54.20
15.19B3	64-66	72-76	41-42	56.5
16.11	66-68	40-44	42-44	47.7
16.20	66-68	76-80	41	44.5
17.10	68-70	36-40	42-44	47.5
21FA	76-78	80-84	53.9	49.6
21FB	76-78	80-84	58-59	52.7
23.16	80-82	60-64	57.8	52.2
23F	80-82	80-84	57-58	60
25F1	84-86	80-84	67-69	56.8
25F2	84-86	80-84	67-69	60

2) Hindgut inhibiting fractions

Fraction	A	B	C	D
15.14	64-66	52-56	44.5	45.9
15.15	64-66	56-60	44.5	46
17.8A	68-70	28-32	43-44	41
17.8B	68-70	28-32	44-46	42
19.15	72-74	56-60	57-58	42
20.16	74-76	60-64	56	44.5
20.17	74-76	64-68	57	43.2
21.17A	76-78	64-68	57.5	47
21.17B	76-78	64-68	58-59	44
21.20	76-78	76-80	57	47.5
23.13	80-82	48-52	60.5	43.9
23.15	80-82	56-60	63.8	43.8

29.7 nmole of this peptide, termed locustapyrokinin or Lom-PK were purified from the extract.

The sequences have been verified by chemical synthesis of peptides according to the solid phase methodology of Merrifield [5]. After synthesis, the peptide was cleaved from the resin and the side chain protecting groups were removed as previously described [3]. Biological and chemical identities of native and synthetic factors were demonstrated.

Evaluation of synthetic analogs of Lom-MT and Lom-SK in the acid and amide forms revealed that these peptides were C-terminally amidated.

Identification of Locust Peptides
by Means of Recombinant DNA Technology

Screening of the locust brain cDNA library was performed using antibody probes. Antisera raised against the vertebrate hormones, bovine growth hormone (bGH) and amphibian-melanocyte stimulating hormone and against the invertebrate hormones HTF (the hypertrehalosaemic factor of *Carausius morosus* [6]) and the molluscan peptide CDCH (the caudo-dorsal cell hormone of *Lymnea stagnalis*) produced immunopositive staining within several distinct plaques of the library. Moreover, all these antisera revealed certain immunoreactive neurons on locust brain tissue sections using immunocytochemical methods.

The immunopositive cDNA clones are currently being analyzed by restriction enzyme digestions, hybridization experiments and dideoxy-sequencing. So far, sequence determination has not demonstrated any appreciable base or amino acid sequence homologies with the original antigens. Some very limited stretches of amino acid sequence, however, were found to occur within the authentic peptide sequences as well. These small similarities may explain the reactivity of the antisera with these locust brain factors (Vanden Broeck et al., this volume).

DISCUSSION

The structures of the three myotropic peptides of *Locusta migratoria* presented here, indicate that they belong to two distinct families. One peptide, designated as locustasulfakinin or Lom-SK [7] exhibited sequence homologies with leucosulfakinin (a myotropic peptide isolated from *Leucophaea*) and with the vertebrate hormones, gastrin and cholecystokinin (Schoofs et al., this volume).

The structures (Lom-MT) of the other two peptides termed locustapyrokinin (Lom-PK) and locustamyotropin (Lom-MT) have a common C-terminal amino acid sequence namely, Phe-X-Pro-Arg-Leu-NH_2. From the roach, only one peptide has been isolated with such a C-terminal amino acid sequence. This peptide was designated leucopyrokinin [8]. Nachman et al. [9] examined the structural requirements of the LPK receptor by preparing a series of octapeptide analogs with single amino acid substitution and synthetic peptides of shorter length. The biological activity of the peptide was relatively insensitive to amino acid substitution in the N-terminal as opposed to the C-terminal region. The most critical sequence within LPK for biological activity was Phe-Thr-Pro-Arg-Leu-NH_2. Hence, LPK and Lom-MT, which has this C-terminal sequence in common with LPK, make up a new family of myotropic neuropeptides. According to the structure-function study of Nachman et al. [9], the fourth to last amino acid may change without substantial loss of biological activity. Therefore, the locust LPK, which has a valine residue instead of threonine at position four from the C-terminal may bind to the same receptor on the *Leucophaea* hindgut as LPK and may also belong to the same family as LPK.

The isolation of 44 peptides from *Locusta migratoria* that influence the motility of the cockroach hindgut is a reflection of the tremendous advances in peptides isolation and purification that have been realized during the last decade. This rapid progress in the isolation of such neuropeptides can be attributed directly to the detection method, namely a very efficient bioassay system [2]. Our strategy for the isolation of these pure peptides in such small quantities depended also on the surgical removal of the brain-corpora cardiaca-corpora allata-suboesophageal ganglion complexes from about 9000 locusts to avoid the need for extraction of whole heads. To find an organism with such an array of peptides of similar biological activity (inhibition or stimulation of hindgut contractions) is perplexing if one assumes that each peptide has only one single action in the organism. Obviously, such will not be the case. Multiple functions are of common occurrence with most peptides of both vertebrates [10] and invertebrates [11-13]. At the moment, we have discovered but one of perhaps many biological activities of a given peptide.

In summary, the search for biologically active peptides in the brain complex of the locust with the use of a heterologous (cockroach) bioassay has been very rewarding as it has led to the systematic discovery of many new neuropeptides which, as already stated, belong to distinct families. Some of them have counterparts in vertebrate tissues, especially the brain and the gut, where they display essential functions (gastrin and cholecystokinin). An alternative approach to identify locust brain peptides, makes use of recombinant DNA technology. By means of these methods, we have attempted to select substances from the locust brain, which are immunoreactive to several antisera, directed against distinct peptide hormones.

By using recombinant DNA techniques many vertebrate and some invertebrate peptide hormone precursors have already been characterized using corresponding nucleic acid sequences. This kind of approach may be particularly helpful in the identification of larger peptides and proteins, even peptide precursor molecules. This method is also useful for small peptides, especially when no appropriate bioassay is available. Therefore, the immunological screening method of an insect brain cDNA library may offer an opportunity to characterize yet unknown factors that may function as transmitters, modulators or neurohoromones.

ACKNOWLEDGEMENTS

The Onderzoeksfonds K. U. Leuven, the NFWO, the Koningin Elizabethstichting and NATO are gratefully acknowledged for financial support. We are indebted to L. Paemen, K. Spittaels, M. Verstuyf, I. De Loof, K. Wilna, A. Tips, P. Verhaert, G. Paesen, E. Fias, M. Amelickx, K. Hendrickx, W. Meulemans and R. Huybrechts for their help in the manual dissection of locust brains. We also thank Dr D. Hard (Mousanto) for a kind gift of bovine somatotropin, used to raise an antiserum.

REFERENCES

1. G. J. Ashby, Locusts, *in*: "The UFAW Handbook on the Care and Management of Laboratory Animals", Churchill Livingstone, Edinburgh, London, pp 582-587 (1972).
2. B. J. Cook and G. M. Holman, Comparative pharmacological properties of muscle functions in the foregut and the hindgut of the cockroach, *Leucophaea maderae*, *Comp. Biochem. Physiol.*, 61C:291-295 (1978).
3. G. M. Holman, B. J. Cook and R. J. Nachman, Isolation, primary structure and synthesis of two neuropeptides from *Leucophaea maderae*: members of a new family of cephalomyotropins, *Comp. Biochem. Physiol.*, 84C:271-276 (1986a).
4. F. Sanger, A. R. Coulsen, B. G. Barrell, A. J. H. Smith and B. A. Roe, Cloning in single-stranded bacteriophage as an aid to rapid DNA sequencing, *J. Mol. Biol.*, 143:161-178 (1980).
5. R. B. Merrifield, Solid phase peptide synthesis, *Adv. Enzymol.*, 32:221-296 (1969).
6. G. Gäde and K. L. Rinehart, Jr., Primary structure of the hypertrehalosaemic factor II from the corpus cardiacum of the Indian stick insect *Carausius morosus* determined by fast atom bombardment mass spectrometry, *Biol. Chem. Hoppe-Seyler*, 368:67-75 (1987).
7. Raina and G. Gäde, Insect peptide nomenclature, *Insect Biochem.*, 18:785-787 (1988).
8. G. M. Holman, B. J. Cook and R. J. Nachman, Primary structure and synthesis of a blocked myotropic neuropeptide isolated from the cockroach, *Leucophaea maderae*, *Comp. Biochem. Physiol.*, 85C:219-224 (1986b).
9. R. J. Nachman, G. M. Holman and B. J. Cook, Active fragments and analogs of the insect neuropeptide leucopyrokinin: structure-function studies, *Biochem. Biophys. Res. Comm.*, 137:936-942 (1986).
10. V. Erspamer, The tachykinin peptide family, *Trends in Neuroscience*, 4:267-269 (1981).
11. R. M. Scarborough, G. C. Jamieson, F. Kalish, S. J. Kramer, C. A. Miller and D. A. Schooley, Isolation and primary structure of two peptides with cardioaccellatory and hyperglycemic activity from the corpora cardiaca of *Periplaneta americana*, *Proc. Nat. Acad. Sci. USA*, 81:5575-5579 (1984).
12. J. L. Witten, M. H. Schaffer, M. O'Shea, J. C. Cook, M. E. Hemling and K. L. Rinehart, Jr., Structures of two cockroach neuropeptides assigned by fast atom bombardment mass spectrometry, *Biochem. Biophys. Res. Comm.*, 124:350-358 (1984).
13. K. Siegert and W. Mordue, Elucidation of the primary structures of the cockroach hyperglycemic hormones I and II using enzymatic techniques and gas-phase-sequencing, *J. Physiol. Entomol.*, 11:205-211 (1986).

ADSORPTION CHROMATOGRAPHY OF SMALL C-TERMINAL PEPTIDE AMIDES ON DIHYDROXYALKYL BONDED SILICA HIGH PERFORMANCE LIQUID CHROMATOGRAPHY COLUMNS AND APPLICATION TO PURIFICATION OF INSECT NEUROPEPTIDES

T. K. Hayes, A. A. Strey and K. W. Beyenbach*

Laboratories for Invertebrate Neuroendocrine Research
and Biotechnology Support Laboratory of the Texas
Agricultural Experiment Station, Department of Entomology
Texas A & M University, College Station, TX 77843, USA, and
*Section of Physiology, VRT 826
Cornell University, Ithaca, NY 14853, USA

SUMMARY

The chromatography of small neuropeptides (< 1400 mw) was studied on a high performance liquid chromatography (HPLC) sized exclusion column designed with an aqueous-compatible dihydroxyalkyl bonded silica support. The major mechanism of retention for small peptide amides on this column was clearly not size exclusion but adsorption. Peptide amides were nearly always retained beyond the calculated salt volume of these columns. This retention was nearly five column volumes for selected peptides like substance P. Under the same conditions, peptide C-terminal acid analogs always eluted much closer to retention times predicted by their molecular weight. Changes in the composition of the mobile phase indicated that the selective retention of the peptide amides was primarily the result of selective yet weak hydrophobic interactions. However, most peptide amides will elute from these supports without the aid of organic solvents, an ideal characteristic for direct coupling of this method to reversed-phase HPLC. A pilot two-dimensional (dihydroxyalkyl HPLC/reversed-phase HPLC) HPLC purification of a diuretic peptide from mosquito head extracts showed that the coupling of these two forms of chromatography could be used to quickly isolate a minor component of a complex mixture.

INTRODUCTION

Many peptide neurotransmitters and hormones undergo post-translational processing to produce C-terminal primary amides. More than half of the naturally occurring peptide hormones with molecular weights less than 4000 are amidated [1]. Vertebrate examples include bombesin, calcitonin, cholecystokinin (CCK), corticotropin releasing factor (CRF), gastrin, growth hormone releasing-hormone or luteinizing hormone releasing hormone (GnRH, LHRH), alpha-melanocyte stimulating hormone (MSH), oxytocin, secretin, substance P, the tachykinin family, thryotropin-releasing hormone (TRH), vasoactive intestinal peptide (VIP) and casopressin. Amidated invertebrate peptides include the *Aplysia* egg laying hormone, molluscan cardioexcitatory neuropeptide (FMRF amide), crustacean pigment concentrating and dispersing hormones, the insect adipokinetic hormone (AKH) family, and the insect leucokinin family. A recent review illustrates that 27 of the 31 then known insect neuropeptides were C-terminal amides [2].

Clearly, peptide amidation is important to many peptides involved in molecular communication. Amidation radically changes the chemical presentation of the C-termini since the negative charge is removed. Replacement of the amide with the acid most often causes major losses in biological activity [3]. Furthermore, amidation reportedly blocks or slows peptide degradation by exopeptidases [1] and even some endopeptidases [4].

Here the selectivity of a type of dihydroxyalkyl high performance liquid chromatography (HPLC) column for the retention of small peptide amides over free acids is reported

Chromatography and Isolation of Insect Hormones and Pheromones
Edited by A. R. McCaffery and I. D. Wilson
Plenum Press, New York, 1990

and the use of this chromatographic system for the isolation and characterization of neuropeptides from natural sources is described.

MATERIALS AND METHODS

Peptide Synthesis

Invertebrate peptides were synthesized by solid phase peptide synthesis on a Biosearch 9500 automated peptide synthesizer by standard tBoc chemistry as recommended by the manufacturer [5]. Hypertrehalosemic hormone acid was synthesized on a Merrifield resin with Thr preattached and the peptide amides were synthesized on a methylbenzhydrylamine (MBHA) resin. The major exception to the standard chemistry was for the addition of pyroglutamic acid (pGlu) to the N-termini of these peptides. pGlu was dissolved in dimethylformamide (DMF) with 1.5 equivalents of hydroxybenztriazole to form the activated ester of pGlu when mixed with diisopropylcarbodiimide. pGlu was coupled with two, two-hour cycles. Each peptide was cleaved with hydrofluoric acid-anisole (9:1 v/v) for one hour at 0°C. The resin was washed with three 20 ml aliquots of diethylether, and the peptide was extracted with two equivalents of 50% acetic acid. After dilution and freeze-drying, each peptide was purified by semipreparative reversed-phase HPLC. PTC-amino acid analysis and analytical reversed-phase HPLC indicated a homogeneous peptide of the appropriate amino acid composition. The remainder of the peptides were purchased from Peninsula Laboratories (Belmont, CA, USA) and Sigma Chemical Company (St Louis, MO, USA).

Mosquito Head Extract

Lyophilized mosquito (*Aedes aegypti*) heads were collected as described by Petzel et al. [6]. Head batches (250 mg) were homogenized in 25 ml of ice cold 15% trifluoroacetic acid (TFA) with a 7K-1 polytron (Brinkmann Instruments, Westbury, NY, USA). The homogenate was centrifuged for 10 min at 11,000 g. The pellet was re-extracted twice. The supernatants were combined and filtered through #1 Whatman filter paper and collected in a flask at 4°C. The filtered supernatant was then loaded onto prepared 50 mg standard size C-18 cartridges (Waters Associates, Milford, Mass., USA; 50 mg mosquito head equivalents/cartridge) and washed with 10 ml of water. Diuretic activity was eluted from the cartridges with 1.5 ml of 30% acetonitrile-5 mM TFA and 60% acetonitrile-5 mM TFA (v/v). Solvent was removed from the combined eluent by vacuum centrifugation.

Analytical Reversed-Phase HPLC

For analytical and small scale preparative reversed-phase HPLC a 100 mg equivalent of mosquito head extract was reconstituted in 2 ml 0.1 M phosphate buffer (pH 6.8). The sample was injected using a WISP automatic injector (Waters Associates, Milford, MA, USA) onto a Vydac C-18 reversed-phased column (250 x 4.6 mm i.d. Catalog #218TP104) equilibrated in water. The aqueous solvent was converted to 5 mM TFA. The elution gradient was as shown in Figure 3 with a flow rate of 1ml/min at a temperature of 31°C. Fractions were collected from 5 to 35 min. Solvent was removed from each fraction by vacuum centrifugation for bioassay analysis.

Two-Dimensional HPLC

The fractions containing the activity (determined by transepithelial assay) associated with factor 2 were reconstituted in 200 µl of 0.1 M phosphate buffer (pH 6.8) and injected with a manual Rheodyne injector (model 7125) onto a dihydroxyalkyl bonded silica column (Waters Protein Pak 60 gel filtration column, 300 x 7.8 mm, Catalog #084601) equilibrated in 0.1 M phosphate buffer (pH 6.8) with a flowrate of 0.5 ml/min. Twenty-five minutes into the run the eluent was directed from the dihydroxyalkyl column through 2 Valco (Houston, TX, USA, model #C6W) zero dead volume high pressure switching valves to a Cel Associates (Houston, TX, USA; 35 x 4.6 mm, Catalog #1707-035) C-18 reversed-phase catch column (equilbrated in water) for 45 min. The catch column is the reversed-phase column placed in line to catch components eluting in the dihydroxyalkyl effluent. Upon completion of the dihydroxyalkyl run, a gradient (see Figure 5) at a flow rate of 200 µl/min was used to elute the sample from the C-18 catch column. One-half minute fractions were collected from 30 to 85 min. Solvent was removed from the fractions by vacuum centrifugation and each was saved for the transepithelial voltage assay. When the dihydroxyalkyl column was run independent from the catch column for extracts or synthetic peptides, we used the conditions described above.

Bioassay

The transepithelial bioassay has been described previously [7]. Changes in the transepithelial voltage of an isolated perfused Malpighian tubule from *Aedes aegypti* served as a rapid, qualitative assay for effects on transepithelial ion transport. Malpighian tubules were isolated and perfused *in vitro* such that Ringer solutions bathed basolateral and apical membranes of the epithelium. The transepithelial voltage (V_T) was measured through the perfusion pipette lodged in the tubule lumen; voltage was measured with respect to ground in the bath. The tubule lumen was perfused at rates between 5 and 10 nl/min. The peritubular bath (volume of 280 µl) was perfused by gravity feed at a rate of 3 ml/min providing for continuous bath flow. Samples were injected into the flowing bath for detection of components that depolarize or hyperpolarize the measured transepithelial voltage.

RESULTS

Dihydroxyalkyl bonded silica columns are designed to separate small proteins and peptides by size exclusion. However, we found in our initial studies of the properties of insect hypertrehalosemic hormone (HTH) that molecular weights for the hormone and other related peptides could not be estimated by HPLC size exclusion when dihydroxyalkyl columns were equilibrated in 0.1 M sodium phosphate (pH 7.2) [8]. These peptides were retained by the column beyond the salt volume, as indicated by phenylalanine elution (Figure 1 and Table 2). Thus, an adsorptive phenomenon and not gel filtration was the major factor responsible for the extended retention of these small peptides on the dihydroxyalkyl columns. Example proteins, larger peptides (such as insulin), and amino acids eluted from the column according to their molecular weights (Figure 1 and Table 1).

The retention of several small peptides (less than twelve amino acids) on dihydroxyalkyl columns were compared (Tables 1 and 2) to help identify any structural features that might be associated with the adsorption. A C-terminal amide was always retained longer than an analogous C-terminal acid. In every case where there was an analogous C-terminal acid peptide (TRH, Leuenkephalin amide, LHRH, vasopressin, substance P and HTH) to test, it approached the retention time of the salt volume much better than the C-terminal amide. For purification purposes, a seven minute retention beyond the salt volume ensured near base-line separation for selectively retained peptides from very small gel-filtered sample components. Eleven out of twelve peptides that were retained for at least seven minutes post salt volume were C-terminal amides (Tables 1 and 2). Some peptide amides were retained for very long time periods. For example, LHRH and substance P were retained for approximately three and five column volumes respectively. Broad peaks with tailing characteristics were noted for adsorbed compounds (Figures 1 and 2).

Changes in solvent composition were used to investigate what physical forces might predominate for the selective retention of peptide amides over peptide acids by dihydroxyalkyl columns. Leu-enkephalin and Leu-enkephalin amide were used as a model analogous pair. An increase from zero to 0.5 M sodium chloride in the buffer did not appreciably change the retention of either form of the peptide (Table 3). Addition of isopropanol to the buffer moved the retentions of Leu-enkephalin and Leu-enkephalin amide together, and closer to the salt volume of the column (Table 3).

Chromatography on dihydroxyalkyl columns provides a complimentary technique to reversed-phase HPLC for the purification of peptide amides from natural sources. Peptide samples were loaded and eluted from dihydroxyalkyl columns in an all aqueous, high-salt, buffer. These solvent conditions were ideal for retention of most peptides on reversed-phase media. As a result, most peptides could be concentrated, desalted and further purified by a reversed-phase step immediately after a dihydroxyalkyl purification.

We have examined the separation of diuretic components from mosquito head extracts as a means of evaluating the coupling of the selectivity of dihydroxyalkyl HPLC with the high resolving-desalting power of reversed-phase HPLC. A transepithelial voltage assay was used to monitor major diuretic activity eluting from both chromatography systems. Separation of head extracts by dihydroxyalkyl HPLC yielded three regions of activity (Figure 2). The bioassay characteristics of these separate activity regions match earlier reported activity regions that elute in the same order from reversed-phase HPLC [6]. The activity labeled factor 1 eluted from the column prior to the salt volume and did not

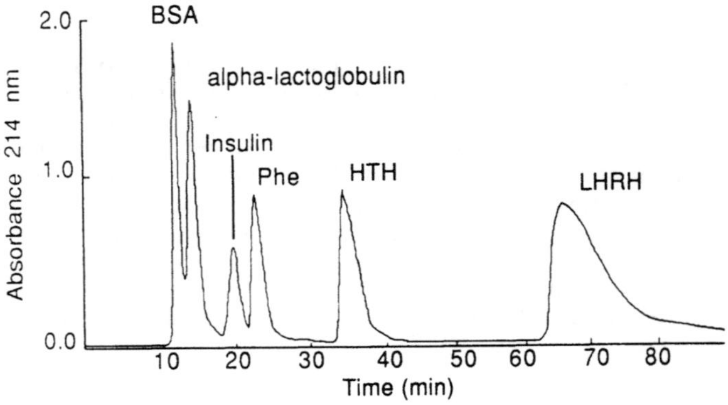

Fig. 1. Dihydroxyalkyl chromatography of example proteins and peptides. Chromatographic conditions are described in the experimental section. Key: BSA, Bovine serum albumin; Phe, phenylalanine; HTH, hypertrehalosemic hormone; and LHRH, leutinizing-hormone-releasing hormone.

Table 1. Retention of peptides and proteins on protein Pak 60 columns.

Sequence	Molecular weight	Retention time (min)
Phenylalanine	165	22.2
Asp-Phe-NH_2	279	22.1
TRH: pGlu-His-Pro-NH_2	362	26.2
TRH-Acid pGlus-His-Pro-OH	363	21.5
Leu-Enkephalinamide: Tyr-Gly-Gly-Phe-Leu-NH_2	554	39.9*
Leu-Enkephalin Tyr-Gly-Gly-Phe-Leu-OH	553	24.2
Trp-Met-Asp-Phe-NH_2*	596	42.3*
LHRH: pGlu-His-Trp-Ser-Tyr-Gly-Leu-Arg-Pro-Gly-NH_2*	1,182	65.8*
LHRH-Acid: pGlu-His-Trp-Ser-Tyr-Gly-Leu-Arg-Pro-Gly-OH	1,183	28.9
(Arg8) Vasopressin (bovine): Cys-Tyr-Phe-Gln-Asn-Cys-Pro-Arg-Gly-NH_2*	1,084	29.1*
(Arg8) Vasopressin (free acid): Cys-Tyr-Phe-Gln-Asn-Cys-Pro-Arg-Gly-OH	1,085	24.1*
Substance P: Arg-Pro-Lys-Pro-Gln-Gln-Phe-Phe-Gly-Leu-Met-NH_2*	1,347	119.6*
Substance P (free acid): Arg-Pro-Lys-Pro-Gln-Gln-Phe-Phe-Gly-Leu-Met-OH*	1,348	34.4*
Insulin	6,000	19.9
Alpha-Lactoglobulin	14,200	14.4
BSA	66,000	12.1

* Retained on the column at least seven minutes longer than phenylalanine and the salt volume.

Table 2. Retention of several natural analogues of insect hypertrehalosemic hormone on protein Pak 60 HPLC columns.

Sequence	Molecular weight	Retention time (min)
RPCH: pGlu-Leu-Asn-Phe-Ser-Pro-Gly-Trp-NH_2*	930	61.9*
AKH_2s: pGlu-Leu-Asn-Phe-Ser-Pro-Gly-Trp-NH_2*	933	49.4*
CC1: pGlu-Val--Asn-Phe-Ser-Pro-Asn-Trp-NH_2*	973	35.7*
CC2: pGlu-Leu-Thr-Phe-Thr-Pro-Asn-Trp-NH_2*	988	46.8*
HTH: pGlu-Val-Asn-Phe-Ser-Pro-Gly-Trp-Gly-Thr-NH_2*	1,073	35.6*
HTH-Acid: pGlu-Val-Asn-Phe-Ser-Pro-Gly-Trp-Gly-Thr-OH	1,074	25.2
Locust AKH: pGlu-Leu-Asn-Phe-Thr-Pro-Asn-Trp-Gly-Thr-NH_2*	1,159	37.7*

* Retained on the column at least seven minutes longer than phenylalanine and the salt volume.

gain much purification from the separation. However, both factors 2 and 3 were retained beyond 7 min post salt volume where few other compounds eluted.

Analytical reversed-phase HPLC demonstrated that the diuretic activity was associated with minor components of the crude mosquito head extracts (Figure 3). Three areas of activity were observed to elute from the HPLC column between 20 and 30% acetonitrile. No major UV absorbing peaks were observed to elute in this region of the chromatogram. Repetition of the injection with collection of individual peaks and valleys directly from the detector outlet, revealed that diuretic activity only eluted with the smallest UV peaks and shoulders of this region. The minor nature of the diuretic factors from mosquito extracts suggests that the purification methods employed would benefit from minimum sample handling and dilution.

A two-dimensional (2D) HPLC system was developed to limit these sample handling problems and to improve the recovery of mosquito diuretic factors (Figure 4). The system consisted of an isocratic HPLC for dihydroxyalkyl chromatography, a set of high pressure valves, and a gradient HPLC for reversed-phase separations. The high pressure valving system allowed the effluent from the dihydroxyalkyl column to be directed to a fraction collector or directly to a small C-18 catch column. Mosquito head extract was first prepurified by reversed-phase HPLC as described in the methods section. The pooled fractions containing the activity associated with factor 2 were injected on the dihydroxyalkyl column and effluent from the column was collected in the fraction collector until 7 min post salt volume (approximately 29 min post injection). After that time period, the effluent was directed to the C-18 catch column that had been equilibrated with water. Sufficiently hydrophobic components of the effluent were concentrated on the catch column until 70 min post injection. The valving system was then realigned to begin pumping from the gradient HPLC system. The catch column was first washed with water to remove excess phosphate buffer (pH 6.8) followed by 5 mM TFA. Finally, the adsorbed components of the dihydroxyalkyl column effluent were eluted from the catch column with an acetonitrile gradient at 200 μl/min (Figure 5). A symmetrical, well-resolved peak eluted from the catch column in approximately 100 μl of 20% acetonitrile in 5mM TFA. This peak was associated with biological activity characteristic of the proposed diuretic factor 2 (Figure 5).

DISCUSSION

We have provided information in this manuscript that demonstrates that dihydroxyalkyl bonded silica columns selectively retain small peptide amides over peptide

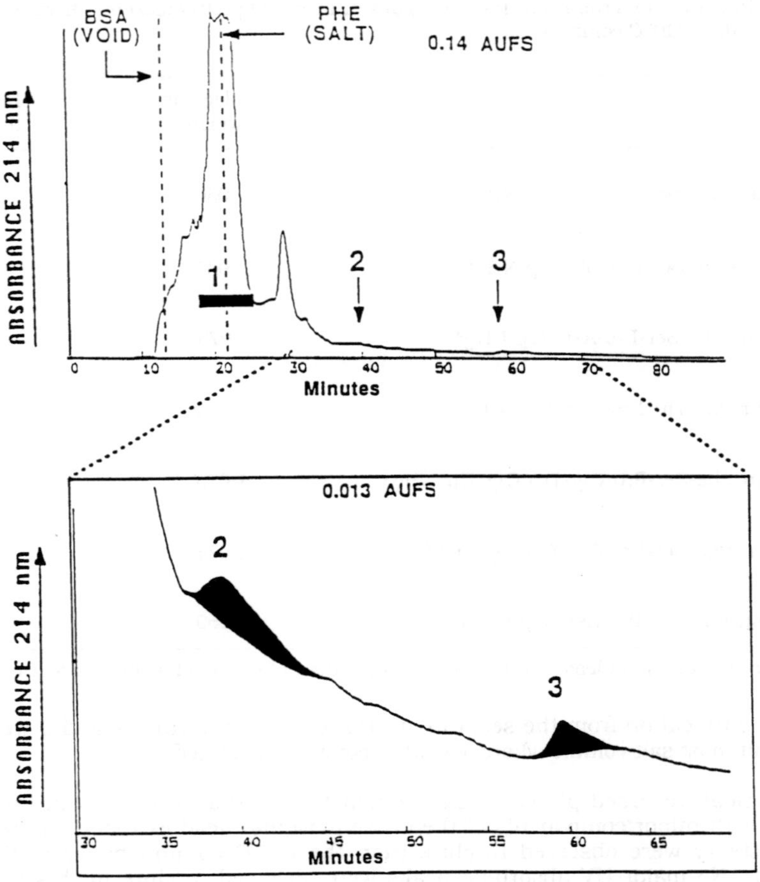

Fig. 2. Separation of diuretic activity from mosquito head extract by dihydroxylalkyl adsorption HPLC. BSA marks the void volume and phenylalanine elution marks the salt volume of the column. The black bar marked 1 represents depolarizing activity that eluted at the salt volume. The peak marked 2 had depolarizing activity associated with the indicated area of elution. Depolarizing and hyperpolarizing activity were associated with elution of the peak marked 3. The bottom panel sensitivity is more than 10 x the top panel.

acids. Hydrophobic forces appear to be an important part of this selective retention. For example, addition of an organic modifier like isopropanol to the mobile phase removed approximately 75% of the selective retention of leu-enkephalin amide over leu-enkephalin (Table 3). However, the major difference between an analogous peptide acid and amide is ionic. The acid has a negative charge at the C-terminus and the amide does not. A large change in the ionic strength of the buffer had little influence on the retention of the model amide-acid pair (Table 3). This indicates that ionic bonds are not a main force in the

Table 3. The influence of moble-phase ionic strength and organic solvent content on the retention of leucine enkephalin and leucine enkephalin amide by protein Pak 60 HPLC columns.

Buffer	Leu enk	Leu enk amide
0.1 M Sodium phosphate (pH 7.2)	24.2	39.9
+0.5 M Sodium chloride	24.4	40.0
+2% Isopropanol	22.6	29.2
+5% Isopropanol	22.2	28.7
+10% Isopropanol	21.3	25.6

selective retention of the amide. Then how could the lone negative charge of a C-terminal peptide acid inhibit retention on a dihydroxyalkyl column?

The answer to the above question could reside in the structure of the dihydroxyalkyl surface. The exact structure of this surface and the chemistry to produce it are proprietary. However, the dihydroxyalkyl silica name implies a general type of structure. The stationary phase consists of a silica backbone, a hydrophobic alkyl extension and a hydrophilic diol-surface. Complete coverage of the silica by the bonded phase will result in a hydrophilic surface with the silica backbone protected from aqueous solvent components by a hydrophobic layer. On the other hand, incomplete coverage of the silanol attachment sites with the bonded phase will result in surface "pits" over each free silanol. Silanols are weakly acidic and are negatively charged at the pH used for this selective chromatography. The surface pits would have hydrophobic walls and a floor that has a negative charge. C-terminal acids would not be bound as well as C-terminal amides in these pits because the negative charge at the acid C-terminus would be repelled by the charged pit bottom. The hydrophobic walls of the pit could then have better interaction with the hydrophobic components of the C-terminal peptide amide.

The adsorptive properties of dihydroxyalkyl columns for peptides and proteins has been previously noted by Meyerson and Abraham [9] who observed that most proteins and peptides above 3000 molecular weight eluted from these columns (equilibrated in 50 mM

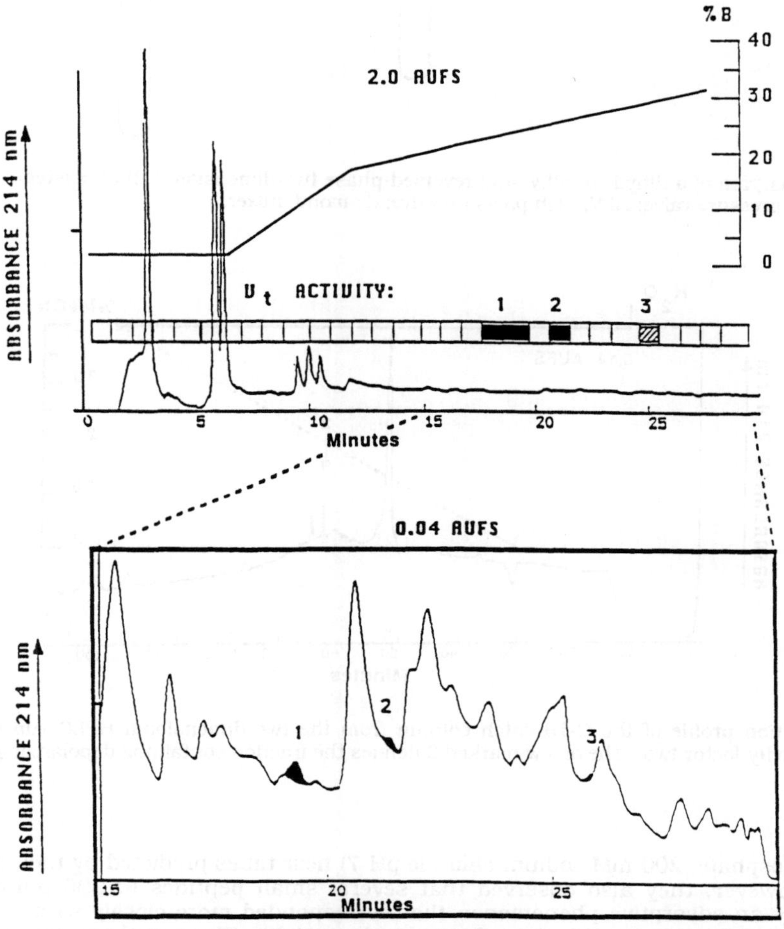

Fig. 3. Separation of diuretic activity from mosquito head extract by analytical reversed-phase HPLC. V_t activity is transepithelial voltage activity measured by the bioassay described in the methods section. Solid black bars denote depolarizing activity. Striped bar denotes depolarization followed by hyperpolarization. The top panel sensitivity is 50 x less sensitive than the bottom panel.

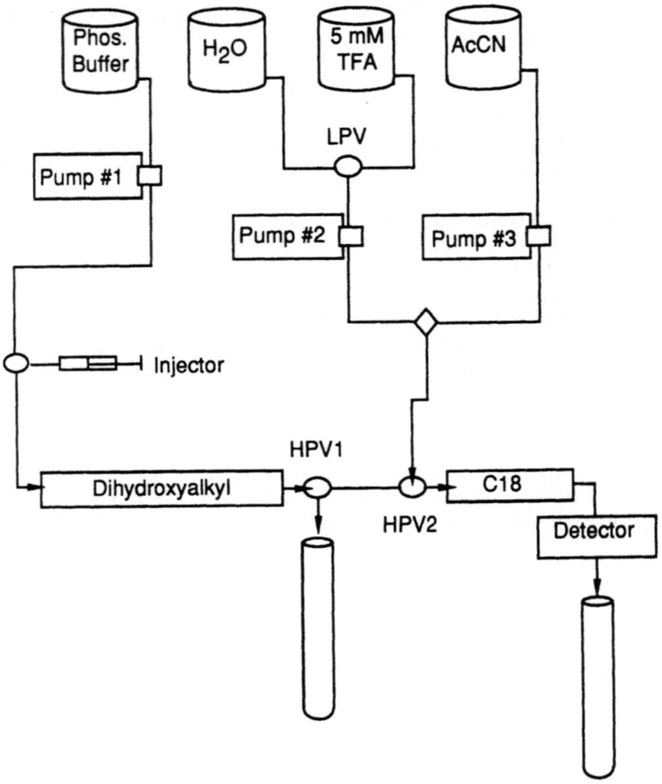

Fig. 4. A diagram of a dihydroxyalkyl and reversed-phase two-dimensional HPLC system. Key: LPV, low pressure valve; HPV, high pressure valve; diamond, mixer.

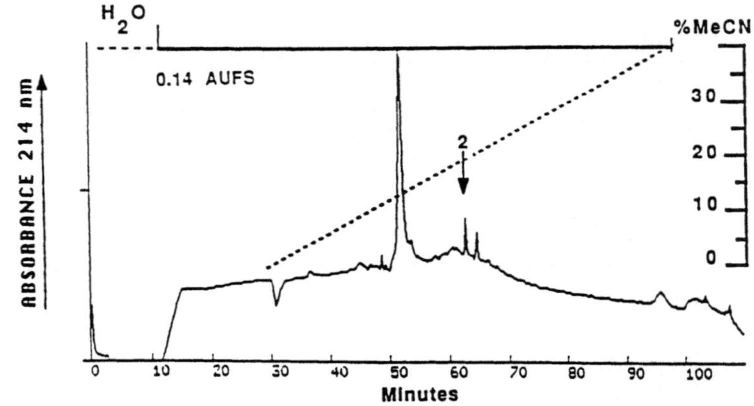

Fig. 5. Elution profile of the C-18 catch column from the two-dimensional HPLC run of diuretic activity factor two. The arrow marked 2 defines the fraction containing depolarizing activity.

sodium phosphate, 200 mM sodium chloride pH 7) near times predicted by their molecular weight. However, they also observed that several small peptides (< 900 daltons) were retained by an adsorptive phenomenon that corresponded more closely with the relative hydrophobicity of each peptide than the molecular weight. These authors did not address the selectivity of the column for C-terminal amide peptides.

The peak shape of adsorbed peptide amides (Figures 1 and 2) on the dihydroxyalkyl phase is reminiscent of that found for basic peptides chromatographed on suboptimally end-capped reversed-phase silica supports [10]. The similar peak shapes on both chromatographic systems were associated with residual silanol groups that failed to react during the synthesis of the bonded phase. On the reversed-phase systems the ionic and/or polar interactions of silanol groups predominated. The number and placement of positive charges on the peptide sequence influence the degree of peak broadening. Likewise, the most highly retained samples in our studies with dihydroxyalkyl columns were the basic peptide amides (LHRH and substance P). Unlike the reversed-phase example, distorted peak shape and adsorption on dihydroxyalkyl columns did not depend as strongly on charge (e.g., see the retention of uncharged peptides in Table 2).

Useful adsorptive properties of size exclusion supports are not limited to HPLC. Classical size exclusion chromatography has the ability to selectively retain specific peptides and was used in isolations of several peptide hormones prior to the wide spread use of HPLC. For example, hydrophobic interaction seems important for the selective adsorption of aromatic peptides to highly crosslinked dextrans (i.e., Sephadex G-10, G-15 and G-25) [11]. Increased support crosslinks, increased number of peptide aromatic residues, and increased ionic strength resulted in increased adsorption of aromatic peptides on these Sephadex supports.

The selective retention of C-terminal peptide amides is, to our knowledge, a unique property of dihydroxyalkyl HPLC. We have demonstrated that the technique is a useful and selective complement to reversed-phase HPLC for the purification of certain small peptides from complex mixtures. The 2D HPLC purification of diuretic factor 2 from mosquito heads allowed us to demonstrate that this coupling of dihydroxyalkyl and reversed-phase HPLC drastically reduced the dilution and handling of the sample. For example, the factor eluted from the dihydroxyalkyl column in 3.5 to 4 mls of buffer (Figure 2). Direct collection on a C-18 catch column allowed concentration and desalting of the factor without exposure to air, light or unnecessary surface adsorption. The factor was eluted in approximately 100 µl instead of 3.5 to 4 mls (Figure 5) of solvent. Trace enrichment of several dihydroxyalkyl runs on a single catch column prior to a single elution could further increase the dilution advantage of the 2D system (i.e., elution in 100 µl instead of 20 mls). We speculate that the peptide associated with factor 2-like activity is a C-terminal amide. However, this prediction will not be confirmed until after structural analysis of the isolated material. We have previously isolated insect hypertrehalosemic hormone from cockroach head extracts by dihydroxyalkyl HPLC coupled with reversed-phase HPLC [8] and assigned this structure as a C-terminal amide (Table 2) on the basis of the structural analysis of the isolated material. Because a large percentage of neuropeptides are of small size and C-terminal amides, adsorptive dihydroxyalkyl HPLC has potential as an important purification and analytical tool for neuroendocrinology.

ACKNOWLEDGEMENTS

We thank Mary Ford and Dr D. H. Petzel for technical assistance. This work was supported in part by funds from NIH grant RO1 NS20137, USDA grant 86-CRCR-1-2008, and NSF grant DCB-891848 to TKH; and NSF grants DCB89040979 and DCB8403305 to KWB.

REFERENCES

1. A. F. Bradbury and D. G. Smyth, Carboxyl-terminal amide formation in peptide hormones, *in*: "Biogentics of Neurohormonal Peptides", R. Håkanson and J. Thorell, eds., Academic Press, London, pp 171-186 (1985).
2. G. M. Holman, M. S. Wright and R. J. Nachman, Insect neuropeptides: coming of age, *in*: "ISI Atlas of Science: Animal and Plant Sciences", 1:129-136 (1988).
3. J. M. Stewart, Peptide Analogs, *Trends Pharm. Sci.*, 3:300-303 (1982).
4. A. C. Martins De Camargo, M. J. Vieira Da Fonseca and H. Caldo, Influence of the carboxyl terminus of LHRH and bradykinin on hydrolysis by brain endo oligopeptidedases, *J. Biol. Chem.*, 257:9265-9267 (1982).
5. D. Hudson, D. Kain and D. Ng, High yielding fully automatic syntheses of cecropin A amide and analogues, *Peptide Chem.*, 413-418 (1986).
6. D. H. Petzel, H. H. Hagedorn and K. W. Beyenbach, Preliminary isolation of mosquito natriuretic factor, *Amer. J. Physiol.*, 249:R379-R386 (1985).
7. T. K. Hayes, T. L. Pannabecker, D. J. Hinckley, G. M. Holman, R. J. Nachman, D. H. Petzel and K. W. Beyenbach, Leucokinins, a new family of ion transport stimulators and inhibitors in insect malpighian tubules, *Life Sci.*, 44:1259-1266 (1989).

8. T. K. Hayes, L. L. Keeley and D. K. Knight, Insect hypertrehalosemic hormone: isolation and primary structure from *Blaberus discoidalis* cockroaches, *Biophys. Biochem. Res. Comm.*, 140:674-678 (1986).
9. L. R. Meyerson and K. I. Abraham, HPLC properties of peptides and proteins on dihydroxyalkyl bonded silica stationary phase, *Peptides*, 7:481-489 (1986).
10. C. T. Mant and R. S. Hodges, Monitoring free silanols on reversed-phase supports with peptide standards, *Chromatographia*, 24:805-814 (1987).
11. R. K. Bretthauer and A. M. Golichowski, Adsorption chromatography of phenylalanine peptides on Sephadex, *Biochem. Biophys. Acta*, 155:549-557 (1968).

ISOLATION AND STRUCTURAL CHARACTERIZATION OF NEUROPARSINS,

POLYTROPIC NEUROHORMONES OF THE AFRICAN LOCUST

J. Girardie, J. C. Pernollet*,
J. C. Huet* and A. Girardie

Laboratoire de Neuroendocrinologie, URA CNRS 1138
Université de Bordeaux I, Avenue des Facultés
33405 Talence Cedex, France, and
*Laboratoire d'Etude des Protéines, INRA
Route de Saint-Cyr, 78026 Versailles Cedex, France

SUMMARY

Neuroparsins A and B, two dimeric sulfur-containing polytropic neurohormones of the African locust, were obtained sufficiently pure to permit their sequencing following two liquid chromatography steps. The N-terminal sequence of neuroparsin B was unique, in contrast to that of neuroparsin A which showed three distinct sequences. The three N-terminal sequences of neuroparsin A included the neuroparsin B sequence itself and extended to five or two supplementary N-terminal residues. The complete sequence of the polypeptide chain of the neuroparsin B homodimer was obtained. It is composed of 78 residues including 2 tyrosines, 1 methionine and 12 cysteines.

INTRODUCTION

Neuroparsin A (NPA) and neuroparsin B (NPB) are two dimeric sulfur-containing proteins of about 16 KD. They are produced by the protocerebral median Al type neurosecretory cells of the African locust, *Locusta migratoria* [1]. The parent molecule NPA is synthesized in the perikarya of the pars intercerebralis [2]. Neuroparsin B originates from NPA by enzymatic cleavage at axonal endings in the corpora cardiaca [2,3]. Both neuroparsins inhibit actions of the juvenile hormone [1], stimulate the fluid reabsorption of isolated recta [4] and induce an increase in hemolymph lipid and trehalose levels [5]. We have determined the N-terminal or complete amino acid structure of neuroparsins, isolated using combined anion exchange, chromatofocusing and reversed-phase liquid chromatography (RP-HPLC).

MATERIALS AND METHODS

Neurosecretory storage lobes of the corpora cardiaca of adult *Locusta migratoria* were crushed in 70% methnaol. The solution was centrifuged at 33,000 g for 30 min and then the supernatant was freeze-dried.

Freeze-dried samples were redissolved in 25 mM Tris-HCl pH 8 and chromatographed on an anion exchange column (MonoQ HR5/5, Pharmacia) using a 33 min linear gradient from 0 to 50% of 1 M NaCl at a flow rate of 1 ml/min. Fractions were manually collected following detection at 225 nm.

Each neuroparsin fraction, identified by its electrophoretic mobility in 7.5% PAGE [6] was desalted and purified on a reversed-phase column (ProRPC HR5/10 Pharmacia). Elution was carried out using a gradient of 20 to 60% acetonitrile with a flow rate of 0.5 ml/min and 0.1% trifluoroacetic acid (TFA) as ion-pair reagent.

Chromatography and Isolation of Insect Hormones and Pheromones
Edited by A. R. McCaffery and I. D. Wilson
Plenum Press, New York, 1990

Purified NPA was chromatofocused on a polybuffer exchange column (MonoP HR5/20 Pharmacia) equilibrated with two types of start buffer: 25 mM bis Tris-HCl pH 6.7 and 25 mM Tris-acetate pH 7 containing 8 M urea desionized with 5% amberlite. Two types of eluent were used to obtain the pH gradient: polybuffer 74 (15% in water) pH 5 with 1 M HCl for 6 to 5 pH gradient, polybuffer 96 (9.5%) and 74 (0.5%) pH 6 with 1 M acetate for 7 to 6 pH gradient. The purified NPA was dissolved in start buffer and injected just before switching the eluent. Chromatofocusing was run at 0.5 ml/min. Neuroparsin A was monitored at 280 nm and pH values were measured on 2 ml fractions using a pH meter until the lowest pH of the gradient was obtained.

Purified NPB was denaturated and reduced (2 h at room temperature in the dark) with a 250 mM Tris-HCl buffer pH 8.5 containing 6 M guanidine chloride and 0.5% 2-mercaptoethanol. Alkylation was achieved with 5% of 4-vinylpyridine for 2 h with the same conditions as above. The mixture was directly applied to the ProRPC column with the conditions previously described.

Reduced and alkylated NPB was cleaved with 250 µg cyanogen bromide (CNBr) in 70% TFA. The mixture was diluted by 3.5 with water and evaporated under nitrogen to 50% of its volume, and then directly applied to the ProRPC column as described above.

The reduced and alkylated NPB was digested with endoprotease Glu-C in a 0.1 M ammonium acetate buffer pH 4 for 24 h at 37°C. The mixture was then applied to an Aquapore RP 300 column (4.6 x 30 mm) and eluted for 30 min gradient of 0 to 40% acetonitrile in 0.1% TFA at a flow rate of 0.5 ml/min.

Purified neuroparsins and all peptide fragments were sequenced using an Applied Biosystems Model 470 A gas phase protein sequencer.

RESULTS

Crude extracts of neurosecretory storage lobes of corpora cardiaca were separated on the MonoQ column into 15 to 20 fractions (Figure 1). Electrophoresis revealed that NPA was in one or two fractions and NPB always in one fraction. After reversed-phase chromatography NPA and NPB fractions gave essentially a single peak, eluting at 40% acetonitrile (Figure 2).

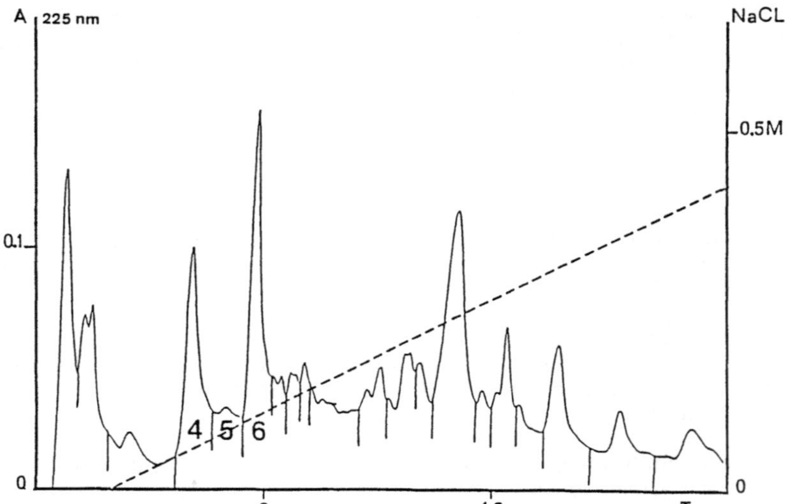

Fig. 1. Separation of crude extract of 60 pairs of neurosecretory storage lobes of corpora cardiaca. The flow rate was 1 ml/min through an anion exchange column MonoQ HR5/5 Pharmacia. Fractions 4 and 5 contained neuroparsin A, fraction 6 contained neuroparsin B.

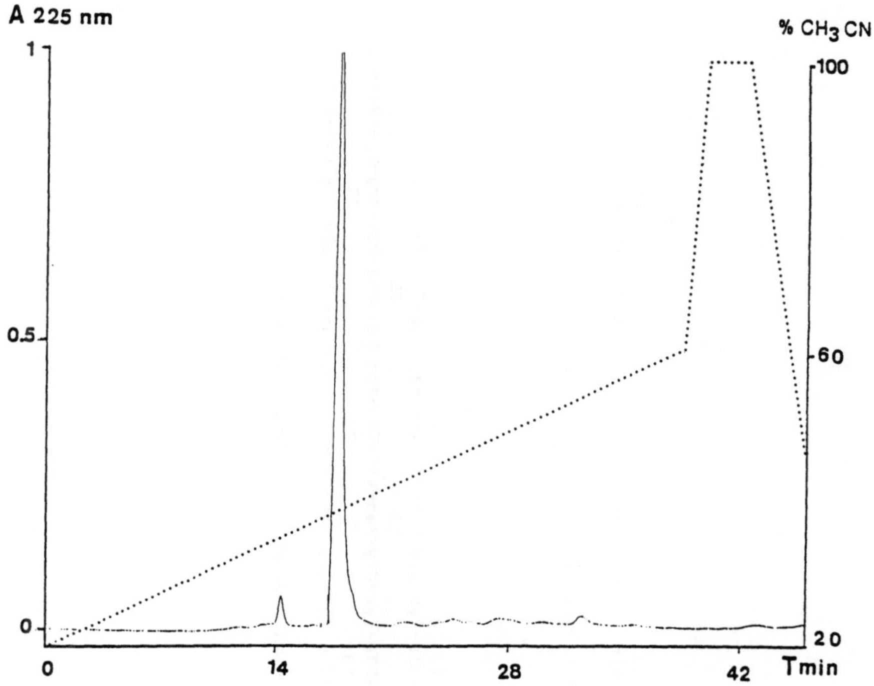

Fig. 2. Purification of fraction 6 obtained from 300 pairs of neurosecretory storage lobes of corpora cardiaca. The flow rate was 0.5 ml/min through a reversed-phase column ProRPC HR5/10 Pharmacia.

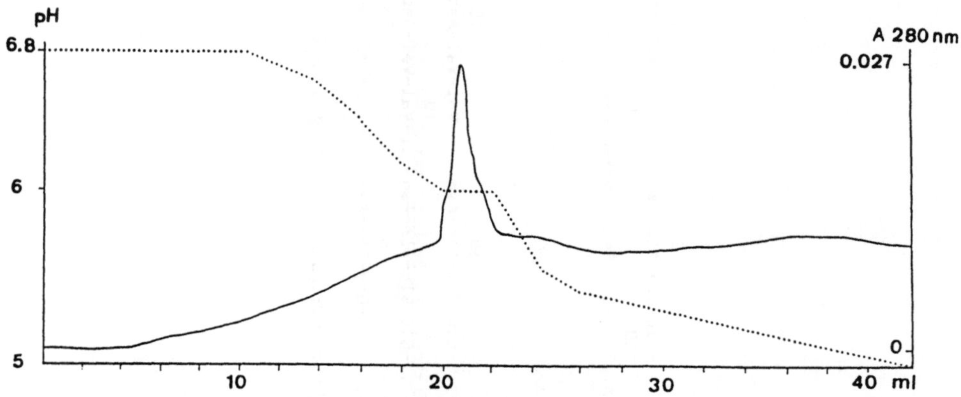

Fig. 3. Separation of the purified neuroparsin A dimers obtained from 360 pairs of neurosecretory storage lobes of corpora cardiaca. The flow rate was 1 ml/min through a polybuffer exchange column MonoP HR5/20 Pharmacia.

Whatever the pH gradient used, chromatofocusing of native pure NPA did not succeed in separating the two dimers expected on the basis of results from anion exchange chromatography. Native NPA was eluted at pH 5.89 during the 6 to 5 pH gradient in a peak with two points of slight inflexion: one in front and one behind (Figure 3). The urea treated NPA was eluted with excellent peak shape at pH 6.8 during the 7 to 6 pH gradient.

Native pure NPB was sequenced up to 44 residues (only one residue at every step of Edman degradation) including five undetermined residues. Thus, NPB is a homodimer. In contrast, native pure NPA, sequenced only up to 29 residues, appeared to be composed of

```
                     1       5          10         15          20          25
NPB:                 HSer-...-Glu-Gly-Ala-Asn-...-Val-Val-Asp-Leu-Thr-Arg-...-Glu-Tyr-Gly-Asp-Val-Thr-Asp-Phe-Phe-Gly-Arg-Lys-Val-
                         30         35         40
                      ...-Ala-Lys-Gly-Pro-Gly-Asp-Lys-...-Gly-Gly-Pro-Tyr-Glu-Leu-His-Gly-...

       1       5          10         15          20          25
NPAα: HAsn-Pro-Ile-Ser-Arg-Ser-...-Glu-Gly-Ala-Asn-...-Val-Val-Asp-Leu-Thr-Arg-...-Glu-Tyr-Gly-Asp-Val-Thr-Asp-Phe-Phe-Gly-...
               1       5          10         15          20          25
NPAβ:          HIle-Arg-Ser-...-Glu-Gly-Ala-Asn-...-Val-Val-Asp-Leu-Thr-Arg-...-Glu-Tyr-Gly-Asp-Val-Thr-Asp-Phe-Phe-Gly-Arg-Lys-Val-..
                     1       5          10         15          20          25
NPB:                 HSer-...-Glu-Gly-Ala-Asn-...-Val-Val-Asp-Leu-Thr-Arg-...-Glu-Tyr-Gly-Asp-Val-Thr-Asp-Phe-Phe-Gly-Arg-Lys-Val-..
```

Fig. 4. Edman amino acid sequencing of the N-terminal portion of the native dimeric neuroparsins B and A.

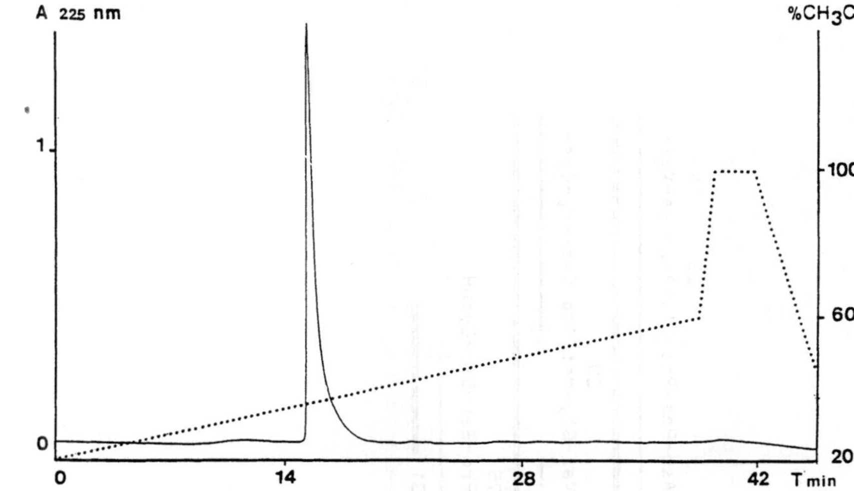

Fig. 5. Separation of the reduced and alkylated pure neuroparsin B obtained from 630 pairs of neurosecretory storage lobes of corpora cardiaca. The flow rate was 0.5 ml/min through a reversed-phase column ProRPC HR5/10 Pharmacia.

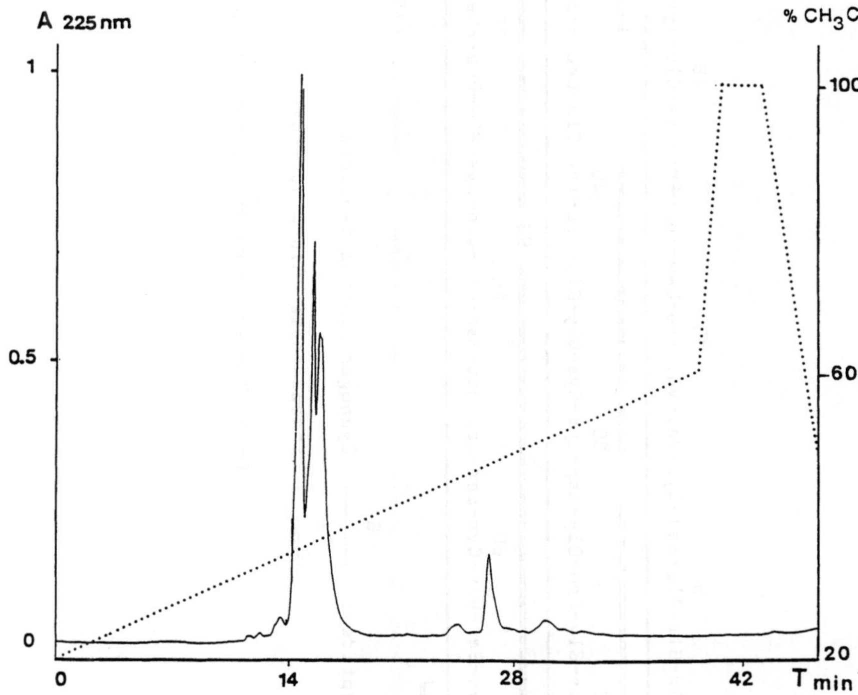

Fig. 6. Separation of peptides after treatment with cyanogen bromide of the neuroparsin B monomer obtained from 1340 pairs of neurosecretory storage lobes of corpora cardiaca. The flow rate was 0.5 ml/min through a reversed-phase column ProRPC HR5/10 Pharmacia.

3 polypeptide chains. One of these accounting for 30% of the total was identical to NPB, while the others, representing 50% and 20% respectively, had five and two supplementary N-terminal residues (Figure 4), respectively.

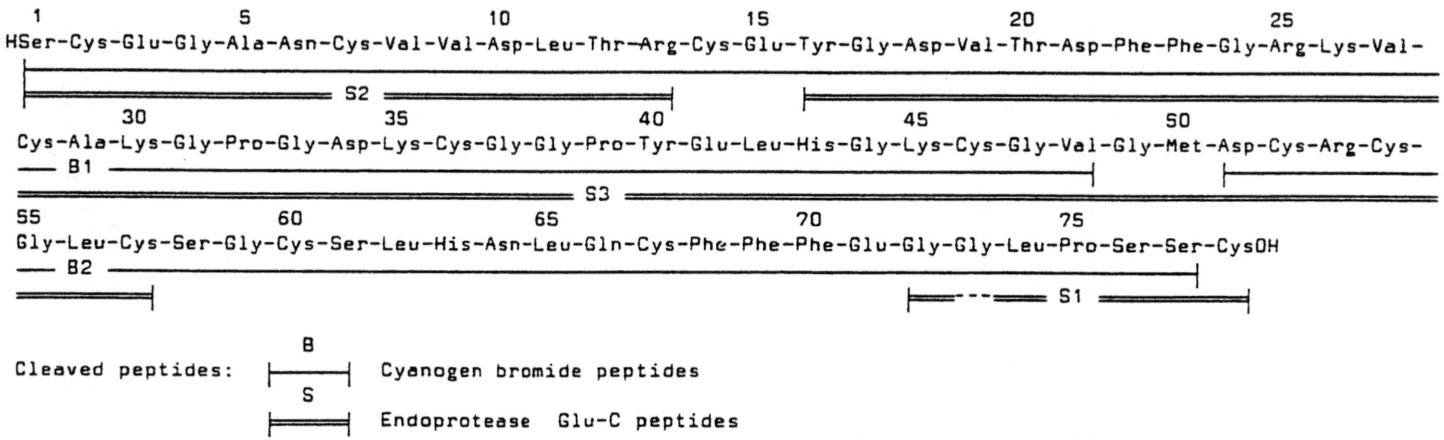

Fig. 7. Complete amino acid sequence of the neuroparsin B monomer.

Reduced and alkylated NPB was eluted at 37% acetonitrile as a single peak (Figure 5). Sequencing of the NPB monomer confirmed the polypeptide chain of 44 residues previously obtained and indicated that the 5 undetermined residues were cysteine as expected.

The monomer NPB, cleaved by CNBr, was separated into three peaks (Figure 6). According to 10 to 20% PAGE results, the first peak corresponded to a peptide of less than 6 KD, the second peak was generally unrevealed and the third peak represented the uncleaved monomer. The two cleaved peptides were sequenced. The biggest (B1) was composed of the 48 residues of the N-terminal portion and the second (B2) was composed of 27 residues that we expected to be the C-terminal portion of the NPB polypeptide chain. Finally, using endoprotease Glu-C, three peptides were obtained overlapping with the native N-terminal and the two peptides B1 and B2. No microheterogeneity was found at any step of the sequencing of these peptides and the C-terminal was free. Thus, the NPB monomer is composed of 78 residues without tryptophan and including 12 cysteines. These residues were usually preceded by a positive charged amino acid and followed by Gly or Glu (Figure 7).

DISCUSSION

Pure neuroparsins were obtained following just two steps of liquid chromatography. The three biological activities of neuroparsins previously demonstrated [1,4-5] cannot be attributed to contaminant molecules but are real effects caused by the neuroparsins.

The sequencing of NPB establishes that the NPB dimer is composed of two identical chains, each containing 12 cysteines. This high cysteine content is in agreement with the intense labelling of neuroparsins following *in vivo* injection of ^{35}S cysteine [2]. The calculated NPB monomer molecular weight is 8188 D and its calculated isoelectric point is 5.98. Both values are in agreement with previous experimental determinations [6,7]. The hydropathy profile of NPB indicates that only the central region of the sequence from residue 30 to 45 appears hydrophilic [8]. The N-terminal sequence of NPB has clear homology with the preliminary and incomplete sequence of a peptide, isolated from the corpora cardiaca of locust and arbitrarily named locust corporis cardiacum peptide (LCCP) by Mordue et al. [9]. The locust corporis cardiacum peptide and NPB are identical from residues 4 to 42 and the 6 unknown residues of LCCP were determined as four cysteines and two aspartic acids in NPB. In spite of a molecular weight estimated to be 4500 D for LCCP, we think that the LCCP, present as neuroparsins in large amounts in the neurosecretory storage lobes of the corpora cardiaca, and neuroparsins are the same molecule.

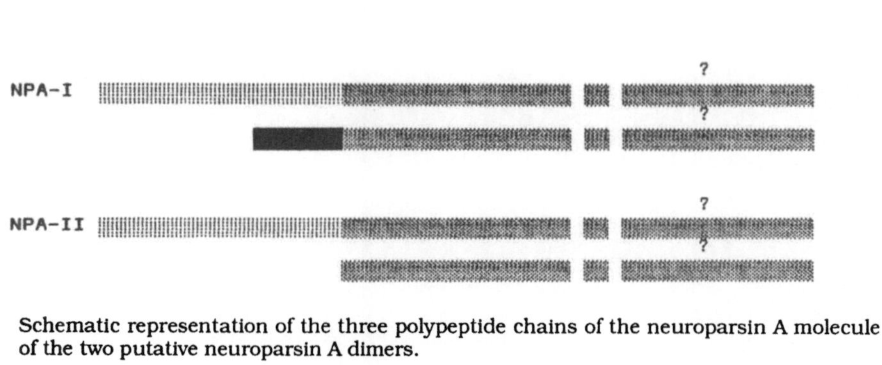

Fig. 8. Schematic representation of the three polypeptide chains of the neuroparsin A molecule and of the two putative neuroparsin A dimers.

The partial sequencing of NPA demonstrates that NPA is heterogenous. The two NPA polypeptide chains (NPAα, NPAβ) appeared to contain NPB. NPB also seems to be a component of native NPA. The arrangement of these three distinct polypeptide chains to form NPA dimers leads to eight putative dimers: two homodimers NPAα-NPAα, NPAβ-NPAβ, three heterodimers NPAα-NPAβ, NPAα-NPB, NPAβ-NPB and their three symmetrical dimers. Unfortunately, chromatofocusing did not allow the separation of the putative NPA dimers. According to the quantitative evaluation of the NPAα, NPAβ and NPB polypeptide chains and from the anion exchange chromatography results, at least two putative dimers are likely, NPA-I (NPAα-NPAβ) and NPA-II (NPAα-NPB) (Figure 8). The post-translational processing which leads from the parent molecule NPA to the final NPB homodimeric molecule previously demonstrated [3], may involve only two steps with, in both cases, the cleavage after an arginyl residue. This hypothesis must be corroborated by completing the sequence of the NPA dimers, hopefully in the near future.

REFERENCES

1. J. Girardie, D. Bourême, F. Couillaud, M. Tamarelle and A. Girardie, Anti-juvenile effect of neuroparsin A, a neuroprotein isolated from locust corpora cardiaca, *Insect Biochem.*, 17:977-983 (1987).
2. J. Girardie, D. Bourême and A. Girardie, Production sites of the three neurosecretory proteins characterized in the corpora cardiaca of the migratory locust, *Insect Biochem.*, 17:29-36 (1987).
3. D. Bourême and J. Girardie, Les neuroparsines du criquet (Othoptera): micro-hétérogénéité d'origine enzymatique, *Ann. Soc. Entomol. France*, 24:390 (1988).
4. B. Fournier and J. Girardie, A new function for the locust neuroparsins: stimulation of water reabsorption, *J. Insect Physiol.*, 34:309-313 (1988).
5. R. Moreau, L. Gourdoux and J. Girardie, Neuroparsin: a new energetic neurohormone in the African locust, *Arch. Insect Biochem. Physiol.*, 8:135-145 (1988).
6. D. Bourême and J. Girardie, Présence de protéines à ponts disulfure dans les corpora cardiaca d'insectes, *Comptes rendus de l'Académie des Sciences Paris*, 18:655-660.
7. J. Girardie, A. Faddoul and A. Girardie, Characterization of three neurosecretory proteins from the A median neurosecretory cells of *Locusta migratoria* by coupled chromatographic, electrophoretic and isoelectrofocusing methods, *Insect Biochem.*, 15:85-92 (1985).
8. J. Girardie, A. Girardie, J. C. Huet and J. C. Pernollet, Amino acid sequence of locust neuroparsins, *Fed. European Biochem. Soc. Lett.*, 245:4-8 (1989).
9. W. Mordue, J. P. Morgan and K. J. Sieghert, Strategies for the isolation of insect peptides, *Peptides*, 6:407-410 (1985).

ISOLATION AND IDENTIFICATION OF A SULFAKININ-LIKE PEPTIDE, WITH SEQUENCE HOMOLOGY TO VERTEBRATE GASTRIN AND CHOLECYSTOKININ, FROM THE BRAIN OF *LOCUSTA MIGRATORIA*

L. Schoofs, M. Holman*, T. Hayes** and A. De Loof*

Zoological Institute of the University
Naamsestraat 59, B-3000 Leuven, Belgium
*Veterinary Toxicology and Entomology Research Laboratory
Agricultural Research Service, US Department of Agriculture
College Station, TX 77841, USA, and
**Laboratories for Invertebrate Neuroendocrine Research
Department of Entomology, Texas A & M University
College Station, TX 77843, USA

SUMMARY

A neuropeptide (pGlu-Leu-Ala-Ser-Asp-Asp-Tyr-Gly-His-Met-Arg-Phe-NH2) with a blocked N-terminus and related to leucosulfakinins I and II from the Madeira roach has been isolated from an extract of 9000 brain complexes of *Locusta migratoria*. The peptide stimulates the motility of the cockroach hindgut. Biological activity was monitored during purification by high performance liquid chromatography by observing the myotropic effect of column fractions on the isolated hindgut of *Leucophaea maderae*.

The peptide, designated as locustasulfakinin or Lom-SK shares a common C-terminal decapeptide fragment with leucosulfakinin II and a common heptapeptide fragment with leucosulfakinin I, indicating the importance of the constituent amino acids for biological activity. Locustasulfakinin is the third invertebrate neuropeptide which exhibits sequence homologies with the hormonally active fraction of the vertebrate hormones cholecystokinin, human gastrin II and caerulin. Its intestinal myotropic activity is analogous to that of gastrin and cholecystokinin. The sequence homologies as well as the analogous myotropic activities suggest that gastrin/cholecystokinin-like peptides have a long evolutionary history, and probably play an important role in physiology.

INTRODUCTION

Immunological studies have demonstrated the presence of various mammalian-like hormones in invertebrate tissues (for reviews see Refs. 1,2). The presence of gastrin or cholecystokinin (CCK)-like immunoreactivity has been demonstrated in nervous and intestinal tissues of several invertebrate species. However, identification of peptides by immunological techniques is of limited value in identifying the molecular structure that is crossreacting with that particular antibody. Nachman et al. [3,4] were the first to report the amino acid sequence of a myotropic neuropeptide in the cockroach *Leucophaea maderae*, with sequence homology with the active C-terminal portion of the vertebrate hormones human gastrin II and cholecystokinin. Moreover, the gastrointestinal myotropic activity of these leucosulfakinins (LSKs) were analogous to that of gastrin and cholecystokinin.

Here we describe the isolation of locustasulfakinin, a third member of the invertebrate gastrin/CCK peptide family, using the same detection method after purification using high performance liquid chromatography (HPLC) (observation of myotropic effect of column fractions on the isolated hindgut of *Leucophaea maderae*).

Chromatography and Isolation of Insect Hormones and Pheromones
Edited by A. R. McCaffery and I. D. Wilson
Plenum Press, New York, 1990

MATERIALS AND METHODS

Locusta migratoria was raised under laboratory conditions as described previously [5]. About 9000 brain-corpora cardiaca/corpora allata-suboesophageal ganglion complexes of *Locusta migratoria* were individually dissected over a period of six months.

The culture of *Leucophaea maderae*, and preparation of the *Leucophaea* hindgut bioassay have been described elsewhere [6].

Extraction and Sep-pak Purification

Groups of 50 brain complexes were extracted in centrifuge tubes, each containing 30 ml of a methanol-water-acetic acid (90:9:1 v/v/v) solution. The contents of each tube were homogenized for 3 min and subsequently sonicated for 1 min. The homogenates were held in the refrigerator at 4°C. Following centrifugation (30 min, 9820 g, 5°C), the supernatants from 10 tubes were decanted into a 1 l round bottom flask. 25 ml of extraction solvent was added to each centrifuge tube and the pellet was resuspended by vigorous shaking. The tubes were recentrifuged and both supernatants were removed by rotary evaporation at 30°C until only the water and a portion of the acetic acid remained. This viscous solution was divided between 10 centrifuge tubes and the final volume in each tube was brought to 15 ml with distilled water. Then 15 ml of ethylacetate was added to each tube. After centrifugation and removal of the upper layer, 15 ml of n-hexane was added to each tube and centrifuged. The lower aqueous portions of each centrifuge tube were pooled, frozen and freeze-dried. All water and acetonitrile-water solutions used for Sep-pak purification and HPLC-analysis contained 0.1% trifluoracetic acid (TFA) as an ion pairing agent. Procedures for Sep-pak purification have been previously described [7]. The dry residue of the extracted material of 250 brain equivalents was dissolved in 50 ml of 0.1% TFA. Aliquots of 5 ml were applied to the Sep-pak and eluted each time with 5 ml of 30% acetonitrile. The Sep-pak was rinsed with 10 ml of 75% acetonitrile solution. All of the 30% (10) and 75% acetonitrile (10) fractions were pooled, evaporated and lyophilized.

High Performance Liquid Chromatography

HPLC analysis was performed on a Waters ALC 100 Liquid Chromatograph equipped with two pumps, a solvent programmer, a Rheodyne septumless injector and a variable wavelength UV detector, set at 214 nm (Waters Assoc. Inc., Milford, MA).

Initially, separation of the biologically active compounds was carried out on a Waters μ-Bondapak Phenyl column (4.6 mm x 30 cm) at ambient temperature and with a solvent rate of 1.5 ml/min. The dry residue in a sample of 250 brain equivalents was dissolved in 0.5 ml of 0.1% TFA and transferred to the injector after passing through a Millipore 0.45 μm filter. The sample tube was rinsed twice with 0.5 ml aliquots of 0.1% TFA, which were also transferred to the injector. Eight minutes after injection a linear gradient from 0 to 50% acetonitrile (made to 0.1% TFA) was started. Final concentration was achieved after 2 hours. Fractions containing 3 ml were arbitrarily collected every 2 min starting at 36 min until 100 min. Thirty six HPLC runs, each containing 250 brain equivalents were required to process the entire extract through μ-Bondapak fractionation. The various fractions containing myotropic compounds were identified by bioassay and retained. One fraction (64 to 66 min) was chosen for further purification. All 36 of the appropriate fractions were pooled and concentrated on Sep-pak. The 80% acetonitrile-water eluate was taken to dryness and further fractionated through a series of three HPLC columns, having different separation characteristics, in the following sequence.

(1) Rainin Microsorb Cl (4.6 mm x 25 cm). Conditions: Solvent A: 0.1% TFA in water. Solvent B: 25% acetonitrile in water, made to 0.1% TFA. Initial conditions: 100% solvent A for 8 min, then a linear program to 100% solvent B over 2 hours. Flow rate 1.5 ml/min. Temperature: ambient. Detector range: 2.0 absorbance units full scale (AUFS) at 214 nm. Fractions were manually collected every 4 min starting at 0 min.

(2) Supelco-C8 Dibasic (4.6 nm x 15 cm). Solvent A: 0.1% TFA in water. Solvent B: 50% acetonitrile in water, made to 0.1% TFA. Initial conditions: 25% solvent B for 8 min, then linear program to 100% solvent B over 80 min. Flow rate: 1.5 ml/min. Detector range: 1.0 AUFS at 214 nm.

(3) Waters-I-125 Protein-Pac column (7.8 nm x 30 cm). Solvent A: 95% acetonitrile made to 0.01% TFA. Solvent B: 50% acetonitrile made to 0.01% TFA. Initial conditions:

100% solvent A for 8 min then linear program to 100% solvent B over 80 min. Flow rate: 1.5 ml/min. Detector range: 0.2 AUFS at 214 nm.

Amino Acid Analysis

Amino acid contents of peptides isolated from the HPLC were determined with the Waters Picotag amino acid analysis system. Samples were hydrolyzed *in vacuo* with gaseous HCl at 150°C for one hour, neutralized with 20 µl of a methanol-water- triethanolamine (2:1:1 v/v/v) solution, vacuum lyophilized and converted to the phenyl-isothiocyanate (PITC) derivatives with 20 µl of an ethanol-triethanolamine-water- phenylisothiocyanate (7:1:1:1 v/v/v/v) solution, vacuum lyophilized and analyzed with the Waters ALC-100 HPLC according to the instructions supplied by the manufacturer. Quantification of the PITC amino acid derivatives was performed by comparing peak areas with the peak areas determined from analyses of an amino acid standard solution also supplied by the manufacturer.

Enzymatic Analysis

Residues of HPLC purified fractions of about 5% of pure peptide were taken up in 0.5 ml of immobilized aminopeptidase M (Pierce Chemical Company) gel suspension (containing 0.1 unit of APM). Following incubation for 2 hours at 37°C, the gel was removed from the suspension by centrifugation and the supernatant was removed for bioassay.

Pyroglutamineaminopeptidase (PGAP) (Sigma) was dissolved in distilled water in a ratio of one activity unit per 20 µl water. Two activity units of enzyme solution (40 µl) were added to at least 500 pmol of the peptide. Following incubation (37°C for 30 min) 0.5 ml of 0.1% TFA was added to the tube and the sample was fractionated with the Supelco C8 column. Peaks were manually collected and stored for amino acid analyses.

RESULTS

After one pass of the 64 to 66 min fraction of the µ-Bondapak-phenyl column through Microsorb Cl column, two fractions, eluting between 36 and 40 min and between 48 and 52 min contained myotropic activity (Figure 1). The 36 to 40 min fraction was further purified on the Supelco C8 column and eluted as one peak at 37.8 min (Figure 2). Final purification of this peak on Waters Protein-Pac revealed a peak at 49.9 min (Figure 3). The purified peptide was subjected to hydrolysis and subsequent amino acid analysis. The molar ratio composition of the isolated peptide was Asp (1 or 2), Glx (1 or 2), Ser (1), Gly (1 or 2), His (1), Arg (1), Tyr (1), Met (1), Phe (1), Leu (1), Ala (1).

Incubation of pure peptide with aminopeptidase M revealed that the biological activity was completely unaffected. The inability of aminopeptidase M to destroy biological activity and the presence of glutamic acid in the amino acid composition suggested that the N-terminal might be blocked by pyroglutamic acid (pGlu). Incubation of at least 500 pmol of the pure peptide with PGAP followed by HPLC-fractionation on the Supelco C8 column, resulted in the isolation of both a small amount of the undigested peptide and a fragment eluting at 18 min (Figure 4). The barely detectable peak of the parent peptide indicated that almost all of it had been digested by the PGAP. Gas-phase sequence analysis of the larger peak showed that it contained all of the amino acids except glutamic acid, indicating that PGAP had removed the pGlu from the N-terminal position: Leu-Ala-Ser-Asp-Asp-Tyr-Gly-His-Met-Arg-Phe. Evaluation of the enzyme digestion data led to the following primary structure:

pGlu-Leu-Ala-Ser-Asp-Asp-Tyr-Gly-His-Met-Arg-Phe

Evaluation of biological and chemical characteristics of the sulfated synthetic replicas in the amidated and acid form will elucidate whether the purified peptide is sulfated and C-terminally amidated.

DISCUSSION

The amino acid sequence of the purified locust peptide is entirely homologous to the sequence of leucosulfakinin II of *Leucophaea maderae*, except for two amino acids, leucine and alanine on positions 2 and 3, which are not present in LSK II. Therefore, we have

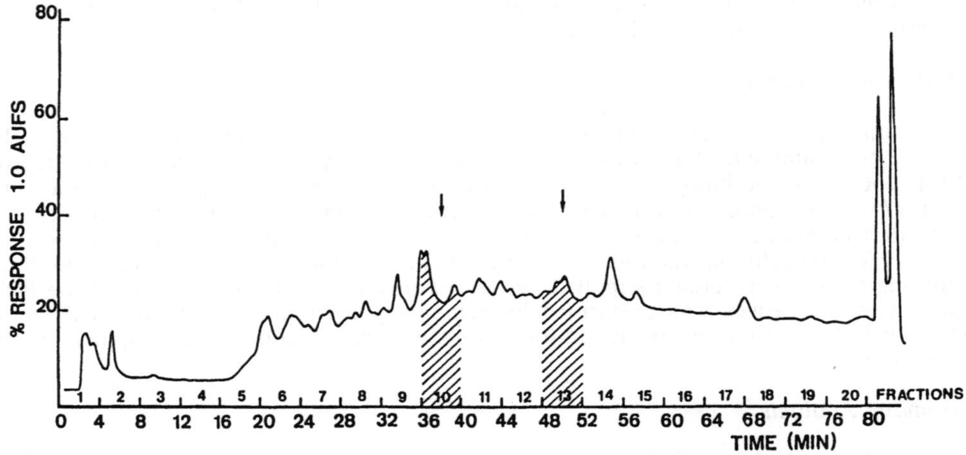

Fig. 1. Fractionation on Microsorb Cl of area eluting at 62 to 64 min from μ-Bondapak phenyl. Sample contained 9000 brain-corpora cardiaca-corpora allata-suboesophageal ganglion equivalents. Myotropic activity is indicated. HPLC conditions are given in the text.

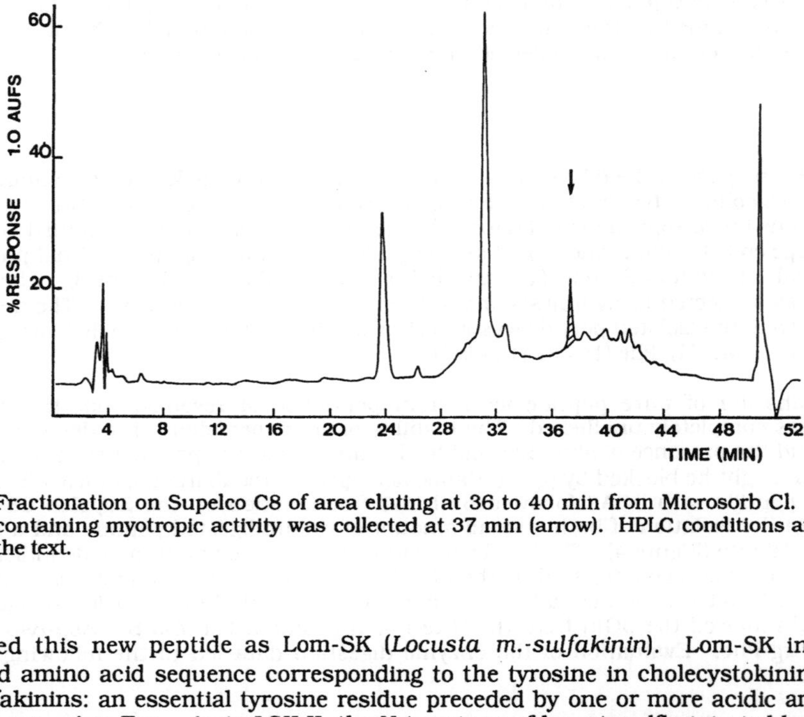

Fig. 2. Fractionation on Supelco C8 of area eluting at 36 to 40 min from Microsorb Cl. One peak containing myotropic activity was collected at 37 min (arrow). HPLC conditions are given in the text.

designated this new peptide as Lom-SK (*Locusta m.-sulfakinin*). Lom-SK includes a conserved amino acid sequence corresponding to the tyrosine in cholecystokinin and the leucosulfakinins: an essential tyrosine residue preceded by one or more acidic amino acid residues, e.g., -Asp-Tyr-. As in LSK II, the N-terminus of locustasulfkainin is blocked by a pyroglutamyl residue, a structural feature found in all insect adipokinetic hormones and some vertebrate hormones (gastrin and cholecystokinin). It may serve the function of protecting the neuropeptide from degradation by aminopeptidase during transport *in vivo* [8]. Investigation of the sulfated synthetic analog will reveal whether the tyrosine residue in Lom-SK is sulfated.

The fact that Lom-SK contains 2 additional amino acids more than the roach LSK at the N-terminus, but is similar to LSK II at the C-terminus, is in agreement with the conclusion of Nachman et al. [3], indicating that the active core of the LSK's is at the C-terminus; the biological active core of the LSK's of the cockroach can be no longer than Asp-Tyr-Gly-His-Met-Arg-Phe-NH2, which constitutes also the C-terminus of locusta-sulfakinin.

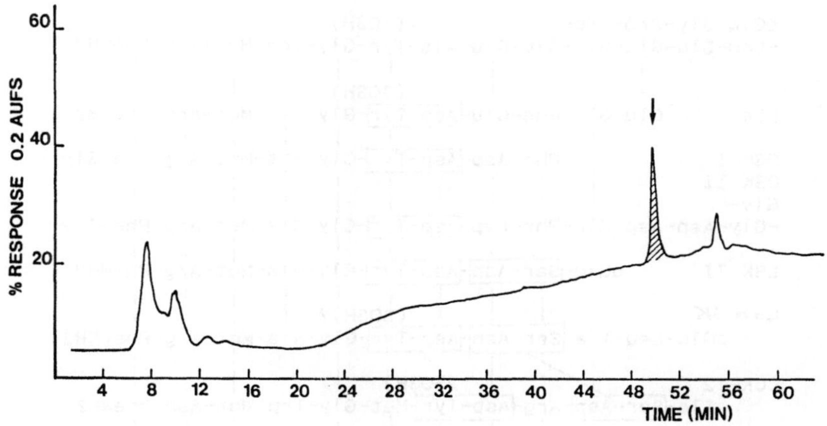

Fig. 3. Fractionation on Waters I-125 Protein-Pac column of active peak collected from Supelco C8. Peak containing myotropic activity eluted at 49.9 min (arrow).

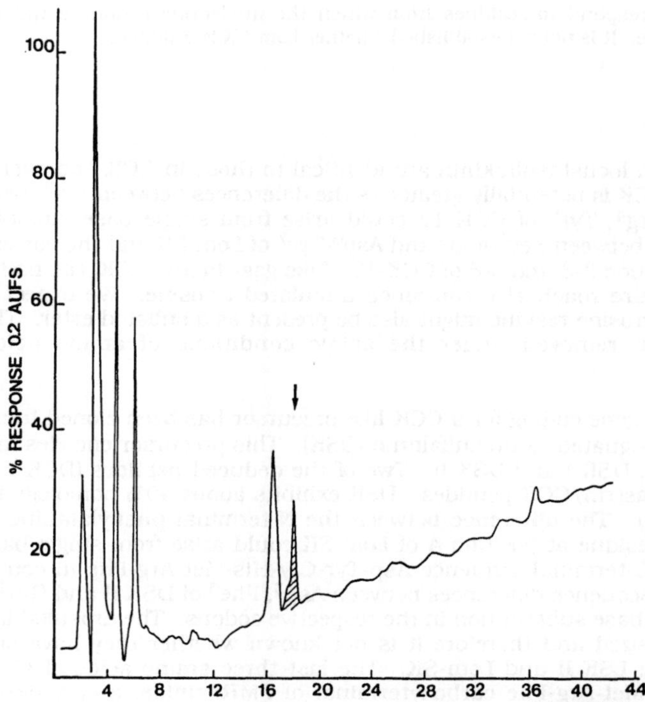

Fig. 4. HPLC fractionation of 500 pmol of Lom-SK incubated with two activity units of pyroglutamate aminopeptidase. Column and conditions are the same as in Figure 2. The peak eluting at 18 min was sequenced and contained all the amino acids minus Glu.

The leucosulfakinins (LSK and LSK II) exhibit sequence homologies with the active portion of the vertebrate hormones cholecystokinin (CKK), human gastrin II and caerulin [3,4]. This is also true for Lom-SK (Figure 5). Residues Tyr^{7}, Gly^{8}, Met^{10} and Phe^{12} of locustasulfakinin are identical to positions 12, 13, 15 and 17 of human gastrin II. The blocked pGlu of Lom-SK is structurally related to Glu^{6} in the analogous position of gastrin II, the former being a cyclic analog of the other. Moreover, the differences between Asp^{5}, Asp^{6} of locustasulfakinin and Glu^{10}, Ala^{11} of gastrin II could arise from single base substitution in the respective codons. Like LSK II, locustasulfakinin is even more closely related to cholecystokinin than to gastrin. Residues Asp^{5}, Gly^{8}, Met^{10} and Phe^{12} are identical to those in the analogous 5, 8, 10 and 12 positions of active CCK 12. Four of twelve

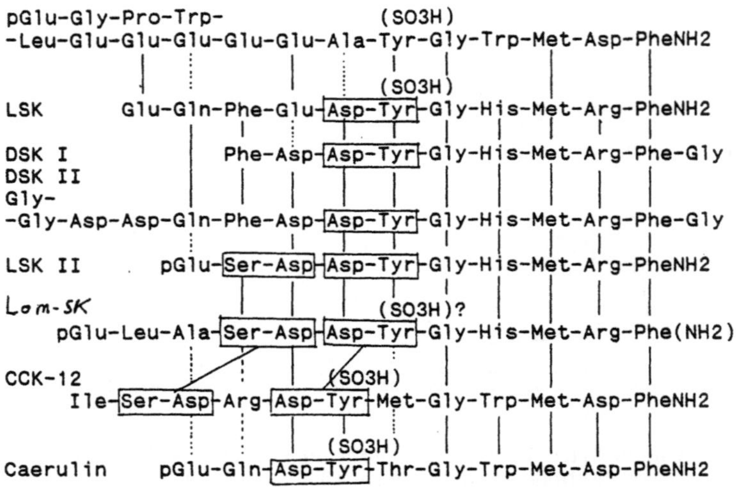

Fig. 5. Sequence homologies between Lom-SK, LSK, LSK II, DSK I, DSK II, human gastrin II, cholecystokinin 12 and caerulin. Solid lines indicated identical amino acids and dashed lines correspond to residues from which the nucleotide codons could differ by a single nucleotide. It is not yet established whether Lom-SK is sulfated.

(30%) residues in locustasulfakinin are identical to those in CCK and caerulin. Sequence homology with CCK is potentially greater as the differences between Ala^3, Ser^4, Asp^6 of Lom-SK and Asp^3, Arg^4, Tyr^6 of CCK 12 could arise from single base substitution. Lateral homologies exist between Ser^4-Asp^5 and Asp^6-Tyr^7 of Lom-SK and the corresponding amino acid pairs in position 2-3 and 5-6 of CCK-12. Like gastrin and CCK, the native LSK and LSK II from the Madeira roach also contained a sulfated tyrosine. We expect that the locust native Lom-SK tyrosine residue might also be present as a sulfated ester. The sulfate group could have been removed under the acidic conditions of amino acid analysis and sequencing.

Recently, a gene coding for a CCK-like precursor has been cloned from *Drosophila* [9] and has been designated as drosulfakinin (DSK). This precursor encodes for three putative peptides: DSK O, DSK I and DSK II. Two of the deduced peptides (DSK I and DSK II) are homologous to gastrin/CCK peptides. DSK exhibits about 90% homology to LSK II and to locustasulfakinin. The difference between the N-terminal phenylalanine residue of DSK and the serine residue at position 4 of Lom-SK could arise from single base substitution. DSK II has the C-terminal sequence Asp-Tyr-Gly-His-Met-Arg-Phe in common with Lom-SK and LSK II. Sequence differences between Asp^3, Phe^6 of DSK II and Glu^1 and Ser^4 could arise from single base substitution in the respective codons. The drosulfakinins have not as yet been synthesized and therefore it is not known whether they have similar biological activities to LSK, LSK II and Lom-SK. The last three amino acids of all sulfakinins are identical to the Met-Arg-Phe carboxyterminal of FMRFamide, a cardioexcitatory peptide isolated from molluscs [10].

In vertebrates, CCK has been identified in every species in which it has been sought, from agnathans to mammalians [11,12]. CCK/Gastrin immunoreactivity has already been reported in the intestinal and in the nervous system of several invertebrate species [13]. In *Diphyllobothrium dendriticum*, gastrin immunoreactive fibers and perikarya were found in and along the main nerve cords, from where fibers extend towards the main excretory ducts [14].

Gastrin/CCK-like peptides occurred in the nervous system of various genera of insects (see Table 1). The gastrin-CCK-like immunoreactivity in the corpora cardiaca of *Leucophaea maderae* [15] may be attributed to the presence of the leucosulfakinins. In *Locusta migratoria*, CCK-8-like neurons and nerve fibers were present in various parts of the brain and the ventral nerve cord [16] and this immunoreactivity is also probably due to the presence of locustasulfakinin. In the light of the homology between locustasulfakinin and gastrin/CCK on one hand, and reports that the locust central nervous system and midgut contain CCK/gastrin-like immunoreactive components, it will be interesting to

Table 1. The occurrence of gastrin/chlolecystokinin-related substances in invertebrates. (CC-CA: corpora cardiaca-corpora allata, NT: nervous tissue)

Protostomia		
Cnidaria		
Metazoa		
Hydra attenuata	NT	[23]
Tealia felina	NT	[23]
Anthozoa		
Anthopleura xanthogrammicca	gut	[24]
Platelminthes		
Diphyllobothrium dendriticum	NT	[14]
Annelida		
Oligochaeta		
Nereis diversicolor	NT	[25]
Lumbricus terrestris	NT	[24]
	gut	[26]
Polychaeta		
Abarenicola pacifica	gut	[24]
Mollusca		
Gastropoda		
Busyon canaliculatus	gut	[24]
	hepatopancreas	
	kidney	
	haemolymph	
Aplysia californica	gut	[24]
	NT	[27]
Helix aspersa	NT	[28]
Lymnea stagnalis	NT	[29]
		[30]
Helix pomatia	NT	[24]
Arthropoda		
Crustacea		
Malacostraca		
Cancer magister	gut	[24, 31]
	NT	
Upogebia pugettensis	gut	[24]
Merostomata		
Limulus polyphemus	gut	[24]
	NT	
Insecta		
Dictyoptera		
Periplaneta americana	gut	[22]
Blaberus craniifer	gut	[21]
Leucophaea maderae	CC-CA	[15]
Orthoptera		
Locusta migratoria	NT	[22]
	gut	
Phasmida		
Carausius morosus	gut	[22]
Coleoptera		
Leptinotarsa decemlineata	NT	[32]
Odonata		
Aeschna cyanea	get	[20]
Lepidoptera		
Manduca sexta	NT	[33]
		[34]
Bombyx mori	NT	[35]
Diptera		
Calliphora erythrocepala	NT	[19]
		[36]
Calliphora vomitoria	NT	[17, 19]
Eristalis aeneus	NT	[18]

Table 1. (continued)

Deuterostomia		
Protohordata		
Urochordata		
Styuela clava	NT	[37]
Ascidiella aspersa	NT	[37]
Ciona intestinalis	gut	[38]
Styela montereyensis	gut	[24]
Styela plicata	gut	[39]
Cephalochordata		
Amphioxus lanceolatus	gut	[40]

examine the ability of a C-terminally directed gastrin/CCK antiserum to bind the locustasulfakinin.

There is substantial evidence from the work of Duve and Thorpe [17] that gastrin/CCK immunoreactive material in the blowfly supplies the abdomen via the abdominal nerves as well as via the dorso-posterior extension of the cardiacal recurrent nerve. They suggested that this peptidergic innervation could be related to a digestive function such as stimulation of secretion of the gut, or that it could be interpreted in terms of control of muscular movement of the gut and associated organs. Therefore, this immunoreactive material could be very similar to locustasulfakinin and to LSK, which show many similarities with gastrin and CCK.

Due to the homologies between CCK-caerulin, gastrin, LSK and Lom-SK, it is reasonable to hypothesize that the peptides arose by gene duplication events followed by subsequent independent evolution, although other mechanisms are possible. However, this hypothesis is more acceptable than the alternative of multiple independent origins followed by convergent evolution. Especially the sequence of active site of the gastrin/CCK family has been highly conserved whereas the remaining portion of the gene has been free to accept a high rate of nucleotide substitution. Indeed, even in vertebrates there are major differences between all CCKs in the N-terminal portion of the molecule. It is not surprising that the amino acid sequences in the latter region of the molecule have diverged during the course of evolution, since full biological potency resides in the CCK-8 portion of the molecule [12]. Also in insects (e.g., *Manduca sexta*), immunoreactive neurosecretory cells were found with antisera against the gastrin C-terminus, but not with antisera against the CCK 9-20 and 9-25 midportion [18]. The material detected in the neuroendocrine system of *Calliphora vomitoria* reacted with antisera specific for the common COOH-terminus of gastrin and CCK but not with the N-terminal directed antisera [11,19]. CCK-8-like material was also demonstrated in basal endocrine cells of the midgut of *Aeschna cyanea* [20], *Carausius morosus, Blaberus craniifer, Locusta migratoria* and *Periplaneta americana* [21, 22]. The antiserum used was directed against the C-terminal non-sulfated cholecystokinin octapeptide.

These immunocytochemical studies as well as the present identification study indicate that the carboxyterminals of the peptides of the gastrin family have been conserved during evolution. The concept has arisen that the C-terminal pentapeptide fragment in caerulin, gastrin, CCK, leucosulfakinins and locustasulfakinin is more invariant, because it is crucial for the biological effect of these peptides [3].

Based on immunocytochemical studies and the present identification study, it is clear that the gastrin/CCK family has a long evolutionary history, extending at least as far back as the primitive Metazoa. It also seems that regulatory peptides with such a long evolutionary history probably play an important role in physiology. The presence of a neuropeptide in the brain-corpora cardiaca-corpora allata-suboesophageal complex, its myotropic activity at the distant hindgut and homology with vertebrate hormones suggest that locustasulfakinin may have a hormonal function in the migratory locust. However, the biological activity of the Lom-SK still has to be evaluated on the visceral muscles of the locust itself.

The intestinal myotropic activity of the LSK's and Lom-Sk is analogous to that of gastrin and CCK. The involvement of CCK-gastrin-like peptides in the stimulation of muscle contractions appears to be a general characteristic in both vertebrates and insects (invertebrates) and apparently constitutes an evolutionary primitive role for CCK-gastrin-like peptides.

Analogous to the increased blood circulation and enzyme secretion in vertebrate intestine, studies on these attendant biological activities on haemolymph circulation and possible involvement in the secretion of digestive enzymes of the gastrin-CCK-like insect peptides will be performed.

A synthetic peptide with the proposed sequence has yet to be prepared. This synthetic peptide must behave in a similar manner on the isolated roach hindgut and should elute at identical retention times on the HPLC columns, used for purification of the native Lom-SK. Once these studies have been performed, the sequence will have been established.

ACKNOWLEDGEMENTS

The authors gratefully acknowledge the Koningin Elizabeth Stichting, the NFWO and het Onderzoeksfonds van de K. U. Leuven for financially supporting this research. L. S. was a visiting scientist at the VTERL, USDA, College Station, Texas and acknowledges the NFWO, NATO and K. U. Leuven for travel scholarships. The authors are indebted to their colleagues, L. Paemen, M. Verstuyf, E. Fias, K. Spittaels, A. Tips, J. Vandenbroeck, P. P. Verhaert, I. De Loof, K. Wilna, M. Amelinckx, W. Meulemans, K. Hendrickx and R. Huybrechts of the Zoological Institute for their help in the manual dissection of locust brains. Peptide sequencing was conducted at the TAES Biotechnology Support Laboratory. Special thanks to Ms. M. Miller for assistance with amino acid sequence analysis; USDA grant nr. 85-CRCD-rz008 and USPHS grant nr. ROINS202137 to TRA for contributions towards sequence costs.

REFERENCES

1. A. De Loof, The impact of the discovery of vertebrate-type steroids and peptide hormone-like substances in insects, *Entomol. Exp. Appl.*, 45:105-113 (1987).
2. L. Schoofs, S. Jègou, H. Vaudry, P. Verhaert and A. De Loof, Localization of melanotropin resembling peptides in the central nervous system of two insect species, the migratory locust *Locusta migratoria* and the fleshfly, *Sarcophaga bullata, Cell Tissue Res.*, 248:25-31 (1987).
3. R. J. Nachman, G. M. Holman, W. F. Haddon and N. Ling, Leucosulfakinin, a sulfated insect neuropeptide with homology to cholecystokinin and gastrin, *Science*, 234:71-73 (1986).
4. R. J. Nachman, G. M. Holman, B. J. Cook, W. F. Haddon and N. Ling, Leucosulfakinin II, a blocked sulfated insect neuropeptide with homology to cholecystokinin and gastrin, *Biochem. Biophys. Res. Comm.*, 140:357-364 (1986).
5. G. J. Ashby, Locusts, *in*: "The UFAW Handbook on the Care and Management of Laboratory Animals", UFAW, ed., Churchill Livingstone, Edinburgh, London, pp 582-587 (1972).
6. B. J. Cook and G. M. Holman, Comparative pharmacological properties of muscle functions in the foregut and the hindgut of the cockroach, *Leucophaea maderae, Comp. Biochem. Physiol*, 61C:291-295 (1978).
7. G. M. Holman, B. J. Cook and R. J. Nachman, Isolation, primary structure and synthesis of two neuropeptides from *Leucophaea maderae*: members of a new family of cephalomyotropins, *Comp. Biochem. Physiol.*, 84C:205-211 (1986).
8. H. Gregory, P. M. Hardy, D. S. Jones, G. W. Kenner and R. C. Sheppard, The antral hormone gastrin, *Nature*, 204:931-933 (1964).
9. R. Nichols, A. Stephen, A. Schneuwly and J. E. Dixon, Identification of a *Drosophila* homologue to the vertebrate neuropeptide, cholecystokinin, *J. Biolog. Chem.*, 263:12167-12170 (1988).
10. D. A. Price and M. J. Greenberg, Structure of a molluscan cardioexitatory neuropeptide, *Science*, 189:670-671 (1977).
11. G. J. Dockray, H. Duve and A. Thorpe, Immunochemical characterization of gastrin/cholecystokinin-like peptides in the brain of the blowfly, *Calliphora vomitoria, Gen. Comp. Endocrinol.*, 45:491-496 (1981).
12. L. I. Larsson and J. F. Rehfeld, Evidence for a common evolutionary origin of gastrin and cholecystokinin, *Nature*, 269:335-338 (1977).
13. N. Dhainaut-Courtois, G. Tramu, R. Marcel, J. Malecha, M. Vergerbocquet, J. C. Andriès, H. Massom, L. Selboum, G. Belemtougri and J. C. Beauvillain, Cholecystokinin in the nervous system of invertebrates and protochordates. Immunohistochemical localization of a cholecystokinin-8-like substance in annelids and insects, *Ann. New York Acad. Sci.*, 448:167-187 (1985).

14. M. K. S. Gustafsson, M. A. I. Lehtonen and F. Sundler, Immunocytochemical evidence for the presence of mamalian neurohormonal peptides in neurones of the tapeworm, *Diphyllobothrium dendriticum*, *Cell Tissue Res.*, 243:41-49 (1986).
15. G. N. Hansen, B. L. Hansen and B. Scharrer, Gastrin/CCK-like immunoreactivity in the corpora cardiaca-corpora allata complex of the cockroach *Leucophaea maderae*, *Cell Tissue Res.*, 248:595-598 (1987).
16. M. Tamarelle, M. Romoeuf and J. J. Vanderhaeghen, Immunohistochemical localization of gastrin/CCK-like molecules in the central nervous system of the migratory locust, *Histochemistry*, 89:201-207 (1988).
17. H. Duve and A. Thorpe, Immunocytochemical mapping of gastrin/CCK-like peptides in the neuroendocrine system of the blowfly, *Calliphora vomitoria* (Diptera), *Cell Tissue Res.*, 237:309-320 (1984).
18. M. El-Salhy, R. Abou-El-Ela, S. Falkmer, L. Grimelius and E. Wilander, Immunohistochemical evidence of gastro-entero-pancreatic neurohormonal peptides of vertebrate-type in the nervous system of larva of a dipteran insect, the hoverfly, *Eristalis aeneus*, *Reg. Peptides*, 1:187-204 (1980).
19. H. Duve and A. Thorpe, Gastrin/cholecystokinin (CCK)-like immunoreactive neurones in the brain of the blowfly, *Calliphora erythrocephala* (Diptera), *Gen. Comp. Endocrinol.*, 43:381-391 (1981).
20. J. C. Andriès and G. Tramu, Distribution patterns of mammalian-like peptide immunoreactive cells in the midgut of *Aeschna cyanea* (Insecta, Odonata), *Experientia*, 41:500-503 (1985).
21. J. C. Andriès and G. Tramu, Ultrastructural and immunohistochemical study of endocrine cells in the midgut of the cockroach *Blaberus craniifer* (Insecta, Dictyoptera), *Cell Tissue Res.*, 240:323-332 (1985).
22. J. C. Andriès and J. C. Beauvillain, Ultrastructural study of cholecystokinin-like immunoreactivity in endocrine cells of the insect midgut, *Cell Tissue Res.*, 254:75-81 (1988).
23. C. J. P. Grimmelikhuijzen, J. F. Sundler and Rehfeld, Gastrin/CCK-like immunoreactivity in the vervous system of coelenterates, *Histochem.*, 69:61-68 (1980).
24. B. A. Larson and S. R. Vigna, Species and tissues distribution of cholecystokinin/gastrin-like substances in some invertebrates, *Gen. Comp. Endocrin.*, 50:469-475 (1983).
25. R. P. Engelhardt, N. Dhainaut-Courtois and G. Tramu, Immunohistochemical demonstration for a cholecystokinin-like peptide in the nervous system of a marine annelid worm, *Nereis diversicolor* O.F. Muller, *Cell Tiss. Res.*, 227:401-411 (1982).
26. R. Rzara, K. V. Kaloustian and E. K. Prokop, Immunochemical evidence for a gastrin-like peptide in the intestinal tissues of the earthworm *Lumbricus terrestris*, *Com. Biochem. Physiol. A*, 71:631-634 (1982).
27. E. Straus, R. S. Yalow and H. Gaines, Molluscan gastrin: concentration and molecular forms, *Science*, 190:687-689 (1975).
28. N. N. Osborne, A. C. Cuello and G. J. Dockray, Substance P and cholecystokinin-like peptides in Helix neurons and cholecystokinin and serotonin in a giant neuron, *Science*, 216:409-410.
29. B. P. Gesser and L. I. Larsson, Localisation of alpha-MSH-, endorphin- and gastrin/CCK-like immunoreactivities in the nervous system of invertebrates, *in*: "Integrated Neuronal Mechanisms", L. Angelucci, E. deWied and E. Endröczi and U. Scapagnini (eds), Elsevier Biomedical Press, Amsterdam (1983).
30. L. P. C. Scot, H. H. Boer, D. F. Swaab and S. Van Noorden, Immunocytochemical demonstration of peptidergic neurons in the central nervous system of the pond snail *Lymneaa stagnalis*, with antisera raised to biologically active peptides of vertebrates, *Cell Tiss. Res.*, 216:273-291 (1981).
31. B. A. Larson and S. R. Vigna, Gastrin-cholecystokinin-like immunoreactive peptides in the Dungness crab, *Cancer magister* (Dana), Immunochemical and biological characterization, *Regulatory Peptides*, 7:155-170 (1983b).
32. J. A. Veenstra, H. M. Romber-Privee, H. Schooneveld and J. M. Polak, Immunocytochemical localisation of peptidergic neurones and neurosecretory cells in the neuro-endocrine system of the colorado potato beetle with antisera to vertebrate regulatory peptides, *Histochemistry*, 82:9-18 (1985).
33. K. J. Kramer, R. D. Speirs and C. N. Childs, Immunochemical evidence for a gastrin-like peptide in insect neuroendocrine systems, *Gen. Comp. Endrocrin.*, 32:423-426 (1977).
34. M. El-Salhy, S. Falkmer, K. J. Kramer and R. D. Speirs, Immunohistochemical investigations of neuropeptides in the brain, corpora cardiaca, corpora allata of an adult lepidopteran insect *Manduca sexta* L. *Cell Tiss. Res.*, 232:295-317 (1983).

35. R. Yui, R. I. Fujita and S. Ito, Insulin-, gastrin-, pancreatic polypeptide-like neurons in the brain of the silkworm, *Bombyx mori*, *Biomed. Res.*, 1:41-46 (1980).
36. R. Cantera, Serotonin and gastrin/cholecystokinin-like immunoreactive neurons in the larval retrocerebral complex of the blowfly *Calliphora erythrocephala*, *Cell Tiss. Res.*, 253:425-433 (1988).
37. M. C. Thorndyke, Cholecystokinin (CCK)/gastrin-like immunoreactive neurons in the cerebral ganglion of the protochordate ascidians, *Styela clava* and *Asciedella aspersa*, *Regulatory Peptides*, 3: 281-288 (1982).
38. H. A. R. Fritsch, S. Van Noorden and A. G. E. Pearse, Localisation of somatostatin and gastrin-like immunoreactivity in the gastrointestinal tract of *Ciona intestinalis* L. *Cell Tiss. Res.*, 186:181-185 (1978).
39. M. Pestarino, J. C. Garaud and F. Ravera, Localization of immunoreactive gastrin-like cells in the alimentary tract of the ascidian, *Styela plicata*, *Gen. Comp. Endrocrin.*, 69:13-18 (1980).
40. S. Van Noorden and A. G. E. Pearse, The localization of immunoreactivity to insulin, glucagon and gastrin in the gut of Amphioxus (Branchiostomata) lanceolatus, *in*: "The Evolution of Pancreatic Islets, T.A.I. Grillo, L. Leibson and A. Epple (eds), pp. 163-178, Pergamon, Oxford (1976).

CHARACTERIZATION AND SEQUENCE DETERMINATION OF LOCUST BRAIN cDNA CLONES SELECTED BY ANTISERA RAISED AGAINST VERTEBRATE PEPTIDE HORMONES

J. Vanden Broeck, H. Smet, E. Fias and A. De Loof

Zoological Institute of the University
Naamsestraat 59, B-3000 Leuven, Belgium

SUMMARY

By means of immunocytochemical studies, performed with antisera directed against known (vertebrate) peptide hormones, numerous immunopositive substances have been detected within insect nervous tissue and gut endocrine cells. The isolation and characterization of these immunoreactive substances, however, may become a very laborious task. Therefore, an alternative and complementary approach was tried. A locust (*Locusta migratoria*) brain cDNA expression library was constructed. This recombinant DNA approach enabled the use of antisera as probes for the selection of specific nucleic acid sequences. In this study bovine growth hormone- (bGH-) and melanotropin- (MSH-), as well as hypertrehalosemic factor- (HTF- from *Carausius morosus*) and caudo-dorsal cell hormone- (CDCH- from *Lymnea stagnalis*) immunopositive plaques have been isolated, purified and amplified. These phage clones were then further analyzed. The sequence of cDNA inserts has been determined.

This immunological screening approach, combined with hybridization analysis of the isolated clones, can lead to the characterization of nucleic acid sequences coding for insect tissue immuno-reactive substances.

INTRODUCTION

Over the past decade many positive immunocytochemical stainings have been obtained in insect tissue sections using antisera directed against many vertebrate and some invertebrate peptide hormones [1-9].

Unfortunately, antibodies do not recognize the entire structure of a peptide but only parts of it (epitopes). Therefore, positive immunoreactions may be due to the recognition of only partly similar substances. Moreover, immunological crossreactions indeed do occur as shown by a number of groups (e.g., [9-13]). Different antisera directed against the same peptide sometimes even yield different results. These observations become very difficult to interpret if all immunoreactive material is considered to be homologous to the original immunogenic peptide(s). This means that, from immunocytochemical studies, one cannot conclude that insects indeed contain all peptides against which the reactive "heterologous" antisera were raised.

More structural data are needed in order to explain the immunological reactions and to allow further functional studies. The objective of this study was to characterize locust (*Locusta migratoria*) immunoreactive substances by means of an alternative and complementary approach. Recombinant DNA technology allows clonal amplification and rapid analysis of selected nucleic acid sequences. Selection of the appropriate nucleic acid sequence(s) can be obtained by immunological screening of a cDNA expression library (see [14]).

Chromatography and Isolation of Insect Hormones and Pheromones
Edited by A. R. McCaffery and I. D. Wilson
Plenum Press, New York, 1990

MATERIALS AND METHODS

Immunocytochemistry

Brains, suboesophageal ganglia, together with the adjacent corpora cardiaca/corpora allata complexes were microdissected from *Locusta migratoria* and *Periplaneta americana* and fixed in Bouin-Hollande's 10% sublimate solution. Tissues were dehydrated in ethanol, cleared in xylene and embedded in Paraplast. Series of alternating tissue sections (4 μm) were processed in the peroxidase-antiperoxidase (PAP) method according to Vandesande and Dierickx [15]. Peroxide and diaminobenzidine were used to localize immunoreactive antigens present in the tissue sections.

Synthesis of a Locust Brain cDNA-Expression Library

Locust brain RNA was prepared according to the method described by Arrand [16]. Poly-A^+ RNA was obtained by oligo-dT-cellulose chromatography [17]. The RNAs were tested by translation *in vitro* in a nuclease treated rabbit reticulocyte lysate.

Double stranded cDNA was synthesized following the method of Gübler and Hoffman [18] and Eco RI restriction sites were methylated by the enzyme Eco RI methylase. Eco RI linkers were ligated to the blunt ended, double stranded cDNA in order to introduce Eco RI cleavage sites. After Eco RI digestion, the cDNA was inserted into the dephosphorylated Eco RI arms of the phage vector λgt 11 (Promega Biotec). Infective phage particles were obtained by *in vitro* packaging of the vector ligation mixtures. For more details see Ref. [19].

Immunological Screening

Recombinant phages (about 10^6 plaque forming units per screening) were mixed with Y1090- bacteria (Promega Biotec), kept for 10 min at 37°C and plated out on culture dishes. Nitrocellulose filter replicas were screened by antisera raised against the vertebrate hormones, bovine growth hormone (bGH) and melanotropin (αMSH), against the hypertrehalosemic factor from the stick insect *Carausius morosus* (HTF) and against the oviposition hormone CDCH of the snail *Lymnea stagnalis*. The filters were incubated with alkaline phosphatase conjugated secondary antiserum (Promega Biotec). Detection of immunopositive plaques was obtained by an enzymatic reaction with nitro-blue tetrazolium (NBT) and 5-bromo-4-chloro-3-indolyl phosphate (BCIP) as substrates, resulting in a dark blue staining. A scheme of the technique is shown in Figure 1.

Amplification of Immunopositive Clones and Sequencing

Immunopositive clones were picked up and purified by further plating and screening. These clones were then amplified in order to prepare recombinant phage DNA [20]. Phage DNA was further analyzed by restriction enzyme digestion, hydbridization experiments and dideoxy-sequencing. Inserts were ligated into another vector (M13 mp18 or mp19) in order to prepare suitable templates for the dideoxy-sequencing reactions. Sequencing was performed according to Sanger et al. [21].

RESULTS

Screening of the cDNA library by means of antisera directed against the vertebrate hormones bGH and αMSH, against HTF of the stick insect *Carausius morosus* and against the molluscan peptide CDCH (*Lymnea stagnalis*) yielded several distinct, positively stained plaques. Moreover, all these antisera revealed certain immunoreactive neurons within insect brain tissue sections (immunocytochemical approach, e.g., Figures 2 and 3).

The immunopositive plaques were collected and purified until homogeneous recombinant phage stocks were obtained. The results of "tertiary" screening tests are shown in Figures 4 and 5. These tests were performed on almost entirely purified phage stocks.

The recombinant phages were amplified and phage DNA was prepared. The cDNA inserts are currently being further analyzed by restriction enzyme digestions, hybridization tests and sequencing. A sequencing result obtained after polyacrylamide electrophoresis of radioactive dideoxy-terminated sequencing mixtures is shown in Figure 6.

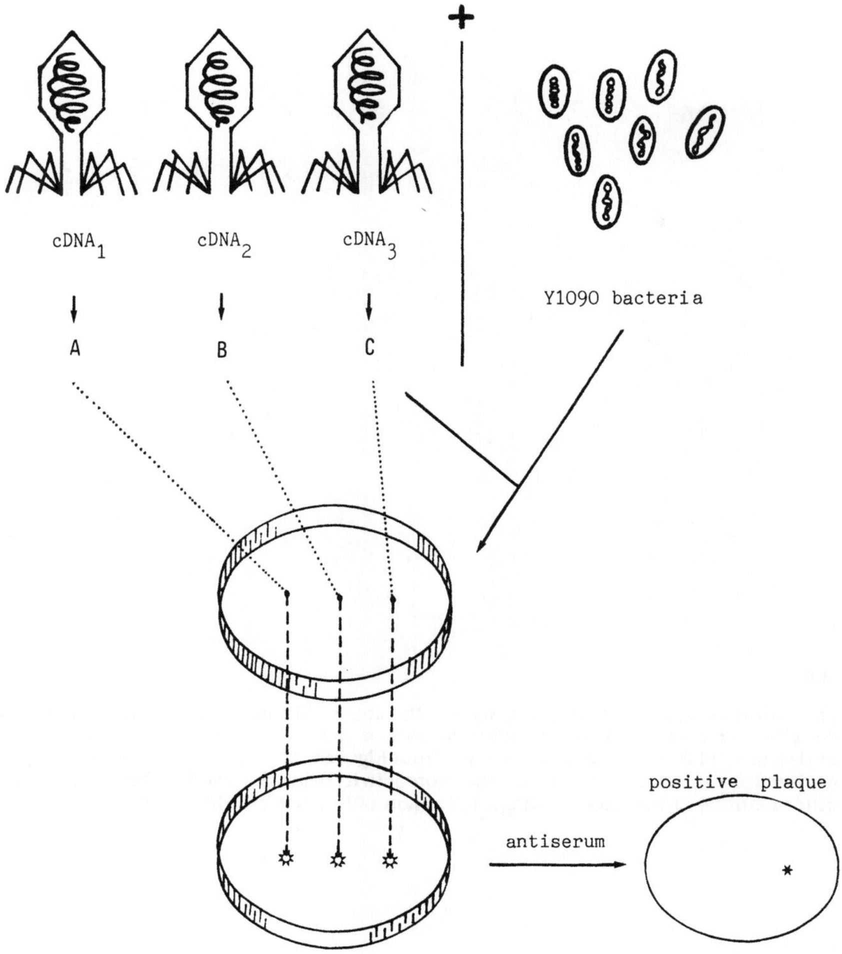

Fig. 1. Scheme of the immunological cDNA-library screening technique. Recombinant phages were mixed with bacteria and incubated for 10 to 15 min at 37°C. The mixture was then spread on agar culture plates. After culture, nitrocellulose filters, prerinsed in a solution containing IPTG (isopropyl β-D-thiogalactopyranoside:lac inducer), were laid on the plates. Filter replicas were incubated with an antiserum and further processed in order to allow detection of immunoreactive plaques.

So far, sequence determination has not demonstrated any appreciable degrees of amino acid or nucleotide sequence homology with the original immunogenic peptides or their corresponding genes. Nevertheless, some very limited similarities occurred between the sequences of these "authentic" peptide hormones and the deduced amino acid sequences of several clones. Parts of the sequences of two distinct bGH-immunopositive clones and of one HTF-immunoreactive clone are shown in Figure 7.

CONCLUSIONS

A locust brain cDNA-expression library was made in order to allow selection of nucleic acid sequences by means of antisera. Locust brain cDNAs indeed were selected by several anti-peptide hormone antisera (anti-bGH, αMSH, HTF, CDCH). Some of the immunoreactive phage clones were further analyzed, e.g., by sequence determination of their cDNA insert. The deduced amino acid sequences did not seem to share any spectacular degree of homology with the peptide hormones against which the antisera were raised. The limited similarities observed between the original immunogenic peptides and the cDNA-derived substances, however, may explain the cross-reactivity of the antisera. Of course,

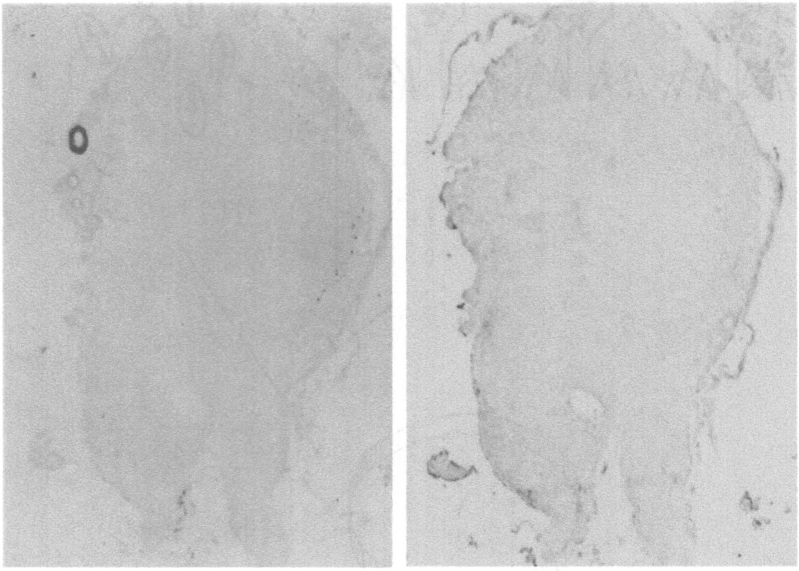

(a) (b)

Fig. 2. a) Immunopositive PAP-staining within the locust (*Locusta migratoria*) suboesophageal ganglion as a result of the incubation with a *first* anti-bGH (bovine growth hormone) antiserum. b) Alternating tissue section from Figure 2 a) incubation with a *second* anti-bGH antiserum. No clear overlap of immunoreactivity was observed. This antiserum stained other neurons of the suboesophageal ganglion not present in this section.

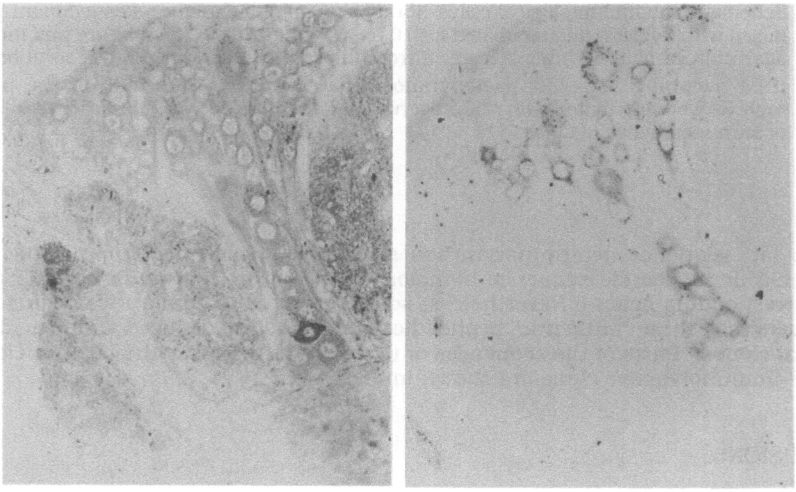

(a) (b)

Fig. 3. a) Positive staining obtained within the suboesophageal ganglion of the American cockroach (*Periplaneta americana*) by means of an anti- MSH (melanotropin) antiserum.
b) Adjacent section which was incubated withan anti-ACTH$_{4-10}$ (adrenocorticotropin amino acid 4-10 sequence-fragment) monoclonal antibody. ACTH$_{4-10}$ is an evolutionary conserved part that is present in most vertebrate melanotropins (MSHs).

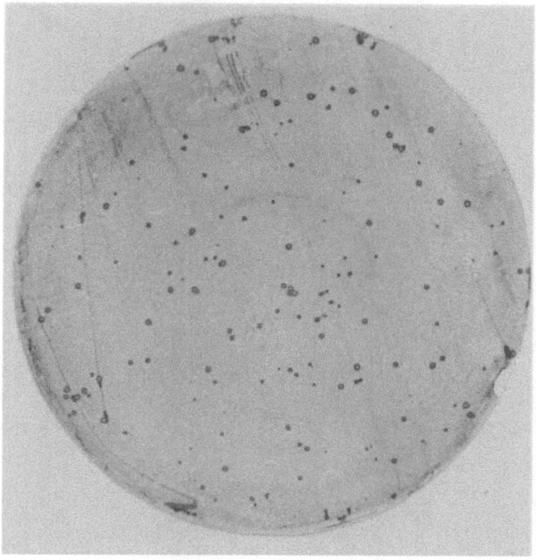

Fig. 4. Result of a tertiary screening of a locust cDNA expression library by an antiserum raised against αMSH (melanotropin).

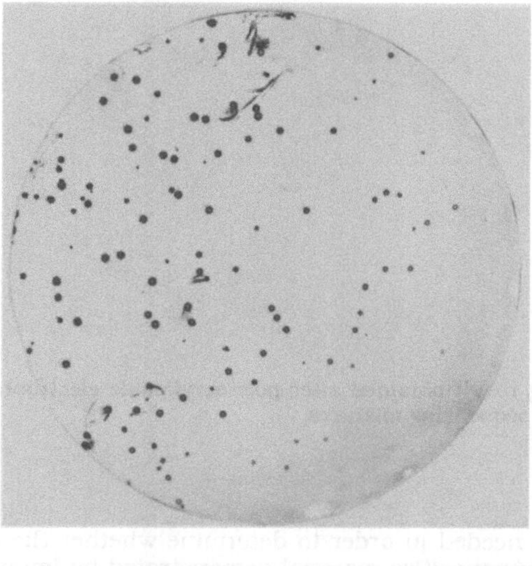

Fig. 5. Detection of an HTF- (hypertrehalosemic factor-) immunopositive clone in a tertiary screening test.

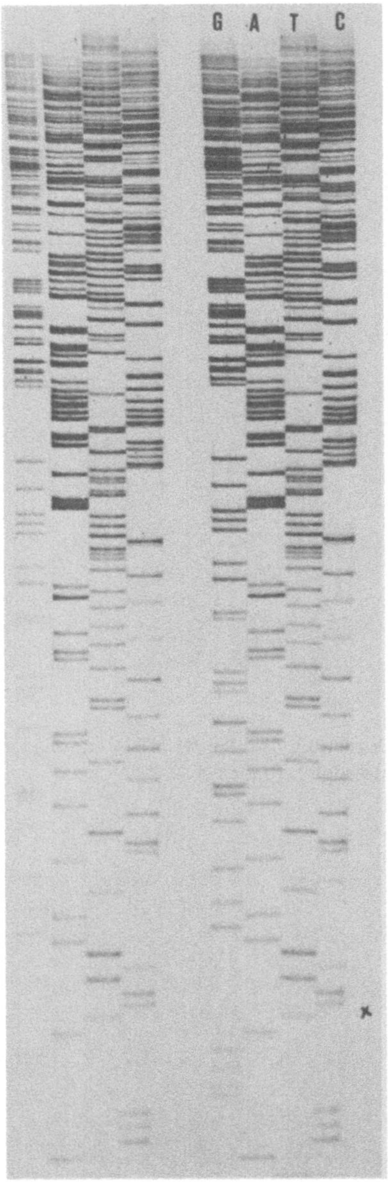

Fig. 6. Sequencing result obtained after poly-acrylamide electrophoresis of radio-active dideoxy-terminated sequencing mixtures.

further analysis is needed in order to determine whether these purified cDNAs really are coding for the immunopositive material demonstrated by immunocytochemical studies. A combination of hybridization histochemistry (*in situ* hybridization) and immunohistochemical methods will be employed in order to perform these further investigations.

By means of recombinant DNA techniques many vertebrate and some invertebrate (e.g., [22,23]) peptide hormone precursors have already been characterized. These methods may be particularly helpful to identify larger peptides or even protein hormones (De Loof et al., this volume). The immunological screening technique may offer an opportunity to characterize as yet unknown insect brain factors that function as transmitters, modulators or neurohormones.

```
(a)   1) -GTG GAC GAG GCG CCG AGG CCA GTC GTG CCG CTG GTC ATC ACC
         -Val-Asp-Glu-Ala-Pro-Arg-Pro-Val-Val-Pro-Leu-Val-Ile-Thr

        GAC TAC TCC TCC GTG GCT CCG GAC GAT CGA AGA AGG AGG CGA CAC
       -Asp-Tyr-Ser-Ser-Val-Ala-Pro-Asp-Asp-Arg-Arg-Arg-Arg-Arg-His

        GTG AGG TCC CCG TGA AGACCGAGACGCCGGCAGCCGCGCCGCAACGGGCCGA...
       -Val-Arg-Ser-Pro OPAL (stop)

      2)  -GCG GCC GGT GCG AGC GTC GGT GGG ACC TCT CCT TCG AGT CGG
          -Ala-Ala-Gly-Ala-Ser-Val-Gly-Gly-Thr-Ser-Pro-Ser-Ser-Arg

        GTT GCA CTC CGG AAT TCC GGG  GCACAGGCGTGCACACTGCTGTGCGTCCAT
       -Val-Ala-Leu-Arg-Asn-Ser-Gly- ...

        GCATGCATGCACAGACTAAAGCATGCACACTGCTGTGGACATGTATTTGAGATAATCC...

b)     -GTT TCG GCG ACT GAC GAG ATG GTC GCC GCC GCG GGA ACA GAG TGG
       -Val-Ser-Ala-Thr-Asp-Glu-Met-Val-Ala-Ala-Ala-Gly-Thr-Glu-Trp

        TCT TCT ACC CCC GGC CGA GTC GTT CCG CCG GTC GGC CCG CAC CCG
       -Ser-Ser-Thr-Pro-Gly-Arg-Val-Val-Pro-Pro-Val-Gly-Pro-His-Pro

        CAC CGA GCG CTT CTG CCA GAC GGT GGC CTC GAG GCC GGA ATC GCA
       -His-Arg-Ala-Leu-Leu-Pro-Asp-Gly-Gly-Leu-Glu-Ala-Gly-Ile-Ala

        CTC TCC TTC ACG CTA CCC GGA ATT CGA ATG TGG GAC TGT ...
       -Leu-Ser-Phe-Thr-Leu-Pro-Gly-Ile-Arg-Met-Trp-Asp-Cys-...


       HTF = pGlu-Leu-Thr-Phe-Thr-Pro-Asn-Trp-Gly-Thr.
```

Fig. 7. Nucleotide and derived amino acid sequences of two bGH- (a) and one HTF- (b) immunopositive clone(s) are shown. In these examples stretches of amino acid sequence shared with the "authentic" peptides are underlined.

ACKNOWLEDGEMENTS

The authors thank the NFWO of Belgium, the Belgian Queen Elisabeth Medical Foundation and the Research Foundation of the University for financial support. We are grateful to F. Clottens (Leuven), L. Schoofs (Leuven), A. White (Manchester, UK), K. Swinnen (Leuven), J. van Minnen and Prof. J. Joosse (Amsterdam, The Netherlands) for their kind gifts of antisera. We thank Dr Haed (Monsanto) for a gift of bovine somatatropin.J. Vanden Broeck is a research assistant of the NFWO of Belgium.

REFERENCES

1. H. Duve and A. Thorpe, Immunofluorescent localization of insulin-like material in the median neurosecretory cells of the blowfly *Calliphora vomitoria* (Diptera), *Cell Tissue Res.*, 200:187-191 (1979).

2. T. Fujita, R. Yui, T. Iwanaga, J. Nishiitsutsuji-Uwo, Y. Endo and N. Yanachara, Evolutionary aspects of "Brain-gut peptides": an immunohistochemical study, *Peptides*, 2:123-131 (1981).
3. B. L. Hansen, G. N. Hansen and B. Scharrer, Immunoreactive material resembling vertebrate neuropeptides in the corpus cardiacum and corpus allatum of the insect *Leucophaea maderae*, *Cell Tissue Res.*, 225:319-329 (1982).
4. C. Rémy, Parentés immunochimiques entre produits de neurosécrétion d'Invertébrés et neuropeptides de Vertébrés, *J. Physiol., Paris*, 78:514-522 (1982).
5. M. El-Salhy, S. Falkmer, K. J. Kramer and R. D. Speirs, Immunohistochemical investigations of neuropeptides in the brain, corpora cardiaca and corpora allata of an adult lepidopteran insect *Manduca sexta*, *Cell Tissue Res.*, 232:295-317 (1983).
6. K. J. Kramer, Vertebrate hormones in insects, *in*: "Comprehensive Insect Physiology, Biochemistry and Pharmacology", G. A. Kerkut and L. I. Gilbert, eds., Pergamon Press, Oxford, 7:Endocrinology I:511-536 (1985).
7. P. Verhaert, R. Huybrechts, D. Schols, J. Vanden Broeck, A. De Loof and F. Vandesande, Characterization of vertebrate peptide hormone-like materials in the American cockroach.Different methods employed to investigate these substances, *in*: "Progress in Insect Neurochemistry and Neurophysiology", A. B. Borkovec and D. B. Gelman, eds., The Humana Press, pp 229-233 (1986).
8. A. De Loof, The impact of the discovery of vertebrate-type steroids and peptide hormone-like substances in insects, *Entomol. Exp. Appl.*, 45:105-113 (1987).
9. J. A. Veenstra, Immunocytochemical demonstration of vertebrate peptides in invertebrates: the homology concept, *Neuropeptides*, 12:49-54 (1988).
10. D. Grube and E. Weber, Immunoreactivities of gastrin (G) cells.Dilution-dependent staining of G-cells by antisera and non-immune sera, *Histochem.*, 65:223-237 (1980).
11. E. A. Nigg, G. Walter and S. J. Singer, On the nature of crossreactions observed with antibodies to defined epitopes, *Proc. Nat. Acad. Sci. USA*, 79:5939-5943 (1982).
12. P. Verhaert, C. P. J. Grimmelikhuijzen, and A. De Loof, Distinct localization of FMRFamide- and bovine pancreatic polypeptide-like material in the brain, retrocerebral complex and suboesophageal ganglion of the cockroach *Periplaneta americana* L., *Brain Res.*, 348:331-338 (1985).
13. F. Berkenbosch and F. J. H. Tilders, A quantitative approach to crossreaction problems in immunocytochemistry, *Neuroscience*, 23:823-826 (1987).
14. B. D. Young and R. W. Davis, Efficient isolation of genes by using antibody probes, *Proc. Nat. Acad. Sci. USA*, 80:1194-1198 (1983).
15. F. Vandesande and K. Dierickx, Immunocytochemical demonstration of separate vasotocinergic and mesotocinergic neurons in the amphibian hypothalamic magnocellular neurosecretory system, *Cell Tissue Res.*, 175:289-296 (1976).
16. J. E. Arrand, "Nucleic Acids Hybridization - A Practical Approach", B. D. Hames and S. J. Higgins, eds., IRL Press (1985).
17. H. Aviv and P. Leder, Purification of biologically active globin messenger RNA by chromatography on oligothymidylic acid-cellulose, *Proc. Nat. Acad. Sci. USA*, 69:1408 (1972).
18. U. Gubler and B. J. Hoffman, A simple and very efficient method for generating cDNA libraries, *Gene*, 25:263-269 (1983).
19. T. V. Huynh, R. A. Young and R. W. Davis, "DNA Cloning Volume I - A Practical Approach", D. M. Glover, ed., IRL Press (1985).
20. T. Maniatis, E. F. Fritsch and J. Sambrook, "Molecular Cloning - A Laboratory Manual", Cold Spring Harbor Laboratory (CSH) (1982).
21. F. Sanger, A. R. Coulson, B. G. Barrell, A. J. H. Smith and B. A. Roe, Cloning in single-stranded bacteriophage as an aid to rapid DNA sequencing, *J. Mol. Biol.*, 143:161-178 (1980).
22. A. B. Smit, E. Vreugdenhil, R. H. M. Ebberink, W. P. M. Geraerts, J. Klootwijk and J. Joosse, Growth-controlling molluscan neurons produce the precursor of an insulin-related peptide, *Nature*, 331:535-538 (1988).
23. L. E. Schneider and P. H. Taghert, Isolation and characterization of a *Drosophila* gene that encodes multiple neuropeptides related to FMRFamide, *Proc. Nat. Acad. Sci. USA*, 85:1993-1997 (1988).

A NOTE ON A COMPARATIVE IMMUNOCYTOCHEMICAL STUDY ON THE PRESENCE OF Cam-HrTH-II- AND Lom-AKH-I-LIKE MOLECULES IN THE CENTRAL NERVOUS SYSTEM OF THREE STICK INSECT SPECIES: *CARAUSIUS MOROSUS, SIPYLOIDEA SIPYLUS* AND *EXTATOSOMA TIARATUM*

F. Clottens, G. Gäde*, L. Schoofs, R. Huybrechts and A. De Loof

Zoological Institute of the University
Naamsestraat 59, B-3000 Leuven, Belgium, and
*Institut für Zoologie IV
Universität Düsseldorf, FRG

SUMMARY

Antisera directed against the hypertrehalosaemic hormone II of *Carausius morosus* (Cam-HrTH-II) and against the adipokinetic hormone I of *Locusta migratoria* (Lom-AKH-I), both members of the AKH/RPCH peptide family, were used to localize immunoreactive molecules in the brain and corpora cardiaca of three species of stick insects. In *Carausius* as well as in *Extatosoma*, both antisera stained cells in the glandular part of the corpora cardiaca and axon fibers in the neuropile of the brain. In Sipyloidea however, molecules reactive to Cam-HrTH-II were located in the cells of the glandular part of the corpora cardiaca and in axon fibers of the neuropile of the brain. Lom-AKH-I-like molecules were situated in cells of the pars intercerebralis, in axon fibers and in granules in the storage part of the corpora cardiaca. From the results, obtained by immunostaining on alternating sections and by pre-absorption tests, it seems probable that in *Carausius* and in *Extatosoma*, only one type of reactive molecule was present. On the other hand, at least two different molecules must have been present in *Sipyloidea*, one which was similar to Cam-HrTH-II and another which was similar to Lom-AKH-I.

INTRODUCTION

After Fernlund and Joseffson identified the red pigment concentrating hormone (Pab-RPCH) in the shrimp, *Pandalus borealis* [1], many other adipokinetic hormone (AKH)/RPCH related peptides were isolated and sequenced (e.g., [2]). In this peptide family, an interesting conservation of pGlu (1), Phe (4), Trp (8) and the C-terminal amide is found [3]. One of these peptides is the hypertrehalosemic hormone II (Cam-HrTH-II), first isolated from *Carausius morosus* [4] which increases the concentration of carbohydrates in the haemolymph of ligated adult *Carausius morosus* [5].

The application of antisera, directed against different peptides of this family, in immunocytochemistry, may yield information on the distribution and the number of peptides present. A first anti-AKH/RPCH antiserum, raised against Lom-AKH-I (adipokinetic hormone I), was produced by Schooneveld et al. [6]. Recently we have prepared an antiserum against Cam-Hr TH-II [7].

In this study, we compared the distribution of Cam-HrTH-II-like molecules in three different stick insect species: *Carausius morosus*, the species where this peptide was first identified, and two related species, *Extatosoma tiaratum* and *Sipyloidea sipylus*.

We have also compared the reaction pattern obtained by anti-HrTH-IIB1, to that obtained by anti-AKH-I241.

Chromatography and Isolation of Insect Hormones and Pheromones
Edited by A. R. McCaffery and I. D. Wilson
Plenum Press, New York, 1990

MATERIALS AND METHODS

Experimental Animals

Adults of three stick insect species of the order of Phasmida (Cheleutoptera) were used, namely *Carausius morosus* MacLeay, *Extatosoma tiaratum* Brunner and *Sipyloidea sipylus* Westwood, respectively belonging to the subfamilies of Lonchodinae, Podocanthinae and Necrosciinae.

Immunocytochemistry

Dissected brains and corpus cardiacum/corpus allatum complexes were fixed *in situ* in a Bouin Hollande Sublimate solution for 24 h. For immunocytochemistry, the indirect peroxidase-anti-peroxidase (PAP) method of Sternberger [8] was used.

3,3'-diaminobenzidine (DAB, Sigma) was used as substrate. The alternating section technique was used to compare the reaction patterns of different antisera.

The anti-AKH-I241 was kindly provided by Dr Schooneveld (Agricultural University, Wageningen, The Netherlands). The anti-HrTH-IIB1 was produced in our laboratory [7].

Method specificity was tested by sequentially omitting the different steps in the immunocytochemical staining method. Antibody specificity was tested by liquid phase pre-absorptions of the antisera.

RESULTS

Carausius morosus

Cam-HrTH-II-like material could be detected in cells of the glandular part of the corpora cardiaca (Figure 1a). In the brain, only some axon fibers in the neuropile and in the central body, but no cells, were found to contain immunoreactive material. No reaction could be observed in other parts of the brain or in the corpora allata.

When the anti-AKH-I241 was applied as primary antiserum, a similar reaction pattern was obtained, suggesting either a co-localization of the two molecules in the same cells or, a cross-reaction of the two antisera with the same molecule(s). Pre-absorption tests by liquid phase absorption, revealed that cross-reactivity occurred. Indeed, pre-absorption of the anti-HrTH-IIB1 with native Lom-AKH-I did not abolish the immunocytochemical reaction in the brain and in the glandular part of the corpora cardiaca. On the other hand, when the anti-AKH-I241 was pre-absorped with the native Cam-HrTH-II, no reaction was found in the cells of the corpora cardiaca or in the axon fibers in the brain (Figure 2).

Extatosoma tiaratum

The results obtained in *Extatosoma* were compared to those in *Carausius*. The anti-HrTH-IIB1 serum stained material in cells of the glandular part of the corpora cardiaca, in axon fibers in the neuropile of the protocerebrum and the central body, in the neuropile of the tritocerebrum and in the suboesophageal ganglion (Figure 1b).

As seen in *Carausius*, a similar reaction pattern was found with anti-AKH-I241. Once again, pre-absorption revealed that there was cross-reactivity between the two antisera. Pre-absorption of the anti-HrTH-IIB1 with the native Lom-AKH-I did not abolish the reactions found with the unpre-absorped antisera, whilst pre-absorption of the anti-AKH-I241 with the native Cam-HrTH-II abolished the staining.

Sipyloidea sipylus

In *Sipyloidea*, completely different results (Figure 1c) were observed compared to *Carausius* and *Extatosoma*. The reaction of anti-HrTH-II, was similar to that in the other two phasmids studied, with immunopositive cells localized in the glandular part of the corpora cardiaca. Immunopositive material was also present in axon fibers of the neuropile of the protocerebrum and tritocerebrum and in the central body. Pre-absorption of this antiserum with the native Lom-AKH-I, did not abolish the positive staining.

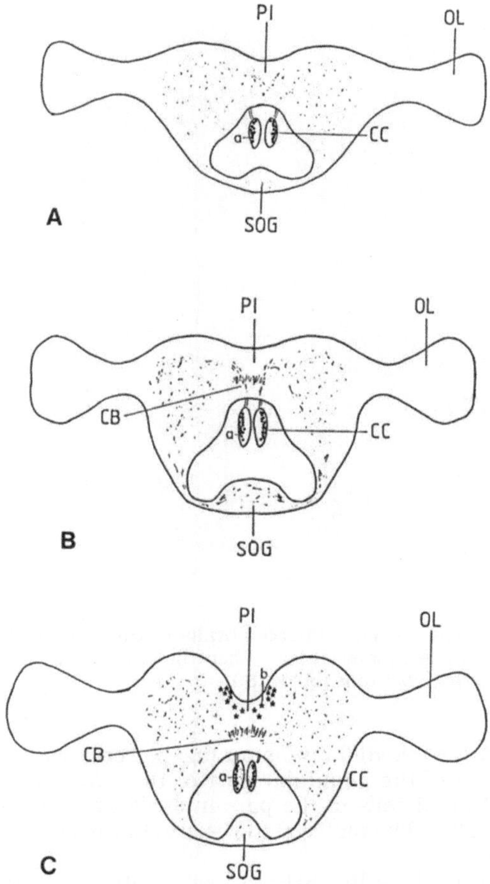

Fig. 1. Schematic representation of the results obtained in the brain and corpora cardiaca of *Carausius morosus* (A), *Extatosoma tiaratum* (B), and *Sipyloidea sipylus* (C), showing Cam-HrTH-II-like (a) and Lom-AKH-I-like (b) neurons. Key: PI pars intercerebralis; OL optic lobe; SOG suboesophageal ganglion; CC corpora cardiaca.

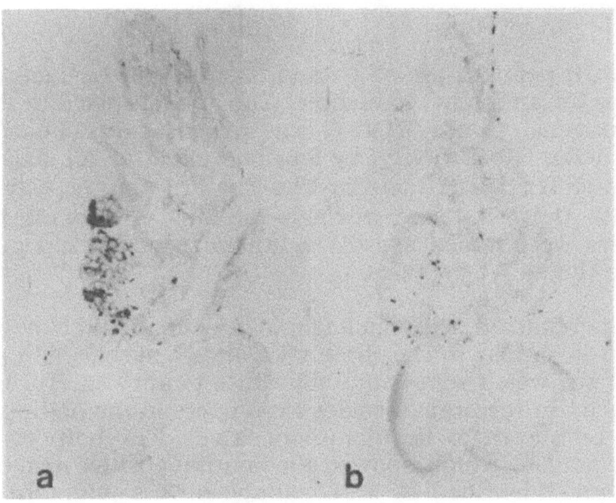

Fig. 2. Cross-sections through the corpora cardiaca of *Carausius morosus*, showing neurons, immunoreactive to anti-AKH-I241 (a). This positive reaction was abolished when the antiserum was pre-absorped with native Cam-HrTH-II (b), indicating a cross-reactivity of the anti-AKH-I241 with the Cam-HrTH-II-like material in the tissue. x 180.

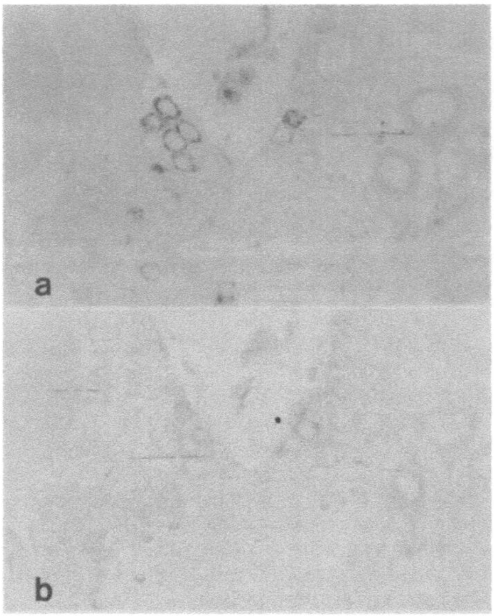

Fig. 3. Cross-sections through the pars intercerebralis of *Sipyloidea sipylus*, showing neurons, containing Lom-AKH-I-like material (a). When the anti-HrTH-IIB1 was applied, no reaction could be detected in these neurons (b). x 280.

With the anti-AKH-I241 a different staining pattern was observed. Besides the reaction found in the cells of the glandular part of the corpora cardiaca, we also found immunopositive material in 28 cells in the pars intercerebralis (Figure 3). In the corpora cardiaca, we found Lom-AKH-I-like material in vesicles in the storage lobe.

After pre-absorption of the anti-AKH-I241 with native Cam-HrTH-II, the reaction in the cells in the pars intercerebralis and in the vesicles of the storage part of the corpora cardiaca was still positive, whilst the reaction in the cells of the glandular part had disappeared. This result indicates that the molecule(s), found in the cells of the pars intercerebralis must be different from those found in the glandular part of the corpora cardiaca.

DISCUSSION

Native AKH/RPCH peptides are all C- and N-terminally blocked and do not therefore provide good antigens for antibody production. The antisera used in this study were made against synthetic antigens. Anti-AKH-I241 was raised against Lom-Tyr-AKH-I (Tyr-Leu-Asn-Phe-Thr-Pro-Asn-Trp-Gly-Thr-NH_2) by Schooneveld et al. [6], and anti-HrTH-IIB1 was raised against Cam-HrTH-IIB (H-Leu-Thr-Phe-Thr-Pro-Asn-Trp-Gly-Thr-NH_2)) [7]. The results obtained with these antisera in cross-sections of nervous tissue showed that, although the antisera were raised against synthetic peptide-analogues, they still could specifically recognize the native peptides.

From the results obtained, the conclusion can be made that the distribution of Cam-HrTH-II-like material is similar in the three stick insect species. In all the three species neurons could be shown which were immunopositive to anti-HrTH-II in the glandular part of the corpora cardiaca. In the brain, besides axon fibers in the neuropile, no neurons were found which could be shown to be immunopositive. The corpora cardiaca factors of *Extatosoma* and *Sipyloidea*, which showed an immunopositive reaction to anti-HrTH-II, have been isolated and identified by HPLC-analysis. As might be expected from our immunocytochemical results, this material was identified as Cam-HrTH-II (Gäde, unpublished results).

A second aim of our study was to seek immunoreactive material with another anti-AKH/RPCH-antiserum (anti-AKH-I241) in order to find out whether more than one

AKH/RPCH-peptide could be localized in the brain of these species of stick insect. An earlier study had demonstrated an immunopositive reaction in the glandular cells of the corpora cardiaca in both *Carausius* and *Extatosoma* with the anti-AKH-I241 [9]. These results were obtained before the Cam-HrTH-II was characterized and could therefore not exclude the possibility of cross-reactivity or co-localization with this compound.

The results, obtained by Schooneveld [9] with anti-AKH-I241, and our results, obtained with anti-HrTH-IIB1, suggest indeed a co-localization of two molecules in the same cells. However, pre-absorption experiments demonstrate that the reactions found in the glandular cells of the corpora cardiaca of the three stick insect species were all Cam-HrTH-II-like. We can conclude that the results described earlier with the anti-AKH-I241, in *Carausius* and *Extatosoma* were due to a cross-reaction of this antiserum with Cam-HrTH-II. This is also the case in *Sipyloidea*, at least concerning the glandular neurons in the corpora cardiaca.

In *Sipyloidea* however, the presence of two different peptides was demonstrated. The Cam-HrTH-II-like material was mainly present in the glandular neurons of the corpora cardiaca, and in axon fibers in the neuropile of the brain. The second peptide, at this moment not yet isolated or characterized, was present in neurons of the pars intercerebralis, and is more Lom-AKH-I-like. It is not necessarily identical to Lom-AKH-I. Indeed, one of the disadvantages of immunocytochemistry is that an antiserum recognizes antigenic sites of only a few amino acids. Other peptides, containing the same amino acid sequence at the antigenic site (but which are not identical to the antigen to which the antiserum was raised) can give a false positive.

Although the immunocytochemical method is of great value in the localization of different substances in various regions of the nervous system, subsequent isolation and characterization of the compounds of interest are still necessary to elucidate the physiological and chemical properties.

ACKNOWLEDGEMENTS

We thank the NFWO of Belgium for financial support and the IWONL of Belgium for providing a research fellowship to F.C. G.G. was supported by a grant from the Deutsche Forschungsgemeinschaft (Ga 241/6-2) and by a Heisenberg Fellowship awarded by the Deutsche Forschungsgemeinschaft (Ga 241/5-2). We thank Mr R. Smisdom for technical assistance.

REFERENCES

1. P. Fernlund and L. Joseffson, Crustacean color-change hormone: amino acid sequence and chemical synthesis, *Science*, 177:173-175 (1972).
2. G. Gäde, New structures of insect neuropeptides, *in*: "Endocrinological Frontiers in Physiological Insect Ecology", F. Sehnal, A. Zabza and D. L. Denlinger, eds., Wroclaw Technical University Press, Wroclaw, pp 635-650 (1988).
3. M. H. Schaffer and B. E. Noyes, The potential of adipokinetic hormone to teach us about neuropeptides, *BioEssays*, 7:67-71 (1987).
4. G. Gäde and K. L. Rinehart, Jr., Primary structure of the hypertrehalosaemic factor II from the corpus cardiacum of the Indian stick insect, *Carausius morosus*, determined by fast atom bombardment mass spectrometry, *Biological and Chemical Hoppe Seyler*, 368:67-75 (1987).
5. G. Gäde and P. Lohr, Restricted specificity of a hyperglycaemic factor from the corpus cardiacum of the stick insect, *Carausius morosus*, *J. Insect Physiol.*, 28:805-811 (1982).
6. H. Schooneveld, G. I. Tesser, J. A. Veenstra and H. M. Romberg-Privee, Adipokinetic hormone and AKH-like peptide demonstrated in the corpora cardiaca and nervous system of *Locusta migratoria* by immunocytochemistry, *Cell Tissue Res.*, 230:67-76 (1983).
7. F. Clottens, G. Gäde, R. Huybrechts and A. De Loof, Immunohistochemical localization of the hypertrehalosaemic hormone II (Cam-HrTH-II) in the nervous system of *Carausius morosus* and *Sarcophaga bullata*, *Cell Tissue Res.*, in press (1989).
8. L. A. Sternberger, "Immunocytochemistry", Wiley Medical Publications, New York (1979).

9. H. Schooneveld, H. M. Romberg-Privee and J. A. Veenstra, Phylogenetic differentiation of glandular cells in the corpora cardiaca as studied immunocytochemically with region-specific antisera to adipokinetic hormone, *J. Insect Physiol.*, 33:167-176 (1987).

PHEROMONES

INSECT TRAIL PHEROMONES: A PERSPECTIVE OF PROGRESS

E. David Morgan

Department of Chemistry
University of Keele, Keele
Staffordshire ST5 5BG, UK

SUMMARY

A comparative description is given of the purpose and glandular source of odour trails in the three insect orders in which trail-following phenomena have been recognized, Isoptera, Lepidoptera and Hymenoptera. The different types of bioassay and the chromatographic methods which together are required for isolation are outlined. Examples are given of the chemical structures of trail pheromones from the three orders, with attention to the non-specific nature of some pheromones active across a range of species and an example of species-specificity in *Tetramorium* ants. Examples of structure-activity relationships are given. Trail pheromones may vary from simple single compounds to complex mixtures containing recruitment and orientation components.

THE CONCEPT OF ODOUR TRAILS

The principle of following a trail is a simple one that appeals even to childhood imagination and is attractive to scientific curiosity. The following is a survey of our present position of knowledge from a chemical point of view, but without an attempt to be wholly comprehensive, of trail pheromones of insects and their isolation and identification.

If one considers the best strategy for a primitive community to forage for food, it is advantageous for individuals to forage singly and communicate to others their food finds. Clearly, sound is not an efficient way, it is too non-specific a means of communication. A shout will bring members of another village, baboons, vultures or crows. A specific and secret code to mark the route to the food site is a better solution. Humans, with their highly developed sense of sight, would use visual clues. Hänsel did in the fairytale. When his parents took him and his sister Gretel deep into the forest to abandon them, he marked the trail as they went with white pebbles, which shone in the moonlight and enabled the children to find their way home.

Insects, with their highly developed chemical sense, might be expected to mark a trail with an odour, and indeed, we find this characteristic has developed many times, and is known today in at least three insect orders, Isoptera, Lepidoptera and Hymenoptera.

Bonnet in 1779 [1] described the phenomenon of trail-following in ants and provided the first crude evidence that the trail was chemical in nature, by drawing his finger across the trail to cause a break in it. The ants stopped at the missing section and had to search about to regain it after the break and then proceeded to re-mark the missing section.

The general principle of discovery of food by foragers, the laying of a trail back to the nest, the recruitment of other workers which follow the trail to the food, reinforcement of the trail as they return, exhaustion of the food and eventual evaporation of the trail are illustrated in Figure 1. Trails are also used for marking a route to a new nest site. I have

Chromatography and Isolation of Insect Hormones and Pheromones
Edited by A. R. McCaffery and I. D. Wilson
Plenum Press, New York, 1990

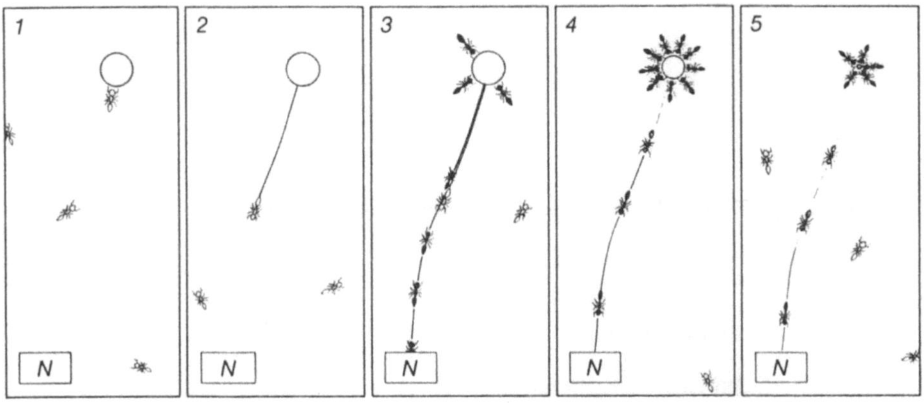

Fig. 1. The nature of trail-laying and trail-following in ants. 1. Foraging workers (in outline) find a food source, feed and returning to the nest (N), lay down a trail. 2. Within or around the nest, other workers (black) are recruited and follow the trail to the food. 3. Returning workers reinforce the trail, 4, until the food is exhausted. 5. No trail pheromone is deposited as workers return without food (stippled ants), the trail evaporates and the colony resumes the pattern of foraging.

observed such an event when a colony of *Tetramorium caespitum* migrated from a bottle in which it had been sent through the post, to a new comodious nest made from laboratory glassware and plaster of Paris, with darkened brood chambers. When the cap of the transporting bottle was removed, some workers swarmed out, and began to explore the large metal tray in which their old and prospective nests were placed, along with test-tubes of sugar-water for feeding. For some time the movement of ants seemed to be chaotic, but gradually fewer workers were observed on the tray and more were moving between the two nests in both directions. A definite trail developed between the two sites, not just a direct route, because that was blocked by a test tube, and the mouth of the bottle and the entrance to the new nest did not face each other. Eventually the large queen appeared, dragged or urged along by numerous workers. Within half an hour activity around the old nest had ceased and the queen and most of her workers had disappeared inside the darkened brood chambers. There may be different messages to recruit workers to food or to a new nest site. In the ant *Camponotus socius*, Hölldobler [2] has shown that recruitment is by antennal contact, and "waggle" and "jerking" movements are used to recruit to food and nest moving respectively.

TERMITE TRAILS

Primitive termites may also lay odour trails to breaches in their tunnel walls, to recruit nest mates to repair them, even though they may never forage in the open. In higher termites, trails are also used to food sources as in ants [3].

The source of the trail in primitive species is a series of three glands under the third, fourth and fifth abdominal sternites. In higher families, there is a single gland under the fourth or fifth sternite [3]. In *Nasutitermes lujae*, Pasteels [4] says that the glands are most developed in older workers which, with the soldiers, do most of the foraging and trail laying. The trail is laid by dragging the surface of the abdomen along the ground [5] (Figure 2).

CATERPILLAR TRAILS

The only group of trail-following Lepidoptera that has so far been studied in detail is the tent-forming genus *Malacosoma* [6]. In this group, trails may serve to maintain aggregation, to facilitate exploration, to recruit to a new bivouac, or to lead to a new source of food [7]. The larvae deposit a thread of silk and the trail substance from an incompletely identified sternal secretary organ. There is a single report of predatory stink bugs being able to follow trails made from the frass of *Malacosoma* sp. [8].

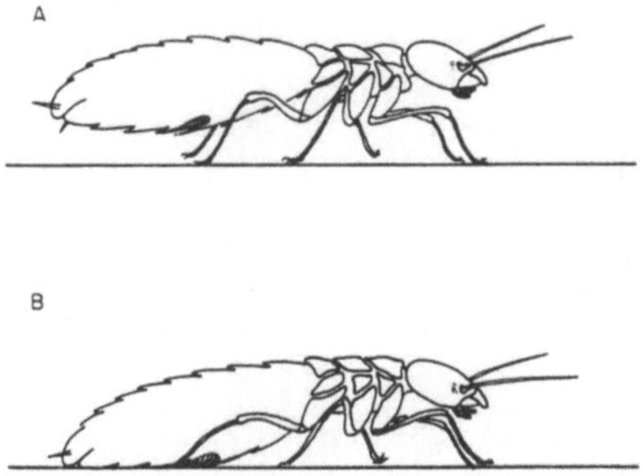

Fig. 2. Trail-laying from the inter-segmental sternal gland of a termite. A. Abdomen in normal position. B. Trail-laying (from Ref. [5]).

BEE AND ANT TRAILS

Among the Hymenoptera, trail laying is found widely among the Formicidae, but a related phenomenon of spot marking of good sources of nectar by male bumblebees is known, in a number of species of *Bombus* [9], using the labial gland secretion [3]. The secretions are species-specific and composed of fatty acid derivatives and terpenoids [10]. A similar behavior has also been observed in stingless bees, *Trigona* species lay "aerial" trails with their mandibular gland secretion [11]. That of *I. subterranea* is dominated by geranial and neral [12]. The honey bee has a rudimentary form of odour trail, called the "footprint substance" which has not yet been further investigated [13]. Ants may use any of a large number of glands in the abdomen or legs (Figure 3), but thus far, the majority have been isolated from the poison gland reservoir, and next to that in frequency is the Dufour gland. The glandular sources have been reviewed by Attygalle and Morgan [14].

BIOASSAY

Wilson, in 1959 [15], showed that it was possible to induce recruitment to trail following in ants with artificial trails made from extracts or a smear of a single gland containing the pheromone, and thereby opened the door to the study of the nature and chemical structure of the trail pheromones.

The two great tools in the chemical study of trail pheromones are a simple bioassay and a palette of chromatographic methods. A number of trail-following tests have been developed from Wilson's observations, either using a straight line [16], a Y-shaped choice test [17], or a circular trail [18] or variations on these as suits the investigation. The straight line test at least possesses simplicity, and resembles a natural situation. A length of paper or plastic, marked with a gland extract or model compound can be inserted into an existing natural trail and the behavior of the insects is observed when they reach the inserted section. The Y-shaped choice test is particularly useful in comparing older and newer trails, or to compare subtle differences, e.g., between two species which apparently share a common substance, or to compare trails made with one substance and those to which a second compound has been added, or to compare the relative effect of two glands, where, as in some species, both Dufour gland and poison gland secretions induce trailing (cf. Ref. [19]). The circular trail may be thought unnatural, since many insects also use the sun or daylight to orient themselves when foraging, but it has the advantage of being quantitative when the circumference is marked off in approximately 1 cm arcs. The trail is laid as a thin line along the circumference. The insects can begin their contact with it at any point, and the number of arcs followed by an individual before it deviates or loses the trail are recorded. However, the realization that in some species, the pheromone system may be complex [20,21], and consist of recruitment and orientation components may require more subtle bioassays in the future.

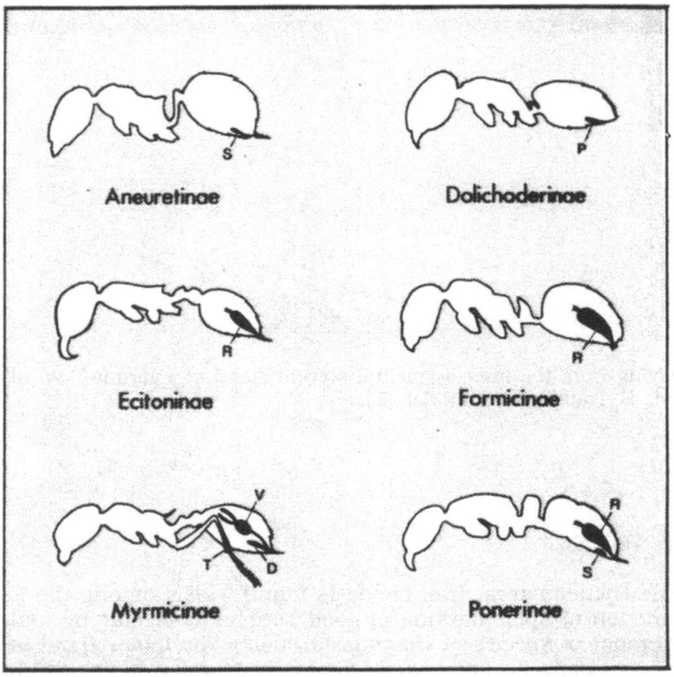

Fig. 3. Common sub-families of ants showing the glandular sources of trail pheromones. S. Sternal Gland. P. Pavan's gland. R. Rectum. V. Venom reservoir. D. Dufour gland. T. Tibial gland.

CHROMATOGRAPHY

The chromatographic methods may be any that the ingenuity of the investigator can devise or apply. The general principles of thin-layer chromatography (TLC) and recovery of sample from the plate and trapping from a gas chromatographic (GC) separation, that we favor, were used by Moore [22] and possibly earlier. With a bioassay, the anatomical source of the pheromone can be located, and with chromatographic methods, the substance can be purified toward identification. In the course of this purification, something will be learned of the polarity and the molecular mass of the compound. By the time it is obtained sufficiently pure for identification by mass spectrometry some structural clues will already be known.

The bioassay functions as the chromatographic detector, since the amount of pheromone is frequently far too little to detect chemically among other substances. An extract of glands or bodies can be placed at the origin of a TLC plate (Figure 4). The silica on the developed plate is divided into sections, and each section is scraped off, placed into a small chromatography tube and the organic substances eluted with acetone or methanol. This extract is then applied to the bioassay trail. Such a simple test is advisable, first, to be sure the pheromone can be recovered from chromatography (e.g., Ref. [23]), secondly to see if only one substance is active (cf. Ref. [24] for an example of a two-component pheromone) and thirdly to obtain some preliminary information about the polarity of the pheromone. It is also possible to carry out simple chemical reactions, either before chromatography or on the plate before elution to learn something of the chemical properties of the pheromone (cf. [25]).

Greater progress can be made with GC because of its much greater resolution and greater sensitivity of detection. It has been our practice, when nothing was known of the pheromone, to chromatograph a few glands at a time and, using an effluent splitter, to collect very broad range fractions (Kovats indices 0 to 1000, 1000 to 2000 etc). When one fraction was found to be active, this chromatography was repeated with fresh glands, dividing the active region into narrower and narrower fractions until its elution properties are known to the nearest carbon number (Figure 5). Doing this on polar and non-polar columns gives the approximate molecular mass and a measure of its polarity. In the case of

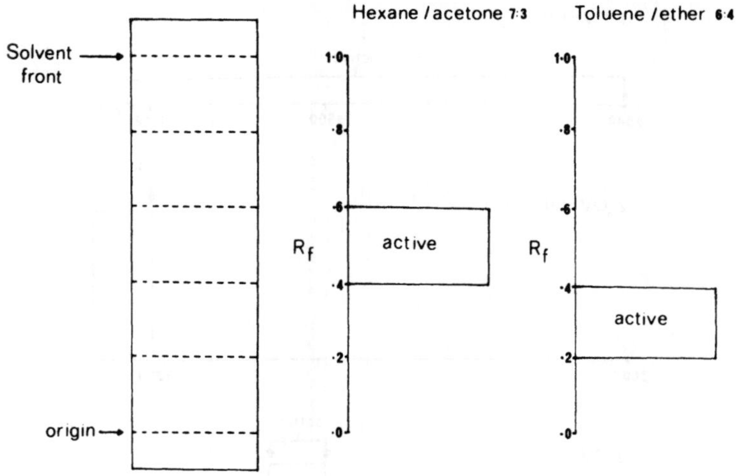

Fig. 4. Thin-layer chromatography of glandular extracts in the isolation of trail pheromones. On the left is shown the thin-layer plate.After development of the plate the silica is cut into R_f bands and each one tested in the bioassay. In this example, activity was confined to R_f 0.4 to 0.6 in hexane-acetone and R_f 0.2 to 0.4 in toluene-ether.

Myrmica rubra, it was not until all this was done, and the gas chromatograph tuned to maximum sensitivity that any peak was seen in the flame detector [26].

With the solid-sampling technique of Morgan and Wadhams [27] it is possible to dissect out one or a few pheromone glands and chromatograph them, without the introduction of solvents or contaminants and obtain a quantitative value for mean amount of pheromone per individual. This can be important in comparing the activity of the synthetic compound at physiological concentration with a gland, to see if the full potency of pheromone is reproduced (Figure 6).

STRUCTURE IDENTIFICATION

Identification is essentially made with the aid of mass spectrometry, because of the small amounts of pheromone commonly available. However, the mass spectrum frequently does not provide sufficient information alone to define a structure and a large number of micro-chemical methods are available for structure determination [28]. Nuclear magnetic resonance spectrometry has been used but requires far more insect material. GC linked to Fourier transform infrared spectroscopy (GC-FTIR) may be helpful in some cases but is not yet sufficiently widely available.

CHEMICAL STRUCTURES OF PHEROMONES

The first trail pheromone isolated was from an Australian termite, *Nasutitermes exitiosus* [22]. The compound was originally called Neocembrene A but is now called (E)-6-cembrene A (Figure 7), and the same compound appears to be used by *N. walkeri* and *N. graveolus* [22], and by *Trinervitermes bettonianus* [29]. Workers of *N. exitiosus* followed trails containing 10^{-8} to 10^{-5} g/ml. Hexanoic acid induces trail-following in *Zootermopsis nevadensis*, where 20 ng corresponds to "1 trail unit", i.e., where 3 out of 10 workers followed the trail for 10 cm [30]. These authors partially identified another substance from *Z. nevadensis* with formula $C_{11}H_{20}$ which was more active. Dodecatrienol (Figure 7), the pheromone of *Reticulitermes virginiacus* [31] is much more potent, less than 1 pg consistently inducing trail following on a 10 cm trail, equivalent to 3 pg per circumference of our circular trail. Other *Reticulitermes* species also follow trails of this compound, though it is not known if it is the true pheromone for them [32]. Other termite species of *Coptotermes, Trinervitermes, Amitermes* and *Schedorhinotermes* also follow this dodecatrienol and natural *Reticulitermes* trails [33], which may indicate a widely shared pheromonal compound. Although termite trail pheromones were the first to be successfully identified, activity in this area has lagged in recent years, and the above appears to be the total list known.

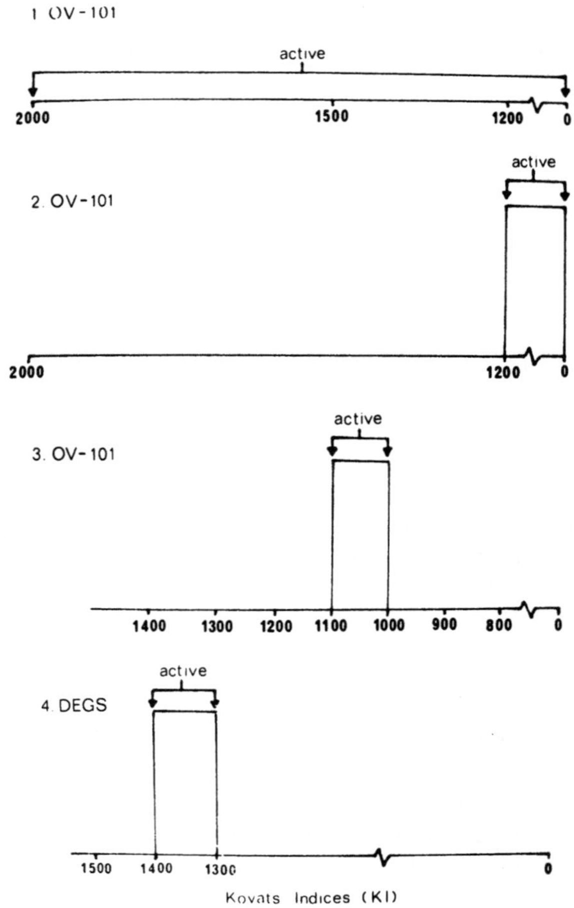

Fig. 5. Gas chromatography of glandular extracts in the isolation of trail pheromones. Using packed columns and an effluent splitter, the effluent was trapped from Kovats index (KI) 0 to 2000 and shown to be active in bioassay (top). Repeating the experiment with successively narrower bands, located the pheromone between KI 1000 and 1100 on a non-polar stationary phase (OV-101) giving information about its molecular mass, and then on a polar phase (diethylene glycol succinate) giving information about its degree of polarity.

The identification of the first lepidopteran trail-pheromone has only recently been completed. Crump et al. [16] have shown that 5β-cholestan-3,24-dione (Figure 8) is present in the trails of *Malacosoma americanum*. The synthetic compound is active at 10^{-8} to 10^{-10} g/mm, equivalent to 3 μg to 30 ng per circular trail. However, this much less volatile substance is detected by surface contact, and the limit of detection will depend upon the surface and the way it is applied. Added to the silk deposited at the same time by the larva (the silk has no trail-following effect alone) the limit of detection may be much lower. It is interesting that two other species, *M. neustria*, and *M. disstria* also follow trails of the same compound, though whether they secrete it is not yet known.

Far more is known today of ant pheromones than of any other group and the subject has been recently reviewed [14]. The pheromones so far identified have been relatively low-mass, volatile substances of specific molecular shape, and generally speaking, they are substances with a distinct odour for the human nose (Figure 9).

The first isolation and identification of an ant trail pheromone was from the leaf-cutting species *Atta texana* [34]. Extraction of 3.7 kg of workers eventually led to the identification of methyl 4-methylpyrrole-2-carboxylate (Figure 9). The substance had a threshold of detection of 0.08 pg per cm, equivalent to 2.5 pg per circular trail.

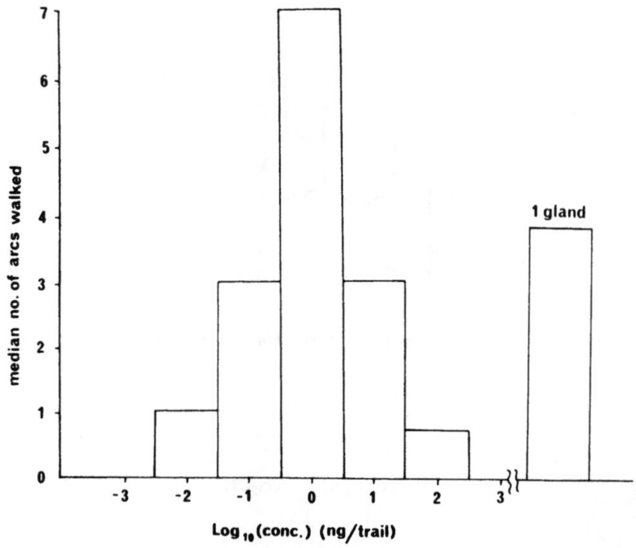

Fig. 6. Comparison of synthetic trail pheromone (3-ehtyl-3,5-dimethylpyrazine) at various concentrations in a trail-following bioassay with a single worker's venom gland, to see if the substance could fully reproduce the activity of the gland (from Ref. [26]).

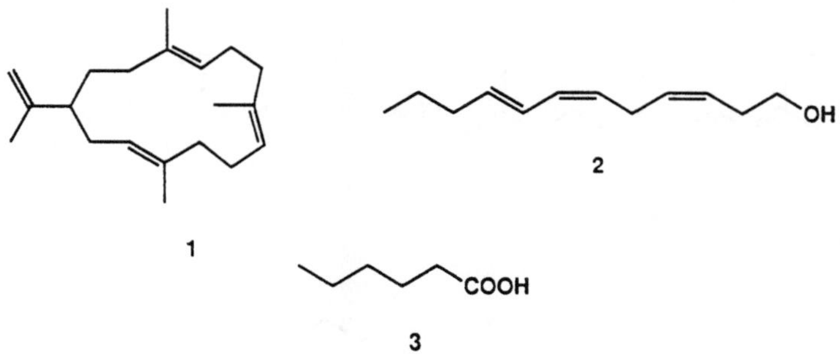

Fig. 7. Known trail pheromones of termites, 1 (E)-6-cembrene; 2, (Z,Z,E)-dodeca-3,6,8-triene-1-ol; 3, hexanoic acid.

In contrast to the species specificity of most insect pheromones it is apparent, even at this stage of discussion, that trail pheromones can be shared by more than one species. 3-Ethyl-2,5-dimethylpyrazine (EDMP) was first identified as the trail pheromone of *Atta sexdens rubropilosa* [35], but subsequently it has been shown to be the pheromone of *Atta sexdens sexdens* [36], of 13 species of *Myrmica* [25,37] and of *Manica rubida* [38] and to be located in the venom of the poison reservoir in each case.

Traces of hydrocarbons from the Dufour gland caused problems in the examination of *Manica rubida* poison glands. However, use of the nitrogen-phosphorus specific GC detector eliminated hydrocarbons and revealed four nitrogen compounds identified as methylpyrazine, 2,5-dimethylpyrazine, trimethylpyrazine and EDMP. On testing each of these, only EDMP showed activity and this activity was not enhanced or synergized by the other substances.

Atta, Myrmica and *Manica* all belong to the subfamily Myrmicinae, as does the genus *Tetramorium*, which does not follow this non-specific trail pheromone rule. Two species, *T. caespitum* and *T. impurum* are morphologically almost identical, but they use distinctly different trail pheromones. *T. caespitum* uses a mixture of two substances, EDMP and 2,5-dimethylpyrazine in a 3:7 ratio mixture, and the synthetic compounds, combined in this ratio give maximum activity in bioassay [39]. *T. impurum* uses methyl 6-methylsalicylate

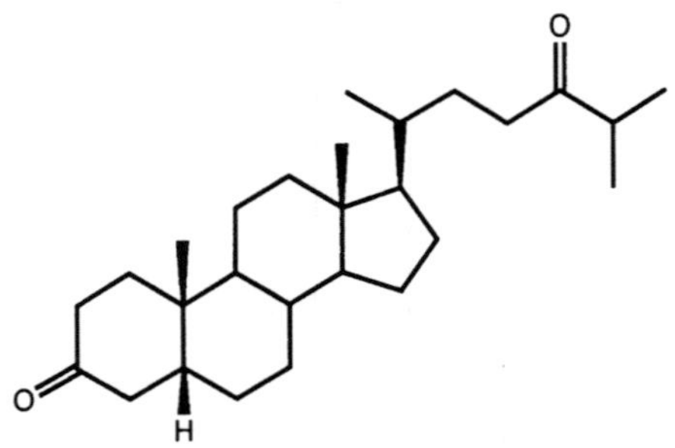

Fig. 8. Trail pheromone of the tent caterpillar *Malacosoma americanum*, 5β-cholestan-3,24-dione.

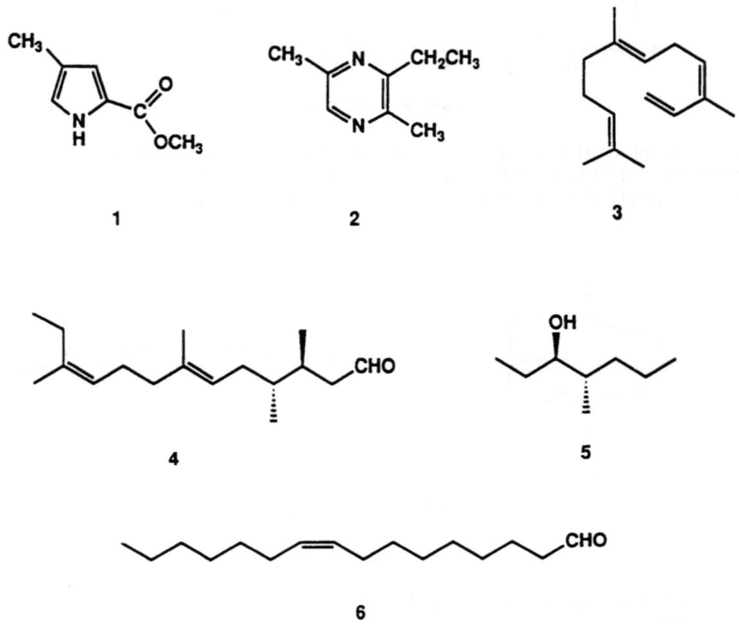

Fig. 9. Some selected trail pheromones of ants. 1. Methyl 4-methylpyrrole-2-carboxylate from the venom reservoir (*Atta texana*, *Atta cephalotes* and *Acromyrmex octospinosus*); 2. 3-ethyl-2,5-dimethylpyrazine from the venom reservoir (numerous myrmicine species); 3. (Z,E)-α-farnesene from the Dufour gland (*Solenopsis invicta*); 4. Faranal from the Dufour gland (*Monomorium pharaonis*); 5. 4-methylheptan-3-ol from the venom reservoir (*Leptogenys diminuta*); and 6. (Z)-hexadec-8-en-1-al from Pavan's gland (*Iridomyrmex humilis*).

(Figure 10) as its pheromone [40] and neither species contains detectable quantities of the other's pheromone in its venom or shows any tendency to follow it. Another *Tetramorium* species, *T. meridionale*, follows neither *T. caespitum* or *T. impurum* trails, but has another pheromone, of which indole forms part (Jackson et al., unpublished).

SPECIFICITY OF STRUCTURE

How specific do these trail pheromones have to be in molecular structure? The pyrazine trail pheromones seem to be highly specific in their activity, and small changes of structure give complete loss of activity [24]. Sonnet and Moser [41] studied a number of analogues of methyl-4-methylpyrrole-2-carboxylate and were able to make conclusions about requirements for activity. The simplicity of the structure of methyl 6-methylsalicylate and the availability either commercially, or through synthesis, of a number of closely related compounds encouraged us to make some structure-activity studies to see how specific the antennal receptors are for this pheromone substance [42]. The compounds selected in Figure 11 show that the presence and position of the methyl group are essential for activity, that the methyl ester is also vital, but surprisingly, the phenolic group is much less important. Activity is retained when it is absent or replaced by an amino group.

COMPLEX PHEROMONES

It is becoming clear that not all cases of trail pheromone are as simple as these examples. Traniello [20] reported that the trail pheromone of the termite *Nasutitermes costalis* consists of an ephermeral recruitment component and an extremely persistent orientation cue. Vander Meer and others have been studying the trail pheromone of the fire ant *Solenopsis invicta*. They identified (Z,E)-α-farnesene from the Dufour gland as the orientation pheromone [21], but have since isolated and partially identified a second component, described as a recruitment pheromone, also from the Dufour gland, in very small quantity [43].

Attygalle et al. [44] have identified (3R,4S)-4-methyl-3-heptanol from the poison gland of *Leptogenys diminuta*, a ponerine ant, as the trail orientation pheromone and isogeraniol as the recruitment component from the pygidial gland. EDMP was identified as present in the poison glands of *Pheidole pallidula* workers, and the pure compound induced trail following, but not as well as a gland extract. It was noted that during isolation of the EDMP that the activity diminished as purification proceeded [45]. Evidently a second component, either a recruitment substance or a powerfully synergistic compound, with similar chromatographic properties, was being removed during purification. In *Messor bouvieri* we identified EDMP again, present in the Dufour gland and active as a pheromone, but here discovered that workers follow trails from the poison gland and Dufour gland equally well. No enhancement of activity was found by combining both gland extracts [46].

The trail pheromones from the poison apparatus, i.e., present either in the venom or the Dufour gland secretion, are laid on the substratum with the sting lance [3]. *Crematogaster* species have a quite different system. Their abdomens are so shaped they cannot bring the sting lance in contact with the substratum. Fletcher and Brand [47] noticed that they walk with a curious gait when laying trails and recognized that the pheromone was coming from the metathoracic legs. Leuthold [48] located the gland in the tibia, with a duct running to the feet. By TLC and GC experiments, the pheromone is known to be a single substance present in the tibial glands in *Crematogaster scutellaris*. However, a synthetic sample of the identified compound showed no activity at all (Morgan et al., unpublished).

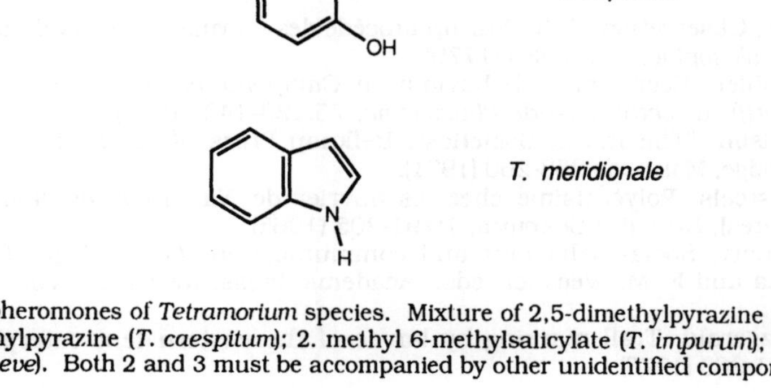

Fig. 10. Trail pheromones of *Tetramorium* species. Mixture of 2,5-dimethylpyrazine and 3-ethyl-2,5-dimethylpyrazine (*T. caespitum*); 2. methyl 6-methylsalicylate (*T. impurum*); and 3. indole (*T. semilaeve*). Both 2 and 3 must be accompanied by other unidentified components.

ACTIVE INACTIVE

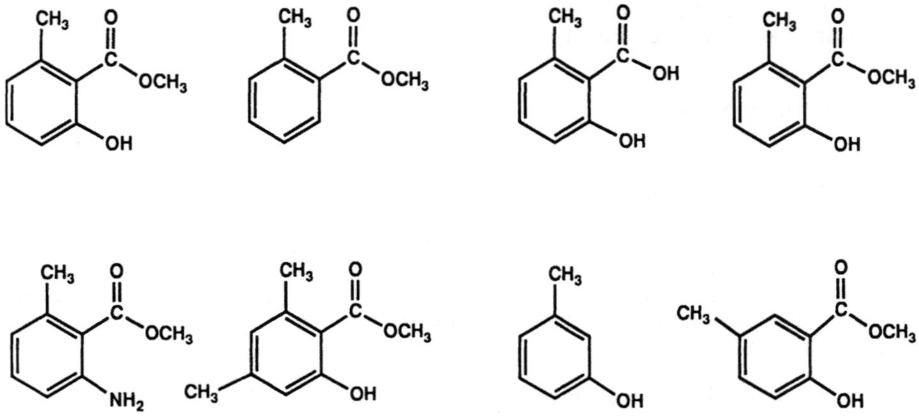

Fig. 11. Methyl 6-methylsalicylate, trail pheromone of *T. impurum* and some compounds of closely related structure, separated to show those active and inactive in the trail following bioassay with this species.

Traniello [49] has discovered a case of colony specificity in the trail pheromone of *Lasius neoniger*. In a Y-shaped choice test, with its own and another colony's pheromone, it has a definite preference of its own. At least two examples are known where the trail is specific to an individual. Foragers of *Pachycondyla tesserinoda* search individually for food and prefer their own over other individual's trails [50], and individuals of *Leptothorax affinis* search for their own trails during colony migration [51]. Wilson [52] has drawn attention to two examples where two species of ant, apparently amicably, share the same trails for food gathering.

The trail pheromones of insects contain many surprises and curiosities still to be revealed by a combination of the ethological insight of entomologists and the chromatographic ingenuity of chemists. Further investigations should help us to understand how the phenomenon of a pheromonal communication evolves in animals.

ACKNOWLEDGEMENTS

I thank all my collaborators, at Keele and elsewhere who have contributed to this work, not forgetting those at great distance who have kindly sent live ants for examination, but particularly Drs R. P. Evershed, A. B. Attygalle, M. C. Cammaerts and R. Cammaerts, Professor J. P. L. Billen, B. D. Jackson, D. G. Ollet, G. W. Sales and Vivienne K. Lancaster. I also thank Professor Billen for the use of his drawings in Figures 1 and 2. Part of this work was carried out under Ministry of Agriculture, Fisheries and Food licence No. PHF 244/35(98).

REFERENCES

1. C. Bonnet, Observation LXIII. Sur un procédé des fourmis, *Oeuvres d'Histoire Naturelle et de Philosophie*, 1:535-536 (1779).
2. B. Hölldobler, Recruitment behaviour in *Camponotus socius* (Hym. Formicidae), *Zeitschrift für vergleichende Physiologie*, 75:123-142 (1971).
3. E. O. Wilson, "The Insect Societies", Belknap Press of Harvard University Press, Cambridge, Mass., pp 258-260 (1971).
4. J. M. Pasteels, Polyéthisime chez les ouvrier de *Nasutitermes lujae* (Termitidae Isotoperes), *Biologica Gabonica*, 1:191-205 (1965).
5. A. M. Stuart, Social behaviour and communication, *in*: "Biology of Termites", K. Krishna and F. M. Weesner, eds., Academic Press, New York, Vol. 1, pp 193-232 (1969)..
6. T. D. Fitzgerald, Trail marking by larvae of the eastern tent caterpillar, *Science*, 194:961-963 (1976).

7. S. C. Peterson, Chemical trail marking and following by caterpillars of *Malacosoma neustria*, *J. Chem. Ecol.*, 14:815-824 (1988).
8. D. K. McLain, Terrestrial trail-following by three species of predatory stink bug, *Florida Entomologist*, 62:152-154 (1979).
9. B. G. Svenson and G. Bergström, Volatile marking secretions from the labial glands of North European Pyrobombus D. T. males (Hymenoptera, Apidae), *Insectes Sociaux*, 24:213-224 (1980).
10. G. Bergström, Chemical aspects of insect exocrine signals as a means for systematic and phylogenetic discussions in aculeate Hymenoptera, *Entomol. Scand., Supp.* 15:173-184 (1981).
11. M. Lindauer and W. E. Kerr, Communication between workers of stingless bees, *Bee World*, 41:29-41, 65-71 (1960).
12. M. S. Blum, R. M. Crewe, W. E. Kerr, L. H. Keith, A. W. Garrison and M. M. Walker, Citral in stingless bees: isolation and functions in trail-laying and robbing, *J. Insect Physiol.*, 16:1637-1648 (1970).
13. C. G. Butler, D. J. C. Fletcher and D. Watler, Nest entrance marking with pheromones by the honey bee *Apis mellifera* L., and by a wasp, *Vespula vulgaris* L., *Animal Behaviour*, 17:142-147 (1969).
14. A. B. Attygalle and E. D. Morgan, Ant trail pheromones, *Advances in Insect Physiol.*, 18:1-30 (1985).
15. E. O. Wilson, Source and possible nature of the odor trail of fire ants, *Science*, 129:643-644 (1959).
16. D. Crump, R. M. Silverstein, H. J. Williams and T. D. Fitzgerald, Identification of trail pheromone of larva of eastern tent caterpillar *Malacosana americanum* (Lepidoptera: Lasiocampidae), *J. Chem. Ecol.*, 3:397-402 (1987).
17. B. Hölldobler, Trail communication in the dacetine ant *Orectognathus versicolor* (Hymenoptera: Formicidae), *Psyche*, 88:245-257 (1981).
18. J. M. Pasteels and J. C. Verhaeghe, Dosage biologique de la phèromone de piste chez les fourrageuses et les reines de *Myrmica rubra*, *Insectes Sociaux*, 21:167-180 (1974).
19. F. J. Ritter, I. E. M. Brüggemann-Rotgans, P. E. J. Verwiel, E. Talman, F. Stein, J. La Brijn and C. J. Persoons, Faranal, a trail pheromone from the Dufour's gland of the pharaoh's ant, structurally related to juvenile hormone, Proceedings of the Eighth International Congress of the International Union for the Study of Social Insects, Pudoc, Wageningen, pp 41-43 (1977).
20. J. F. A. Traniello, Colony specificity in the trail pheromone of an ant, *Naturwissenschaften*, 67:361-362 (1980).
20. R. K. Vander Meer, F. D. Williams and C. S. Lofgren, Hydrocarbon components of the trail pheromone of the red imported fire ant *Solenopsis invicta*, *Tet. Lett.*, 22:1651-1654 (1981).
22. B. P. Moore, Isolation of the scent-trail pheromone of an Australian termite, *Nature, London*, 211:746-747 (1966).
23. M. C. Cammaerts-Tricot, E. D. Morgan and R. C. Tyler, Isolation of the trail pheromone of the ant *Myrmica rubra*, *J. Insect Physiol.*, 23:511-515 (1977).
24. A. B. Attygalle and E. D. Morgan, Identification of trail pheromone of the ant *Tetramorium caespitum* L. (Hymenoptera: Myrmicinae), *J. Chem. Ecol.*, 10:1453-1468 (1984).
25. R. P. Evershed, E. D. Morgan and M. C. Cammaerts, 3-Ethyl-2,5-dimethylpyrazine, the trail pheromone from the venom gland of eight species of *Myrmica* ants, *Insect Biochem.*, 12:383-391 (1982).
26. R. P. Evershed, E. D. Morgan and M. C. Cammaerts, Identification of the trail pheromone of the ant *Myrmica rubra* L. and related species, *Naturwissenschaften*, 67:374-376 (1981).
27. E. D. Morgan and L. J. Wadhams, Gas chromatography of volatile compounds in small samples of biological materials, *J. Chromatographic Sci.*, 10:528-529 (1972).
28. A. B. Attygalle and E. D. Morgan, Pheromones in nanogram quantities: structure determination by combined microchemical and gas chromatographic methods, *Angew. Chem., Int. Ed. (Eng)*, 27:460-478 (1988).
29. P. G. McDowell and G. W. Oloo, Isolation, identification and biological activity of trail-following pheromone of termite *Trinervitermes bettoniacus* (Sjostedt) (Termitidae: Nasutitermitinae), *J. Chem. Ecol.*, 10:835-851 (1984).
30. P. Karlson, M. Lüscher and H. Hummel, Extraction und biologische Auswertung des Spürpheromons der Termite *Zootermopsis nevadensis*, *J. Insect Physiol.*, 14:1763-1771 (1968).
31. F. Matsumura, H. C. Coppel and A. Tai, Isolation and identification of termite trail-following pheromone, *Nature, London*, 219:963-964 (1968).

32. R. Howard, F. Matsumura and H. C. Coppel, Trail following pheromones of the Rhinotermitidae: approaches to their authentication and specificity, *J. Chem. Ecol.*, 2:147-166 (1976).
33. G. D. Prestwich, Chemical systematics of termite exocrine secretions, *Ann. Rev. of Ecology & Systematics*, 14:287-301 (1983).
34. J. H. Tumlinson, R. M. Silverstein, J. C. Moser, R. G. Brownlee and J. M. Ruth, Identification of the trail pheromone of a leaf-cutting ant *Atta texana*, *Nature, London*, 234:348-349 (1971).
35. J. H. Cross, R. C. Byler, U. Ravid, R. M. Silverstein, S. W. Robinson, P. M. Baker, J. S. de Oliveira, A. R. Jutsum and J. M. Cherrett, The major component of the trail pheromone of the leaf cutting ant *Atta sexdens rubropilosa* Forel: 3-Ethyl-2,5-dimethylpyrazine, *J. Chem. Ecol.*, 5:187-204 (1979).
36. R. P. Evershed and E. D. Morgan, The amounts of trail pheromone substances in the venom of workers of four species of Attine ants, *Insect Biochem.*, 13:469-474 (1983).
37. B. D. Jackson, E. D. Morgan and C. A. Collingwood, The chemical secretions of *Myrmica specioides* Bondroit and *Myrmica gallieni* Bondroit (Myrmicinae), *Actes des Colloques Insectes Sociaux*, 5:307-313 (1989a).
38. A. B. Attygalle, M. C. Cammaerts, R. Cammaerts, E. D. Morgan and D. G. Ollett, Chemical and ethological studies of the trail pheromone of the ant *Manica rubida* (Hymenoptera: Formicidae), *Physiol. Entomol.*, 11:125-132 (1986).
39. A. B. Attygalle and E. D. Morgan, Trail pheromone of the ant *Tetramorium caespitum* L., *Naturwissenschaften*, 70:364 (1983).
40. E. D. Morgan and D. G. Ollett, Methyl 6-methylsalicylate, trail pheromone of the ant *Tetramorium impurum* (Formicidae: Myrmicinae), *Naturwissenschaften*, 74:596-597 (1987).
41. P. E. Sonnet and J. C. Moser, Trail pheromone: response of the Texas leaf-cutting ant, *Atta texana* to selected halo- and cyanopyrrole-2-aldehydes, ketones and esters, *Environmental Entomol.*, 2:851-854 (1973).
42. E. D. Morgan, B. D. Jackson and G. W. Sales, Trail pheromone of the ant *Tetramorium impurum* and model compounds: structure-activity comparisons, submitted (1990).
43. R. K. Vander Meer, F. Alvarez and C. S. Lofgren, Isolation of the trail recruitment pheromone of *Solenopsis invicta*, *J. Chem. Ecol.*, 14:825-838 (1988).
44. A. B. Attygalle, O. Vostrowsky, H. J. Bestmann, S. Steghaus-Kovac and U. Maschwitz, (3R,4S)-4-Methyl-3-heptanol, the trail pheromone of the ant *Leptogenus diminuta*, *Naturwissenschaften*, 75:315-317 (1988).
45. M. F. Ali, E. D. Morgan, C. Detrain and A. B. Attygalle, Identification of a component of the trail pheromone of the ant *Pheidole pallidula* (Hymenoptera: Formicidae), *Physiol. Entomol.*, 13:257-265 (1988).
46. B. D. Jackson, P. J. Wright and E. D. Morgan, 3-Ethyl-2,5-dimethylpyrazine, a component of the trail pheromone of the ant *Messor bouvieri*, *Experientia*, 45:487-489 (1989).
47. D. J. C. Fletcher and J. M. Brand, Source of the trail pheromone and method of trail laying in the ant *Crematogaster peringueyi*, *J. Insect Physiol.*, 14:783-788 (1968).
48. R. H. Leuthold, A tibial gland scent-trail and trail-laying behaviour in the ant *Crematogaster ashmeadi* Mayr, *Psyche*, 75:233-248 (1968).
49. J. F. A. Traniello, Colony specificity in the trail pheromone of an ant, *Naturwissenschaften*, 67:361-362 (1980).
50. K. Jessen and U. Maschwitz, Orientation and recruitment behaviour in the ponerine ant *Pachycondyla tesserinoda* (Emery): laying of individual-specific trails during tandem running, *Behavioural Ecol. and Sociobiol.*, 19:151-155 (1986).
51. U. Maschwitz, S. Lenz and A. Buschinger, Individual specific trails in the ant *Leptothorax affinis* (Formicidae: Myrmicinae), *Experientia*, 42:1173-1174 (1986).
52. E. O. Wilson, Trail sharing in ants, *Psyche*, 72:1-7 (1965).

GAS CHROMATOGRAPHY LINKED TO ELECTROANTENNOGRAPHY: A VERSATILE TECHNIQUE FOR IDENTIFYING INSECT SEMIOCHEMICALS

A. Cork, P. S. Beevor, A. J. E. Gough and D. R. Hall

Overseas Development Natural Resources Institute
Central Avenue, Chatham Maritime
Chatham, Kent ME4 4TB, UK

SUMMARY

The efficient separating power of high resolution gas chromatography (HRGC) has been combined with the sensitivity and selectivity of electroantennography to produce a versatile technique for identifying insect olfactory stimulants. These chemicals are potentially useful as control agents by modifying insect behaviour. Here we describe the equipment currently used in our laboratory to conduct this work and discuss the scope and limitations of the technique experienced in the identification of a range of insect semiochemicals.

INTRODUCTION

Chemical communication in insects is highly developed. They can perceive and, under the correct physiological conditions, respond to remarkably small quantities of chemicals. In the study of olfactory stimuli which influence these behavioural responses, it is first necessary to identify the particular chemicals involved. Classically this would be undertaken by physically or chemically separating a behaviourally active natural extract and testing the fractions by means of a laboratory bioassay or field test. These procedures are time-consuming, and, moreover, the behavioural response may well be dependent on the presence of a specific blend of chemicals, so that fractionation could result in an apparent loss of activity.

Insects generally perceive odors by means of specially adapted sensillae on the antennae. In 1957 Schneider [1] showed that by inserting microelectrodes into the base and tip of an insect's antenna it was possible to record a slow depolarization across the antenna in response to stimulation by volatile compounds. This technique was termed electroantennography (EAG), and the EAG response is considered to represent a summation of depolarizations of all the stimulated sensillae on the antenna.

The electrophysiological response is essentially independent of other environmental and neural factors which affect behavioural responses, so for a given dose of a particular chemical an individual insect of a particular species will respond in a predictable manner. Conversely, the fact that a compound elicits an EAG response from an insect does not mean that it will cause, or even contribute to causing, any particular behavioural response.

Olfactory stimulants are by definition volatile and perceived in the gaseous phase, so ideally a chemical mixture should be separated and presented to an EAG preparation in this medium. This is most effectively achieved using high resolution gas chromatography (HRGC), and by combining the separating ability of HRGC with the sensitivity and selectivity of the EAG as a detector, it is possible to detect the chemicals perceived by an insect in a complex mixture.

Chromatography and Isolation of Insect Hormones and Pheromones
Edited by A. R. McCaffery and I. D. Wilson
Plenum Press, New York, 1990

MATERIALS AND METHODS

The EAG Preparation

Securing the Insect: Intact live insects were used in the EAG preparation so as to maximize their useful life, and such EAG preparations could last at least one working day without significant deterioration.

EAG preparations made with male moths or diptera, such as tsetse flies, were mounted on their dorsal surface in a groove cut in a block of plasticine and restrained with a strip of polystyrene secured in place with pins. Moth antennae were laid out in a V-form with the ventral surface uppermost, and held in place with a series of U-shaped wires (0.2 mm diameter) made from lengths of copper taken from the screening in coaxial cables. Tsetse fly antennae were prised from their resting position and the arista buried in the plasticine before securing the funiculus with a similar U-shaped wire.

For small coleoptera, such as the larger grain borer, *Prostephanus truncatus*, it was found to be more satisfactory to restrain the insect by placing its abdomen and thorax in a hole drilled into a perspex disc whose diameter was slightly larger than the insect's cross-section. The head was allowed to protrude out of the hole and was restrained with a metal plate. The surface of the perspex disc was coated with a thin layer of plasticine onto which the antennae were laid.

Microelectrodes: Glass microelectrodes were fashioned from capillary glass tubing, typically 50 mm length by 2 mm diameter, by pulling the tubing under a heat source. Ideally the process should be controlled by using a microelectrode puller (e.g., Havard Apparatus Ltd.) to ensure reproducibility between batches. In order to facilitate the filling of the microelectrode with saline it is advisable to use a capillary glass tubing with an inner filament (e.g., Kwik-Fil, Clark's Medical Ltd.).

The microelectrodes were filled with saline solution to make electrical continuity between the haemolymph in the antennae and the silver/silver chloride (Ag/AgCl) junction of the recording system. Ideally this should be isotonic with the haemolymph of the insects as is that described by Reolofs and Comeau [2] for lepidoptera. We have employed this solution in all of our EAG work, but usually add 0.2% agar to reduce losses by evaporation. Other workers have achieved satisfactory results with 3 M potassium chloride solution.

Location of Microelectrodes in the Antenna: In order to record EAGs from an insect antenna it is normally necessary to insert microelectrodes into the antenna. Olfactory sensillae are generally located on the antennal flagellum so that the maximum amplitude of response will be obtained by inserting the recording electrode as near to the distal end of the flagellum as practicable, and the reference or indifferent electrode as near to the proximal end of the flagellum as possible. In principle the reference electrode can be inserted anywhere in the insect, but we prefer to locate it near the basal scape on the same antenna used for recording in order to reduce the interference from extraneous neuromuscular activity.

The microelectrodes were mounted in a holder located on micromanipulators (Leitz) where they were in contact with a Ag/AgCl electrode. In order to monitor movement of the microelectrodes with the micromanipulators it was necessary to work under low power magnification using a standard binocular microscope (x 64). Ideally the EAG set-up should be placed on a vibration-free table, separate from the GC, and shielded with an earthed Faraday cage.

Microelectrodes are most easily inserted into the antennae of lepidoptera by bending the antenna slightly at the appropriate positions to expose the interstitial membranes between the annuli. These are soft enough to allow easy penetration by glass microelectrodes. However, the antennae of coleoptra are not so easily punctured, so that it may be necessary to pre-puncture the cuticle with a micro-knife (e.g., Rudolph Beaver Inc., USA) or a sharpened needle before inserting the microelectrodes.

With micro-lepidoptera it was often easier to remove completely one antennal flagellum and position the reference electrode in the haemocoel exposed in the pedicel. The recording electrode was then inserted in the distal end of the other antenna using one of the several methods. The microelectrode could be inserted through the interstitial membrane as above, or into the haemocoel exposed by removing two of the distal segments.

Alternatively, the tip could be broken from the microelectrode and the distal end of the antenna inserted into the microelectrode.

EAG recordings from tsetse flies were produced by inserting the reference electrode in the proximal end of the funiculus of one antenna and the recording electrode into the distal end of the funiculus of the other antenna.

These methods ensured good electrical contact between the haemolymph of the antenna and the recording equipment, but in fact the electrodes did not necessarily have to puncture the antenna in order to record EAG responses. Nagai [3] successfully recorded responses from the European corn borer, *Ostrinia nubilalis*, by wrapping thin strips of filter paper saturated with saline solution at intervals along the antenna. Each paper strip acted as an electrode making electrical contact with the haemolymph through the pores on the sensilla despite the increased resistance that would be expected from such a procedure. A similar phenomena was observed by Den Otter and Saini [4] who showed that not only could a tsetse fly EAG be recorded by merely placing the tip of the recording electrode on the surface of the antenna, but that this also increased the amplitude of the response by a factor of five, and we have confirmed this [5]. However, the situation may be further complicated in the diptera with the finding that the amplitude of the EAG response to different compounds varies according to the location of the microelectrode on the surface of the funiculus of the antenna of *Dacus oleae* [6]. It was suggested that the olfactory sensilla eliciting the responses are not evenly distributed across the antenna and receptor density maps have been reported in support of these findings. This result may well turn out to be more general for diptera, but numerous studies with lepidoptera indicate that the different types of olfactory sensilla are generally evenly distributed on the ventral surface of the antennal flagellum, and so should not affect the EAG in this manner.

Amplification and Recording of the EAG Response: Electrical depolarization across the antenna was recorded after suitable amplification with a high impedance (10^7 ohms) micro-amplifier, such as a Grass P16 or similar instrument in the DC mode.

Electrical contact between the microelectrodes and the amplifier was achieved using a Ag/AgCl junction. Silver wire soldered onto one end of a screened coaxial cable was coated with AgCl by electrolysis in an aqueous solution of sodium chloride (ca. 1 M) or dilute hydrochloric acid (ca. 5%). The coaxial cable was coupled to the amplifier with DIN plugs, and the EAG signal typically amplified 100 times and monitored on an oscilloscope (e.g., Tektronix 5110 or similar) before recording. At present we store the EAG responses on a Racal Store 4, FM tape recorder and simultaneously print both EAG and GC-FID traces on a Linseis dual pen recorder (L2045). However, the oscilloscope and tape recorder are not essential for linked GC-EAG analyses and could be replaced by a good quality, dual-channel recorder such as that described above, which has a full scale response time of 80 ms. Computer-based integrating packages are now available and have been used by other workers (J. C. Dickens, USDA Boll Weevil Research Unit, personal communication).

The GC-EAG Link

Simultaneous recording of EAG responses from male antennae to GC column effluent was first described by Moorhouse et al. [7]. The system involved the collection of a portion of the volatiles eluting from a packed GC column in the vapor phase in a 10 cm^3 glass reservoir placed in the GC detector oven of a Varian Aerograph 204 1B. The reservoir was periodically emptied by a secondary nitrogen flow into a constant nitrogen flow which both cooled and diluted the GC effluent before passing it over the EAG preparation. The periodic pulsing of GC column effluent resulted in electroantennographic responses which were recorded on an ink jet recorder (Siemens Oscillomink) as a series of depolarizations, typically -0.2 to -10 mV from the resting potential.

Our current system utilizes essentially the same technique, modified to exploit the higher efficiencies afforded by capillary GC columns (Figure 1). The system was designed around a modified Carlo Erba 5300 Mega series instrument fitted with two Grob type split/splitless injectors (a). Two fused silica capillary columns (b) (25 m x 0.33 mm ID) coated with a polar, chemically-bonded Carbowax 20M equivalent and a non-polar, chemically-bonded methyl silicone were connected to the two injectors. The column outlets were combined by means of a low dead volume, push-fit Y tube (c) and passed along a short length (10 cm) of fused silica capillary tubing coated with chemically-bonded methyl silicone to another Y tube (d). There the column effluent was split 50:50 into two identical lengths (0.5 m) of coated fused silica capillary tubing. One of these was connected to a flame ionization detector (FID) (h) and the other delivered column effluent to a larger

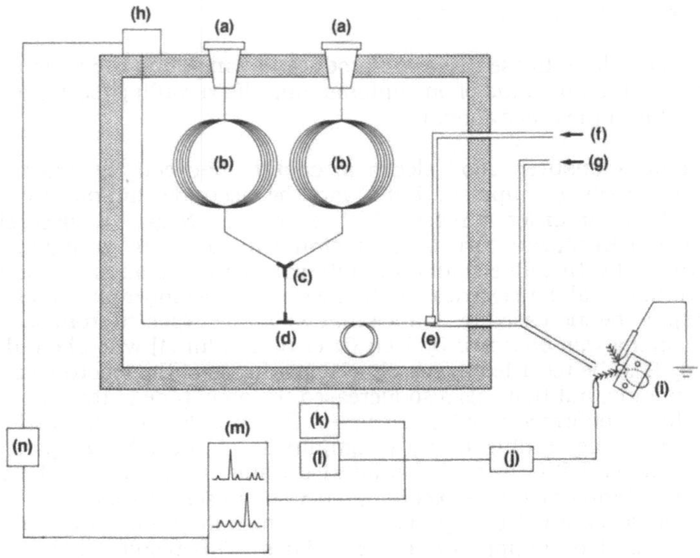

Fig. 1. Diagram of coupled GC-EAG system: (a) GC Grob type split/splitless injectors; (b) capillary columns; (c) and (d) Y tube connectors; (e) GC column outlet to EAG preparation; (f) inlet for pulsed nitrogen flow (100 cm^3/min); (g) inlet for constant bypass nitrogen flow (250 cm^3/min); (h) GC column outlet to flame ionization detector (FID); (i) EAG preparation; (j) high impedance DC amplifier; (k) oscilloscope; (1) FM tape recorder; (m) dual pen recorder for FID and EAG outputs; (n) FID amplifier.

bore glass tube (e) which could be evacuated either periodically or continuously over the EAG preparation (i).

One of the GC columns was used for the analysis with helium carrier gas at a flow rate of 1 ml/min, while the other column provided make-up at 2 ml/min to minimize peak broadening through the Y connectors. The flow rates were interchanged to use the other GC column for analysis.

Normally the GC column effluent was pulsed over the EAG preparation using nitrogen delivered through an electronically-controlled valve. Column effluent was collected in the glass tube (10 cm long, 2 mm ID) (e) for 15 sec then displaced by a 3 sec pulse of nitrogen (100 cm^3/min) into another glass tube (g) which was constantly evacuated with a nitrogen flow of 250 cm^3/min over the EAG preparation. This constant nitrogen flow was not essential for cooling the GC effluent, but was used for micro-lepidoptera, such as the cocoa moth, *Conopomorpha cramerella*, which are prone to dessication. It was also used for species which give large EAG responses to changes in air flow, such as the cotton bollworm, *Heliothis armigera*. This contribution to the EAG response by the mechanoreceptors could in some instances obscure the response by the chemoreceptors. A constant nitrogen flow through tube (g) was also preferred when the GC was operated in the temperature-programmed mode because the amplitudes of the EAG responses tended to increase in response to the increasing temperatures experienced by the preparation. Ideally the basic response should remain constant throughout a linked analysis ensuring that any increase in EAG response due to the presence of an eluting compound is not obscured. For this reason the signal was not filtered by any electronic means to reduce noise levels or ensure a consistent baseline. Both were useful indicators of the condition of the EAG preparation.

RESULTS AND DISCUSSION

The GC-EAG output consisted of a normal GC-FID trace with a synchronized EAG trace showing a series of responses to the portions of GC column effluent collected and delivered to the EAG preparation every 15 sec (Figure 2). Eluting compounds were considered to have caused a significant EAG response if the magnitude of the associated response was at least 10% greater than the average of the responses to the GC fractions eluting immediately before and after it.

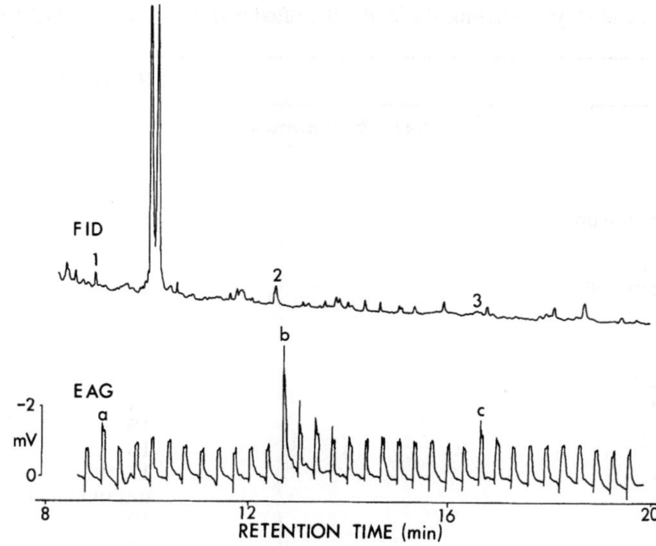

Fig. 2. A typical combined GC-EAG analysis of solvent extracted female moth volatiles entrained on charcoal. Top trace is the gas chromatogram recorded from the flame ionization detector. GC conditions: polar CP Wax 52CB fused silica capillary column (25 m x 0.33 mm internal diameter), temperature programmed from 70°C at 20°C/min to 120°C then 4°C/min to 210°C with helium as carrier gas (0.6 kg/cm^3). EAG trace (bottom) of male moth is characterized by a series of EAG responses resulting from periodic collection and expulsion of the GC column effluent over the EAG preparation. The trace indicates the presence of a homologous series of three electrophysiologically active compounds which give rise to responses a, b and c. The major EAG response b was elicited in response to 0.4 ng of compound 2.

The discontinuous delivery of GC effluent to the EAG preparation had two advantages. Firstly, it provided a concentration step by collecting the effluent over 15 sec and then passing it over the EAG preparation in less than a second. In our hands, this gave greater sensitivity compared with continuous delivery systems.

Secondly, it avoided fatigue of the antennal receptors due to continuous exposure to both electrophysiologically active and inactive compounds. The EAG response consisted of a rapid negative depolarization over the first 0.5 sec of the nitrogen pulse followed by a slight but generally significant repolarization over the next 2.5 sec of the pulse. After the nitrogen pulse, the EAG returned to the resting potential at a rate that was dependent upon the chemical nature and quantity of material that eluted from the GC column in the previous fraction. In order to optimize both chromatography and electrophysiology, it was generally aimed to keep the quantities of material chromatographed well below 100 ng per component, and under these conditions the EAG resting potential was normally easily restored in the interval between stimuli. Thus EAG responses to each fraction could be measured accurately without having to compensate for a fluctuating resting potential or adaptation due to prolonged over-stimulation.

However, the resolution of the pulsed delivery system was limited by the interval between stimuli which was set at 15 sec to enable the EAG to attain the resting potential before the next pulse. As the peak width of compounds eluting from capillary GC columns under temperature-programmed conditions is typically less than 10 sec, two components resolved by the GC may not be resolved by the EAG system. If more than one component was thought to be eluting in the EAG-active GC fraction, the identity of the electrophysiologically active compound could be established more clearly by changing the GC conditions so that the compounds were better separated, e.g., by using appropriate isothermal conditions, or by switching to a GC column of different polarity. With the dual-column arrangement described above it was possible to carry out successive analyses on columns of different polarity with the same EAG preparation. Alternatively, the valve system controlling delivery of the GC fractions to the EAG could be operated manually so that components that were resolved on the GC could be presented to the EAG individually.

Thus, in GC-EAG analyses of pheromone gland extracts from the African armyworm moth, *Spodoptera exempta*, on a polar GC column, an EAG response was observed to a GC

Table 1. Behaviour Modifying Chemical Blends Identified with the Aid of the GC-EAG Technique

Species name	Reference
Sex Pheromones	
LEPIDOPTERA	
Gelechiidae	
Scrobipalpopsis solanivora	11
Gracillariidae	
Conopomorpha cramerella	12
Noctuidae	
Busseola fusca	9
Diparopsis castanea	13
Diparopsis watersi	14
Earias insulana	15
Earias vittella	16
Heliothis armigera	17, 18
Heliothis assulta	unpublished
Heliothis punctiger	19
Sesamia inferens	20
Spodoptera exempta	21, 8
Spodoptera littoralis	22
Pyralidae	
Chilo auricilius	23
Chilo partellus	24
Chilo suppressalis	25
Chila sacchariphagus	26
Corcyra cephalonica	27
Ephestia cautella	28
Scirpophaga incertulas	29
Tortricidae	
Cryptophlebia batrachopa	30
Cryprophlebia leucotreta	30
Yponomeutidae	
Prays citri	31
Prays oleae	32
Aggregation Pheromones	
COLEOPTERA	
Bostrichidae	
Prostephanus truncatus	33
Kairomones	
DIPTERA	
Glossinidae	
Glossina morsitans morsitans	34, 5
Glossina pallidipes	34, 5

fraction which contained two completely resolved compounds at 15.01 and 15.16 equivalent carbon lengths (ECL) relative to the retention times of saturated, straight-chain acetates. In order to show which, if any, of the compounds was EAG active the analysis was repeated on a GC column with a non-polar liquid phase. On this GC phase the compounds were easily separated, eluting at 13.91 and 12.51 ECL respectively, and both were found to be electrophysiologically active. Relative GC retention times together with GC-MS data were then used to establish the structures of these compounds as (*Z*)-9-tetradecenol and (*Z,E*)-9,12-tetradecadienyl acetate respectively [8].

However, the separating power of the GC-EAG is ultimately limited by that of the HRGC, and a GC fraction may contain several unresolved compounds. For example, GC-EAG analyses of volatiles collected from cultures of *P. truncatus* on Porapak Q indicated that the main EAG response was coincident with one of three well-resolved GC peaks on both polar and non-polar GC columns. In this case the GC peaks were shown to be electrophysiologically inactive diethylbenzene impurities from the Porapak, and the main pheromone component was totally obscured by them. Separation was only achieved by liquid chromatography (Cork et al., in preparation).

The EAG provides a specific, generally very sensitive detector for olfactory stimuli. As it records the summed depolarizations of all the chemoreceptors on the antenna, a single preparation can in principle detect all the olfactory stimulants in a complex mixture, and is ideal for use in detecting multi-component blends of unknown structure. The system was originally developed for use in the identification of sex pheromones of lepidoptera, but has subsequently proved equally valuable for aggregation and sex pheromones of coleoptera such as *P. truncatus*, and for host-produced kairomones for tsetse flies, as shown in Table 1.

However, in all this work, the limitations of the EAG technique must be recognized. On the one hand, compounds which are perceived by the insect and which are behaviourally very important may only elicit weak EAG responses. In this case they could go undetected if only small amounts of material are available for analysis. For example, in the case of the sex pheromone of the maize stalk borer, *Busseola fusca*, three pheromone components are vital for attraction of male moths, but one of these, (*Z*)-9-tetradecenyl acetate, caused virtually no EAG response [9]. This was probably due to the small number of chemoreceptors on the antenna which respond specifically to this compound, and the problem could be avoided by using single-cell recording. However, if these chemoreceptors are present at low density then the probability of locating one of these sensillae for recording is correspondingly low.

On the other hand, the production of an EAG response by a compound gives no indication of the behavioural significance of that compound. This is particularly apparent outside the pheromone field, for example with tsetse flies where numerous compounds have been detected by EAG, some of which act as attractants, e.g., 1-octen-3-ol, 4-methylphenol and 3-propylphenol, some as repellents, e.g., acetophenone, 2-methoxyphenol [5] and some of which have no apparent behavioral activity, e.g., 2,6,10,10-tetramethyl-1-oxaspiro[4.5]dec-2-en-8-one [10].

CONCLUSIONS

The GC-EAG has many advantages as a method for screening mixtures for potential semiochemicals. Providing its limitations are recognized and accommodated, the technique provides a sensitive means for selecting out potential candidate compounds for further study either by laboratory bioassay or in the field. The value of the technique is well illustrated by the number of semiochemicals identified to date as shown in Table 1.

REFERENCES

1. D. Schneider, Elecktrophysiologische untersuchungen von chemo- und mechanorezeptoren der antenne des seidenspinners *Bombyx mori* L. *Z. Vergl. Physiol.*, 40:8-41 (1957).
2. W. Roelofs and A. Comeau, Sex pheromone perception: electroantennogram responses of the red banded leafroller moth, *Insect Physiol.*, 17:1969-1982 (1971).
3. T. Nagai, Electroantennogram response gradient on the antenna of the European corn borer, *Ostrinia nubilalis, J. Insect Physiol.*, 27:889-894 (1981).
4. C. J. Den Otter and R. K. Saini, Pheromone perception in the tsetse fly, *Glossina morsitans morsitans.*, *Entomol. Exp. Appl.*, 39:155-161 (1985).
5. E. Bursell, A. J. E. Gough, P. S. Beevor, A. Cork, D. R. Hall and G. A. Vale, Identification of components of cattle urine attractive to tsetse flies, *Glossina* spp. (Diptera: Glossinidae), *Bull. Ent. Res.*, 78:281-291 (1988).
6. R. Crnjar, G. Scalera, A. Liscia, A. M. Angioy, A. Bigiani, P. Pietra and I. Tomassini Barbarossa, Morphology and EAG mapping of the antennal olfactory receptors in *Dacus oleae*, *Entomol. Exp. Appl.*, 51:77-85 (1989).
7. J. E. Moorhouse, R. Yeadon, P. S. Beevor and B. F. Nesbitt, Method for use in studies of insect communication, *Nature*, 223:1174-1175 (1969).
8. A. Cork, J. Murlis and T. Megenasa, Identification and field testing of additional components of female sex pheromone of African armyworm, *Spodoptera exempta* (Lepidoptera: Noctuidae), *J. Chem. Ecol.*, 15:1349-1364 (1989).
9. D. R. Hall, P. S. Beevor, A. Cork, R. Lester, B. F. Nesbitt, G. K. C. Nyirenda, D. D. Nota Phiri, B. W. Blair and J. Tannock, The female sex pheromone of the maize stalk-

borer, *Busseola fusca* (Fuller) (Lepidoptera: Noctuidae), *Zimbabwe J. Agric. Res.*, 19:111-122 (1981).

10. A. J. E. Gough, D. R. Hall, P. S. Beevor, A. Cork, E. Bursell and G. A. Vale, Attractants for tsetse from cattle urine, *in*: "Proceedings of the 19th Meeting of the International Scientific Council for Trypanosomiasis Research Control", Lome, Togo, 27th March-4th April 1987 (in press).
11. B. F. Nesbitt, P. S. Beevor, A. Cork, D. R. Hall, R. M. Murillo and H. R. Leal, Identification of components of the female sex pheromone of the potato tuber moth, *Scrobipalpopsis solanivora, Entomol. Exp. Appl.*, 38:81-85 (1985).
12. P. S. Beevor, A. Cork, D. R. Hall, B. F. Nesbitt, R. K. Day and J. D. Mumford, Components of female sex pheromone of cocoa pod borer moth, *Conopomorpha cramerella, J. Chem. Ecol.*, 12:1-23 (1986).
13. B. F. Nesbitt, P. S. Beevor, R. A. Cole and R. G. Poppi, The isolation and identification of the female sex pheromones of the red bollworm, *Diparopsis castanea, J. Insect Physiol.*, 21:1091-1096 (1975).
14. P. S. Beevor, D. G. Campion and J. E. Moorhouse, Cross-attractancy and cross-mating between the red bollworm, *Diparopsis castanea* (Hmps.) and the Sudan bollworm *Diparopsis watersi* (Roths.) (Lepidoptera: Noctuidae), *Bull. Ent. Res.*, 62:439-442 (1973).
15. D. R. Hall, P. S. Beevor, R. Lester and B. F. Nesbitt, (*E,E*)-10,12-Hexadecadienal: a component of the sex pheromone of the spiny bollworm, *Earias insulana* (Boisd.) (Lepidoptera: Noctuidae), *Experientia*, 36:152-153 (1980).
16. A. Cork, D. J. Chamberlain, P. S. Beevor, D. R. Hall, B. F. Nesbitt, D. G. Campion and M. R. Attique, Components of the sex pheromone of spotted bollworm, *Earias vittella* F. (Lepidoptera: Noctuidae): identification and field evaluation in Pakistan, *J. Chem. Ecol.*, 14:929-945 (1988).
17. B. F. Nesbitt, P. S. Beevor, D. R. Hall and R. Lester, Female sex pheromone components of the cotton bollworm, *Heliothis armigera, J. Insect Physiol.*, 25:535-541 (1979).
18. B. F. Nesbitt, P. S. Beevor, D. R. Hall and R. Lester, (*Z*)-9-Hexadecenal: a minor component of the female sex pheromone of *Heliothis armigera* (Hubner) (Lepidoptera: Noctuidae), *Ent. Exp. Appl.*, 27:306-308 (1980).
19. G. H. L. Rothschild, B. F. Nesbitt, P. S. Beevor, A. Cork, D. R. Hall and R. A. Vickers, Studies of the female sex pheromone of the native budworm, *Heliothis punctiger, Ent. Exp. Appl.*, 31:395-401 (1982).
20. B. F. Nesbitt, P. S. Beevor, D. R. Hall, R. Lester and V. A. Dyck, Identification of the female sex pheromone of the purple stem borer moth, *Sesamia inferens, Insect Biochem.*, 6:105-107 (1976).
21. P. S. Beevor, D. R. Hall, R. Lester, R. G. Poppi, J. S. Read and B. F. Nesbitt, Sex pheromones of the armyworm moth, *Spodoptera exempta* (Walker), *Experientia*, 31:22-23 (1975).
22. B. F. Nesbitt, P. S. Beevor, R. A. Cole, R. Lester and R. G. Poppi, Sex pheromones of two noctuid moths, *Nature, New Biology*, 137:208-209 (1973).
23. B. F. Nesbitt, P. S. Beevor, A. Cork, D. R. Hall, H. David and V. Nandagopal, The female sex pheromone of sugarcane stalk borer, *Chilo auricilius*.Identification of four components and field tests, *J. Chem. Ecol.*, 12:1377-1388 (1986).
24. B. F. Nesbitt, P. S. Beevor, D. R. Hall, R. Lester, J. C. Davies and K. V. Seshu Reddy, Components of the sex pheromone of the female spotted stalk borer, *Chilo partellus* (Swinhoe) (Lepidoptera: Pyralidae): identification and preliminary field trials, *J. Chem. Ecol.*, 5:153-163 (1979).
25. B. F. Nesbitt, P. S. Beevor, D. R. Hall, R. Lester and V. A. Dyck, Identification of the female sex pheromones of the moth, *Chilo suppressalis, J. Insect Physiol.*, 21:1883-1886 (1975).
26. B. F. Nesbitt, P. S. Beevor, D. R. Hall, R. Lester and J. R. Williams, Components of the sex pheromone of the female sugarcane borer, *Chilo sacchariphagus* (Bojer) (Lepidoptera: Pyralidae): identification and field trials, *J. Chem. Ecol.*, 6:385-394 (1980).
27. D. R. Hall, A. Cork, R. Lester, B. F. Nesbitt and P. Zagatti, Sex pheromones of rice moth, *Corcyra cephalonica* (Stainton). II. Identification and role of female pheromone, *J. Chem. Ecol.*, 13:1575-1589 (1987).
28. J. S. Read and P. S. Beevor, Analytical studies on the sex pheromone complex of *Ephestia cautella* (Walker) (Lepidoptera: Phycitidae), *J. Stored Prod. Res.*, 12:55-57 (1976).
29. A. Cork, P. S. Beevor, D. R. Hall, B. F. Nesbitt, G. S. Arida and O. Mochida, Components of the sex pheromone of the yellow stem borer, *Scirpophaga incertulas, Entomol. Exp. Appl.*, 37:149-153 (1985).
30. D. R. Hall, P. S. Beevor, A. Cork, B. F. Nesbitt and E. A. S. La Croix, (*Z*)-8-Dodecenyl acetate: the major component of the female sex pheromone of *Cryptophlebia*

batrachopa, a Tortricid pest of macadamia in Malawi *Entomol. Exp. Appl.* , 35:33-36 (1984).
31. B. F. Nesbitt, P. S. Beevor, D. R. Hall, R. Lester, M. Sternlicht and S. Goldenberg, Identification and synthesis of the female sex pheromone of the citrus flower moth, *Prays citri, Insect Biochem.*, 7:355-359 (1977).
32. D. G. Campion, L. J. McVeigh, J. Polyrakis, S. Michaelakis, G. N. Stravrakis, P. S. Beevor, D. R. Hall and B. F. Nesbitt, Laboratory and field studies of the sex pheromone of the olive moth, *Prays oleae, Experientia*, 35:1146-1147 (1979).
33. R. J. Hodges, A. Cork and D. R. Hall, Aggregation pheromone for monitoring the greater grain borer, *Prostephanus truncatus, British Crop Protection Conference - Pests and Diseases* (1984).
34. D. R. Hall, P. S. Beevor, A. Cork, B. F. Nesbitt and G. A. Vale, 1-Octen-3-ol. A potent olfactory stimulant and attractant for tsetse isolated from cattle odors, *Insect Sci. Appl.*, 5:335-339 (1984).

TECHNIQUES FOR ISOLATION AND CHARACTERIZATION OF VOLATILE SEMIOCHEMICALS OF PHYTOPHAGOUS INSECTS

Margaret M. Blight

AFRC Institute of Arable Crops Research
Rothamsted Experimental Station
Harpenden, Hertfordshire AL5 2JQ, UK

SUMMARY

An air entrainment method for collecting volatiles from plants and phytophagous insects is described, and the results of a study of the aggregation pheromone of *Sitona lineatus* are reported.

INTRODUCTION

Multicomponent semiochemical systems, which were shown some 23 years ago to be of importance in bark beetle behavior [1], are being implicated increasingly in intraspecific communication by many insects [2]. In particular, the behavior of phytophagous insects is modified both by insect-produced pheromonal substances and by plant volatiles [3,4]. Besides synergizing responses to insect-produced pheromones, plant volatiles also cue orientation to, and selection of, the host plant for purposes of feeding, mating and oviposition [5].

Although insect-produced pheromones have most often been isolated by extraction of whole insects or insect glands, there are two main disadvantages to this procedure. Firstly, some substances are synthesized and released immediately by the insect instead of being stored in a glandular structure, and secondly, glands may contain precursors and inhibitory substances which may also complicate the isolation and purification of the active chemicals [6]. In contrast, the collection of volatiles by headspace and air entrainment techniques from the air surrounding insects allows information to be gained not only on the qualitative composition but also on the relative concentrations of the components. Ratios of volatiles in a multicomponent semiochemical bouquet are usually critical for biological activity [7].

Entrainment of volatiles from insects alone, particularly the Lepidoptera, may be appropriate, but it is often necessary, and advantageous, to collect volatiles from phytophagous insects when they are established on the host plant. Studies of such systems therefore involve the separation and characterization of very complex mixtures of insect- and plant-produced substances which are often present only in minute quantities.

This paper outlines the general strategy which we follow in studies on volatile semiochemicals of phytophagous insects and describes a method of differential air entrainment which has been modified since being developed originally for the study of bark beetle pheromones [8]. The use of this strategy is exemplified with some results from studies on the aggregation pheromone of the pea and bean weevil, *Sitona lineatus*.

Chromatography and Isolation of Insect Hormones and Pheromones
Edited by A. R. McCaffery and I. D. Wilson
Plenum Press, New York, 1990

MATERIALS AND METHODS

General Approach

The strategy for the study of volatile semiochemicals of phytophagous insects is:

(a) Obtain evidence for semiochemical production in laboratory or field behavioral bioassay experiments.

(b) Concurrently with (a), collect and analyze volatiles produced by the host plant alone and by insect-colonized plants. In some instances it is also appropriate to collect volatiles from insects out of contact with the plant.

(c) Locate the electrophysiologically-active components in the extract using coupled gas chromatography (GC)-electroantennogram (EAG) and coupled GC-single cell recording (SCR) techniques (Wadhams, this volume).

(d) Identify these components using physico-chemical methods (e.g., nuclear magnetic resonance spectrometry and coupled GC-mass spectrometry (GC-MS)) and micro-chemistry. Coupled GC-mass spectrometric techniques are described by Pickett (this volume).

(e) Investigate the behavioral activity of these substances in laboratory and field bioassays.

(f) Devise means of using the behaviorally-active substances in integrated control systems so that the use of pesticides can be minimized.

Isolation of Volatiles

An air entrainment method was used in which the volatiles were collected in an air stream and trapped on Porapak Q.

All glassware was scrupulously cleaned and heated for several hours at 200°C before use. Solvents were rigorously purified. Pentane (1 l) was passed sequentially through silica gel (60-120 mesh, 200 g), activated 4Å molecular sieve (50 g) and basic aluminium oxide (50 g), and then distilled. Dichloromethane (AR, 1 l) was freshly purified with basic aluminium oxide (90 g), followed by distillation.

Porapak Q (50-80 mesh) must be purified and conditioned each time it is used. New samples contained high levels of contaminants, but these became negligible after the Porapak has been recycled a few times. The Porapak was heated in a stream of nitrogen at 190°C for 24 h and then stirred, x 3, in the purified dichloromethane for 2 h periods. After filtration, most of the residual solvent was removed by flushing the flask containing the Porapak with nitrogen. Tubes used for collecting the volatiles were then filled with the Porapak and heated, under nitrogen, for 24 h at 100°C and stored tightly capped until used.

A differential air entrainment method (see Figure 1 for schematic diagram) was used in which volatiles were collected from a number of separate "lines" simultaneously. This arrangement allowed direct comparisons to be made between volatiles produced by plant materials and by the two sexes of the insect (either alone or on host plants). A blank line acted as a control. In each line, air, dried and purified by passage, in series, through activated 5Å molecular sieve and charcoal, was drawn at 1000-2000 ml/min through a glass culture vessel (Quickfit FV range) (or a tube) containing the insects and then through the glass tube containing the conditioned Porapak Q. The tube containing the Porapak had ground-glass joints and fitted directly into a socket on the flat flange lid of the culture vessel. It was connected via a flow meter and a needle valve assembly through a silica gel tower to the pump (Charles Austen, Model B100 SE).

Volatiles were desorbed from the Porapak by eluting with purified solvent. Generally pentane (c. 10 ml/g of Porapak) was used, followed by freshly purified dichloromethane (c. 5 ml/g) but other solvents, e.g., 20% diethyl ether (freshly purified) in hexane, are suitable. Pentane or hexane were normally preferred because they are inert carriers for the volatiles. Extracts of volatiles were concentrated to low volume by slow distillation of the solvent through a 17 cm column of 4 mm Fenske glass helices. Final concentration to 100 to 500 µl

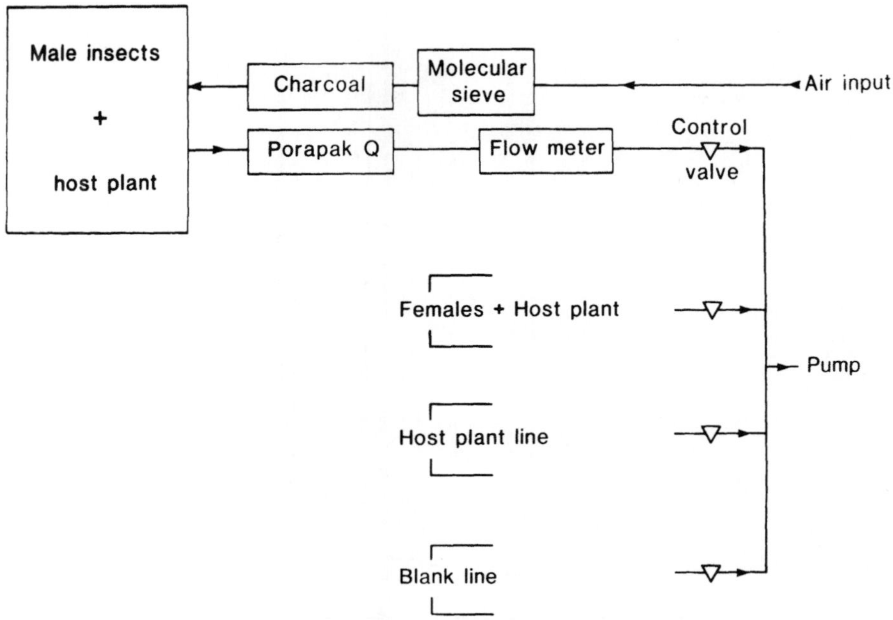

Fig. 1. Schematic diagram of air entrainment apparatus. One line only is drawn in full.

was performed by slow evaporation under a gentle stream of nitrogen. Extracts were stored under nitrogen at -20°C in tightly-capped microvials (1.5 ml) or glass ampoules.

Collection and Analysis of Volatiles from S. Lineatus *on* Vicia Faba *(Broad Bean)*

An air entrainment was carried out as described above, except that a Drechsel gas washing bottle, containing redistilled water, was placed in line between the charcoal tube and a 20 l culture vessel. In each vessel 4 x 150 ml conical flasks were placed containing broad bean shoots in water. The necks of these flasks were sealed with aluminium foil. A total of 400 insects (males, females, or males + females) were added to each "line". Volatiles were collected on 4 g Porapak Q over a period of six days.

The initial gas chromatographic analyses were done on a 25 m x 0.3 mm i.d. BP-1 (methyl silicone) fused silica column with hydrogen as the carrier gas. Subsequent analyses were performed on a 50 m x 0.32 mm i.d. HP-1 (methyl silicone) fused silica column with hydrogen as the carrier gas (see legend to Figure 3 for details). Flame ionization detection was employed in all GC analyses.

RESULTS

The aggregation pheromone of *S. lineatus* was studied using the procedures described above. A field experiment provided clear evidence that male *S. lineatus* produced a pheromone which attracted approximately equal numbers of both sexes from over-wintering sites in the spring [9]. Extracts of volatiles from *V. faba* shoots, and from male and female *S. lineatus* feeding (separately and together) on bean shoots, were analyzed initially on a 25m BP-1 column. A broad peak was detected in the chromatogram of male-produced volatiles. This was identified, by coupled GC-MS on an OV-101 column, as 4-methyl-3,5-heptanedione, although no peak was observed in these studies ([10]; Pickett, this volume).

The GC analysis, on the BP-1 column, of a synthetic sample of 4-methyl-3,5-heptanedione is shown in Figure 2. It eluted as a single abnormally broad peak. Since the column was clearly "active", further analyses of the extracts of volatiles were done using a 50 m HP-1 column. In Figure 3a and b the chromatograms of volatiles produced by male and female weevils maintained separately on beans are shown. The peaks 5, 5' (Figure 3a)

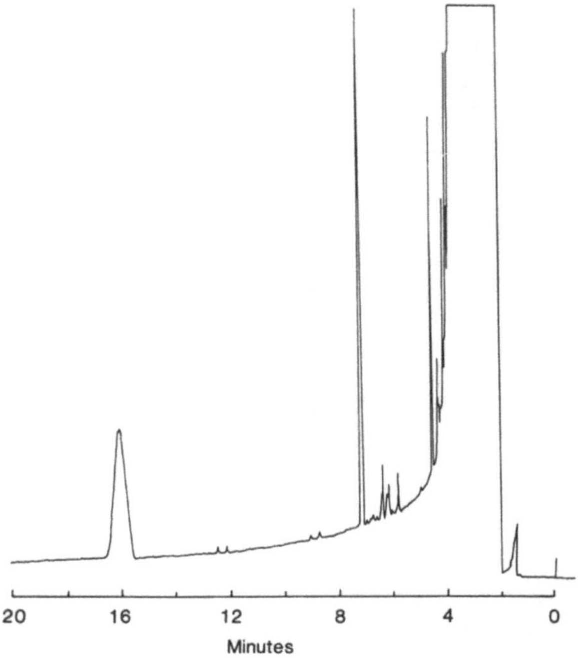

Fig. 2. Gas chromatogram of a synthetic sample of 4-methyl-3,5-heptanedione (Rt ~ 16 min). Operating conditions: splitless (Grob) injection on a 25 m x 0.33 mm i.d. BP-1 (methyl silicone) fused silica column with hydrogen as the carrier gas. Oven temperature: 40°C for 1 min, 40° to 100°C at 4°C/min.

among the male-produced volatiles are the keto- and enol-forms of 4-methyl-3,5-heptanedione. The keto-form (peak 5) co-eluted with limonene (Figure 3a and b, peak 4), a broad bean volatile. There was no evidence that female *S. lineatus* produced any of this substance. Coupled GC-EAG recordings showed that antennae of both male and female weevils responded to the natural 4-methyl-3,5-heptanedione and to a synthetic sample. The response of a single cell which is specialized to 4-methyl-3,5-heptanedione is shown in Figure 4. Comparison of the GC and single cell recordings suggests that the cell is more responsive to the enol (Figure 4, peak 2) than the keto (Figure 4, peak 1) form of the dione.

Coupled GC-EAG recordings also showed that the antennae of both male and female weevils responded to green leaf volatiles produced by *V. faba*. These were identified by GC-MS as (*Z*)-3-hexen-1-ol, (*Z*)-3-hexen-1-yl acetate, and linalool (see Figure 3). Single cell recordings confirmed that these substances elicited responses from both male and female antennae. The response of a cell, on the antenna of a male weevil, which responds to linalool (1) is shown in Figure 5.

Synthetic samples of 4-methyl-3,5-heptanedione and the electrophysiologically-active green leaf volatiles were tested in a factorial field experiment on weevil over-wintering sites in the spring [10]. 4-Methyl-3,5-heptanedione dispensed at two different concentrations (see Table 1) attracted both sexes, although significantly more males than females were trapped. The reason for this is not yet clear. The mixture of green leaf volatiles was not attractive but when a high concentration was released in combination with a high concentration of 4-methyl-3,5-heptanedione it synergized the activity of the dione.

DISCUSSION

In studies on insect-plant relationships it is most important that the characteristic odor of the plant, which attracts and/or arrests insects, should not be destroyed during isolation of the volatiles. Enzymic or chemical reactions can cause changes in composition

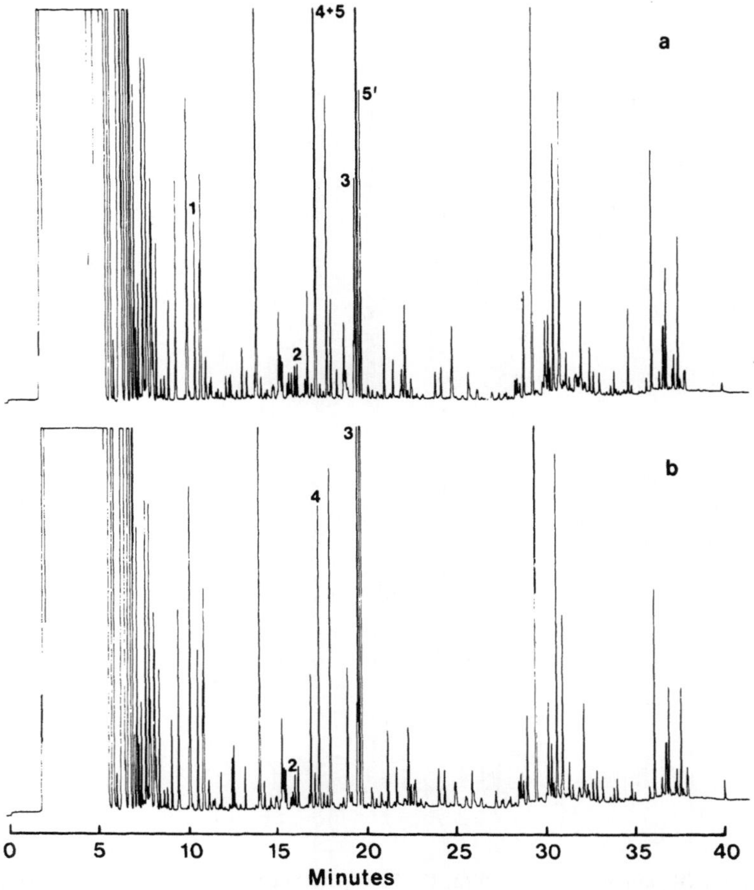

Fig. 3. Gas chromatograms of volatiles produced by (a) male *S. lineatus* on *V. faba* and (b) female *S. lineatus* on *V. faba*. Operating conditions: splitless (Grob) purged injection on a 50 m x 0.32 mm i.d. HP-1 (methyl silicone) fused silica column with hydrogen as the carrier gas. Oven temperature: 40°C for 5 min, 40° to 150°C at 5°C/min, then 150° to 250°C at 10°C/min. 1. (*Z*)-3-hexen-1-ol. 2. (*Z*)-3-hexen-1-yl acetate. 3. linalool. 4. limonene. 5,5'. Diketo and enol forms (respectively) of 4-methyl-3,5-heptanedione.

when some isolation methods are used, e.g., maceration or chopping of the plant tissue, followed by extraction and distillation to separate the volatile components. This procedure can be useful when it is necessary to isolate larger amounts of specific volatile substances for chemical identification, although all heating in air is to be avoided. However, headspace, or dynamic headspace (air entrainment) methods are more likely to produce extracts which contain the volatiles in ratios similar to those found in air around plants. It is important, though, that the growth stage and physiological condition of the plant material used for collection of volatiles should be appropriate for the insect which it supports.

Porapak Q (an ethylvinyl benzene - divinyl benzene copolymer), Tenax GC [poly(2,6-diphenyl-*p*-phenylene oxide] and charcoal are three adsorbents which are widely used to trap entrained volatiles. All have advantages and disadvantages. Contaminants in Porapak Q have presented a problem to some workers, but a careful purification procedure, as described in this paper, minimizes these difficulties. Tenax GC probably contains fewer contaminants than Porapak Q, but it has a much lower capacity as an adsorbent. Charcoal is the most efficient adsorbent but elution of volatiles requires a polar solvent and degradation or isomerization of adsorbed substances may sometimes occur. In general, Porapak Q is the best compromise. Air entrainment procedures are reviewed by Golub and Weatherston [11].

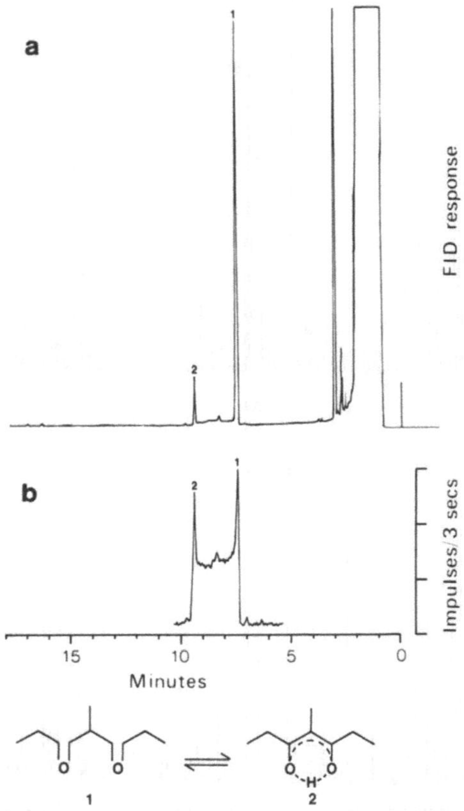

Fig. 4. Coupled GC-single cell recording (SCR) of the response of an antennal cell of male *S. lineatus* to 4-methyl-3,5-heptanedione. (a) Gas chromatogram of a synthetic sample of 4-methyl-3,5-heptanedione. For GC conditions, see legend to Figure 2. (b) Response of the olfactory cell specialized to 4-methyl-3,5-heptanedione: 1. diketo form. 2. enol form of the substance.

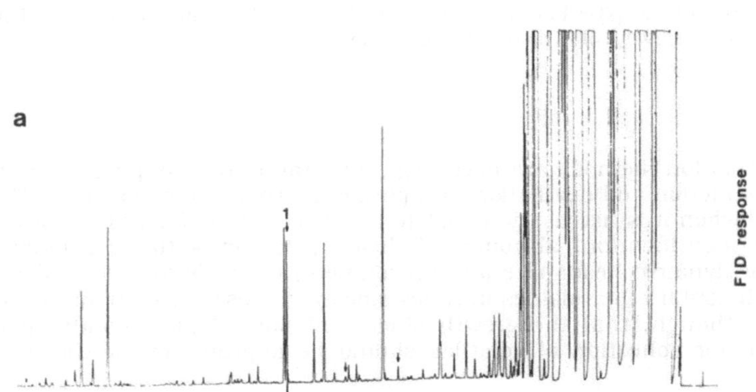

Fig. 5. Coupled GC-SCR of the response of an antennal cell to linalool. (a) Gas chromatogram of *V. faba* volatiles. 1. linalool. (b) SCR from a male *S. lineatus* preparation. The less intense responses to minor components (arrowed) of the mixture of volatiles may have been recorded from other cells.

Table 1. Field response of *S. lineatus* to baits releasing the weevil metabolite, 4-methyl-3,5-heptanedione and/or the bean volatiles, (*Z*)-3-hexen-1-ol, (*Z*)-3-hexen-1-yl acetate and linalool.

Treatment	[e]Mean number of weevils caught per replicate	Sex ratio ♂ : ♀
1. Blank (empty trap)	4.9	1.7:1
2. Bean (1)[a]	6.6	1.2:1
3. Bean (2)[b]	3.9	1.6:1
4. Weevil (1)[c]	39.1	1.9:1
5. Weevil (2)[d]	38.9	2.0:1
6. Weevil (1) + Bean (1)	32.1	2.1:1
7. Weevil (1) + Bean (2)	41.5	1.8:1
8. Weevil (2) + Bean (1)	47.7	2.2:1
9. Weevil (2) + Bean (2)	54.8	2.5:1

The substances were released from polythene vials

[a] 7 mg (*Z*)-3-hexen-1-ol + 5 mg (*Z*)-3-hexen-1-yl acetate + 34 mg linalool

[b] 34 mg (*Z*)-3-hexen-1-ol + 21 mg (*Z*)-3-hexen-1-yl acetate + 3 x 50 mg linalool

[c] 5 mg 4-methyl-3,5-heptanedione

[d] 21 mg 4-methyl-3,5-heptanedione

[e] The data were derived from 2 blocks, with 9 replications per block. The total number of weevils caught was 4796.

Fundamental to the efficient chemical analysis of the complex mixtures of plant and insect volatiles obtained by air entrainment is the use of high resolution GC and coupled GC-MS, together with coupled GC-electrophysiological techniques which allow the identification of physiologically-active substances. The use of these procedures, which make the study of complex insect-plant interactions a feasible proposition, led to this identification of the multicomponent aggregation pheromone of *S. lineatus* and the formulation of a lure which attracts these insects from overwintering sites in the spring.

REFERENCES

1. R. M. Silverstein, J. O. Rodin and D. L. Wood, Sex attractants in frass produced by male *Ips confusus* in ponderosa pine, *Science*, 154:509-510 (1966).
2. J. H. Tumlinson, Contemporary frontiers in insect semiochemical research, *J. Chem. Ecol.*, 14:2109-2130 (1988).
3. P. A. Hedin, G. H. McKibben, E. B. Mitchell and W. L. Johnson, Identification and field evaluation of the compounds comprising the sex pheromone of the female boll weevil, *J. Chem. Ecol.*, 5:617-627 (1979).
4. T. L. Ladd, M. G. Klein and J. H. Tumlinson, Phenethyl propionate + eugenol + geraniol (3:7:3) and japonilure: a highly effective joint lure for Japanese beetles, *J. Econ. Entomol.*, 74:665-667 (1981).
5. J. H. Visser, Host odor perception in phytophagous insects, *Ann. Rev. Entomol.*, 31:121-144 (1986).
6. P. J. Silk, S. H. Tan, C. J. Wiesner, R. J. Ross and G. C. Lonergan, Sex pheromone chemistry of the eastern spruce budworm, *Choristoneura fumiferana*, *Environ. Entomol.*, 9:640-644 (1980).
7. J. H. Visser and D. A. Avé, General green leaf volatiles in the olfactory orientation of the Colorado beetle, *Leptinotarsa decemlineata*, *Ent. Exp. et Appl.*, 24:738-749 (1978).
8. M. M. Blight, L. J. Wadhams and M. J. Wenham, Volatiles associated with unmated *Scolytus scolytus* beetles on English elm: differential production of α-multistriatin and 4-methyl-3-heptanol, and their activities in a laboratory bioassay, *Insect Biochem.*, 8:135-142 (1978).
9. M. M. Blight and L. J. Wadhams, Male-produced aggregation pheromone in pea and bean weevil, *Sitona lineatus* (L.), *J. Chem. Ecol.*, 13:733-739 (1987).
10. M. M. Blight, J. A. Pickett, M. C. Smith and L. J. Wadhams, An aggregation pheromone of *Sitona lineatus*, *Naturwissenschaften*, 71:480-481 (1984).

11. M. A. Golub and I. Weatherston, Techniques for extracting and collecting sex pheromones from live insects and from artificial sources, *in*: "Techniques in Pheromone Research", H. E. Hummel and T. A. Miller, eds., Springer-Verlag, New York, pp 223-285 (1984).

THE USE OF COUPLED GAS CHROMATOGRAPHY: ELECTROPHYSIOLOGICAL TECHNIQUES IN THE IDENTIFICATION OF INSECT PHEROMONES

L. J. Wadhams

AFRC Institute of Arable Crops Research
Rothamsted Experimental Station
Harpenden, Hertfordshire AL5 2JQ, UK

SUMMARY

Use of coupled GC-electrophysiological techniques in the identification of semiochemicals is described for the cabbage stem flea beetle, *Psylliodes chrysocephala*, the European elm bark beetle, *Scolytus scolytus*, and the aphids *Lipaphis erysimi* and *Aphis fabae*.

INTRODUCTION

Semiochemicals, and phermones in particular, have assumed an increasingly important role in integrated pest management strategies by improving the efficacy of insecticides and reducing their environmental impact. However, the isolation and identification of pheromones poses many problems, not least in that they generally occur in trace quantities accompanied by large amounts of biologically inactive materials. Although gas chromatography (GC) is a powerful tool for the analysis of complex natural products, additional techniques are required to locate the biologically active components in the chromatograms. Electroantennogram (EAG) recordings from insect antennae have been used as an assay method for monitoring the activity of fractions collected from the gas chromatograph [1], but this does not take full advantage of the GC resolution capabilities. Direct coupling of the EAG and GC, so that the column effluent is continuously monitored by both GC and EAG detectors, provides an extremely sensitive detection system which has proved invaluable in the identification of a wide range of insect pheromones [2-5]. The EAG technique, however, has limitations in its use as a specific GC detector and this has led to the development of coupled systems based on recordings from individual olfactory cells [6,7]. The advantages and disadvantages of these systems are discussed and their use in the isolation and identification of semiochemicals from Coleoptera and aphids is described.

MATERIALS AND METHODS

A typical coupled GC electrophysiological system is shown in Figure 1. The capillary column effluent is split equally between the flame ionization detector (FID) and the transfer line to the insect preparation by a low volume splitter constructed from glass-lined stainless-steel tubing and deactivated fused silica tubing. One length of tubing leads to the FID whilst the other is taken through the oven wall and directs the remainder of the effluent into a purified airflow passing continuously over the antennal preparation. Components eluting from the capillary column are monitored simultaneously by both FID and insect detectors. N_2 can be added as a make-up gas to reduce residence time in the transfer lines and to maximize FID response.

Chromatography and Isolation of Insect Hormones and Pheromones
Edited by A. R. McCaffery and I. D. Wilson
Plenum Press, New York, 1990

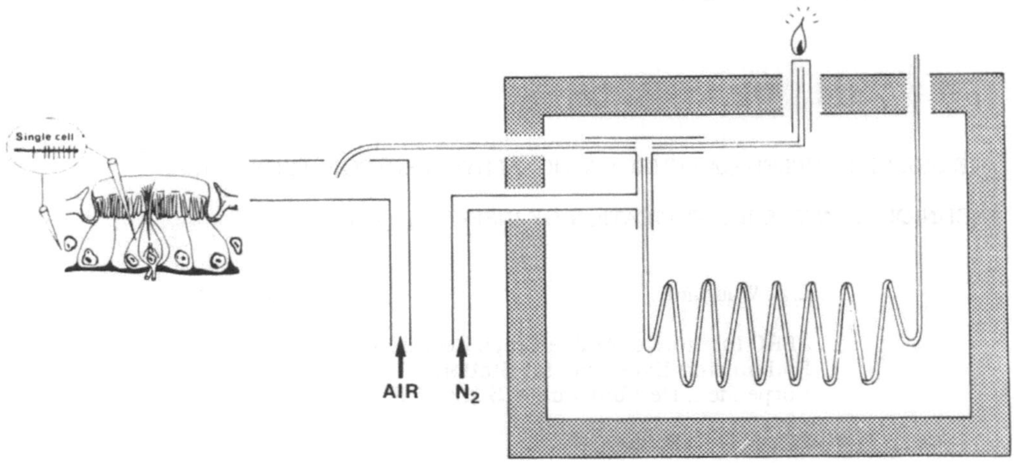

Fig. 1. Diagram of a typical coupled GC electrophysiological system.

Electrophysiology

EAG Recordings: These were made using Ag-AgCl glass electrodes filled with saline (composition as in Maddrell, 1969 [8], but without glucose). The recording electrode was positioned in the distal end of the antenna and the indifferent electrode at the base.

Single-Cell Recordings (SCR): These were made from individual olfactory cells using tungsten microelectrodes [9]. The indifferent electrode was placed in the first antennal segment for aphids, and in either the scape or mouthparts for Coleoptera, and the recording electrode was then brought into contact with the surface of the antenna until impulses were recorded.

The EAG and SCR signals were amplified and recorded by standard methods [10].

Stimulation: The stimulus was delivered into a purified airstream which flowed continuously over the preparation. The delivery system and stimulation protocol have been described previously [10].

RESULTS AND DISCUSSION

The EAG technique was pioneered by Schneider [11], who suggested that the response was essentially the sum of many receptor potentials. As such, it appears to be the ideal detector with which to monitor the GC effluent, since eluting components will be exposed simultaneously to all the receptors on the antenna and in theory, all biologically active compounds in an extract can be located with just one analysis. The application of coupled GC-EAG to the analysis of lepidopteran sex pheromones is particularly useful because of the magnitude of the EAG response and the ease with which the antennae can be prepared [12]. The technique has also proved effective in the isolation and identification of coleopteran pheromones [4,5].

The ability of the coupled GC-EAG to locate biologically active components in complex extracts is demonstrated by the response of the cabbage stem flea beetle, *Psylliodes chrysocephala*, to volatiles entrained from oilseed rape (Figure 2): 10 potential semiochemicals were located in a single analytical run. Although the 3-butenyl (I) and 4-pentenyl (II) isothiocyanates elicited strong responses, many of the other physiologically active compounds showed only weak activity. Such variations in the EAG response to different components can impose severe constraints on the use of the EAG as a specific GC detector. In the identification of lepidopteran sex pheromones, this can in part be overcome by employing the antennae of other moth species to detect minor pheromone components in the extracts [13]. However, this compound-specific detector technique relies on the

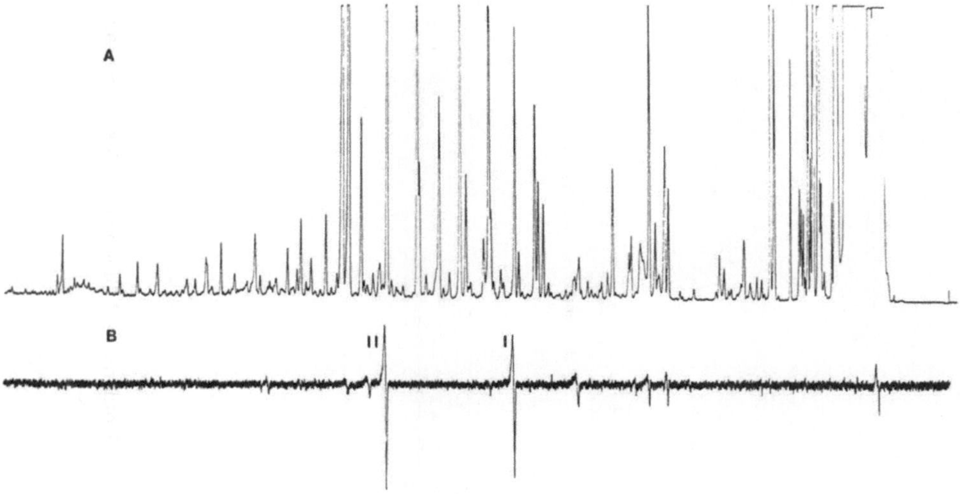

Fig. 2. *P. chrysocephala* coupled GC-EAG. (A) gas chromatogram of oilseed rape volatiles and (B) corresponding EAG response.

generality of sex pheromone components in this Order and is not equally applicable to other Orders of insect.

I

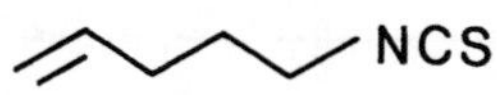

II

The aggregation pheromone of the elm bark beetle, *Scolytus scolytus*, comprises a complex of insect and host produced metabolites. In air entrainment extracts of beetles boring into elm bolts, coupled GC-EAG showed electrophysiological activity corresponding to the beetle metabolites (-)-*threo* and (-)-*erythro*-4-methyl-3-heptanol (III,IV) and (-)-α-multistriatin (V). No EAG response was observed for the host metabolite (-)-α-cubebene (VI), although olfactory cells were found on the antenna that responded specifically to this compound [7]. α-Cubebene also synergized the activity of 4-methyl-3-heptanol in the field [14].

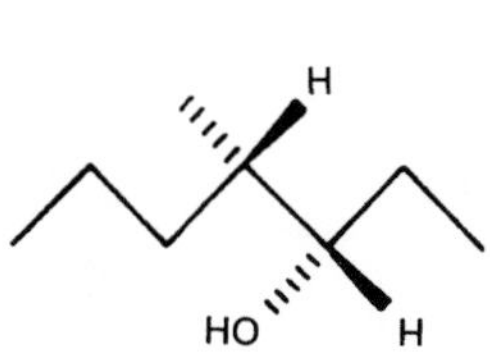

III

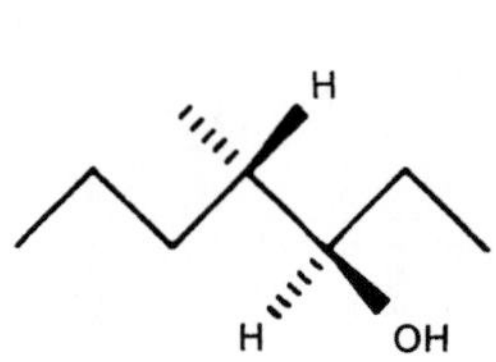

IV

V

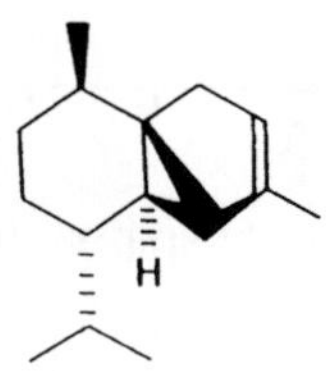

VI

In *Scolytus*, the peripheral receptors are specialized for response to different components of the aggregation pheromone. Although the EAG and SCR threshold

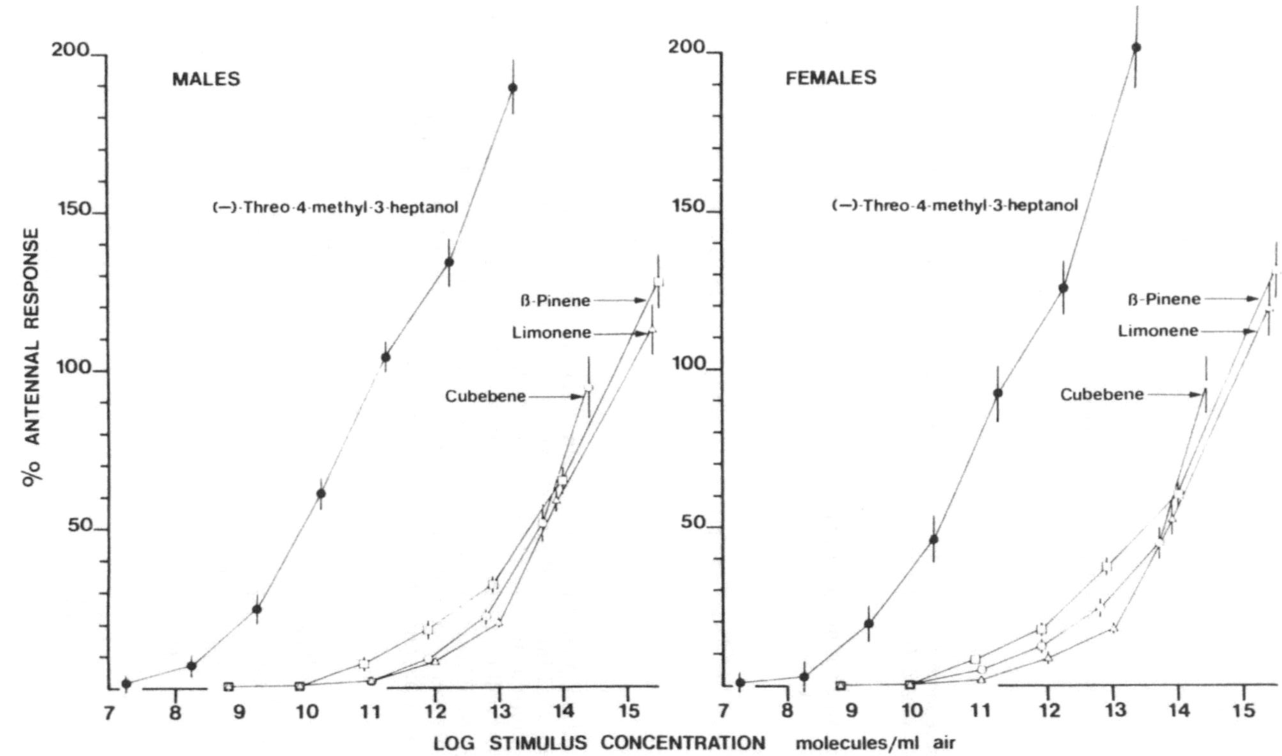

Fig. 3. EAG responses of male and female *S. scolytus* to (-)-*threo*-4-methyl-3-heptanol, (-)-α–cubebene, (-)-β-pinene and limonene. Responses are normalized with respect to a (±)-*threo*-4-methyl-3-heptanol standard (for details, see Ref. [10]).

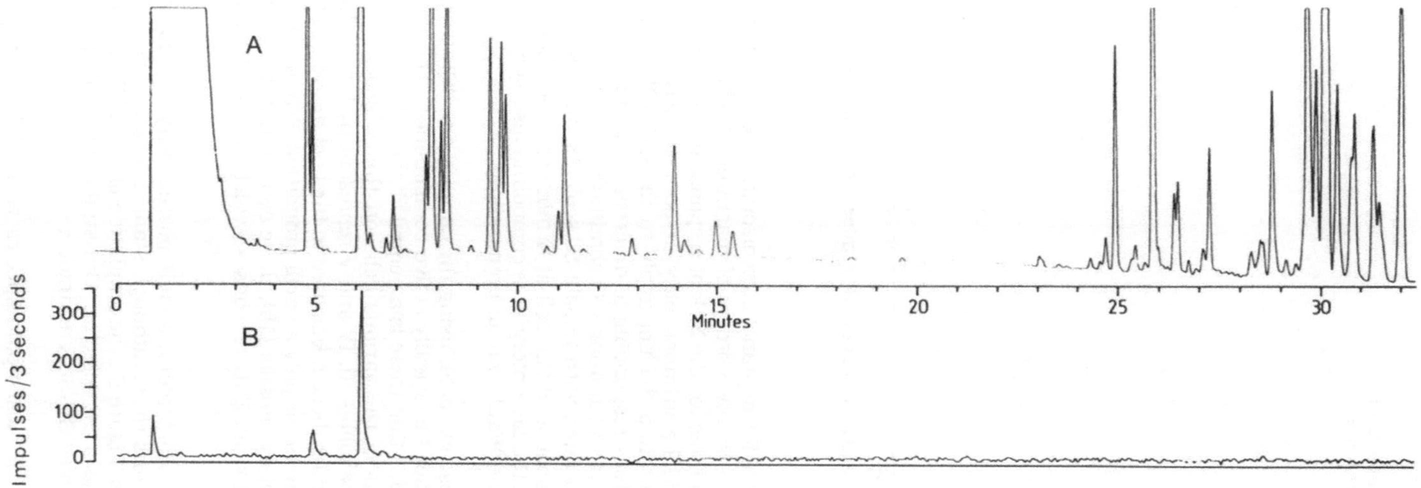

Fig. 4. *S. scolytus* coupled GC-SCR. (A) gas chromatogram of the volatiles from female *S. scolytus* boring into English elm and (B) corresponding action potential frequency from a cell specialized to β-pinene.

Table 1. Proportion of olfactory cells on the antenna of *Scolytus scolytus* responding to different compounds.

Key compound for cell*	%
Insect derived	
(-)-*erythro*-4-methyl-3-heptanol	24.2
(-)-*threo*-4-methyl-3-heptanol	18.9
(-)-α-multistriatin	9.9
4-methyl-3-heptanone	5.3
Host derived	
(-)-β-pinene	5.5
(-)-α-cubebene	3.7
unknown	5.3
"	1.8
"	1.1
"	1.1
"	0.9
"	0.7
"	0.4
"	0.2
Other	
Mechanoreceptors	6.1
Uncharacterised cells	14.8

Total number of cells recorded - 545.
* Where the identity of the key compound was not determined, the cell was characterised from the GC-SCR data.

concentrations for the 4-methyl-3-heptanol stereoisomers were similar, for α-cubebene the EAG threshold concentration was two orders of magnitude higher (Figure 3). The relative insensitivity of the EAG accounted for the lack of response to α-cubebene in the coupled GC-EAG, since responses could be attained only at concentrations which overloaded the capillary column. This limitation did not apply to the GC-SCR system, since threshold concentrations of olfactory cells responding to different compounds were similar and could readily be achieved under high resolution GC conditions. A coupled GC-SCR trace of an extract of the volatiles associated with female *S. scolytus* infestations on English elm is shown in Figure 4. The response of the cell at 6 min corresponds to the elution of (-)-β-pinene from the column. In the laboratory, this compound also synergized the activity of 4-methyl-3-heptanol (Blight and Wadhams, in preparation).

Further analysis of these extracts, using the coupled GC-SCR technique, revealed the presence of a number of other biologically active components. The data from 545 cells is given in Table 1 and, assuming that there is an equal chance of recording from any cell type, must reflect the proportions of the various cells on the *S. scolytus* antenna. The cells specialized to the beetle metabolites III, IV and V represent by far the highest proportion of receptors on the antenna, whilst those specialized to the host derived compounds, such as α-cubebene and β-pinene, account for only a small proportion. Thus, despite the fact that the antenna has 2 to 3000 olfactory sensilla [15], these compounds are perceived by only a few cells and consequently elicit weak EAG responses, which are not detected with the coupled GC-EAG technique.

Unlike Coleoptera and Lepidoptera, aphids possess very limited numbers of olfactory cells on their antennae [16] and thus, although stable preparations can be obtained which will survive for several hours (Figure 5), the amplitude of the EAG response is so small as to render it unsuitable for use as a GC detector. Our studies on aphid pheromones have therefore concentrated on the coupled GC-SCR technique.

The alarm pheromone of most aphid species is (*E*)-β-farnesene (VII) [17,18] and is released when aphids are attacked by predators, causing neighboring aphids to cease feeding and move away rapidly. Although the turnip aphid, *Lipaphis erysimi*, produces VII as an alarm pheromone component, it did not respond well to this compound when presented at levels comparable to that produced by the aphid. An extract of volatiles obtained from *L. erysimi* did, however, give rise to a good alarm response.

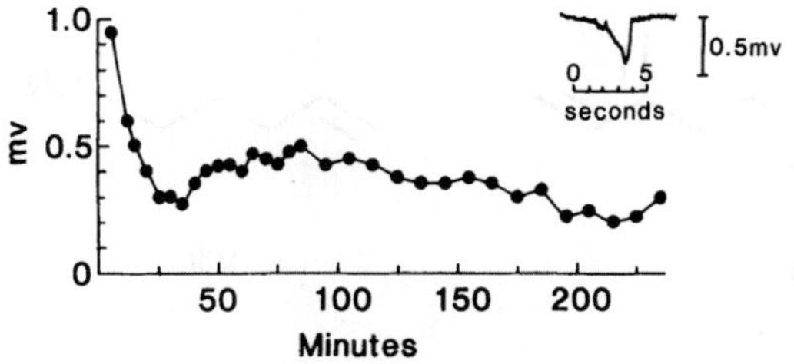

Fig. 5. EAG responses of an alate *Myzus persicae* virginopara to repeated stimulation with 10 ng (*E*)-β-farnesene.

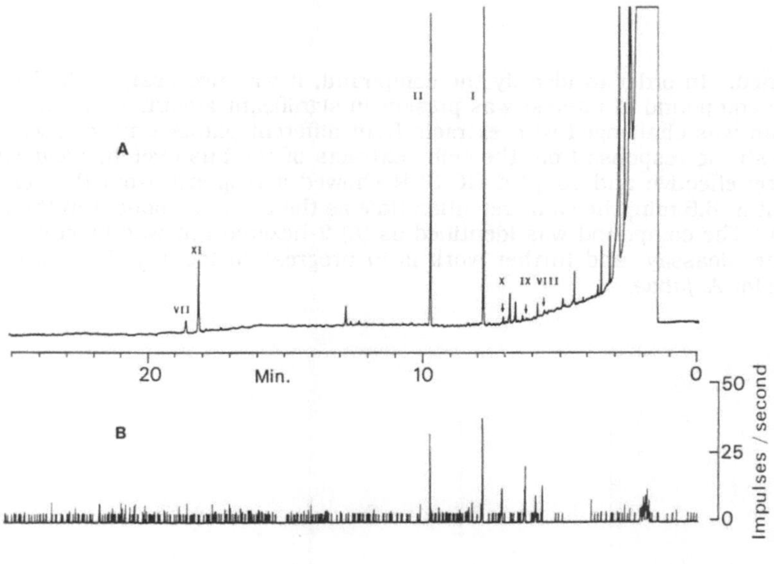

Fig. 6. *L. erysimi* coupled GC-SCR. (A) gas chromatogram of *L. erysimi* volatiles and (B) corresponding action potential frequency of an olfactory cell from the proximal rhinarium.

Cells on the primary rhinarium on the fifth (proximal) antennal segment responded strongly to the *L. erysimi* volatiles and coupled GC-SCR of the extract revealed the presence of a number of biologically active components, all of which were perceived by one cell (Figure 6). These compounds were identified as 4-pentenyl cyanide (VIII), allyl isothiocyanate (IX), 2-butyl isothiocyanate (X), 3-butenyl isothiocyanate (I) and 4-pentenyl isothiocyanate (II). The olfactory receptor exhibited considerable specificity for these compounds and no response was elicited from 2-phenylethyl isothiocyanate (XI). Dose response data for these compounds showed that the receptors were maximally responsive to allyl and 3-butenyl isothiocyanates and although these compounds were inactive on their own in the alarm pheromone bioassay, when they were applied together with VII, all the aphids responded [19]. Isothiocyanates are well-known components of cruciferous plants and are derived from enzymic decomposition of glucosinolate precursors [20]. These components were therefore considered to have arisen from the host plant.

The existence of host-derived alarm pheromone synergists for other aphid species has also been examined. Studies on the black bean aphid, *Aphis fabae*, have shown that the alarm pheromone is a multicomponent mixture and that, as with *L. erysimi*, synergists derived from the host plant are involved. SCRs from the rhinarium on the fifth antennal segment again showed the presence of olfactory cells responding to host plant volatiles and coupled GC-SCR of the bean extract revealed the presence of three electrophysiologically active components [21]. However, the major activity was associated with a minor component on the trailing edge of a major polar peak, so characteristic mass spectra were

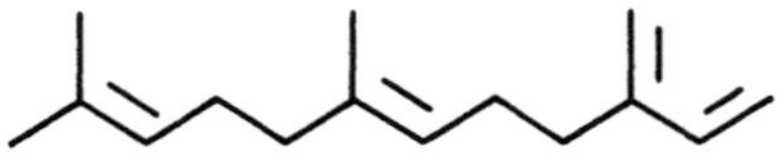

VII

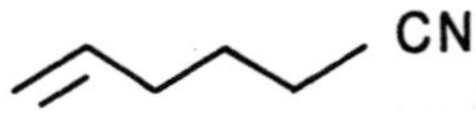

VIII

IX

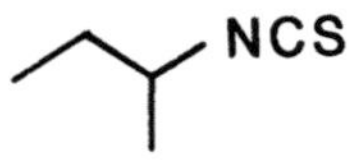

X

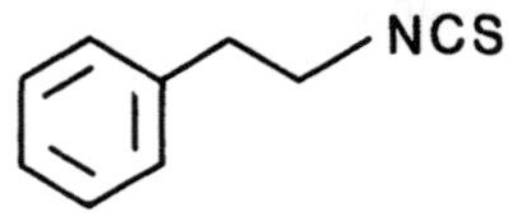

XI

not obtained. In order to identify the compound, it was necessary to find an extract in which the compound of interest was present in significant amounts. To achieve this, the preparation was challenged with extracts from different plants until one was found that elicited a strong response from the cell. Extracts of the bush vetch, *Vicia sepium*, were particularly effective and coupled GC-SCR showed a response from the cell to a major component at 6.5 min, the same retention time as the trace component in the bean extract (Figure 7). The compound was identified as (*E*)-2-hexenal but was inactive in the alarm pheromone bioassay, and further work is in progress to identify the alarm pheromone synergists for *A. fabae*.

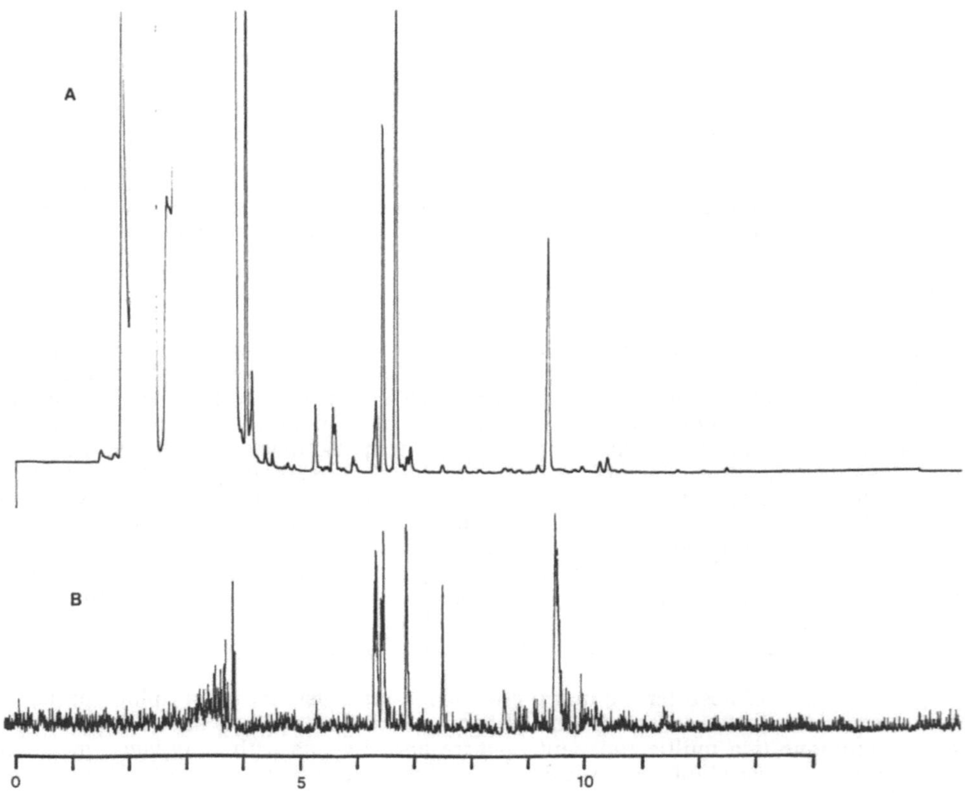

Fig. 7. *A. fabae* coupled GC-SCR. (A) gas chromatogram of *Vicia sepium* volatiles and (B) corresponding action potential frequency from a number of cells on the proximal rhinarium.

The use of olfactory cells to screen for extracts containing significant amounts of biologically active compounds can be particularly useful in the elucidation of insect-host plant interactions, where activity is frequently associated with trace components. However, this approach is not practical with the EAG technique since the extracts also contain a number of other active compounds and it is not possible, without recourse to a full GC-EAG analysis, to determine whether the EAG activity associated with the extract is due to the compound of interest or to another component.

Coupled GC-electrophysiological techniques provide rapid and precise information on the retention times of biologically active compounds in complex natural product extracts, enabling pheromone identification to be readily achieved even when no discernible GC-MS peaks are present (Pickett, this volume). The EAG offers a simple detector system that monitors the column effluent with the full range of olfactory receptors possessed by the insect, although it may lack sensitivity for certain pheromone components. In contrast, the SCR technique is highly sensitive to all compounds but provides information from only a limited number of receptors at one time. In practice, the choice of technique is often determined by the insect's antennal morphology.

REFERENCES

1. W. Roelofs, Electroantennogram assays: rapid and convenient screening procedures for pheromones, *in:* "Techniques in Pheromone Research", H. E. Hummel and T. A. Miller, eds., Springer, New York, pp 131-159 (1984).
2. H. Arn, E. Städler and S. Rauscher, The electroantennographic detector - a selective and sensitive tool in the gas chromatographic analysis of insect pheromones, *Zeitschrift für Naturforschung*, 30C:722-725 (1975).
3. P. M. Guerin, E. Städler and H. R. Buser, Identification of host plant attractants for the carrot fly, *Psila rosae*, *J. Chem. Ecol.*, 9:843-861 (1983).
4. M. M. Blight, J. A. Pickett, M. C. Smith and L. J. Wadhams, An aggregation pheromone of *Sitona lineatus*. Identification and initial field studies, *Naturwissenschaften*, 71:480-481 (1984).
5. P. R. White and M. C. Birch, Female sex pheromone of the common furniture beetle *Anobium punctatum* (Coleoptera: Anobiidae): extraction, identification and bioassays, *J. Chem. Ecol.*, 13:1695-1706 (1987).
6. J. Van Der Pers and C. Löfstedt, Continuous single sensillum recording as a detection method for moth pheromone components in the effluent of a gas chromatograph, *Physiol. Entomol.*, 8:203-211 (1983).
7. L. J. Wadhams, Coupled gas chromatography - single cell recording: a new technique for use in the analysis of insect pheromones, *Zeitschrift für Naturforschung*, 37C:947-952 (1982).
8. S. H. P. Maddrell, Secretion by the Malpighian tubules of *Rhodnius*. The movement of ions and water, *J. Exp. Biol.*, 51:71-97 (1969).
9. J. Boeckh, Electrophysiologische untersuchungen an einzelnen Geruchs-Rezeptoren auf den antennen des Totengräbers (Necrophorus: Coleoptera), *Zeitschrift für Vergleichende Physiologie*, 46:212-248 (1962).
10. L. J. Wadhams, M. E. Angst and M. M. Blight, Responses of the olfactory receptors of *Scolytus scolytus* (F) (Coleoptera: Scolytidae) to the stereoisomers of 4-methyl-3-heptanol, *J. Chem. Ecol.*, 8:477-492 (1982).
11. D. Schneider, Electrophysiologische untersuchungen von chemo- und mechanorezeptoren der antenne des Seidenspinners *Bombyx mori* L., *Zeitschrift für Vergleichende Physiologie*, 40:8-41 (1957).
12. P. Guerin, H. Arn and H. R. Buser, Characterization of some tortricid sex pheromones using the insect antenna as a gas chromatographic detector, *in:* "Semiochemistry: Flavors and Pheromones", T. E. Acree and D. M. Soderlund, eds., Walter de Gruyter & Co., Berlin, pp 239-250 (1985).
13. D. L. Struble, H. R. Buser, H. Arn and G. E. Swailes, Identification of sex pheromone components of redbacked cutworm, *Euxoa ochrogaster*, and modification of sex attractant blend for adult males, *J. Chem. Ecol.*, 6:573-584 (1980).
14. M. M. Blight, C. J. King, L. J. Wadhams and M. J. Wenham, Attraction of *Scolytus scolytus* (F) to the components of multilure, the aggregation pheromone of *S. multistriatus* (Marsham) (Coleoptera: Scolytidae), *Experientia*, 34:1119-1120 (1978).
15. N. C. Henderson and L. J. Wadhams, The morphology of the antennal club of *Scolytus scolytus* (F) (Coleoptera: Scolytidae), *Zeitschrift für Angewandte Entomologie*, 92:477-487 (1981).

16. A. K. Bromley, J. A. Dunn and M. Anderson, Ultrastructure of the antennal sensilla of aphids. I. Coeloconic and placoid sensilla, *Cell & Tissue Res.*, 203:427-442 (1979).
17. W. S. Bowers, L. R. Nault, R. E. Webb and S. R. Dutky, Aphid alarm pheromone: isolation, identification, synthesis, *Science*, 177:1121-1122 (1972).
18. J. A. Pickett and D. C. Griffiths, Composition of aphid alarm pheromones, *J. Chem. Ecol.* 6:349-360 (1980).
19. G. W. Dawson, D. C. Griffiths, J. A. Pickett, L. J. Wadhams and C. M. Woodcock, Plant-derived synergists of alarm pheromone from turnip aphid, *Lipaphis (Hyadaphis) erysimi* (Homoptera, Aphididae), *J. Chem. Ecol.*, 13:1663-1671 (1987).
20. H.-Y. Ju, C. Chong, W. J. Mullin and B. B. Bible, Volatile isothiocyanates and nitriles from glucosinolates in rutabaga and turnip, *J. Amer. Soc. Horticul. Sci.*, 107:1050-1054 (1982).
21. G. W. Dawson, D. C. Griffiths, J. A. Pickett, L. J. Wadhams and C. M. Woodcock, Plant compounds that synergize activity of the aphid alarm pheromone, 1986 British Crop Protection Conference - Pests and Diseases, pp 829-834 (1986).

GAS CHROMATOGRAPHY-MASS SPECTROMETRY IN INSECT PHEROMONE IDENTIFICATION: THREE EXTREME CASE HISTORIES

J. A. Pickett

AFRC Institute of Arable Crops Research
Rothamsted Experimental Station
Harpenden, Hertfordshire AL5 2JQ, UK

SUMMARY

Use of modern gas chromatography-mass spectrometry (GC-MS) for the identification of insect pheromones, using a minimum of insect material, is described for three extreme cases: (i) a mosquito oviposition pheromone: a GC-MS peak but no evidence of activity; (ii) the *Sitona lineatus* aggregation pheromone: no GC-MS peaks in active region; (iii) aphid sex pheromones: GC-MS peaks associated with activity - the ideal situation.

INTRODUCTION

Gas chromatography coupled with mass spectrometry (GC-MS) is a powerful tool for the identification of insect pheromones. GC-MS has played an important role in the identification of an oviposition pheromone for various mosquitoes [1,2], the *Sitona lineatus* aggregation pheromone [3] and aphid sex pheromones [4-6]. Here we describe in detail the GC-MS studies that led to these identifications, which represent three extreme cases where GC-MS has been used to identify insect pheromones.

MATERIALS AND METHODS

Biological extracts were stored at -20°C in glass ampoules sealed under nitrogen, prior to GC-MS analysis.

A Mosquito Oviposition Pheromone

Apical droplets from eggs of *Culex quinquefasciatus*, reared as described previously [7], were collected using a fine glass rod, dissolved in hexane and analyzed by GC-MS. GC was performed using a fused SiO_2 capillary column, 25 m x 0.2 mm, OV101, starting at 50°C (10 min), then 4°/min to 200°C (isothermal); samples were injected into a section of heated (150°C) silanized glass tubing on the front of the column. MS was by electron impact, 70 eV, 200°C or chemical ionization, isobutane (VG Micromass 70-70F with data system); accurate mass at low resolution was as described by Powers et al. [8]. Purification was by thin-layer chromatography (TLC) on silica gel 60, 0.25 mm, with ether as eluent. Components were detected by treating part of the plate with iodine vapor and analyzing extracts from the other part of the plate by GC-MS. The putative pheromone was extracted at R_f 0.39 (17 μg from 50 egg rafts). Portions were treated with various reagents under standard conditions, but within fine test tubes made by drawing out Pasteur pipettes, and the products again analyzed by GC-MS.

The Sitona lineatus *Aggregation Pheromone*

Volatiles, obtained by air entrainment as described previously [3], were analyzed by GC-MS as above, but the programme was 40°C (10 min), then 2°/min to 125°C (isothermal). A similar column was employed for GC-coupled electrophysiological studies [3].

Biological and chemical samples, obtained as described previously [4-6], were analyzed by GC-MS. GC was performed on a fused SiO_2 capillary column, 50 m x 0.32 mm, OV101 bonded, starting at 30°C (2 min), and then 3 /min to 120°C (isothermal), using on-column injection. MS was by electron impact at 70 eV, 230°C (VG Analytical 70-250 integrated data system). A similar column was employed for GC-coupled electrophysiological studies.

RESULTS AND DISCUSSION

A Mosquito Oviposition Pheromone

Species of mosquito in the genus Culex are vectors for a number of diseases. Individual females of these mosquitoes lay their eggs in groups to form egg-rafts and it has been shown that gravid females preferentially oviposit on water where egg-rafts have already been laid, in response to a pheromone produced by the eggs [9]. Although it had also been demonstrated that the pheromone was to some extent volatile, early identification work by other groups concentrated on the largely involatile lipid components of the egg apical droplets from which the pheromone is released. However, by analyzing apical droplets from *Culex quinquefasciatus* using GC, and by allowing the temperature gradient to finish with a long isothermal period at 200°C, it was possible to see a large distorted GC peak (A, Figure 1). The mass spectrum obtained for this peak (Figure 2) showed interesting ions at m/z 252 (3%), 142 (36%), 99 (100%). A very low intensity ion at 312 (0.1%) appeared to be the parent ion (M^+). This was confirmed by chemical ionization GC-MS, which showed an M^+ + 1 ion at m/z 313. Although Wadhams (see this volume) has shown the value of coupling GC with electrophysiological recordings in order to demonstrate the activity of compounds giving GC peaks, such techniques were not available for this study. Since it was not known that biological activity was associated with peak A in the chromatogram (Figure 1), it was necessary that the tentative identification from the GC-MS analysis was substantiated before synthesis and biological testing.

Assuming a molecular weight of 312, loss of 60 mass units to give the fragment ion at m/z 252 is likely to have been caused by loss of acetic acid from an acetate. The ion at m/z 99 could have arisen from a δ-lactone or a methyl-substituted γ-lactone, but it was not possible to suggest how the abundant rearrangement ion at m/z 142 would have occurred. It was conjectured that proximity of the acetate group to the lactone may have allowed the acetyl (CH_3CO) group to migrate onto the lactone fragment (i.e., m/z 99), thus creating the ion with m/z 142. The mass spectrum also showed no evidence of branching anywhere in the molecule. From these considerations, structure (I) was proposed, i.e., the δ-lactone structure. Initially, only small samples were available (about 2 nanograms of compound per egg), so accurate mass determination at high resolution was not possible, but semi-accurate mass data were obtained by measuring the mass at relatively low resolution (and therefore high sensitivity), using tetraiodoethylene as an internal standard (Table 1).

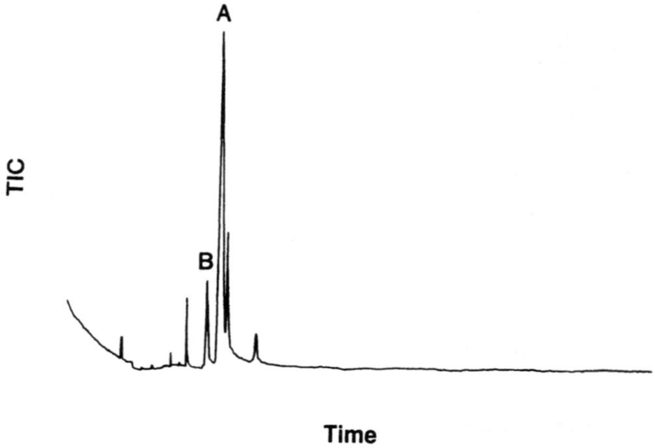

Fig. 1. GC-MS of apical droplet from mosquito egg, total ion current (TIC).

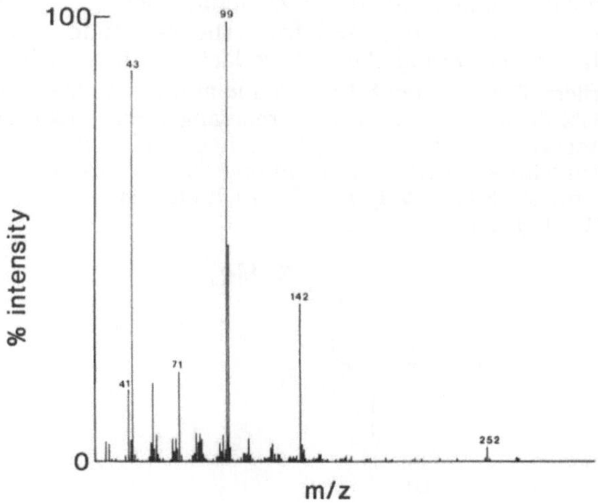

Fig. 2. MS from peak A in GC-MS of the mosquito egg apical droplet.

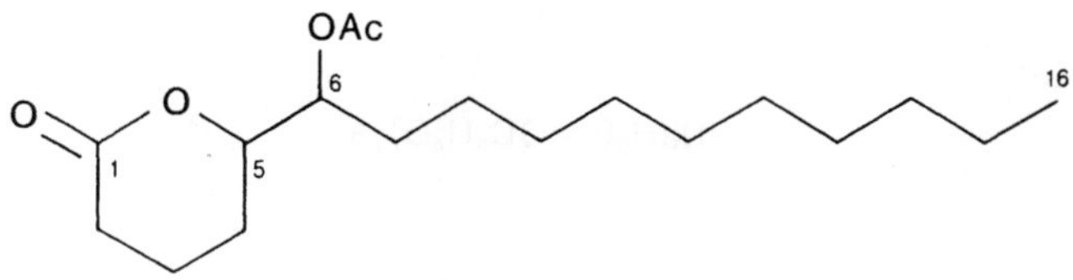

I

Table 1. Accurate mass data obtained at low resolution from peak A in the GC-MS of the mosquito egg apical droplet

Formula	$C_{18}H_{32}O_4$	$C_{16}H_{28}O_2$	$C_7H_{10}O_3$	$C_5H_8O_2$	$C_5H_7O_2$
m/z required	312.2300	252.2089	142.0630	100.0524	99.0446
m/z found	312.2009	252.1992	142.0621	100.0474	99.0432

This confirmed the earlier proposals regarding composition of the main fragment ions, and reaffirmed structure (I) as a strong possibility. To confirm structure (I), a larger biological sample was obtained and purified by TLC, monitoring by GC-MS of samples eluted from the plate. This gave microgram amounts of the putative pheromone, which could be examined by microchemical techniques. The first method involved hydrolysis with 2M NaOH, which on acidification gave a new compound eluting at position B on the pheromone chromatogram (Figure 1). The mass spectrum showed M^+ at m/z 270 (0.05%) and diagnostic ions at m/z 252 (0.5%), 100 (100%) and 99 (25%). The M^+ at m/z 270 was consistent with the hydroxylactone (II), m/z 252 arising from loss of water (i.e., M^+-18).

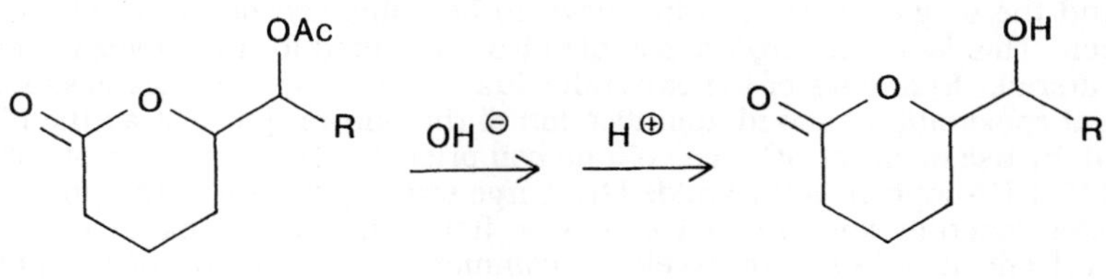

I II

The rearrangement fragment at m/z 100 would have been formed by H replacing acetyl, m/z 99 would have again arisen from the δ-lactone. Treatment with *N,O*-bistrimethylsilylacetamide converted the hydroxylactone (II) into the trimethylsilyl ether (III), which eluted earlier. The expected M^+ was found at m/z 342 (3%) and the presence of an intense ion at m/z 172 (59%) again arose from rearrangement, this time with trimethylsilyl replacing acetyl. Acetylation of the proposed hydroxylactone (II) gave the original compound. Reduction with a hydride soluble in organic solvents gave the 1,5,6-trihydroxy-hexadecane (IV) with m/z 238 (M^+ - $2H_2O$, 2%), which on acetylation was converted into the triacetate, m/z 341 (M^+ - CH_3CO_2, 5%).

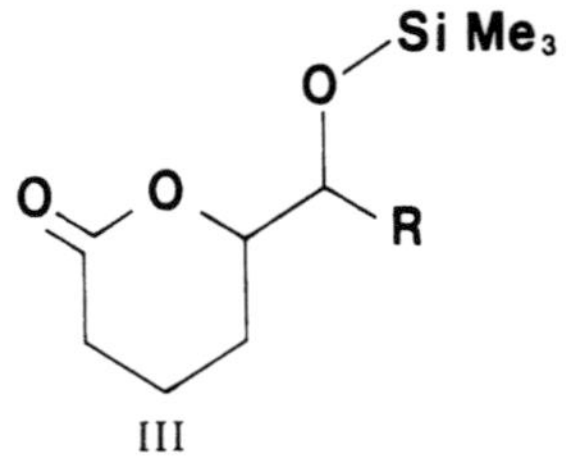

III

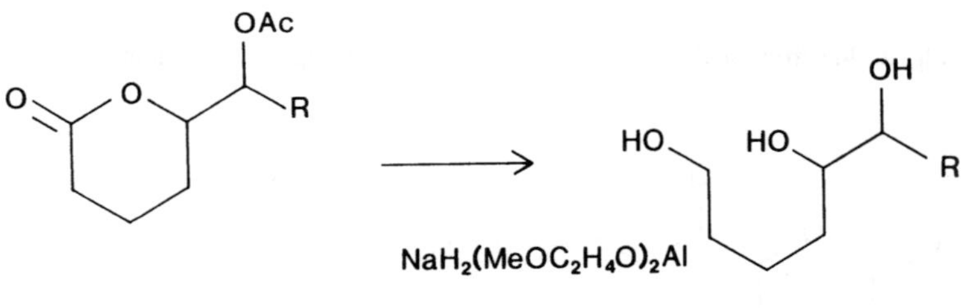

I IV

The above chemistry is consistent with structure (I) for the *Culex quinquefasciatus* pheromone. An extensive search was therefore made of the literature for a precedent for the unusual rearrangement reaction, in order to make final confirmation of this structure. A precedent was found in which acyclic α-glycol diesters were shown to eliminate aldehydes and ketones on mass spectrometry [10]. Using the mechanism proposed for this, the formation of the even-mass fragment from the mosquito pheromone (I) and from the derivatives (II) and (III) could be explained (Figure 3).

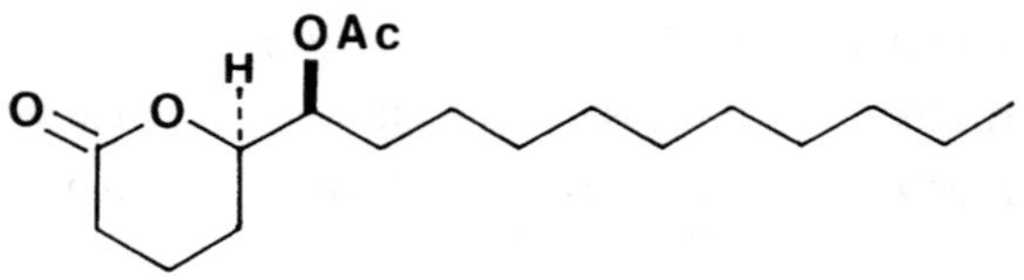

Ia

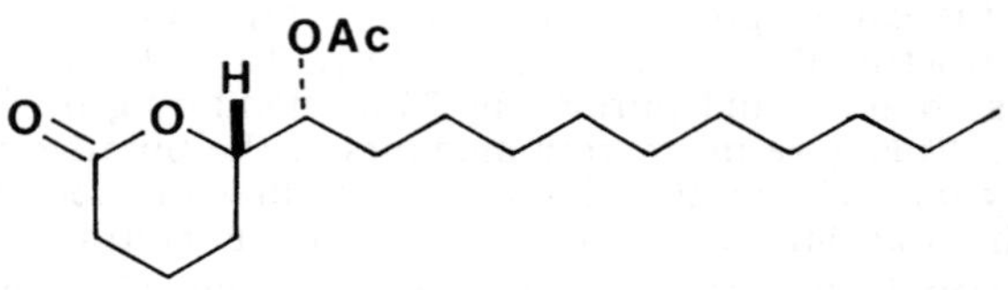

Ib

Structure (I) contains two asymmetric carbons, at positions 5 and 6, which give rise to four isomeric possibilities. A mixture of the two *erythro* enantiomers (Ia and b) was synthesized and the single GC peak was shown to be enhanced on co-injection with the natural product. This synthetic *erythro* sample also accounted for the biological activity of the natural material. Synthesis of the two individual isomers by the Sharpless kinetically controlled chiral epoxidation allowed identification of the natural product as the (-)-(5*R*,6*S*) enantiomer (Ia), by use of chiral GC on a column comprising *O*-(1*S*,3*S*)-chrysanthemoyl- (*R*)-mandelic acid (*S*-1-(1'-naphthyl)ethylamide [2]. Large scale synthesis of the pheromone via a Baeyer-Villiger reaction has enabled extensive field trials in Africa to be conducted successfully [11] and it is hoped to develop a commercial formulation of the pheromone combined with a juvenoid larvicide, which will provide a safe and efficient method for

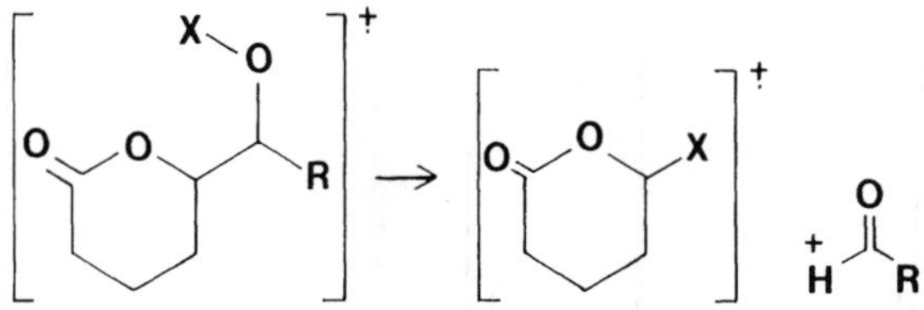

I ,X = AcO	$\frac{m}{z}$ 142,X = OAc
II ,X = X	$\frac{m}{z}$ 100,X = H
III,X = Me_3 Si	$\frac{m}{z}$ 172,X = Si Me_3

Fig. 3. Proposed MS rearrangement based on Sasaki et al. [10].

killing important disease vectors in the *Culex* genus of mosquitoes. Thus, although it must be stressed that electrophysiological recordings combined with GC should be employed wherever possible, in certain cases it is still feasible to identify insect pheromones on the basis of GC-MS studies alone.

The Sitona lineatus *Aggregation Pheromone*

The pea and bean weevil, *Sitona lineatus* (Curculionidae), is a pest of legume crops. The adults notch the leaves, but the larvae cause more important damage by attacking the nitrogen-fixing nodules on the roots. Like many Curculionids, *S. lineatus* produces an aggregation pheromone. This has been identified and synthetic material, together with certain host plant compounds, can be used to trap the adults [3].

GC-coupled electrophysiological studies using antennal preparations showed that volatiles obtained by entrainment from plants on which the insects were feeding contained active material derived from the insects. However, no GC-MS peaks were found, even at maximum sensitivity (Figure 4), in the region of the chromatogram where electrophysiological activity occurred and the search for very small peaks was made more difficult by the fact that volatiles obtained from plants without insects, which showed no activity, were substantially different (Figure 5). The first step was to expand the region where electrophysiologically active peaks were found (Figure 6) and this was repeated for the inactive sample (Figure 7). Peaks A and B were shown to be the same for the two samples and to be from the monoterpene hydrocarbons limonene and (*E*)-β-ocimene. These compounds were used to pinpoint accurately the region in the GC-MS chromatogram where electrophysiological activity was obtained, i.e., between scan number 350 and 360 (Figure 6). Each spectrum between these points was examined and the spectrum at scan number 357 showed some interesting ions, e.g., m/z 86. Ion chromatograms were run for each of these, e.g., Figure 8, and ions at m/z 142, 133, 86 and 57 etc., were shown to peak at scan number 357. Manual background subtraction gave a good spectrum (Figure 9), which suggested a molecular weight of 142. Although this spectrum was not in the library of mass spectra, fragmentation theory suggested that the compound was 4-methyl-3,5-heptanedione (V, Figure 10). This compound can exist both in the keto form (Va) and the enol form (Vb) and it was this isomerization taking place during gas chromatography which caused the GC peak to be so diffuse. Synthesis of the compound allowed confirmation of its identity as the aggregation pheromone [3]. The amount present in the original chromatogram was, at the most, hundreds of picograms. MS ion peaks from this compound were lost in the background, but were available using the MS data system once accurate information on where to look in the chromatogram was obtained from GC-coupled electrophysiological studies. This demonstrates that it is possible, even without a clearly visible GC-MS peak, to make a tentative pheromone identification, provided that GC-coupled electrophysiological recordings are available.

Aphid Sex Pheromones

When the summer or secondary host plant begins to senesce, many species of aphids form winged gynoparae which fly to the primary or winter host, where they give rise to flightless oviparae. It has been shown that, for some species, the oviparae attract winged males by releasing a sex pheromone from their hind tibiae. Although several research

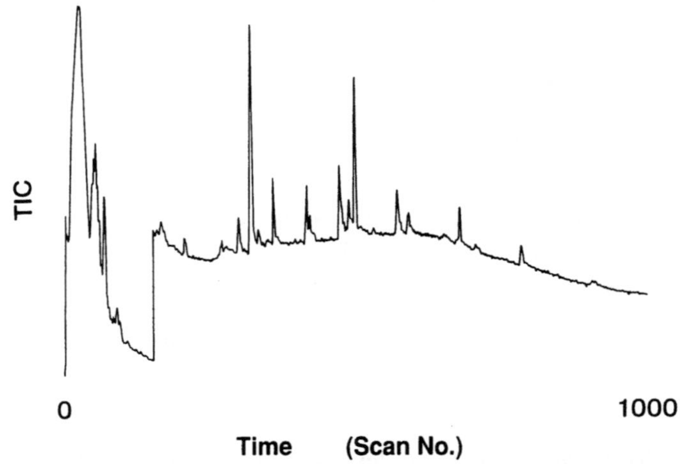

Fig. 4. GC-MS of sample obtained by entrainment from plants with insects (maximum sensitivity, biologically active sample).

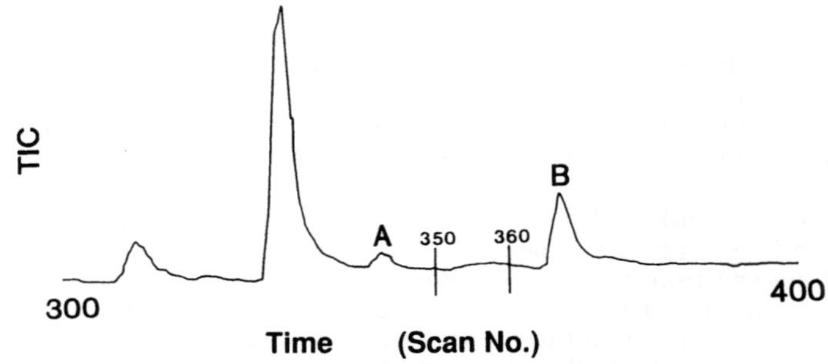

Fig. 5. GC-MS of sample obtained by entrainment from plants only (maximum sensitivity, biologically inactive sample).

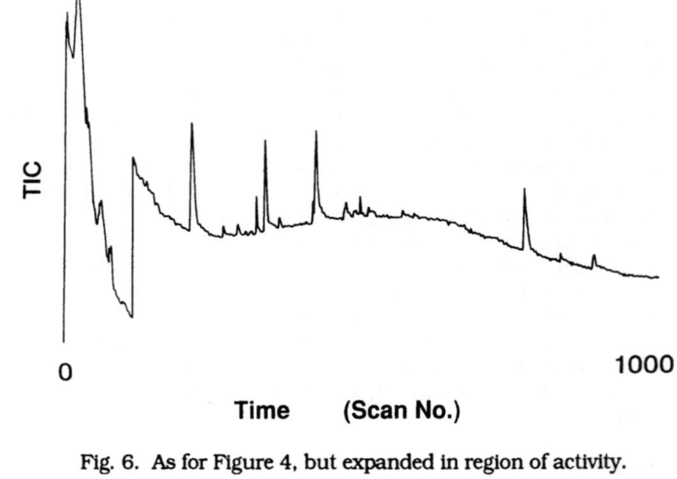

Fig. 6. As for Figure 4, but expanded in region of activity.

groups have put considerable effort into identification of aphid sex pheromones, this has only proved possible by the use of GC-coupled single cell recordings (GC-SCR) from the antennae of male aphids [4].

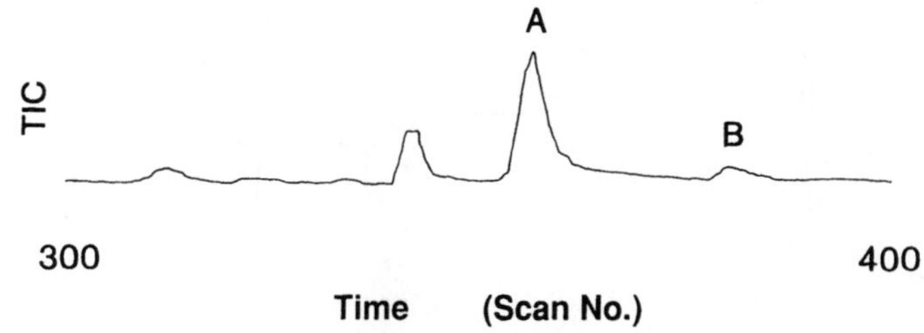

Fig. 7. As for Figure 5, but expanded in region where activity would be.

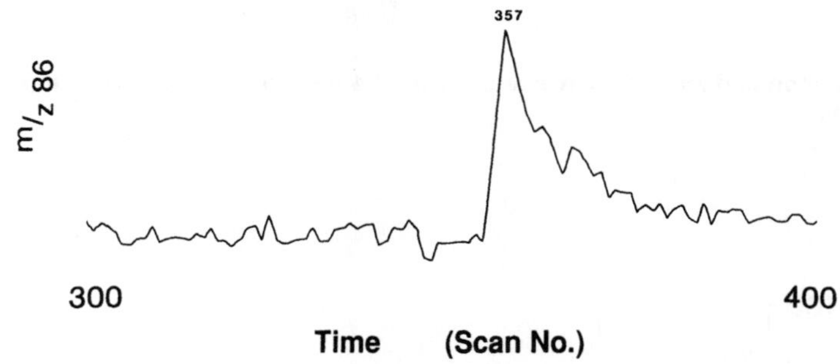

Fig. 8. Ion chromatogram for m/z 86 from biologically active region, i.e., as for Figure 6.

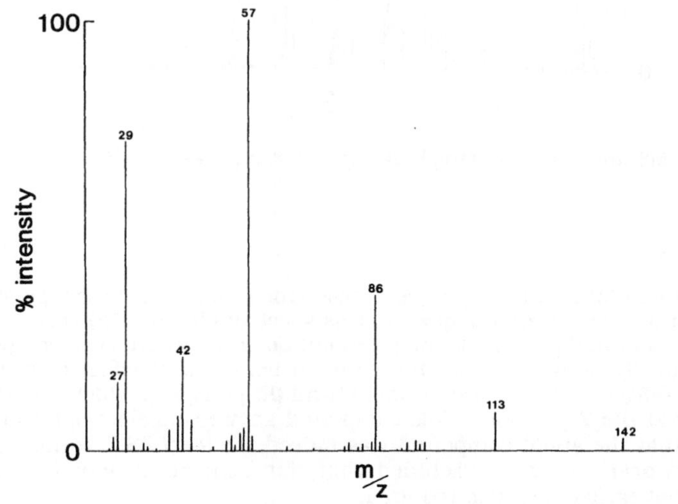

Fig. 9. Spectrum at scan number 357 after manual background subtraction.

Initially, in studies on the vetch aphid, *Megoura viciae*, two GC peaks were shown to coincide with stimulation of cells in the secondary rhinaria on the third antennal segment, which had previously been suggested as the site for reception of the female-produced sex pheromone. The MS of the second component suggested a molecular weight of 166 (Figure 11). Fragment ions in this mass spectrum suggested a monoterpenoid and although the pattern was not similar to any in the mass spectrum library, it appeared that the compound could be a monoterpenoid in the nepetalactone series (VI). The three asymmetric centres at positions 4a, 7 and 7a give rise to eight possible isomers. Three compounds, (VIa), (VIb) and (VIc), were obtained from *Nepeta cataria* and *Nepeta mussinii* by HPLC, with confirmation

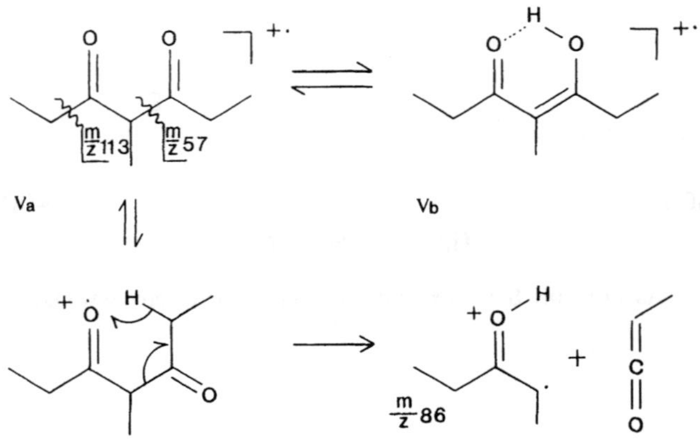

Fig. 10. Fragmentation and McLafferty rearrangement of 4-methyl-3,5-heptanedione (V), m/z 142 (M^+, $C_6H_{14}O_2$).

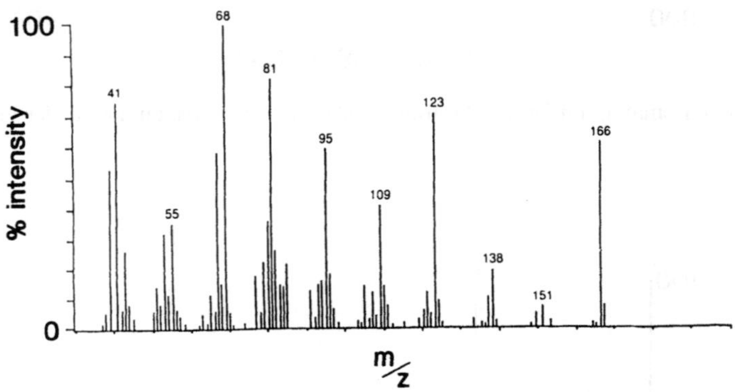

Fig. 11. MS from second electrophysiologically active peak from *M. viciae* oviparae.

of structures by 400 MHz nuclear magnetic resonance (NMR) spectroscopy [4]. Compound (VIa), obtained from *Nepeta cataria*, gave a mass spectrum (Figure 12) very similar to that of the aphid component and gave peak enhancement on co-injection. The compound was also electrophysiologically active, but did not cause a behavioural effect in attracting males. The enantiomer (VId) was synthesized from natural (*R*)-pulegone, which has the appropriate stereochemistry at the 7 position. This compound showed no electrophysiological activity and confirmed that the aphid compound was indeed the (4a*S*,7*S*,7a*R*)-nepetalactone (VIa) from *Nepeta cataria*. It was concluded that, for behavioral activity, the component chromatographing earlier was also required.

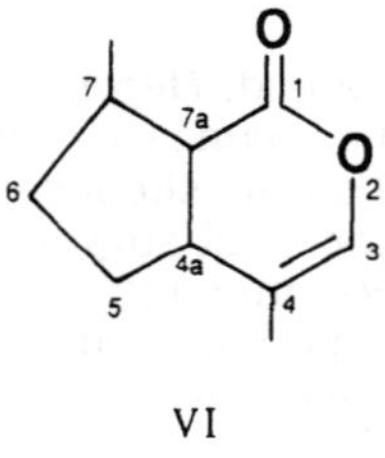

VI

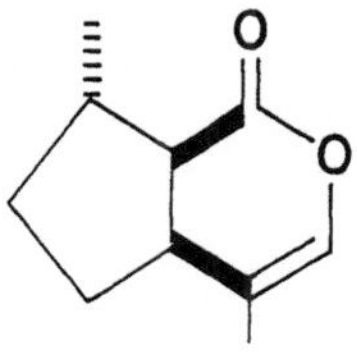

VIa

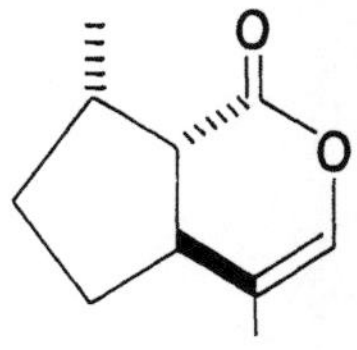

VIb

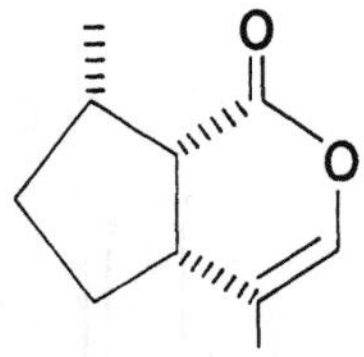

VIc

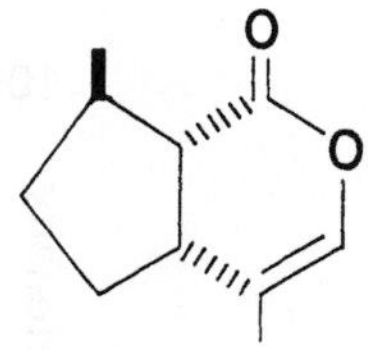

VId

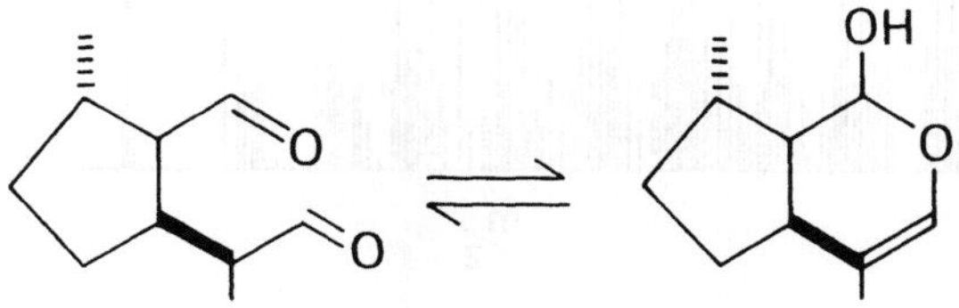

VII VIII

The mass spectrum (Figure 13) indicated a molecular weight of 168 and the fragment at 58 suggested a McLafferty rearrangement arising from structure (VII), an iridodial. Such a compound was obtained from the Staphylinid beetle, *Ocypus olens*, but the spectrum (Figure 14) was substantially different from that obtained from the aphid compound. It has been suggested that iridodial is in equilibrium with its isomer, the nepetalactol (VIII) [12]. Although this lactol was also a candidate for the earlier aphid peak, no iridodial-like compounds could be detected. In addition, such a lactol would be expected to chromatograph after the lactone and not before. Nonetheless, reduction of the nepetalactone (VIa) to the nepetalactol (IX) was attempted. This was successfully achieved using the reducing agent DIBAH. This created a new asymmetric carbon atom with two isomeric possibilities (IXa and IXb), but the product comprised mainly one isomer, which had a similar mass spectrum (Figure 15) to the earlier aphid component and also gave peak enhancement on co-injection. However, its absolute stereochemistry could not be determined, even with use of the 400 MHz NMR instrument in various sophisticated modes. A crystalline derivative was therefore obtained by producing the 3,5-dinitrobenzoate, and X-ray crystallography on this demonstrated that the DIBAH reduction product was mainly the nepetalactol having the structure (IXa) [6]. Although a small amount of the other isomer (IXb) was obtained, this did not give a separate GC peak. The lactol was therefore converted to the acetate, which on re-chromatography allowed separation of the two isomers and full confirmation of the aphid compound as the nepetalactol (IXa), having the stereochemistry of the main DIBAH

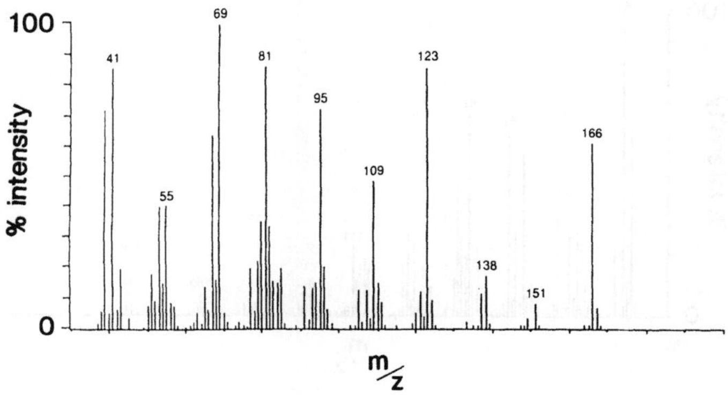

Fig. 12. MS of nepetalactone (VIa) isolated from *N. cataria*.

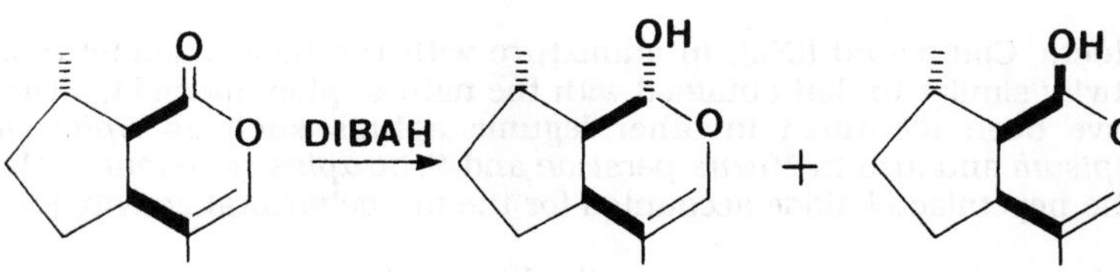

VIa IXa IXb

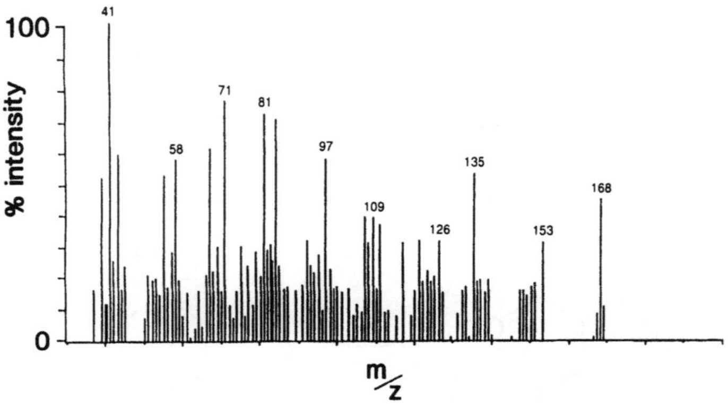

Fig. 13. MS from first electrophysiologically active peak from *M. viciae oviparae*.

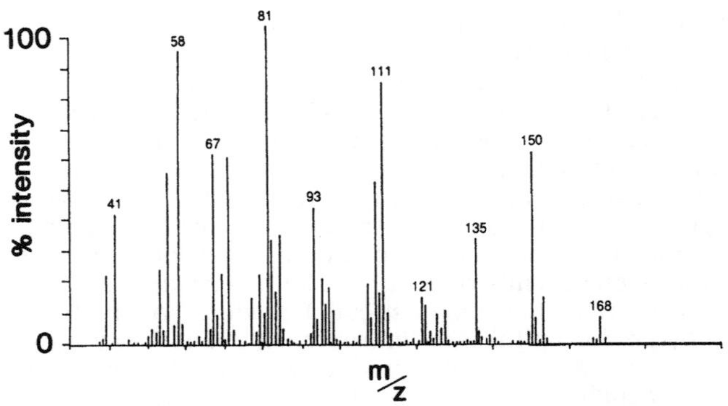

Fig. 14. MS of iridodial isolated from *Ocypus olens*.

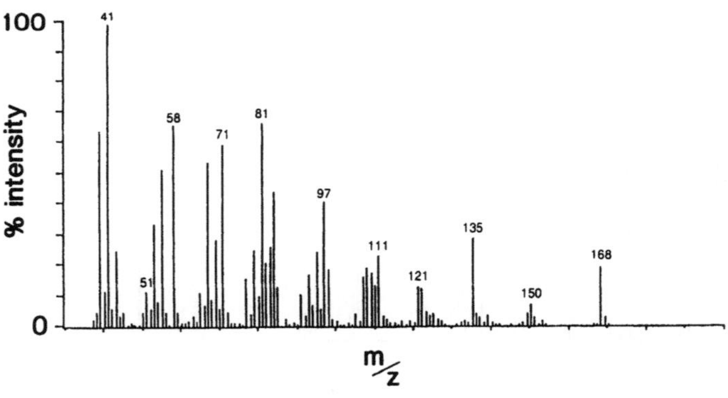

Fig. 15. MS of nepetalactol (IXa) obtained by DIBAH reduction of the nepetalactone (VIa).

reduction product. Compound (IXa), in admixture with the nepetalactone (VIa), showed behavioural activity similar to that obtained with the natural pheromone [4]. The same two compounds have been identified in other legume aphids such as *Aphis fabae* and *Acyrthosiphon pisum* and also in *Myzus persicae* and *Schizaphis graminum*. However, for *S. graminum*, the nepetalactol alone accounted for the full behavioral activity [5].

It is hoped that the sex pheromone for the hop aphid, *Phorodon humuli*, will also be identified. The sex pheromone plays a vital role for this aphid, which must go through the sexual phase to maintain a population after the hop crop is harvested. It may therefore be employed to monitor populations or to disrupt mating. Initial investigations using GC-MS of volatiles entrained from oviparous *P. humuli* showed the presence of cycopentanoid-type monoterpenoids related to the nepetalactol (IXa) and this has now been confirmed by coupled GC-SCR studies. In the case of aphid sex pheromones, it is clear that electrophysiological techniques played a very important role in the GC-MS identification.

REFERENCES

1. B. R. Laurence and J. A. Pickett, *Erythro*-6-acetoxy-5-hexadecanolide, the major component of a mosquito oviposition attractant pheromone, *J. Chem. Soc. Chem. Comm.*, 59-60 (1982).
2. B. R. Laurence, K. Mori, T. Otsuka, J. A. Pickett and L. J. Wadhams, Absolute configuration of mosquito oviposition attractant pheromone, 6-acetoxy-5-hexadecanolide, *J. Chem. Ecol.*, 643-648 (1985).
3. M. M. Blight, J. A. Pickett, M. C. Smith and L. J. Wadhams, An aggregation pheromone of *Sitona lineatus*, *Naturwissenschaften*, 71:S.480 (1984).
4. G. W. Dawson, D. C. Griffiths, N. F. Janes, A. Mudd, J. A. Pickett, L. J. Wadhams and C. M. Woodcock, Identification of an aphid sex pheromone, *Nature*, 325:614-616 (1987).
5. G. W. Dawson, D. C. Griffiths, L. A. Merritt, A. Mudd, J. A. Pickett, L. J. Wadhams and C. M. Woodcock, The sex pheromone of the greenbug, *Schizaphis graminum*, *Entomol. Exp. et Applic.*, 48:91-93 (1988).
6. G. W. Dawson, N. F. Janes, A. Mudd, J. A. Pickett, A. M. Z. Slawin, L. J. Wadhams and D. J. Williams, The aphid sex pheromone, *Pure & Applied Chem.*, 61:555-558 (1989).
7. D. W. Bruno and B. R. Laurence, The influence of the apical droplet of *Culex* egg rafts on oviposition of *Culex pipiens fatigans* (Diptera: Culicidae), *J. Med. Entomol.*, 16:300-305 (1979).
8. P. Powers, P. H. D'Arsey, J. C. Bill and M. J. Wallington, Applications of a new real-time accurate mass facility for organic mass spectrometers, 26th Annual Conference of Mass Spectrometry Allied Topics, St. Louis, Missouri, p 480 (1978).
9. C. E. Osgood, An oviposition pheromone associated with the egg rafts of *Culex tarsalis*, *J. Econ. Entomol.*, 64:1038-1041 (1971).
10. S. Sasaki, H. Abe, Y. Itagaki and K. Nakanishi, Two-step expulsion of aldehydes and ketones in the fragmentation of acyclic α-glycol diesters, *Tet. Lett.*, 25:2357-2361 (1967).
11. W. A. Otieno, T. O. Onyango, M. M. Pile, B. R. Laurence, G. W. Dawson, L. J. Wadhams and J. A. Pickett, A field trial of the synthetic oviposition pheromone with *Culex quinquefasciatus* Say (Diptera: Culicidae) in Kenya, *Bull. Entomol. Res.*, 78:463-470 (1988).
12. S. A. Abou-Donia, L. J. Fish and G. Pattenden, Iridodial from the odoriferous glands of *Staphylinus olens* (Coleoptera: Staphylinidae), *Tet. Lett.*, 43:4037-4038 (1971).

HIGH PERFORMANCE LIQUID CHROMATOGRAPHY-MASS SPECTROMETRY IN THE ANALYSIS OF SEMIOCHEMICALS

A. Mudd

AFRC Institute of Arable Crops Research
Rothamsted Experimental Station
Harpenden, Hertfordshire AL5 2JQ, UK

SUMMARY

The use of high performance liquid chromatography linked to mass spectrometry as a technique for the analysis of semiochemicals is discussed. Examples are given of high performance liquid chromatography-mass spectrometry analysis of insect antifeedants from *Ajuga* species and of kairomones from Lepidoptera.

INTRODUCTION

High performance liquid chromatography linked with mass spectrometry (HPLC-MS) is becoming an important technique for the analysis of natural product mixtures. There are several advantages in using HPLC-MS for the analysis of sensitive biologically active molecules compared with gas chromatography linked with mass spectrometry (GC-MS). For example, most biologically active molecules are more stable in solution at ambient temperatures than in the gas phase at high temperatures and, unlike GC, HPLC will separate involatile and very polar molecules without derivatization. Also since guard columns can be used with HPLC columns, less sample clean-up is normally required than with GC. Thermally unstable and very high molecular weight compounds are obviously better separated by HPLC. However, there is no point in using HPLC-MS if the problem can be conveniently solved by GC-MS, which is a more sensitive and routine technique.

There are three main types of commercially available HPLC-MS interface; the fast atom bombardment (FAB) interface, the thermospray/plasmaspray (or discharge) interface and the moving belt interface. They differ in their sensitivities, the types of sample they will analyze and also in their availability for particular mass spectrometers and their compatibility with different types of HPLC equipment. Given that no interface is equally suitable for all types of sample and that a choice has to be made, that choice will be determined by the nature of the key samples to be analyzed and by the compatibility of the interface with existing equipment.

The belt interface has had the longest period of commercial development and is, in some ways, the most versatile in terms of the different methods of sample ionization that can be used with it. However, it is not as useful as the other interfaces for the more difficult samples; the very involatile, polar, thermally unstable compounds. Comparisons of sensitivity are therefore misleading because they depend very much upon the type of sample being examined.

The thermospray/plasmaspray (discharge) and belt interfaces will handle normal analytical HPLC flow rates (1 to 2 ml/min), so that conventional HPLC pumps, columns, injectors etc. can be used. The continuous or dynamic FAB interface will only handle very low flow rates of about 10 µl/min, which means that capillary or microbore columns are required as well as special pumps, low volume injectors and detectors for optimizing separations "off-line" prior to HPLC-MS.

Chromatography and Isolation of Insect Hormones and Pheromones
Edited by A. R. McCaffery and I. D. Wilson
Plenum Press, New York, 1990

All of the interfaces can use reversed-phase solvents (methanol, water, acetonitrile etc.) and volatile buffers such as ammonium acetate but only the plasmaspray (discharge) and belt interfaces can use the normal phase solvents (hexane, ether, chloroform etc.) required for adsorption chromatography on silica or alumina.

Because we wished to use existing HPLC equipment to study polar, involatile and thermally unstable plant components, we chose the thermospray/plasmaspray interface in preference to the FAB or belt interfaces. For semiochemical analysis, plasmaspray rather than thermospray has been used because it is much more sensitive and fragmentation information is much more easily available, which can be important.

MATERIALS AND METHODS

HPLC-MS analyses were performed with a VG 70-250 mass spectrometer equipped with a VG thermospray/plasmaspray interface operated with a source temperature of 250°C, probe temperature of 220°C and a 450 v discharge voltage. HPLC analyses were performed with Gilson 303 pumps using a 250 x 4.6 mm Spherisorb ODS2 (5 micron) column (Anachem). Methanol-water (60:40, 1.5 ml/min) was used as the eluent in the analysis of *Ajuga remota* extracts, with ultra-violet (UV) detection at 213 nm. Methanol containing 1% trifluoroacetic acid (1.5 ml/min) was used for the analysis of mandibular gland extracts from *Ephestia kuehniella*, with UV detection at 233 nm. Electron impact (EI) mass spectra of ajugarin I and II were obtained by evaporation from a probe inserted directly into the source of the mass spectrometer operated in the EI mode. GC-MS analyses were performed with a Hewlett-Packard 5890 gas chromatograph coupled directly to the VG 70-250 mass spectrometer (70 eV, source 250°C) using "on-column" injection on a J & W fused silica capillary column (30 m x 0.32 mm) coated with DB1 (0.25 micron). After operating isothermally at 30°C for 3 min, the column was programmed from 30 to 280°C at 15°/min, then at 3°C/min to 300°C.

RESULTS AND DISCUSSION

The isolation and identification of unknown components from biologically active plant extracts is a difficult task, especially if the extract is unsuitable for GC-MS analysis. It is made difficult not only by the very large number of components that have to be separated, but also by the fact that many of them are difficult to detect. If components have little or no UV absorption, detection can be a major problem unless refractive index detection can be used, which presupposes that sufficient of the active components are present to be detectable and that gradient HPLC is excluded. Also, extracts unsuitable for GC-MS analysis usually give very poor and uninformative EI mass spectra because they contain involatile and thermally unstable components. HPLC-MS analysis can often overcome the problems of detection and identification in such extracts.

For example, extracts of *Ajuga* species (Labiateae) contain a large number of involatile, thermally unstable insect-antifeedants, which have very similar, clerodane structures. Extracts of *Ajuga remota* have antifeedant activity against lepidopteran larvae, aphids and coleopteran pests. Preliminary experiments suggest that different components of the extract are responsible for the antifeedant activity in different species. Since many of the clerodane components have no UV chromophore, they are difficult to detect and since their EI mass spectra are very similar and show no molecular ions they are hard to identify. HPLC-Plasmaspray was used to overcome problems of detection and identification. Comparison of the HPLC analysis of an extract of *A. remota* using UV detection (Figure 1) with HPLC analysis of the same extract using plasmaspray detection (Figure 2) shows that many more components are detectable by plasmaspray. Also, it is much easier to identify these components from their plasmaspray spectra than their EI mass spectra [1]. The EI mass spectra of ajugarin I (Figure 3a) and ajugarin II (Figure 3b) show no molecular ions at m/z 434 and 392 respectively. Their spectra are very similar and difficult to distinguish between, especially in mixtures where only the component with the higher mass fragments is detectable; the mass spectrum of the lower molecular weight ajugarin II simply merges with the spectrum of ajugarin I since nearly all of their fragments are similar. This is a very common situation in the analysis of plant extracts of this type. The plasmaspray spectra of the ajugarins, however (Figures 4a and b) show strong protonated molecular ions MH^+ m/z 435 and 393 with diagnostic high mass fragments. Ajugarin I shows strong fragment ions at m/z 375 (MH - 60) and m/z 315 (MH - 120) for the losses of the acetoxy groups at C6 and C19 whereas ajugarin II shows strong fragment ions at m/z 375 (MH - 18)

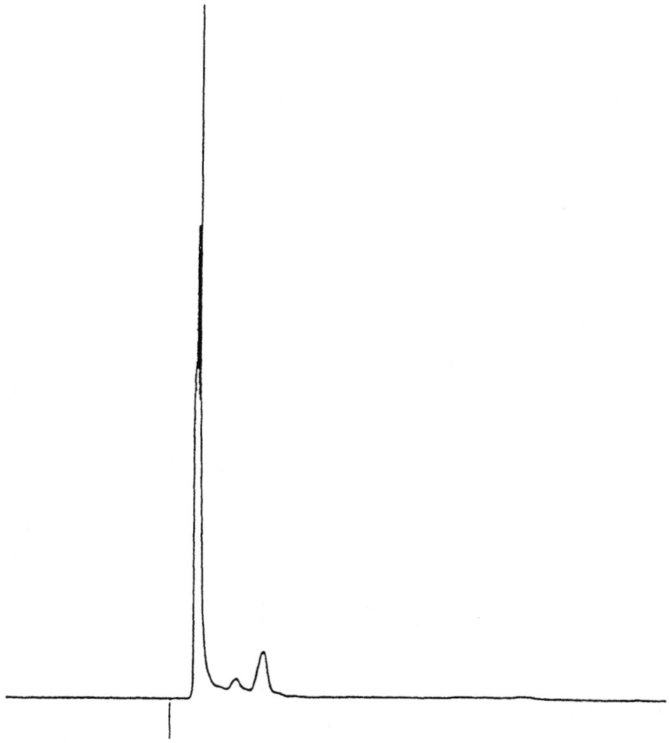

Fig. 1. HPLC of *Ajuga remota* extract with UV detection (213 nm).

and m/z 333(MH - 60) corresponding to losses of the hydroxyl and acetoxy groups at C6 and C9 respectively. A sample of ajugarin I which was thought to be pure because it had the literature m.p., NMR and EI mass spectrum was shown by HPLC-plasmaspray mass spectrometry to be a mixture of ajugarins I and II, which illustrates how valuable the technique can be.

Another, quite different example of the use of HPLC-plasmspray mass spectrometry in the analysis of semiochemicals is provided by examination of the larval manibular gland components of the Mediterranean flour moth *Ephestia kuehniella*. The glands contain components which act as pheromones, by affecting the egg-laying of the moth and as kairomones, by attracting the parasitoids *Nemeritus canescens* (Grav.) [2] and *Bracon hebtor* (Say) [3] which kill the larvae. The active compounds are a mixture of at least sixteen closely related, high molecular weight (ca. 400) β-triketones or 2-acylcyclohexane-1,3-diones and their 4-hydroxy analogues [4,5] (Figure 5). GC-MS using bonded-phase, fused silica capillary columns (Figure 6) did not completely separate the mixture and it was eventually separated using a series of different types of HPLC columns. Comparison of the HPLC separations of the gland contents using UV detection and plasmaspray detection shows that they appear to be very different (Figure 7).

HPLC with UV detection gives a misleading impression of the composition of the extract since only UV absorbing components are detected, and peak areas depend upon the extinction coefficients of the chromophores present [6]. Non-UV absorbing compounds such as sterols and fatty acids that were present remained undetected, whereas components with conjugated dienes in their side-chains (ε_{max} 30,000) appeared much more abundant than similar amounts of components with only the β-triketone chromophore (ε_{max} 10,000). So, HPLC with plasmaspray detection gave a more accurate representation of the gland composition than UV detection because it did not depend on the type or even the presence of a chromophore. Also by generating ion chromatograms, for example of the protonated molecular ion MH^+ (m/z 393) of the major gland component 4-hydroxy-2-acylcyclohexane-1,3-dione, it was possible to find previously undetected isomers that appeared as a single peak using UV detection.

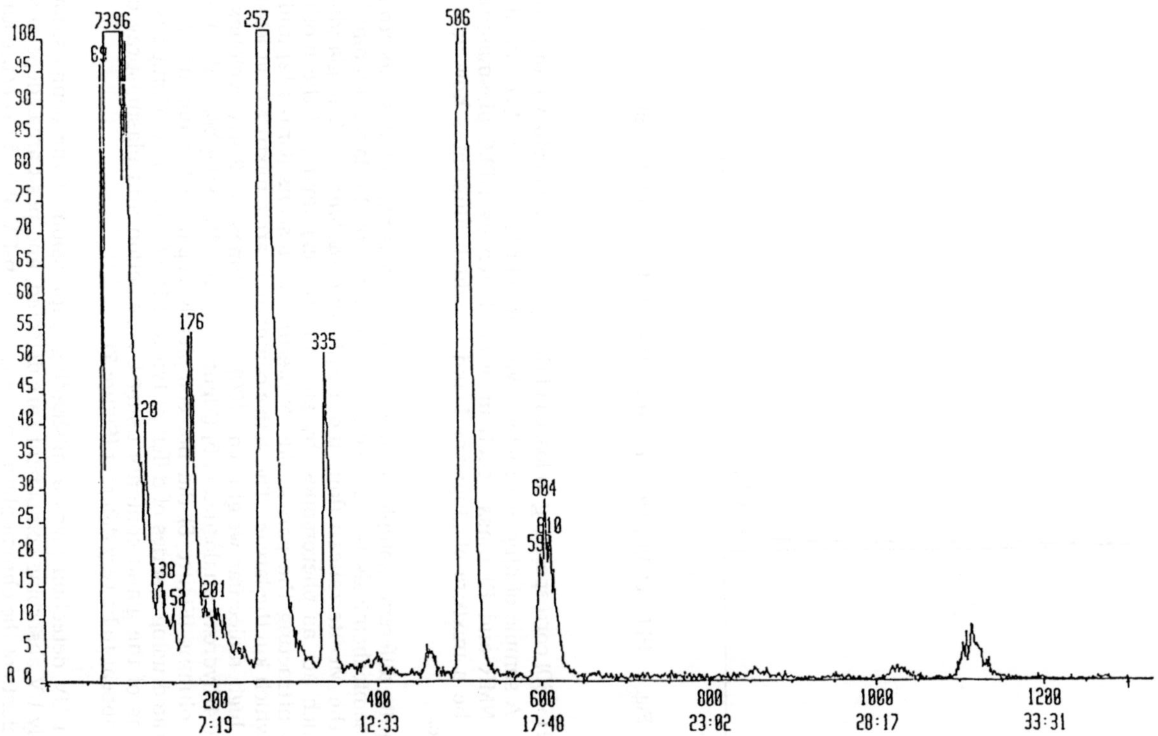

Fig. 2. HPLC of *Ajuga remota* extract with plasmaspray detection.

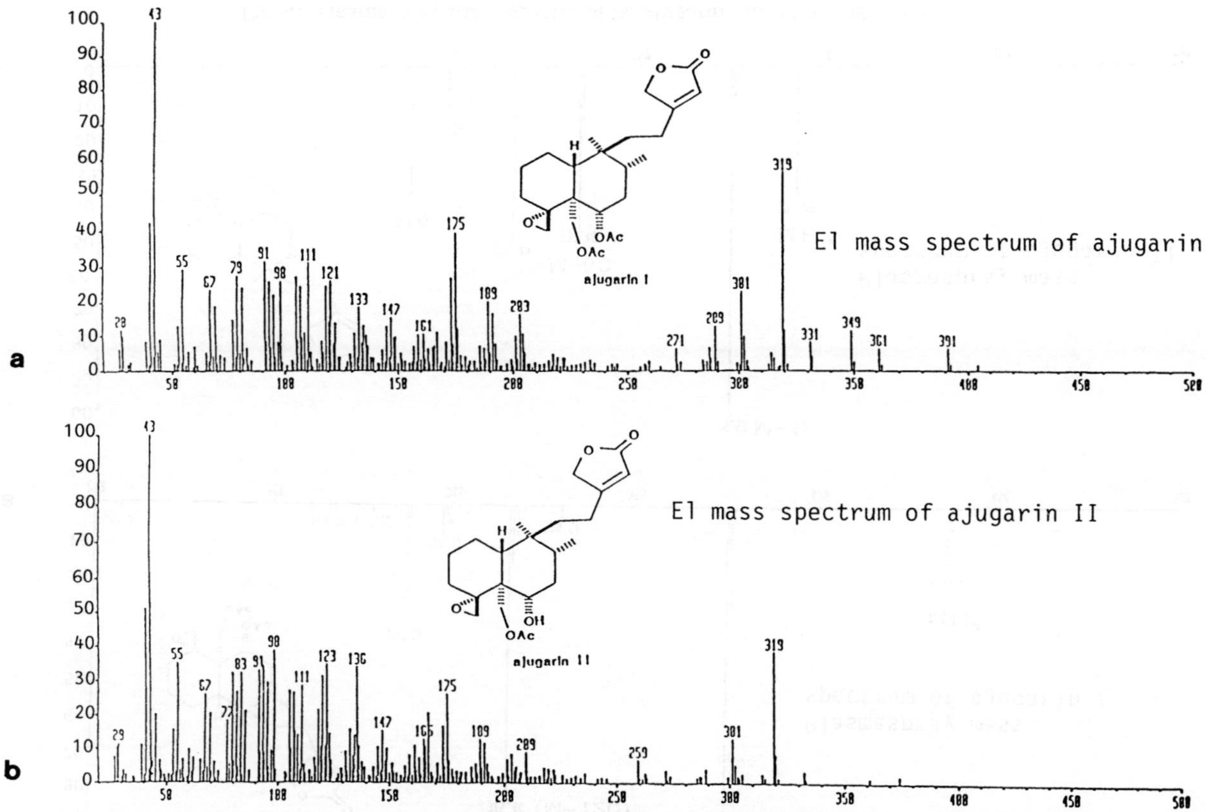

Fig. 3. EI mass spectra of (a) ajugarin I and (b) ajugarin II.

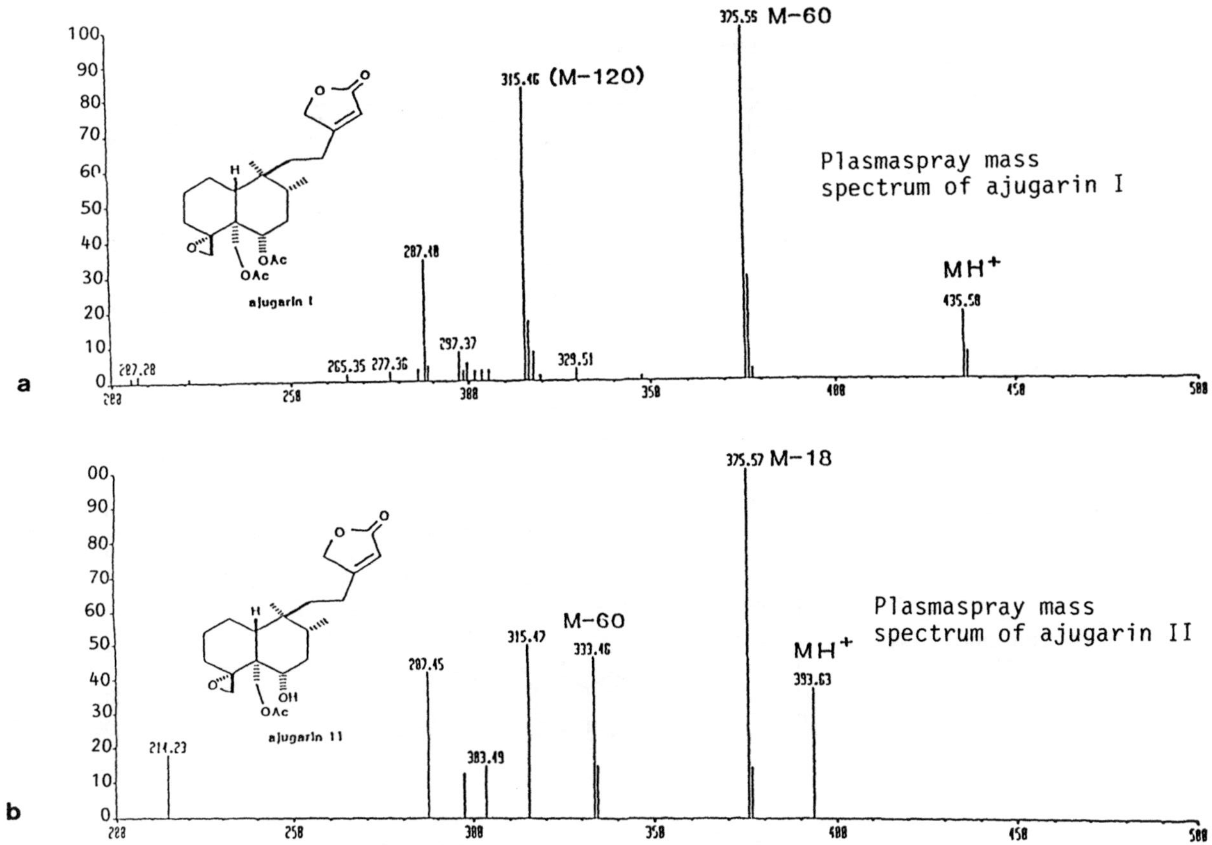

Fig. 4. Plasmaspray mass spectra of (a) ajugarin I and (b) ajugarin II.

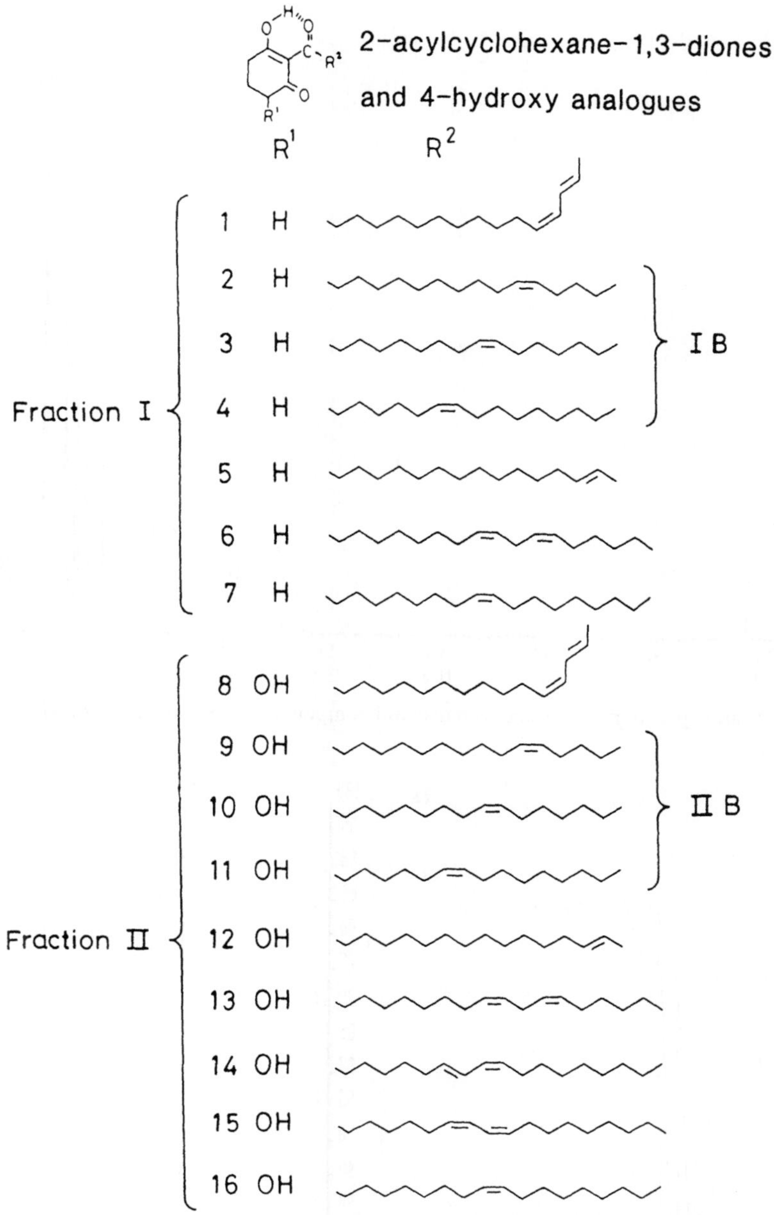

Fig. 5. Major active components from the mandibular glands of *Ephestia kuehniella* larvae.

In summary, HPLC-MS is proving to be a valuable method for studying involatile, thermally unstable semiochemicals. It is not yet as sensitive or routine as GC-MS, but it is most useful for precisely those samples that are unsuitable for GC-MS. It gives us the opportunity to study the composition of very polar, hydrophilic, involatile and thermally unstable natural product mixtures that have always been difficult to analyze. We can expect improvements in the sensitivity and reliability of thermospray and plasmaspray in the next few years. We can also expect developments in capillary and microbore HPLC equipment that can be used with continuous or dynamic FAB. This will, I believe, make HPLC-MS the method of choice for many analyses in the future.

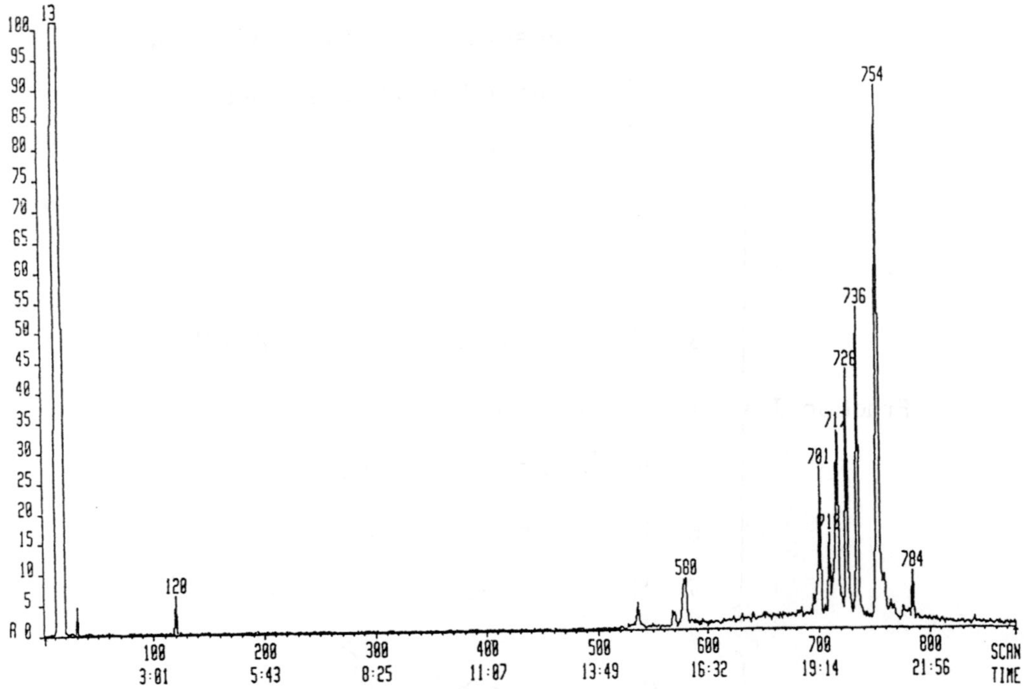

Fig. 6. GC-MS separation of the manibular gland components of *Ephestia kuehniella* larvae.

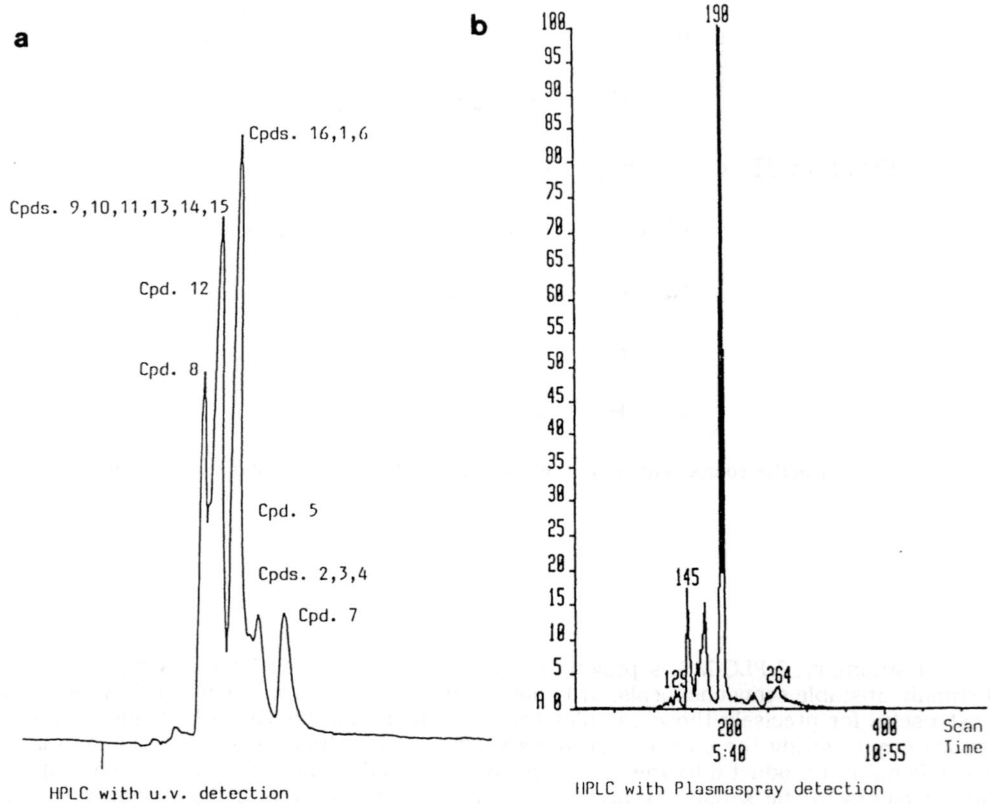

Fig. 7. HPLC separation of the larval mandibular components of *Ephestia kuehniella* larvae using (a) UV detection (233 nm) and, (b) Plasmaspray detection.

REFERENCES

1. A. Mudd, J. A. Pickett and L. H. Wadhams, HPLC-Plasmaspray analysis of natural products, 11th International Mass Spectrometry Conf. Proc., 11B:1252-1253 (1988).
2. A. Mudd and S. A. Corbet, Response of the ichneumonid parasite *Nemeritus canescens* (Grav.) to kairomones of the flour moth *Ephestia kuehniella* Zeller, *J. Chem. Ecol.*, 8:843-850 (1982).
3. M. R. Strand, H. J. Williams, S. B. Vinson and A. Mudd, Arrestment and trail following response of *Bracon hebetor* (Say) to kairomones from *Ephestia kuehniella* Zeller, *J. Chem. Ecol.*, 15:1491-1500 (1989).
4. A. Mudd, Further novel 2-acylcyclohexane-1,3-diones from Lepidopteran larvae, *J. Chem. Soc. Perkin Trans.*, I:2161-2164 (1989).
5. A. Mudd, Synthesis of kairomonal 2-acylcyclohexane-1,3-dione components of the larval mandibular glands of *Ephestia kuehniella*, *J. Chem. Ecol.*, 11:51-57 (1985).
6. A. Mudd, GC-MS analysis of pheromones/kairomones, 11th International Mass Spectrometry Conf. Proc., 11B:1250-1251 (1988).

ANALYSIS OF GLAND SECRETIONS OF PENTATOMOIDEA (HETEROPTERA) BY GAS CHROMATOGRAPHY-MASS SPECTROMETRY TECHNIQUES

C. E. Evans, B. W. Staddon* and D. E. Games

School of Chemistry and Applied Chemistry and
*School of Pure and Applied Biology
University of Wales College of Cardiff
Cardiff CF1 3TB, UK

SUMMARY

Results of analyses by gas chromatography and gas chromatography-mass spectrometry on the semiochemicals of several species of the superfamily Pentatomoidea are given. In addition to the primarily defensive metathoracic gland (MG) the four Acanthosomatidae species studied also possess a pair of active dorsal abdominal exocrine glands (DG1) in the adults. The use of microscale hydrogenations in confirming the identity of the DG1 monoterpenes is illustrated. Two methods of sample collection were used, one involving gland dissection under saline into solvent, followed by injection of the sample into the GC, the other involving trapping the volatiles from the bugs onto a porous polymer, followed by their thermal desorption into the gas chromatograph.

INTRODUCTION

Bugs of a number of families of Heteroptera have scent glands which secrete noxious defensive secretions. Members of the superfamily Pentatomoidea are commonly known as stink bugs and usually possess a dorsal metameric series of three (occasionally four) defensive scent glands in the abdomen during larval stages. The DG1 consists of a minute pair of widely separated half-glands whereas the larger second and third abdominal scent glands (DG2 and DG3) are undivided [1,2]. The DG1 usually retains its activity in the adult pentatomid bugs.

Adults possess another defensive scent gland which is located in the thorax, the metathoracic gland. Sexual dimorphism in the metathoracic gland has not been observed in the pentatomids. In contrast, sexual dimorphism in DG1 has been reported in several species of Pentatomoidea [1,3-6]. These include some but not all predaceous bugs Asopinae (Pentatomidae) and two species of Scutelleridae, namely *Hotea gambiae* and *Sphaerocoris annulus*.

A study of about 250 species from various families revealed that a sexual dimorphism of the abdominal sternal glands exists in the Pentatomoidea [7]. These apparently unicellular glands are found in the abdominal integument of most males but are less abundant or absent in the females, and it is thought that they may be responsible for the secretion of a sex pheromone. Several reports of presumed sex pheromones within the Pentatomoidea have been made [8-13].

The objectives of this work were two fold. Firstly, to seek to extend the knowledge of pentatomid defense chemicals and secondly to investigate the possible emission of sex pheromones by the abdominal sternal glands. Individuals calling a mate may release volatiles from a gland continually, whereas in defensive glands the secretion is stored until needed to deter an aggressor. It is unusual for insects to produce copious quantities of sex pheromone, therefore trapping volatiles over a period of time and concentrating the miniscule quantities seems to be the best approach to sample collection in such cases.

Chromatography and Isolation of Insect Hormones and Pheromones
Edited by A. R. McCaffery and I. D. Wilson
Plenum Press, New York, 1990

MATERIALS AND METHODS

Insects

Adults and larvae of the four British species of the Acanthosomatidae [14] were collected on warm days in parks and woodlands in the Cardiff area from mid July until late October. *Cyphostethus tristriatus* (Fabricius) was captured almost entirely on Lawson's Cypress (*Cupressus Lawsonia*). *Elasmostethus interstinctus* (Linnaeus) and *Elasmucha grisea* (Linnaeus) were usually found on the same host trees in mixed birch-oak-alder woodlands (*Betula: Quercus: Alnus* spp.). *Acanthosoma haemorrhoidale* (Linnaeus) was found on hawthorn (*Crataegus* spp.) together with several species of cultivated bushes possessing juicy berries. Adults of *Agonoscelis versicolor* (Fabricius) were collected in Nigeria.

All the insects were kept in an incubator (photoperiod 14 hr light, 10 hr dark; daily temperature cycle 18°C night and 28°C day).

Instruments

Gas chromatographic (GC) analyses of samples were performed using a Varian Vista 6000 instrument equipped with a FID and a Shimadzu C-E1B integrator. GC-MS analyses were performed with a DANI 3800 gas chromatograph interfaced to a VG 7070E mass spectrometer.

Separation of the scent constituents was achieved on two capillary columns of different polarities; 25 m BP1 and 30 m DB225. A temperature program of 70 to 160°C at 6°C/min was used for all the chromatographic runs except for the separations of monoterpenes on DB225 which were performed isothermally at 65°C.

Mass spectra were generated at electron energy of 70 eV and emission current 200 μA.

GC-MS analyses were carried out in the electron impact (EI) mode on all samples. In addition, spectra in the chemical ionization (CI) mode were acquired for the metathoracic gland secretion of *Agonoscelis versicolor*.

Sample Collection

Method 1. The scent volatiles of the larval DG2 and DG3, the adult DG1 and MG of the Acanthosomatidae and the MG of *Agonoscelis versicolor* were extracted by gland dissection. Excision of the scent glands was effected under 200 mM NaCl and was followed by the extraction of scent volatiles in ca. 5 μl acetone (or ethanol for samples prepared for hydrogenation). Extracts were stored in sealed glass ampoules at -20°C until required for analysis.

Method 2. The second method involved trapping airborne volatiles onto a porous polymer followed by their desorption into a gas chromatograph. The vapor collection system used is illustrated in Figure 1. The components were connected with Swagelock unions and seals were provided by Teflon ferrules. The glass collection chamber was treated with Surfasil Siliconizing Fluid (Pierce) in order to minimize adsorption of volatiles onto the glass surface. Cartridges were packed with 0.2 g of Tenax TA (Alltech) (mesh size 20/35) which was held in place by silanized glass wool. All sampling cartridges were preconditioned by purging with helium for 10 minutes at room temperature followed by 2 hours at 275°C. The cartridges were of equal dimensions to those of the gas chromatograph injector glass liner. Air was drawn over the insects contained in the glass collection chamber at a flow rate of 40 ml/min. Insects were aerated for 10 minutes in defense chemicals investigations and for 3 hours in the pheromone experiments.

Thermal desorption was achieved by a simple method [15-16]. The GC injector port nut and septum were removed, the cartridge was inserted directly into the injector body and then the septum and nut were quickly replaced. The desorbed volatiles were reconcentrated by cryogenic focusing (Figure 2).

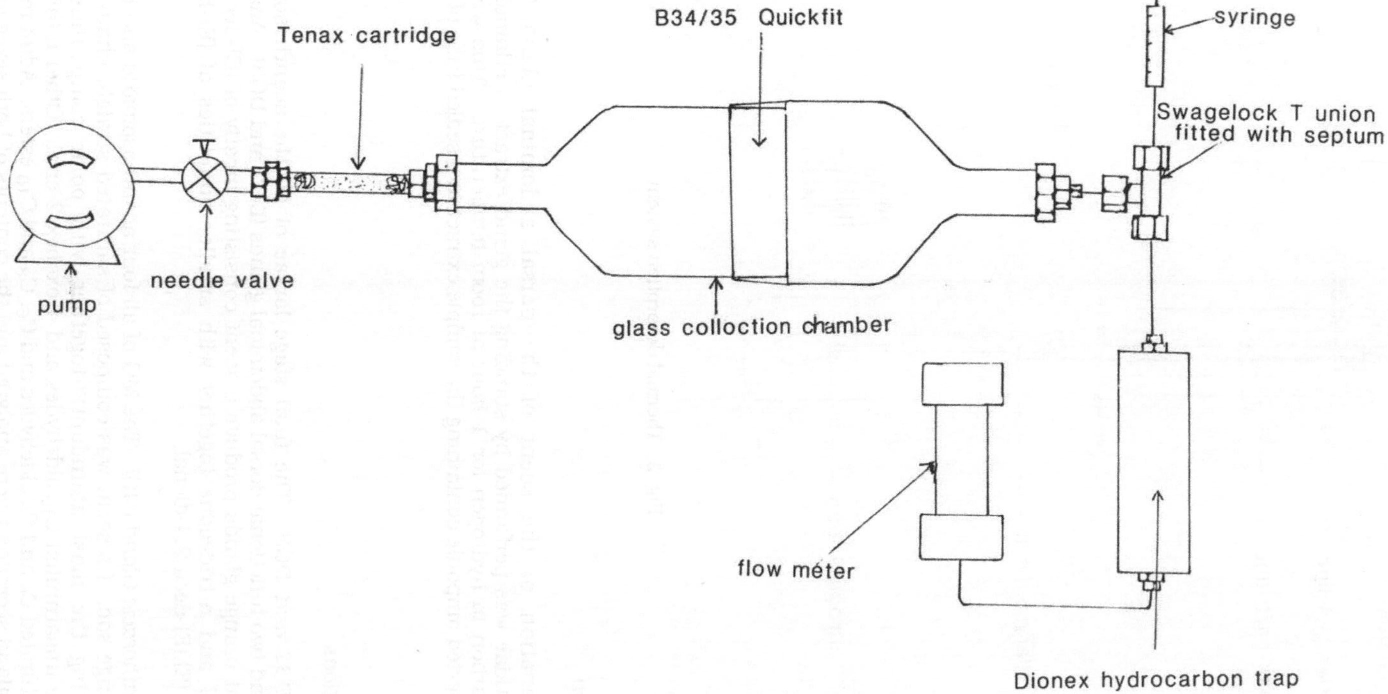

Fig. 1. Vapor collection system.

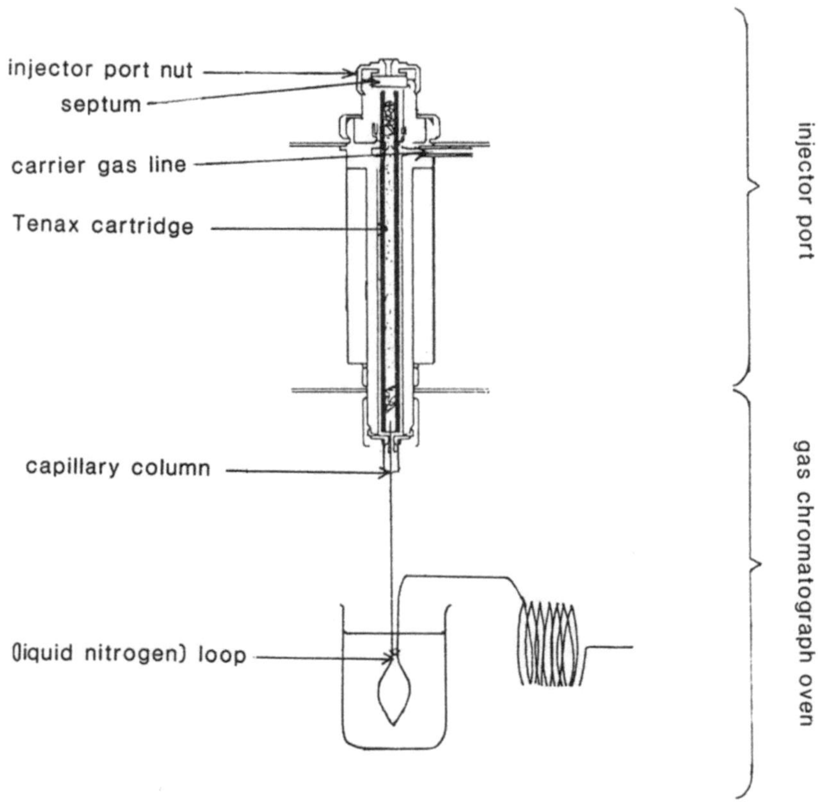

Fig. 2. Thermal desorption system.

Hydrogenation

Hydrogenation of the scent of the dorsal abdominal gland (DG1) of the Acanthosomatidae was performed by standing the gland extract in ethanol with 10% Pd catalyst on carbon in hydrogen for 1 hour at room temperature. This was achieved by placing an inverted ampoule containing the sample extract in a sealed tube of hydrogen.

RESULTS

Gland Dissections

Larval DG2 and DG3. The final stage larvae of all the acanthosomatid species investigated had two functional dorsal abdominal glands (DG2 and DG3). Analyses revealed that the bright orange glands produce a scent consisting mostly of (E)-dec-2-enal, (E)-4-oxohex-2-enal and n-tridecane together with smaller quantities of (E)-hex-2-enal, n-dodecane and (E),(E)-deca-2,4-dienal.

The Metathoracic Gland (MG). The MG of all four acanthosomatid species was a large undivided orange sac. Its scent was composed of saturated straight chain hydrocarbons (tridecane being the most abundant), together with polar compounds which were predominantly unsaturated C_{10} aldehydes and 4-oxohex-2-enal. Other minor components included unsaturated C_6 and C_8 aldehydes and C_6, C_8 and C_{10} esters. Adult male and female metathoracic gland secretions were analyzed and the contents of both sexes were essentially identical. The results are shown in Table 1.

The constituents of the metathoracic gland of *Agonoscelis versicolor* included the common pentatomid carbonyl defense compounds and saturated aliphatic hydrocarbons together with a few unusual components. On the basis of matching retention time data on two different columns and comparison of mass spectra generated in both electron impact

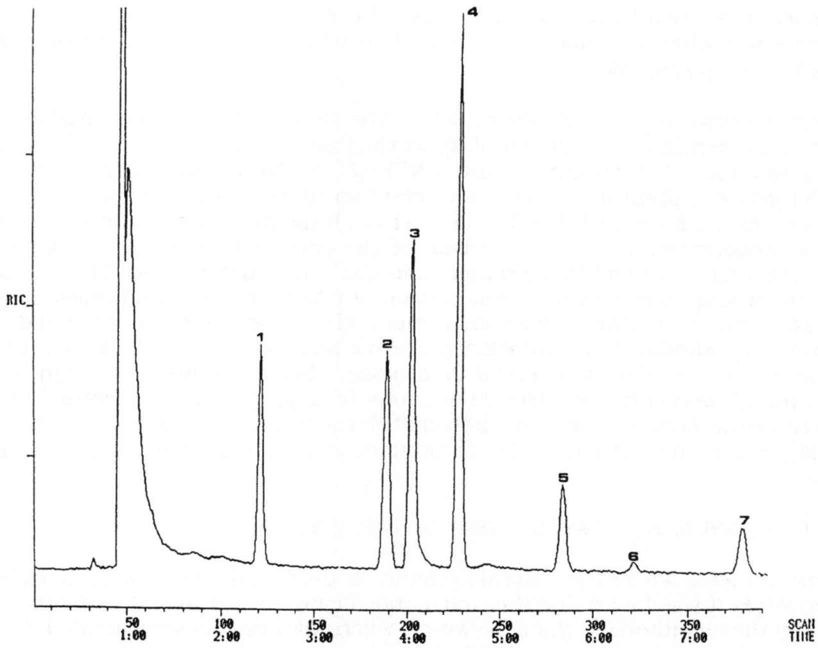

Fig. 3. DG1, female in acetone. Reconstructed total ion chromatogram from GC-MS of the anterior dorsal abdominal gland of a composite sample of three individual female parent bugs, *Elasmucha grisea* on DB225 capillary at 65 C isothermal. Peak numbering corresponds to Table 2.

Table 1. Composition of Metathoracic Scent Glands of Adult Male Acanthosomatidae

		Mean R.t.(min) T.P. 70-160°C at 6°C/min		Percentage composition			
Peak	Scent constituent	DB225	DB1	*A. haemorrhoidale*	*E. grisea*	*E. interstinctus*	*C. tristriatus*
1	n-Undecane	2.9	3.0	-	-	-	-
2	(E)-Hex-2-enal	3.4	1.0	14	1	2	1
3	n-Dodecane	3.9	5.1	TR	TR	TR	TR
4	(E)-Hex-2-enyl acetate	4.6	2.2	3	-	-	TR
5	n-Tridecane	5.4	7.1	52	49	21	62
6	Tridecene	5.8	7.2	-	-	-	TR
7	(E)-Oct-2-enal	7.3	2.6	-	TR	TR	TR
8	(E)-4-Oxohex-2-enal	8.0	1.3	-	5	25	2
9	n-Pentadecane	8.4	11.2	TR	TR	TR	TR
10	(E)-Dec-2-enal	11.2	6.0	13	32	48	21
11	(E),(E)-Deca-2,4-dienal	12.1	6.1	TR	10	3	11
12	(E)-Dec-2-enyl acetate	12.1	9.1	16	TR	TR	TR
13	(E),(E)-Deca-2,4-dienyl acetate	12.9	9.0	-	-	-	TR

TR = Trace.

and chemical ionization (methane) with those of the authentic standard one of the major constituents was characterized as linalool. Two peaks corresponding to dodecenal and decadienal were also recorded.

Anterior Dorsal Abdominal Gland (DG1). The DG1 of the Acanthosomatidae consists of a pair of pigmentless half glands of approximately the same size in both sexes. The reconstructed total ion chromatogram (RIC) of the DG1 constituent compounds of *Elasmucha grisea* is given in Figure 3 and retention times and percentage compositions of the four species are given in Table 2. The DG1 of all the Acanthosomatidae was composed entirely of monoterpenes, and comparison of the composition of the gland in the four species reveals mostly quantitative rather than qualitative differences. The initial identity of the monoterpenes was based on comparison of retention times and mass spectra with those of authentic standards. However, compounds within the same terpenoid structure class have very similar fragmentation patterns and are thus notoriously difficult to identify on the basis of mass spectral data alone. Since the proposed identities of the acanthosomatid monoterpenes were compounds belonging to four different fundamental types, hydrogenations of the acanthosomatid secretion and of individual authentic standards were carried out in order to produce supporting evidence for the proposed identities.

Effluvial Collection of Agonoscelis versicolor *volatiles*

In experiments where the airborne volatiles of disturbed insects were collected on the Tenax cartridges it was found that the compounds being emitted were the defense chemicals produced by the metathoracic gland. However, when calm insects were aerated (in this case in an attempt to trap sex pheromones), defense compounds were not emitted. Due to a decline in the population of sexually mature *Agonoscelis versicolor* males, reproducible results in the investigations of male sex pheromones were not obtained.

DISCUSSION

The results on the larval dorsal glands and the metathoracic gland of the Acanthosomatidae conform to a pattern previously established for other members of the Pentatomoidea. The MG and larval DG secretion of terrestrial Heteroptera are primarily defensive irritants [17]. It is proposed that in the defensive secretion of the Acanthosomatidae, the hydrocarbons function as relatively non-volatile solvents for the active aldehydes and also as spreading and penetrating agents, as has been proposed for the

Table 2. Composition of Anterior Dorsal Abdominal Gland DG1 of 4 Species of Adult Male and Female Acanthosomatidae

		Mean R.t.(min)		Percentage composition							
				A. haemorrhoidale		*E. grisea*		*E. interstinctus*		*C. tristriatus*	
Peak	Scent constituent	DB225 65°iso	BP1 70-160°C at 6°C/min	♂	♀	♂	♀	♂	♀	♂	♀
1	α-pinene	2.5	4.5	4	3	11	13	TR	TR	-	-
2	β-pinene	3.8	5.3	74	66	18	17	17	20	28	27
3	sabinene	4.1	5.4	-	-	33	23	47	40	25	26
4	myrcene	4.5	5.6	15	25	16	30	10	14	23	22
5	limonene	5.6	6.4	7	6	14	11	11	12	17	18
6	unknown	6.4	6.8	-	-	TR	TR	TR	TR	-	-
7	γ terpinene	7.5	7.1	-	-	7	6	13	12	7	7

TR = Trace.

defensive secretions of other Heteroptera [18]. In addition to releasing the usual pentatomid defense secretion constituents, the metathoracic gland of the Nigerian bug *Agonoscelis versicolor* also emits three unusual components. Among the Heteroptera the tertiary monoterpene alcohol, linalool has been reported as a constituent of the metathoracic gland of two *Dysdercus* species [19-20] and in the dorsal abdominal gland (DG1) of the predatory stink bugs, *Podisus* spp. [5,21]. The occurrence of C_{12} aldehydes has not previously been reported in heteropteran secretions. The function of dodecenal and dodecadienal in *Agonoscelis versicolor* is not known but one would suppose that they have a similar function to the lower aldehydes present in the metathoracic gland secretion.

The anterior dorsal abdominal gland (DG1) of adult Acanthosomatidae can be added to the mushrooming list of heteropteran glands which release monoterpenes. Gough et al. [1] and Hamilton et al. [6] were the first to report on the occurrence of monoterpenes as constituents of pentatomid scent glands when they identified monoterpenes in the DG1 and MG of *Hotea gambiae* (Scutelleridae). The biological role of the DG1 constituents of the Acanthosomatidae is not known.

Although information on the emission of male pheromones was not obtained using the method of effluvial collection, the trapping and subsequent desorption of *Agonoscelis versicolor* defensive compounds does indicate that the system described can be used as a viable method of sample collection in heteropteran semiochemical analysis. Thus thermal desorption can be effected using a simple method which requires no modification of the gas chromatograph. The two approaches to sample collection undertaken, namely gland dissection and effluvial collection, both have their relative merits.

REFERENCES

1. A. J. E. Gough, J. G. C. Hamilton, D. E. Games and B. W. Staddon, Multichemical defense of plant bug *Hotea gambiae* (Westwood) (Heteroptera: Scutelleridae), *J. Chem. Ecol.*, 11:343-351 (1985).
2. B. W. Staddon, M. J. Thorne and D. W. Knight, The scent glands and their chemicals in the aposematic cotton harlequin bug, *Tectocoris diophthalmus* (Heteroptera: Scutelleridae), *Australian J. Zool.*, 35:227-234 (1987).
3. C. Dupuis, Donnees nouvelles sur la morphologie abdominale des hemipteres, heteropteres et en particulier des Pentatomoidea, Comptes Rendus. XIIIe Congres International de Zoologie, Paris (1948).
4. C. Dupuis, Notes remarques et observations diverses sur les Hemipteres; premiere serie: notes I-IV, III-Dimorphisme sexuel de la glande dorso-abdominale anterieure de certains Asopinae (Pentatomoidae), *La Feuille des Naturalistes*, 7:1-4 (1952).
5. J. R. Aldrich, N M. S. Blum, H. A. Lloyd and H. M. Fales, Pentatomid natural products chemistry and morphology of III and IV dorsal abdominal glands of adults, *J. Chem. Ecol.*, 4:161-172 (1978).
6. J. G. C. Hamilton, A. J. E. Gough, B. W. Staddon and D. E. Games, Multichemical defense of plant bug *Hotea gambiae* (Westwood) (Heteroptera: Scutelleridae) (E)-2-hexenol from abdominal gland in adults, *J. Chem. Ecol.*, 11:1399-1409 (1985).
7. J. Carayon, Dimorphisme sexuel des glandes, tegumentaires et production de pheromones chez les Hemipteres Pentatomoidea, Comptes Rendus des Seances, *Academie des Sciences (Paris) Serie III*, 299:867-870 (1981).
8. W. C. Mitchell and R. F. L. Mall, Response of the female southern green stink bug and its parasite, *Trichopoda pennipes* to male stink bug pheromones, *J. Economic Entomol.*, 64:856-859 (1971).
9. S. Moriya and M. Shiga, Attraction of the male brown-winged green bug *Plautia stali* (Scott) (Heteroptera: Pentatomidae) for males and females of the same species, *App. Entomol. & Zool*, 19:317-322 (1984).
10. D. W. Knight, B. W. Staddon and M. J. Thorne, Presumed sex pheromone from androconial glands of male Cotton Harlequin bug *Tectocoris diophthalmus* (Heteroptera: Scutelleridae).Identified as 3,5-dihydroxy-4-pyrone, *Zeitschrift fur Naturforschung*, 40:851-853 (1985).
11. C. Pavis and C. Malosse, Physiologieanimale - mise on evidence d'un attractif sexual produit par les males de *Nezara viridula* (L) (Heteroptera: Pentatomidae), C.R. Acad. Sci. Paris, Series III, 7:272-276.
12. J. R. Aldrich, J. E. Oliver, W. R. Lusby and J. P. Kochansky, Identification of male specific exocrine secretions from predatory stink bugs (Hemiptera, Pentatomidae), *Archives Insect Biochem. & Physiol*, 3:1-12 (1986).

13. M. Borges, P. C. Jepson and P. E. Howse, Long-range mate location and close-range courtship behaviour of the Green Stink Bug, *Nezara viridula* and its mediation by sex pheromones, *Entomol. Experimentalis & Applicata*, 44:205-212 (1981).
14. P. Stys and I. Kerzhner, The rank and nomenclature of higher taxa in recent Heteroptera, *Acta. Ent. Bohemoslov.*, 72:65-79 (1975).
15. J. H. Cross, A vapor collection and thermal desorption method to measure semiochemical release rates from controlled release formulations, *J. Chem. Ecol.*, 6:781-785 (1980).
16. C. Kitamura, S. Wakamura and S. Takahashi, Identification and functions of ventral glands secretion of some Heteroptera, *App. Entomol. & Zool.*, 19:33-41 (1984).
17. B. W. Staddon, The scent glands of Heteroptera, *Adv. Insect Physiol.*, 14:351-418 (1979).
18. D. H. Calam and A. Youdeowei, Identification and functions of secretion from the posterior scent gland of fifth instar larva of the bug *Dysdercus intermedius*, *J. Insect Physiol.*, 14:1147-1158 (1968).
19. I. J. Everton, D. W. Knight and B. W. Staddon, Linalool from the metathoracic scent gland of the cotton stainer *Dysdercus intermedius* Distant (Heteroptera: Pyrrhocoridae), *Comp. Biochem.Physiol.*, 63B:157-161 (1979).
20. H. Daroogheh and T. O. Olagbemiro, Linalool from the cotton stainer *Dysdercus superstitiosus* (F.) (Heteroptera: Pyrrhocoridae), *Experientia*, 38:421-422 (1982).
21. J. R. Aldrich, W. R. Lusby, J. P. Kochansky and C. B. Abrams, Volatile Compounds from the predatory insect *Podisus maculiventris* (Hemiptera: Pentatomoidae): male and female metathoracic scent gland and female dorsal abdominal gland secretions, *J. Chem. Ecol.*, 10:561-569 (1984).

A NOTE ON KAIROMONES OF THE GREAT EUROPEAN

SPRUCE BARK BEETLE *DENDOROCTONUS MICANS*

D. R. Kelly and M. Barghian

School of Chemistry and Applied Chemistry
University of Wales College of Cardiff
Cardiff CF1 3TB, UK

INTRODUCTION

The Great European Spruce bark beetle, *D. micans*, is a major pest of spruce from eastern Siberia to the west of Europe. The first reported outbreak of *D. micans* in Britain was in 1982 but the earliest attacks probably occurred in 1973. It breeds under the bark of the spruce tree causing gradual death of the tree and its natural enemy is the bark beetle, *Rhizophagus grandis* Gyl. (*Coleoptera Rhizophagidae*).

The classical model for tree colonization by bark beetles involves three distinct phases [1,2]. Dispersal and random colonization is followed by a mass attack phase in which a large number of insects are attracted by aggregation pheromones [3] and finally late arrivals are deterred by repellent pheromones [4].

Sequences of this kind have been described in detail for *Ips typographus* and *Dendroctonus brevecomis* (the Western Pine Beetle), and to a lesser extent for other species. Nevertheless, it is only a prerequisite of species that are incapable of successfully colonizing a tree by single individuals.

If the host tree is weakened by climatic conditions, low nutrient levels in the soil, pathogenic fungi or prior attack by other species, then it is possible for an attack by an individual to be successful [5]. Naturally, the disadvantage of this strategy for the insect is that the host, because of its weakened condition, will not be able to provide as large a supply of nutrients as a healthy tree [6]. Alternatively, if the bark beetles are particularly large and vigorous, they may be able to overwhelm the defences of healthy trees by single or uncoordinated attacks. This appears to be the case for *Dendroctonus valens* which is found in North America and *D. micans*, the Great European Spruce Bark Beetle.

Unlike most bark beetles, attack by *D. micans* takes place by a single female. It breeds under bark causing destruction of the cambium, which debilitates and, in extreme cases, kills the tree. Because attack occurs by single individuals, there is no aggregation pheromone and, hence, no possibility of using conventional monitoring traps baited with a pheromone to monitor populations. This work is concerned with the identification of semiochemical cues, whereby the predator, *R. grandis*, is able to locate its prey, *D. micans*. It is believed that the volatile components of *D. micans* frass are responsible for the attraction.

MATERIALS AND METHODS

In order to obtain acid, base and neutral extracts, the crude extract (66g of diethyl ether soxhlet extract, from 150g frass) was initially treated with hydrochloric acid (1 M) and extracted with diethyl ether (3 x 150 ml). The ether layer containing the acid and neutral components of the crude extract was kept aside while the aqueous layer was treated with a solution of sodium hydroxide (2 M) (pH = 12 to 14). The aqueous suspension was then extracted with diethyl ether (3 x 150 ml). The ether layers were collected and dried over

Chromatography and Isolation of Insect Hormones and Pheromones
Edited by A. R. McCaffery and I. D. Wilson
Plenum Press, New York, 1990

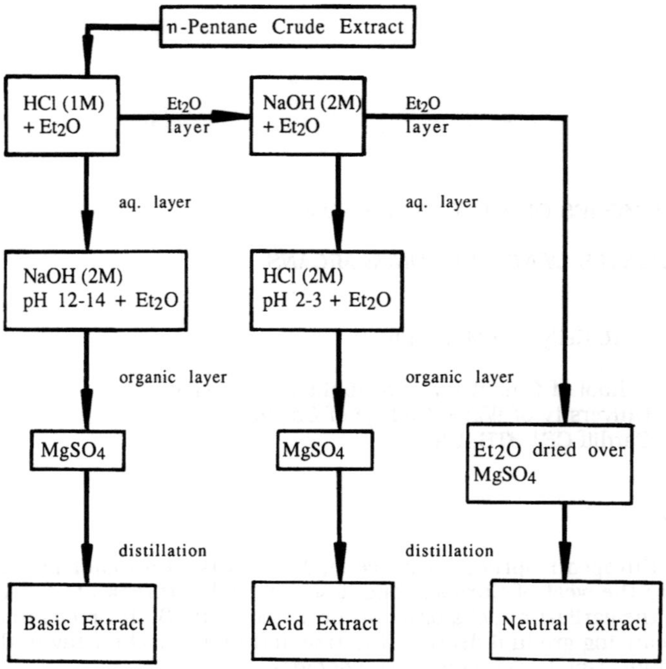

Fig. 1. Fraction scheme.

magnesium sulphate for a few hours. Subsequent filtration and evaporation of the solvent afforded the basic fraction of the crude extract. The ether layer containing the neutral and acidic components of the crude frass extract was further treated with a solution of sodium hydroxide (2 M). This was neutralized with a solution of hydrochloric acid (1 M) and extracted with diethyl ether (3 x 150 ml). The ether layers were dried over magnesium sulphate. Subsequent filtration and distillation of the solvent gave the acidic components of the crude extract. The remaining organic layer was dried over magnesium sulphate. Filtration and evaporation of the solvent gave the neutral extract. As a result, the acidic fractions of the crude extract were obtained (Figure 1). The majority of the material obtained was neutral with small variable amounts of acids and bases.

The neutral extract was separated into several small fractions using flash silica column chromatography with gradient elution from petroleum ether (40 to 60°C) to diethyl ether. The fractions were analyzed by 360 Mhz ^{1}H nuclear magnetic resonance spectroscopy (NMR) and gas liquid chromatography (GLC).

The crude soxhlet extract (17.2 g) was distilled under a water vacuum to 75°C and gave 4 g of distillate, the residue was further distilled under high vacuum to 70°C to give 1.88 g of hydrocarbons. The two distilled fractions were essentially identical. The residue weighing 11.4 g also consisted of hydrocarbons, 99% of the material was accounted for.

RESULTS AND DISCUSSION

Frass was collected from infested Norway or Sitka spruce and stored at -18°C. In some cases, it was possible to obtain several hundred grams from a single tree. The frass was extracted using n-pentane or diethyl ether, in a soxhlet continuous extraction apparatus for a period of 3 to 5 days. The crude extracts normally weighed between 30 to 50% of the original weight of the frass and were partitioned into acid, base and neutral fractions (Figure 1). In most cases, the majority of the material obtained was in the neutral fraction (approximately > 90%) with small and variable amounts of acids and bases. *Rhizophagus grandis* beetles have shown a high response to both crude and neutral extracts but low activity has been observed with acid and base extracts. Further separation of the neutral extract by flash silica gel column chromatography afforded several smaller fractions, which were each tested for biological activity in walking and flying assays. All of the fractions showed some activity, which was a little stronger in the earliest eluting fractions

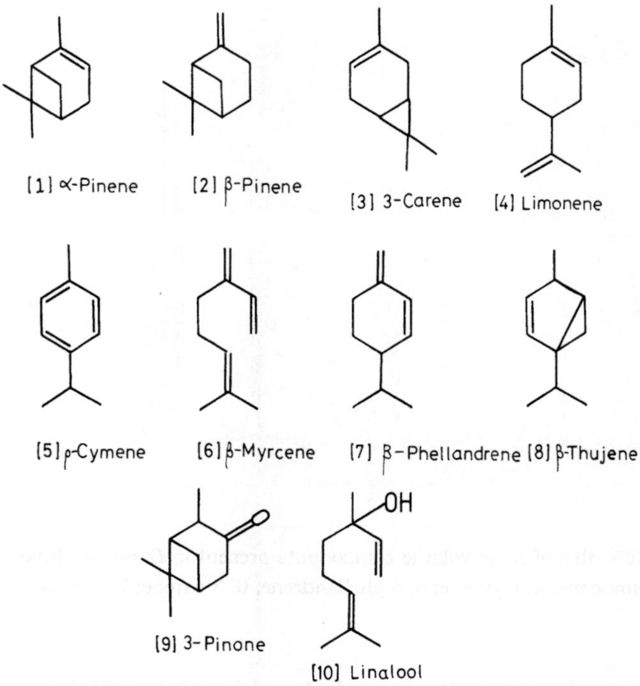

Fig. 2. **Monoterpenes found in *D. micans* frass.**

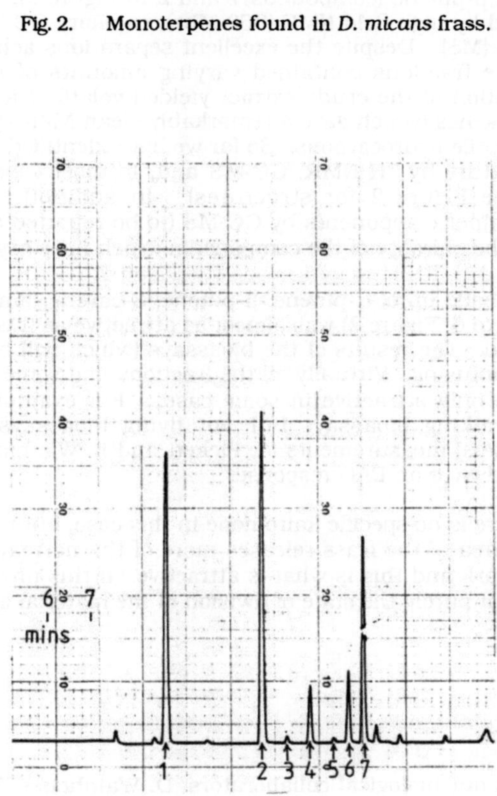

Fig. 3. **GLC trace of crude neutral extract, OB 5 column, 4°/min, ambient to 240°C. 1. α-pinene; 2. β-piene; 3. β-myrcene; 4. 3-carene; 5. p-cymene; 6. β-phellandrene; 7. limonene.**

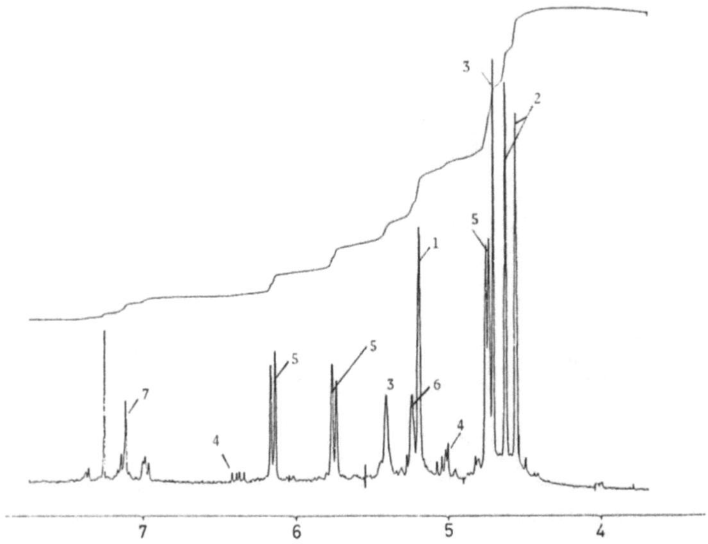

Fig. 4. ^1H NMR (360 Mhz) of some volatile components present in *D. micans* frass. 1. α-pinene; 2. β-pinene; 3. limonene; 4. myrcene; 5. β-phellandrene; 6. 3-carene; 7. p-cymene.

which contained α-and β-pinene (compounds 1 and 2 in Figure 2). The components of the fractions were identified by high field NMR (360 Mhz), capillary GC (Figure 3) and capillary GC-mass spectroscopy (MS). Despite the excellent separations achieved by flash column chromatography all the fractions contained varying amounts of high molecular weight hydrocarbons. Distillation of the crude extract yielded volatile fractions (35% by weight) containing terpenoid alkenes (which gave a remarkably clean NMR spectrum, Figure 4) and a residue (65%) of involatile hydrocarbons. So far we have identified the ten most abundant components of the mixture by ^1H NMR, GC-MS and, in most cases, by coinjection with authentic samples (see Figure 2 for structures). In addition, we have provisionally identified another 20 minor components by GC-MS (to be reported elsewhere). Work is in progress to determine the chirality of the components which have asymmetric centers [7,8].

A synthetic lure made up of α-pinene, β-pinene, 3-carene, limonene , ρ-cymene and myrcene (compounds 1 to 6, Figure 2) was almost as attractive as frass in the bioassays, but less effective in the field. The results of the bioassays (which will be described elsewhere) have been somewhat confusing. Virtually all the fractions had some attractiveness, but the pure components were only attractive in some cases. For example, pure α-pinene was highly attractive in walking bioassays but not flying bioassays. Similarly, GC-MS-electroantennograph (EAG) measurements (J. Pickett and L. Wadhams) demonstrated that many components could elicit an EAG response.

It may be that there is no specific kairomone in this case, but the wounding of the tree and the large surface area of the frass releases more of the normal tree terpenes than an equivalent amount of bark and this is what is attractive. If this proves to be true, this will be the first case in which purely the state of division of the material causes a semiochemical phenomenon.

ACKNOWLEDGEMENTS

We wish to thank our biological collaborators, D. Wainhouse, D. Cross and T. Wyatt, without whom this project would not have been possible and the Forestry Commission for providing funding for MB.

REFERENCES

1. D. L. Wood, The role of pheromones, kairomones and allomones in the host selection and colonization behavior of bark beetles, *Ann. Rev. Entomol.*, 27:411 (1982).
2. M. C. Birch, Aggregation in bark beetles, *in*: "Chemical Ecology of Insects", W. J. Bell and R. T. Carde, eds., Chapman and Hall, New York, pp 331-353 (1984).
3. J. P. Vite and W. Francke, The aggregation pheromones of bark beetles: progress and problems, *Naturwissenschaften*, 63:550 (1976).
4. J. A. Byers, Male-specific conversion of the host plant compound, myrcene to the pheromone (+)-ipsdienol, in the bark beetle, *Dendroctonus brevecomis*, *J. Chem. Ecol.*, 8:363 (1982).
5. A. A. Berryman, Site characteristics and population dynamics: a theoretical perspective, *in*: "Site Characteristics and Population Dynamics of Lepidopteran and Hymenopteran Forest Pests", D. Bevan and J. T. Stoakley, eds., Forestry Commission, Research and Development Paper 135, Edinburgh (1985).
6. D. Wainhouse, *in*: "Forests, Integrated Pest Management", Academic Press, London, pp 361-401 (1987).
7. H. C. Brown and N. N. Joshi, Hydroboration of terpenes. 9. A simple improved procedure for upgrading the optical purity of commercially available α– and β-pinenes. Conversion of (+)-α-pinene to (+)-β-pinene via hydroboration isomerization, *J. Org. Chem.*, 53:4059 (1988).
8. M. Lindstrom, T. Norin, G. Birgersson and F. Schlyter, Variation of enantiomeric composition of α-pinene in Norway Spruce, *Picea abies*, and its influence on production of verbenol isomers by *Ips typographus* in the field, *J. Chem. Ecol.*, 15:541 (1989).

A NOTE ON PYGIDIAL GLANDS OF PRIMITIVE AUSTRALIAN ANTS:

A NEW SOURCE OF ODOROUS CHEMICALS

B. D. Jackson, E. D. Morgan and J. P. J. Billen*

Department of Chemistry, University of Keele
Keele, Staffordshire ST5 5BG, UK, and
*Zoological Institute of the University
B-3000 Leuven, Belgium

INTRODUCTION

The pygidial gland of ants is an abdominal gland associated with the intersegmental membrane between the 6th and 7th abdominal tergites. Large pygidial glands are found in most species of the subfamily Dolichoderinae and in a number of species scattered throughout other subfamilies. However, the only non-dolichoderine pygidial gland to have been studied chemically is that of the ponerine *Rhytidoponera metallica* [1].

The pygidial glands of *Nothomyrmecia macrops* Clark and *Myrmecia nigriceps* Mayr, two primitive Australian ants, are relatively large and occur laterally in pairs (Figure 1). Using gas chromatography-mass spectrometry (GC-MS) and a solventless injection technique, they were found to contain nanogram quantities of volatile substances.

MATERIALS AND METHODS

Workers of *N. macrops* and *M. nigriceps*, identified by Dr R. W. Taylor, were killed by placing in a Biofreezer at -50°C and then dissected under water. Pygidial glands were removed with some cuticle and sealed immediately in soft-glass capillaries (2 mm x 2 cm) in pairs. Samples of cuticle from the 2nd abdominal tergite were similarly sealed to serve as controls. The samples were analyzed by GC-MS on a Hewlett Packard 5890 Gas Chromatograph and 5970B Mass Selective Detector with HP59970C ChemStation. A fused silica capillary column (12 m x 0.2 mm) coated with HP-1 (cross linked methyl silicone gum ≅ OV-1) of 0.33 μm film thickness was used. The carrier gas was helium at 10 psi column head pressure. The capillary tubes containing the samples were kept in the solid injector [2] in the injection port at 200, 170 or 140°C for 2 to 3 min before crushing. A splitless injection was performed with the injection port purge turned on after 0.5 min. The oven temperature was initially 30°C for 2 min then increased at a rate of 4°C/min to 250°C. The Mass Selective Detector was set to monitor m/z 35 to 350 in the scan mode under "Autotune" conditions using 70 eV ionization. Identification was confirmed by comparison of the mass spectra and retention times with those of synthetic specimens either purchased or prepared as indicated by Billen et al. [3] and Jackson et al. [4]. The quantity of each component was determined by external standards.

RESULTS

The pygidial glands of *N. macrops* and *M. nigriceps* were found to contain mixtures of ketones, terpenoid compounds, lactones and indole (Table 1). Typical chromatograms of the contents of the paired glands of single insects are shown in Figures 2 and 3.

M. nigriceps, the larger ant (body length 20-35 mm), contains on average nearly 10 times as much material as *N. macrops*, and a greater number of components were detected.

Chromatography and Isolation of Insect Hormones and Pheromones
Edited by A. R. McCaffery and I. D. Wilson
Plenum Press, New York, 1990

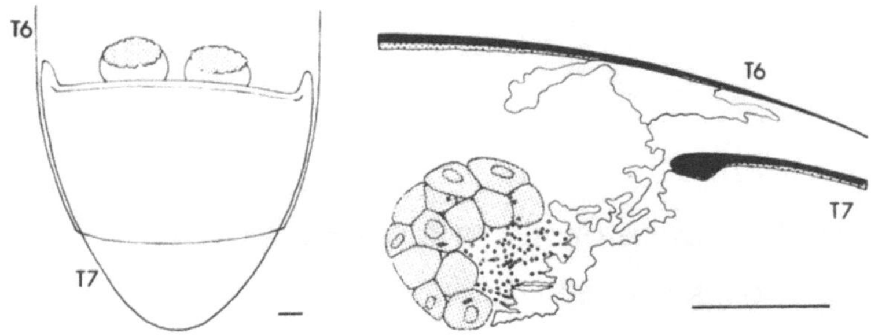

Fig. 1. Dorsal view of the abdomen of *M. nigriceps* showing the position of the pygidial glands under the 6th tergite (T6), and a longitidual vertical section, showing the secretory cells and the invagination of the intersegmental membrane which forms a storage chamber.

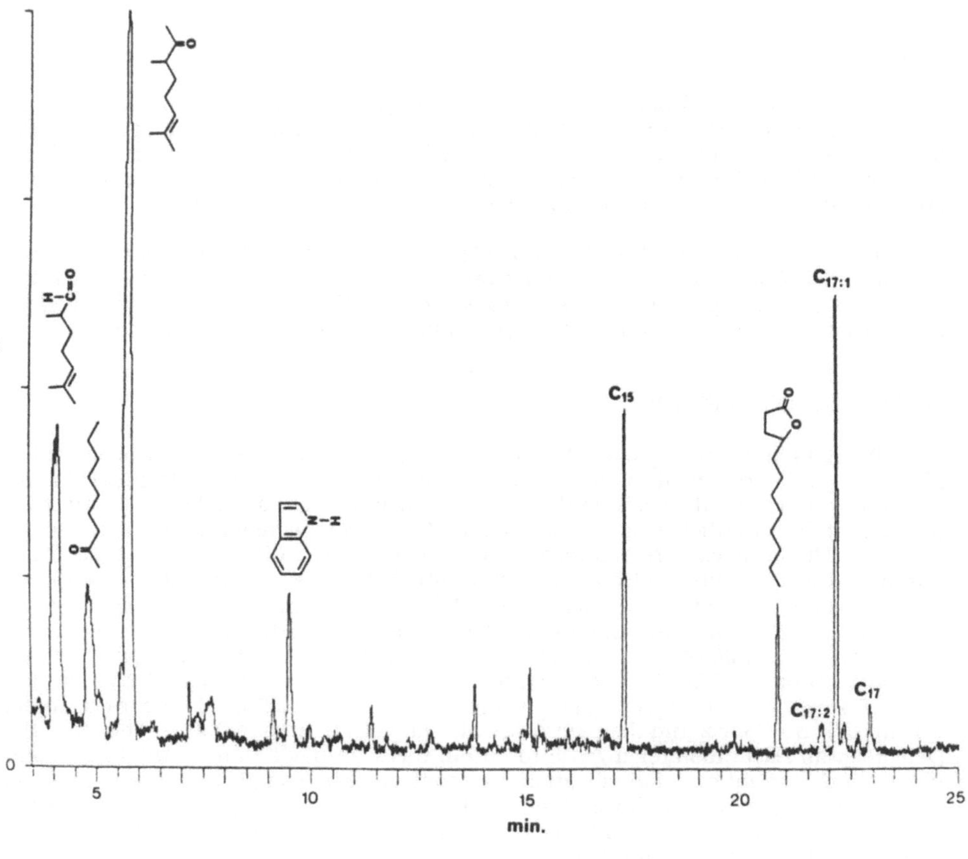

Fig. 2. Typical total ion chromatogram of the contents of the pygidial glands of a single worker of *N. macrops*. The hydrocarbons probably arise from the Dufour secretion.

The principal constituents of *N. macrops* and *M. nigriceps* were 3,7-dimethyloct-6-en-2-one and (E)-nerolidol, respectively, both of which have not previously been identified in insects.

To show that the mixture of farnesene isomers was actually present in the gland, and that the more stable (E,E)-α-and (E)-β-isomers did not arise from isomerization in the injection port, the temperature of the injection port was varied from 140°C to 200°C. Little change in the relative proportions resulted. Further confirmation was provided by

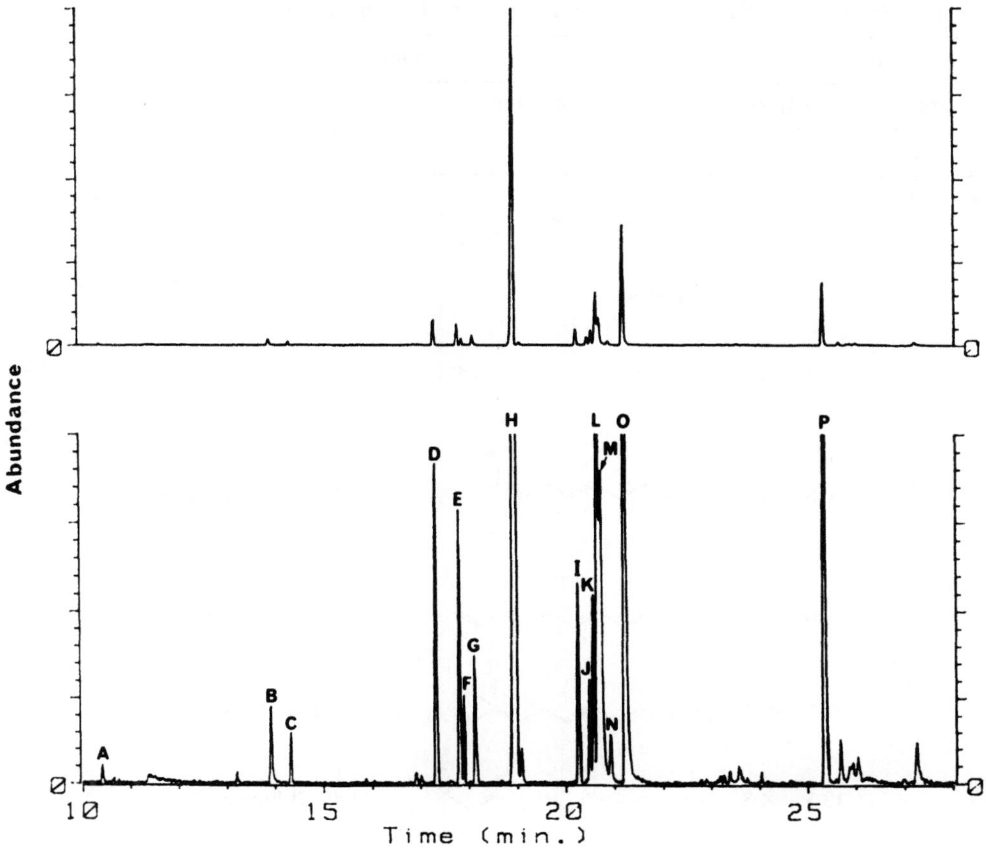

Fig. 3. Typical total ion chromatogram of the contents of the pygidial glands of a single worker of *M. nigriceps*. The lower trace is at higher sensitivity to show the minor components. For identification of the compounds see Table 1.

examining glands that contain only the (Z,E)-α-isomer by the same method. No trace of the more stable isomers was found [5].

DISCUSSION

The compounds present in the pygidial glands of these two primitive Australian ants have previously been found in a variety of insect secretions [6,7], as shown in Table 1. The exceptions are nerolidol, 3,7-dimethyl-6-en-2-one and δ-dodecalactone which have not been reported from insects.

Although most of the components have been found in other ant secretions, only two occur in the ant pygidial glands. 2-Nonanone and 2-undecanone have been reported from *Azteca* species (Dolichoderinae) [7]. The large pygidial glands of most dolichoderine species typically contain iridoids, a group of cyclopentanoid monoterpenes, but some low molecular weight terpenoid and linear ketones have also been found [8]. The only other ant pygidial gland to have been investigated chemically is that of the ponerine *Rhytidoponera metallica*, which contained 3-hydroxybenzaldehyde, isogeraniol, heptadecane and heptadecene [9].

N. macrops and *M. nigriceps* both contain 2-nonanone, γ-dodecalactone and indole. The pygidial glands of these two species thus have common characteristics, but they are both quite different from the dolichoderine and ponerine species that have been studied. We have already drawn attention to the similarity between the Dufour gland secretion of these two species [9,10]. Thus, the chemistry of the secretions of these species provides evidence of

Table 1. Mean proportions of substances found in the pygidial glands of *N. macrops* and

Name				Compound		
N. macrops		*N. nigriceps*		Structure	Name	
%	SD	%	SD			
14	34	-	-	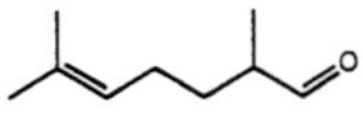	2,6-Dimethylhept-5-enal	
65	57	-	-	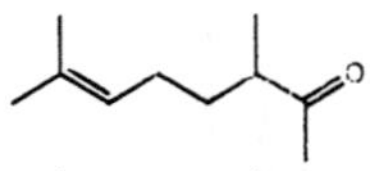	3,7-Dimethyloct-6-en-2-one	
-	-	1.8	0.8	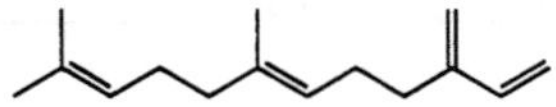	(E)-β-Farnesene	D†
-	-	0.5	0.2	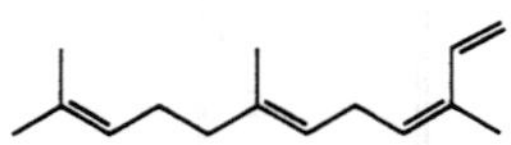	(Z,E)-α-Farnesene	F
-	-	1.0	0.3	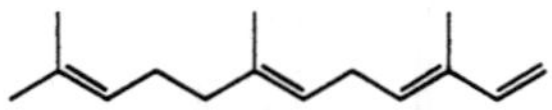	(E,E)-α-Farnesene	G
-	-	41.4	10.0	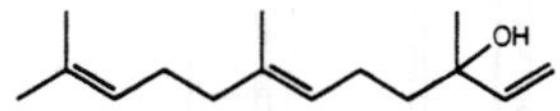	(E)-Nerolidol	H
-	-	13.5	4.8	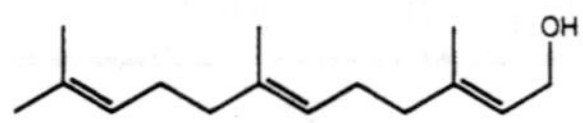	(E,E)-Farnesol	O
-	-	7.4	2.9	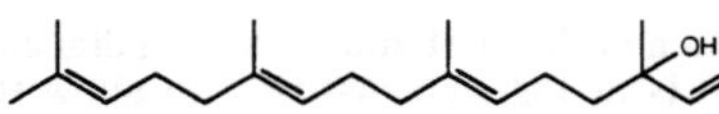	Geranyllinalool	P
12	10	0.4	0.4	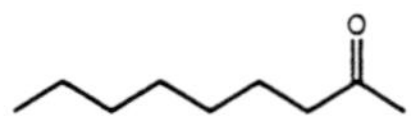	2-Nonanone	A
-	-	1.6	0.7	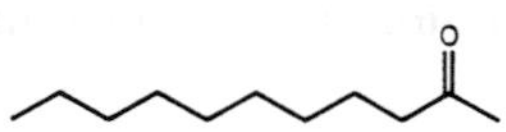	2-Undecanone	C
-	-	8.6	4.5	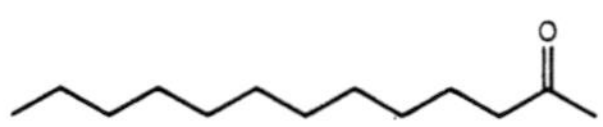	2-Tridecanone	E
-	-	1.2	1.1	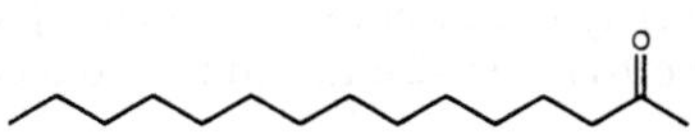	2-Pentadecanone	N
6	4	2.4	0.8	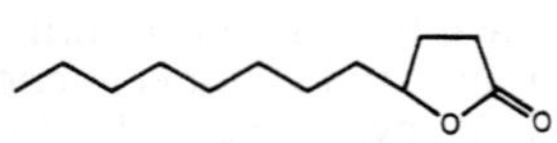	γ-Dodecalactone	I
-	-	9.9	3.9	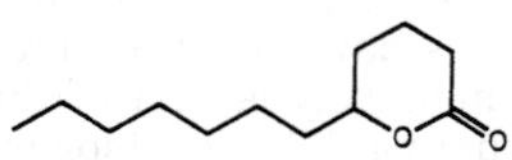	δ–Dodecalactone	L
4	2	0.5	0.3	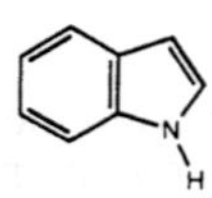	Indole	B
51 ng		442 ng			Mean total amount	

* Wheeler & Duffield [6]; Blum [7].
† Peak identification for Fig. 3.

M. nigriceps, and the occurrence of these substances in other insects.

Occurrence in other insects*	
Ants	**Others**
Subfamily: *Genus* (gland)	Order: *Genus* (gland)
Formicinae: *Acanthomyops, Lasius* (Mandibular)	-
-	-
Myrmicinae: *Manica* (Dufour)	Hymenoptera: *Andrena* (Dufour) Isoptera: *Reticulitermes* (Frontal) Homoptera: Aphidiae - *many genera* (Alarm pheromone)
Formicinae: *Camponotus, Formica, Polyergus* (Dufour) Myrmicinae: *Myrmica, Manica, Solenopsis Pheidole, Aphaenogaster* (Dufour)	Hymenoptera: *Andrena* (Dufour)
Myrmicinae: *Manica, Solenopsis* (Dufour)	-
-	
Myrmicinae: *Atta* (Mandibular)	Hymenoptera: *Andrena* (Dufour) *Apis* (Nasanov); *Bombus, Psithyrus* (Labial)
-	Isoptera: *Reticulitermes* (Frontal)
Dolichoderinae: *Azteca* (Pygidial)	Hymenoptera: *Andrena, Bombus, Trigona* (Mandibular) Hemiptera: *Nezara* (Metasternal)
Dolichoderinae: *Azteca* (Pygidial) Ponerinae: *Bothroponera* (Mandibular) Myrmicinae: *Tetramorium* (Mandibular)	Hymenoptera: *Andrena, Bombus Trigona* (Mandibular) Isoptera: *Armitermes* (Frontal) Lepidoptera: *Heterocampa* (Cervical of Larvae)
Formicinae: *Acanthomyops, Camponotus Formica, Lasius, Notoncus* (Dufour)	Hymenoptera: *Bombus, Trigona* (Mandibular) Isoptera: *Armitermes, Curvitermes, Schedorhinotermes* (Frontal)
Formicinae: *Acanthomyops, Camponotus Gigantiops, Lasius, Notoncus* (Dufour)	Hymenoptera: *Trigona* (Mandibular) Isoptera: *Armitermes, Schedorhinotermes* (Frontal)
-	Coleoptera: *Bledius* (Pygidial)
-	-
-	Trichoptera: *Pycnopsyche* (Sternal)

a closer relationship than the classification into different subfamilies, Nothomyrmeciinae and Myrmeciinae, (cf. Ref. [11]) would suggest.

No behavioral studies have been carried out to determine the function of the pygidial gland secretion in *M. nigriceps*. However, Hölldobler and Taylor [12] suggested that the pygidial gland secretion of *N. macrops* has an alarm-defense function, perhaps in encounters with other ants. γ-Dodecalactone may contribute to this effect, since the homologue γ-decalactone, from the anal exudate of the thrips *Bagnalliella yuccae* (Thysanoptera), is known to repel small predators such as ants [13].

The compounds present in these 2 Australian ants have been found to have a variety of behavioral effects, mainly deterrent and alarm, in other insects, and they are used widely in defensive secretions.

Nerolidol, the major component of *M. nigriceps*, has recently been shown to be highly deterrent and harmful to the leaf-cutting ant *Atta cephalotes* [14]. 2-Undecanone and 2-tridecanone are minor components in the cervical gland opening of the underside of the neck in larvae of *Heterocampa manteo* (Lepidoptera: Notodontidae) which is dominated by formic acid. The ketones are reported to increase considerably the deterrent effectiveness of formic acid [15].

The principle alarm pheromone utilized by several aphid species is (E)-β-farnesene, and it is now known to occur in species from 19 genera in 3 aphid subfamilies [16]. 2-Heptanone, homologous to 2-nonanone found in both *N. macrops* and *M. nigriceps*, is present in *Apis* (Hymenoptera: Apidae) worker mandibular glands, and is known to act as an alarm pheromone in conjunction with sting gland secretions [17].

A widespread function of the mandibular gland secretions of bees is that of defense [17], and 2-alkanones occur in these secretions in species from 3 genera. In the defensive frontal gland secretions of soldier termites geranyllinalool, β-farnesene and 2-alkanones are produced.

The nature of the constituents of the pygidial glands of *N. macrops* and *M. nigriceps* supports the idea of an alarm-defense role, but their function remains to be investigated.

ACKNOWLEDGEMENTS

We thank Dr R. W. Taylor for his help in collecting the ants, the Belgian National Fund for Scientific Research for a travel grant to Australia for J.P.J.B., the Royal Society and the SERC for grants for equipment to E.D.M. and the SERC for a research studentship to B.D.J.

REFERENCES

1. J. Meinwald, D. F. Wiemer and B. Hölldobler, Pygidial gland secretions of the ponerine ant *Rhytidoponera metallica*, *Naturwissenschaften*, 70:46-47 (1983).
2. E. D. Morgan and L. J. Wadhams, Gas chromatography of volatile compounds in small samples of biological materials, *J. Chromatog. Sci.*, 10:528-529 (1972).
3. J. P. J. Billen, B. D. Jackson and E. D. Morgan, The contents of the pygidial gland of the primitive ant *Nothomyrmecia macrops* (Hymenoptera: Formicidae), *Experientia*. 44:794-797 (1988).
4. B. D. Jackson, E. D. Morgan and J. P. J. Billen, Contents of the pygidial gland of the ant *Myrmecia nigriceps* (Hymenoptera: Formicidae), submitted, (1990).
5. M. F. Ali, A. B. Attygalle, J. P. J. Billen, B. D. Jackson and E. D. Morgan, Change of Dufour gland contents with age of workers of *Formica sanguinea* (Hymenoptera: Formicidae), *Physiol. Entomol.*, 13:249-255 (1988).
6. J. W. Wheeler and R. M. Duffield, Pheromones of Hymenoptera and Isoptera, *in*: "Handbook of Natural Pesticides Vol. 4B Phermones", E. D. Morgan and N. B. Mandava, eds., CRC Press, Boca Raton, pp 59-206 (1988).
7. M. S. Blum, "Chemical Defences of Arthropods", Academic Press, New York and London (1981).
8. A. B. Attygalle and E. D. Morgan, Chemicals from the glands of ants, *Chem. Soc. Rev.*, 13:245-278 (1984).

9. J. P. J. Billen, B. D. Jackson and E. D. Morgan, Secretion of the Dufour gland of the ant *Nothomyrmecia macrops* (Hymenoptera: Formicidae), *Experientia*, 44:715-719 (1988).
10. B. D. Jackson, J. P. J. Billen and E. D. Morgan, Dufour gland contents of 3 species of *Myrmecia* (Hymenoptera: Formicidae), primitive ants of Australia, *J. Chem. Ecol.*, 15:2191-2205 (1989).
11. R. W. Taylor, *Nothomyrmecia macrops*: a living-fossil ant rediscovered, *Science*, 201:979-985 (1978).
12. B. Hölldobler and R. W. Taylor, A behavioral study of the primitive ant *Nothomyrmecia macrops* Clark, *Insectes Sociaux*, 30:384-401 (1983).
13. D. F. Howard, M. S. Blum and H. M. Fales, Defense in thrips: forbidding fruitiness of a lactone, *Science*, 220:335-336 (1983).
14. J. J. Howard, J. R. J. Cazin and D. F. Wiemer, Toxicity of terpenoid deterrents to the leafcutting ant *Atta cephalotes* and its mutualistic fungus, *J. Chem. Ecol.*, 14:59-69 (1988).
15. T. Eisner, A. F. Kluge, J. E. Carrel and J. Meinwald, Defense mechanisms of arthropods XXXIV Formic acid and acyclic ketones in the spray of a caterpillar, *Ann. Entomol. Soc. Amer.*, 65:765-766 (1972).
16. L. R. Nault and P. L. Phelan, Alarm pheromones and sociality in pre-social insects, *in*: "Chemical Ecology of Insects", W. J. Bell and R. T. Cardé, eds., Chapman and Hall, London, pp 237-256 (1984).
17. R. M. Duffield, J. W. Wheeler and G. C. Eickwork, Sociochemicals of bees, *in*: "Chemical Ecology of Insects", W. J. Bell and R. T. Cardé, eds., Chapman and Hall, London, pp 387-428 (1984).

A NOTE ON VOLATILES FROM EXOCRINE GLANDS OF SOME GRASSLAND ANTS AND TERMITES FROM TROPICAL WEST AFRICA

T. O. Olagbemiro, K. M. Sani* and B. W. Staddon**

Department of Chemistry, Abubakar Tafawa Balewa
University, PMB 0248, Bauchi, Nigeria
*Department of Chemistry, Nigerian Defence Academy
PMB 2109, Kaduna, Nigeria, and
**School of Pure and Applied Biology
University of Wales College of Cardiff
PO Box 915, Cardiff CF1 3TL, UK

INTRODUCTION

Ants and termites form an important component of the soil fauna of scrub-savannah in Northern Nigeria. The seed gatherer (harvester ant) *Messor galla* (Mayr) and honeydew ant *Camponotus maculatus* (Fabricus) are among the commoner species of ants to be found in such areas. The commonest termites include the grass cutter *Trinervitermes geminatus* (Wassman) and the tree climber *Odontotermes magdalenae* (Grasse). The need to understand the ecology of these social insects is essential, in view of their potential competitive interactions with man through a common requirement for grains and grassland. As a contribution to the chemical ecology of the social insects of tropical grassland we report here the chemical analyses of volatile secretions from Dufour's glands in *M. galla* and *C. maculatus*, from the frontal gland in *T. geminatus*, and from the salivary gland in *O. magdalenae*.

MATERIALS AND METHODS

Workers of *Messor galla* (Mayr), *Camponotus maculatus* (Fabricus) as well as soldiers of *Trinervitermes geminatus* (Wassman) and *Odontotermes magdalenae* (Grasse) were separately collected at night from their colonies in open fields on the campus of Bayero University, Kano, Nigeria and the Institute for Agricultural Research, Ahmadu Bello University, Kano Station. They were chilled prior to the isolation of scent materials. Dufour's scent glands of *Messor* and *Camponotus* ants were excised under a 200 mM sodium chloride solution and the scent volatiles extracted in chromatoquality spectrograde ether which had previously been distilled. Scent materials from the frontal glands of soldiers of *Trinervitermes* were isolated in a saline solution as above by removing the termite head from the body and extraction made by crushing the heads in ether solvent as above. The *Odontotermes* volatiles from the salivary gland of soldiers were again extracted into chromatoquality spectrograde ether solvent after the bilobed sac had been carefully excised from the abdomen under saline. Extracts (sample volume, ca. 5 μl) for each species were stored at low temperature in glass ampoules after sealing in the flame of a gas burner and transported by air for analytical work in the Department of Chemistry at the University of Sussex.

The secretion components were examined by gas chromatography (GC) and gas chromatography-mass spectroscopy (GC-MS). Identification of the components was tentatively based on mass spectral data and confirmed by comparison of the GC retention data and fragmentation pattern of mass spectra with those of authentic samples. In some cases GC identifications were corroborated by co-chromatography with authentic standards.

Chromatography and Isolation of Insect Hormones and Pheromones
Edited by A. R. McCaffery and I. D. Wilson
Plenum Press, New York, 1990

Analysis of the scent volatiles was carried out on a Packard model 427 gas chromatograph equipped with a flame ionization detector and glass capillary column (2 m x 2 mm) packed with 5% OV-225 on 100 to 120 mesh Gas Chrom G. The flow rate of the carrier gas (nitrogen) was 30 ml/min. The oven temperature was 70°C isothermal for 15 min, then programmed to 270°C at 6°C/min. Mass spectra were measured in electron impact (EI) modes at 70 eV and 100 μA ionizing current using a Kratos MS25 mass spectrometer interfaced to a Carlo-Erbo-Fractovap 4200 gas chromatograph. Separations were achieved using a silicon OV column; 2 m x 2 mm 5% OV-225 on 100 to 120 mesh Gas Chrom G, 30 ml/min, 70°C isothermal for 15 min, and then temperature programmed to 200°C at 6°C/min. Data was acquired using a Kratos D555M data system. Authentic standards from Fluka and BDH Chemicals were used where possible, for comparative GC and GC-MS purposes.

RESULTS AND DISCUSSION

The series of hydrocarbons within the range of C_{10} to C_{21} which were present in the Dufour's glands of both the seed gathering ant *M. galla* and the honeydew ant *C. maculatus* are shown in Table 1. Aliphatic hydrocarbons C_{13} to C_{21} (*M. galla*) and C_{10} to C_{17} (*C. maculatus*) were recorded from the two ant species. In general, the Dufour's glands of ants have a remarkable ability to synthesize hydrocarbons and over sixty alkanes and alkenes have been identified and reported in Dufour's gland secretions [1-3]. As in most *formicine* [4], *myrmicine* [1] as well as species of *ponerine* [3], *pseudomyrmecine* [1], *myrmecine* [3], and *dolichoderine* [4], the Dufour's gland of ants are typically but not exclusively filled with linear hydrocarbons with odd numbers of carbon atoms. Hydrocarbons with an even number of carbon atoms are sometimes encountered as minor components [5]. Branched chain hydrocarbons, alkadienes and terpenoids, as well as oxygenated compounds, have been recorded in a number of species of *Myrmica* [6], *Formica* [7], *Lasius* [8], *Camponotus* [9,10], *Notoncus* [11] and *Gigantiops* [12]. In our work we found no branched chain hydrocarbons in either *M. galla* or *C. maculatus*. Pentadecadiene and heptadecadience were detected in our work on the Dufour's gland secretions of *M. galla* and *C. maculatus*. 4,7-Tridecadiene is the only recorded alkadiene to have been detected in the Dufour's gland of workers from the genus *Camponotus* [9]. For all hydrocarbons detected the mass fragmentation patterns were typical of alkadienes, alkenes and alkane hydrocarbons (Table 1). Benzyl alcohol and 4-phenyl-3-buten-2-one (data in Table 1) were the only oxygenated scent volatiles recorded in small quantities from the Dufour's gland of *M. galla*. Although a homologue of benzyl alcohol, 2-phenylethanol, had been recorded from the mandibular glands of males of *Camponotus clavithorax* [13], to our knowledge, benzyl alcohol and 4-phenyl-3-buten-2-one have not been previously recorded in any ant material.

Termites soldier scent volatiles are chemically more novel and variable than in any other insect taxon of comparable species number [14]. Our findings on a study of the frontal gland secretions of soldiers of *T. geminatus* (Table 2) indicated the presence of monoterpenes α-pinene, β-pinene, (Z)-β-ocimine, camphene, limonene and nothing else. The aromatics, toluene and xylene, were detected as minor components from the whole insect solvent extract. The presence of the monoterpenes was confirmed from GC retention times, co-chromatography with authentic standards, as well as supportive mass spectra data from both the eight peak index as well as referenced authentic standards. Monoterpenic hydrocarbons such as α-pinene, β-pinene, limonene, camphene, myrcene, phellandrene, terpinolene and α-terpinene have been reported in the frontal glands of some species of nasute termites [15,16]. Previous studies of the frontal gland contents of *T. geminatus* from the Savannah grassland of East Africa and West Africa reported the presence of fatty acids, alcohols and ketones, monoterpene hydrocarbons, diterpenes as well as esters [14,17,18].

The salivary gland secretions of *O. magdalenae* from the scrubland of Africa, South of the Sahara, on which no chemical work had been previously recorded [19,20], contained 2-methyl-1,4-benzoquinone, 2-ethyl-1,4-benzoquinone, 2-hydroxy-5-methyl-1,4-benzoquinone and 2-methyl-1,4-hydroxy-benzoquinone (Table 3). Toluene was detected as a minor component from the total body solvent extract from this termite species. Quinones have been recorded as the major components of the salivary glands of some *Odontotermes* species from South America [21,22]. Some of the quinones recorded included benzoquinone, toluquinone, 1,4-benzoquinone, 2-methyl-1,4-benzoquinone and 2-ethyl-1,4-benzoquinone.

These results provide yet a further illustration of the diversity of natural product chemistry to be found within insects. Among the various functions assigned to Dufour's

Table 1. Volatile Compounds Found in Dufour's Gland of Worker Ants of *Messor galla* and *Camponotus maculatus*

Compound	Dufour's Gland M.g.	C.m.	mol. wt.	Mass of fragment ion[a], m/z (% abundance)
n-alkane				
C_{10}		+	142	142(M+;3), 85(15), 71(75), 57(70), 56(23), 55(30), 43(100), 41(40)
C_{11}		+	156	156(M+;17), 85(40), 71(80), 57(90), 56(34), 55(35), 43(100), 41(88)
C_{13}	+	+	184	184(M+;5), 85(42), 71(85), 57(98), 56(27), 55(28), 43(100), 47(71)
C_{15}	+	+	212	212(M+;2), 85(43), 71(83), 57(100, 56(23), 55(25), 43(98), 41(65)
C_{19}	+		268	268(M+;5), 85(72), 71(89), 57(100), 56(31), 55(47), 43(97), 41(77)
C_{21}	+		296	296(M+;5), 85(78), 71(100), 57(190), 56(25), 55(43), 43(85), 42(88)
n-alkene				
C_{11}		+	154	154(M+;3), 83(10), 70(19), 69(21), 57(22), 56(25), 55(31), 43(100)
C_{12}		+	168	168(M+;2), 83(27), 69(30), 56(20), 55(35), 41(53), 43(100), 41(50)
C_{15}	+		210	210(M+;3), 83(9), 70(55), 69(82), 57(85), 56(77), 55(60), 43(100)
C_{21}	+		294	294(M+;2), 97(68), 83(75), 71(62), 69(65), 57(100), 55(96), 43(95)
alkadiene[c]				
C_{15}	+		208	208(M+;1), 167(3), 155(5), 137(7), 123(12), 82(18), 68(73), 55(63), 54(100), 41(90), 42(82)
C_{17}		+	236	235(M+;2), 180(4), 165(5), 137(10, 123(13), 81(30), 68(25), 57(100, 41(71)
Miscellaneous				
Acetic acid		v.g.	60	60(M+;20), 45(100)
Benzyl alcohol[b]	+		108	108(M+;85), 107(65), 91(18), 79(100), 77(57), 57(45), 51(32)
4-phenyl-3-buten-2-one[b,c]		+	146	146(M+;36), 131(100), 103(54), 77(38), 57(27), 51(33), 43(45), 41(17)
Limonene[b]	t.ext.		136	136(M+;8), 121(10), 107(7), 105(9), 94(10), 93(42), 92(35), 91(18), 79(30), 77(25), 68(100), 67(35), 59(20), 53(35)

Abbreviations:
M.g., *Messor galla*; C.m., *Camponotus maculatus*; v.g., venom gland; t.ext., total extract;
[a] significant ions above m/z 40; [b] confirmed by mass spectrum and retention time (OV225) of authentic sample; [c] Bond geometry not determined.

Table 2. Monoterpenes from Frontal Gland Secretions of Soldiers of *Trinervitermes geminatus*

Compound	mol. wt.	Mass of fragment ion[a], m/z (% abundance)
α-Pinene[b]	136	136(M⁺;8), 121(14), 107(7), 105(12), 93(100), 92(30), 91(35), 80(10), 79(23), 77(28), 67(10), 65(5), 55(6), 53(10), 51(11), 41(20)
(Z)-β-ocimene[b]	136	136(M⁺;9), 121(21), 107(8), 105(15), 93(100), 92(35), 91(30), 80(10), 79(20), 77(25), 67(8), 65(5), 53(6), 41(6)
Camphene[b]	136	136(M⁺;12), 135(19), 121(60), 107(22), 105(5), 96(12), 94(10), 93(100), 92(10), 90(30), 79(36), 77(24), 67(15), 53(5), 41(8)
β-Pinene[b]	136	136(M⁺;9), 121(13), 107(5), 105(4), 94(12), 93(100), 92(20), 79(21), 77(20), 69(32), 53(10), 41(25)
Limonene[b]	136	136(M⁺;8), 121(12), 107(10), 105(12), 94(15), 93(40), 92(35), 91(16), 79(25), 77(21), 68(100), 67(44), 53(22)

Toluene and p-xylene were identified from the whole body solvent extract.
Abbreviations:
[a] significant ions above m/z 40; [b] confirmed by mass spectrum and retention time (OV225) of authentic sample.

Table 3. Quinones from Salivary Gland Secretions of Soldiers of *Odontotermes magdalenae*

Compound	mol. wt.	Mass of fragment ion[a], m/z (% abundance)
2-Methyl-1,4-benzoquinone[b]	122	122(M⁺; 26), 94(60), 82(35), 67(30), 66(70), 54(100), 53(25), 43(90)
2-Ethyl-1,4-benzoquinone[b]	136	136(M⁺;32), 108(44), 79(50), 71(33), 54(60), 45(31), 43(100), 41(50)
2-Hydroxy-5-methyl-1,4-benzoquinone	138	138(M⁺;8), 91(20), 82(25), 69(100), 53(44), 43(80), 42(30), 41(25)
2-methyl-1,4-hydroquinone[b]	124	124(M⁺;100), 123(39), 95(35), 77(30), 67(40), 55(35), 53(30), 41(30)

Toluene was recorded from the whole body solvent extract.

Abbreviations:
[a] significant ions above m/z 40; [b] confirmed by mass spectrum and retention time (OV225) of authentic sample; [c] Bond geometry not determined.

glands in ants the main one seems to be defense against predation [5]. A defensive role has similarly been assigned to the frontal gland and salivary gland secretions from termites [19]. Perhaps the differences in the secretions found here for termite soldiers of *T. geminatus* and *O. magdalenae* are related to differences in accessibility to termite predators. Open foraging is a feature of the biology of the workers of *T. geminatus* unlike *O. magdalenae* whose workers construct protective foraging galleries from clay. Future work will be directed towards investigation of the effects of components of the secretion on possible predators and on the producers themselves. Multifunctions are to be expected from multicomponent secretions such as we have described [2,5].

ACKNOWLEDGEMENTS

Part funding for this work was provided under the International Foundation for Science Research grant No. F/882-2 and Appl. No. AF/2089 to T. O. Olagbemiro. We thank British Airways for supporting a visit by K. M. Sani to the Chemistry Department, Sussex University, for GC-MS analysis of sample material and the British Council for providing a travel grant to B. W. Staddon and T. O. Olagbemiro to Nigeria and Britain respectively for this work. This article is based in part on previously published material [22,23].

REFERENCES

1. M. S. Blum and H. R. Hermann, Venoms and venom apparatuses of the Formicidae: Myrmeciinae, Ponerinae, Dorylinae, Pseudomyrmecinae, Myrmicinae and Formicinae, *in*: Arthropod Venoms, S. Bettini, ed., Handb. exp. Pharmakol., Springer-Verlag, Berlin, 48:801 (1978a).
2. M. S. Blum, "Chemical Defenses of Arthropods", Academic Press, New York (1981).
3. G. W. K. Cavil and P. J. Williams, Constituents of Defour's gland in *Myrmica gulosa*, *J. Insect Physiol.*, 13:1097-1103 (1967).
4. M. S. Blum and H. R. Hermann, Venoms and venom apparatuses of the Formicidae: Dolichoderinae and Aneuretinae, *in:* Arthropod Venoms, S. Bettini, ed., Handb. exp. Pharmokol., Springer-Verlag, Berlin, 48:871-894 (1978b).
5. A. B. Attygalle and E. D. Morgan, Chemicals from the glands of ants, *Chem. Soc. Rev.*, 13:245 (1985).
6. E. D. Morgan, K. Parry and R. C. Tyler, The chemical composition of the Dufour gland secretion of the ant *Myrmica scrabrinodis*, *Insect Biochem.*, 9:117-121 (1979).
7. C. Bergström and J. Lofqvist, Odour similarities between the slave-making ants *Formica sanguinea* and *Polyergus refescens* and their slaves *Formica fusca* and *Formica rufibarbis*, *J. Insect Physiol.*, 14:995-10118 (1979).
8. G. Bergström and J. Lofqvist, *Camponotus ligniperda* Latr. A model for the composite volatile secretions of Dufour's gland in formicine ants. Pp 195-223, 2nd Int. IUPAC Congr., Tel Aviv, *in*: "Chemical Releasers in Insects", A. S. Tahori, ed., Gordon and Breach, New York (1971).
9. J. J. Brophy, G. W. K. Cavill and J. S. Shannon, Venom and Dufour's gland secretions in an Australian species of *Camponotus*, *J. Insect Physiol.*, 19:791-798 (1973).
10. J. J. Brophy, G. W. K. Cavill, J. A. McDonald, D. Nelson and W. D. Plant, Volatile constituents of two species of Australian *Notoncus* and *Polyrhachis*, *Insect Biochem. Physiol.*, 12:215-219 (1983).
11. M. S. Blum, T. H. Jones, W. L. Overal, H. M. Fales, J. O. Schmidt and N. A. Blum, Exocrine chemistry of the monotypic ant genus *Gigantiops*, *Comp. Biochem. Physiol.*, 75B:15 (1983).
12. H. A. Lloyd, M. S. Blum and R. M. Duffield, Chemistry of male mandibular gland secretion of the ant *Camponotus clavithorax*,*Insect Biochem.*, 5:489 (1975).
13. G. D. Prestwich, Chemical systematics of termite exocrine secretions, *Ann. Rev. Ecol Syst.*, 14:287-311 (1983).
14. J. Vrkoc, J. Krecek and I. Hrdy, Monoterpene alarm pheromones in two *Nasutitermes* species, *Acta Entomol. Bohemoslow*, 75:1-8 (1979).
15. B. P. Moore, Studies on chemical composition and function of the cephalic gland secretion in Australian termites, *J. Insect Physiol.*, 14:33-39 (1968).
16. G. D. Prestwich, Chemical defense by termites, *J. Chem. Ecol.*, 5:459-480 (1979).
17. J. C. Brakeman, D. Daloze, A. Dupont, J. M. Pasteels and G. Josens, Diterpene composition of defense secretion of four West African *Trinervitermes* soldiers, *J. Chem. Ecol.*, 10:1363-1370 (1984).
18. G. D. Prestwich, Defense mechanisms of termites, *Ann. Rev. Entomol.*, 29:201-232 (1984).
19. G. Josens, Le bilan énergêtique de *Trinervitermes geminatus* Wasmann (Termitidae, Nasutitermitinae) 1. Mesure de biomasses, d'équivalents énergêtiques, de longévité et de production en laboratoire, *Insectes Soc.*, 29:295-307 (1982).
20. U. Maschwitz and Y. P. Tho, Chinone als Wehrsubstanzen bei einigen orientalische Macrotermitinen, *Insectes Soc.*, 21:231-234 (1974).
21. W. F. Wood, W. Trackenbrodt and J. Meinwald, Chemistry of the defensive secretion from the African termite *Odontotermes badius*, *Ann. Entomol. Soc. Am.*, 68:359 (1975).
22. T. O. Olagbemiro, K. M. Sani, L. Lajide, M. O. Agoh, B. W. Staddon and C. E. Chukwu, *trans*-4-Phenyl-3-buten-2-one from the African harvester ant Messor galla (Mayr) (Hymenoptera: Formicidae), *Z. Naturforsch*, 43b:339-342 (1987).

23. T. O. Olagbemiro, L. Lajide, K. M. Sani and B. W. Staddon, 2-Hydroxy-5-methyl-1,4-benzoquinone from the salivary gland of the soldier termites *Odontotermes magdalenae*, *Experentica*, 44:1022-1024 (1988).

PATTERN RECOGNITION

A NOTE ON A COMPUTERIZED SYSTEM FOR PATTERN RECOGNITION DESIGNED FOR GAS CHROMATOGRAMS

B. A. Sommerville, M.A. Green and A. J. Bleasby*

Department of Forensic Medicine
Clinical Sciences Building
St James's University Hospital
Beckett Street, Leeds LS9 7TF, UK, and
*Department of Biophysics
University of Leeds, Leeds LS6 9JT, UK

INTRODUCTION

The aim of this research project was to identify a range of odorous volatiles in human sweat which are under genetic influence and might be used as a basis for recognizing an individual in a similar way to fingerprinting. Assuming that such a range of compounds should have a similar pattern in identical twins, we have analyzed sweat samples from ten pairs of twins. These have been run on a gas chromatograph equipped with head space desorption facilities and an outlet splitter allowing successive peak fractions to be collected. The fractions have been presented to a trained dog for scent matching as described elsewhere [1].

A fraction of medium volatility comprising about ten peaks has been selected on the basis that the dog could use it to discriminate between unrelated people but not between identical twins. This fraction has been monitored by the computerized pattern recognition system described below.

MATERIALS AND METHODS

The software was based on "CHAS", an integrative FORTRAN 77 program using IBM FORTVS [2]. It was used with an ATARI 1040 ST computer connected to a video display and printer. The integration data were stored on disc and derived either from the output of a flame ionization detector connected to a gas chromatograph (GC-FID) or, indirectly, from a mass spectrometer (MS). The reconstructed ion current print-out from the MS was digitized and then converted to ATARI mode.

After loading the software, two stored chromatograms, A and B, were loaded and B was modified to make it as similar to A as possible by eliminating variation arising from the running technique and sample size.

Thus Figure 1 demonstrates how the entire chromatogram may be translated vertically (or horizontally) and how a prescribed portion of the area beneath the peaks may be removed.

Editors' Note: This research was presented as a poster at the meeting. Whilst not entomological in content, the techniques reported are clearly applicable to pheromone research on insects. In addition, it seems likely that components of human sweat could act as semiochemicals for insects and therefore the Editors have concluded that the inclusion of this work in this volume is appropriate.

Chromatography and Isolation of Insect Hormones and Pheromones
Edited by A. R. McCaffery and I. D. Wilson
Plenum Press, New York, 1990

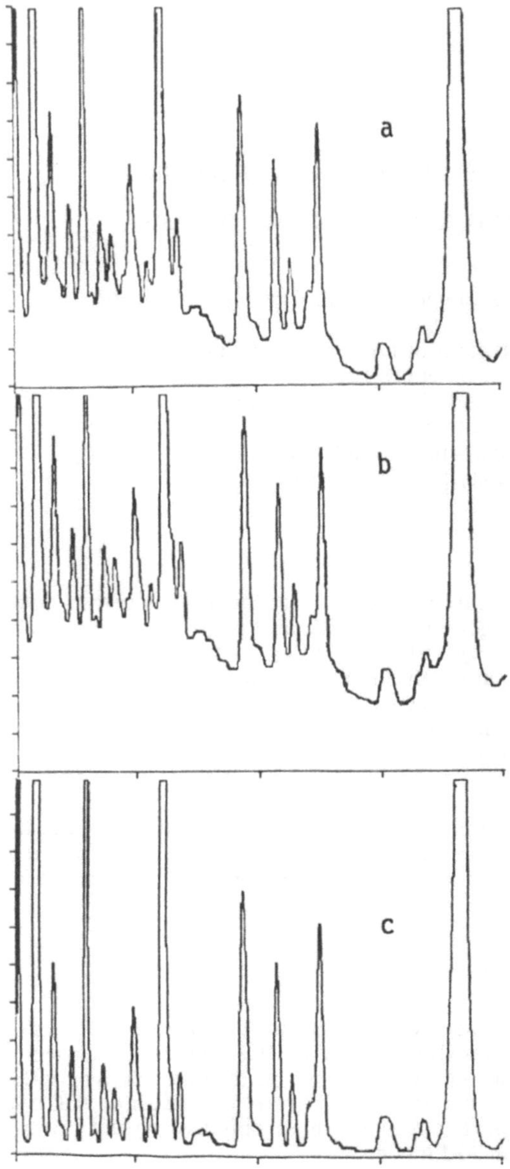

Fig. 1. Chromatograms of human sweat volatiles demonstrating vertical translation (a), unprocessed chromatogram (b), and removal of 90% of the baseline (c).

As shown in Figure 2, an upward linear trend in the baseline may be removed and marker spikes deleted. This may also be carried out for a downward trend.

Any prescribed peak in one chromatogram may be aligned with one in the other as shown in Figure 3. Alignment should normally be restricted to marker spikes in the GC-FID chromatogram but may be applied to any peaks whose common identity has been determined in the MS derived chromatograms. The program remembers up to 99 realignments.

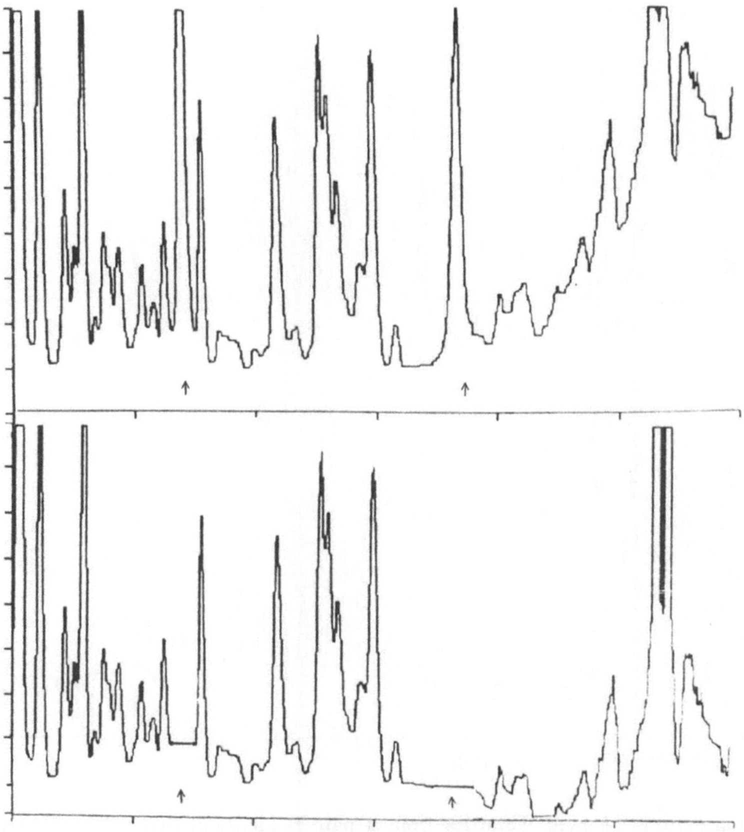

Fig. 2. A chromatogram demonstrating how a linear trend in the baseline may be removed and also any spurious peak, e.g., marker spike.

Similarly Figure 4 shows how chromatogram B may be scaled up to match chromatogram A. Down-scaling is also possible.

Two optional facilities exist in the software. The area of chromatogram B may be normalized to match that of chromatogram A and a peak weighting factor may be applied to enhance the contribution of selected peaks when the match parameters are measured.

Four pattern match parameters can then be measured on the tailored pair of chromatograms:

1. Alignment gives the number of peaks aligned as a ratio of the total number of peaks and the coefficient gives a figure for this ratio (a peak B, in chromatogram B, is considered to be aligned with a peak A, in chromatogram A, if the beginning of the later one precedes the other's apex and if also the later apex precedes the end of the earlier peak). Total alignment gives a value of 1.0.

2. Profile correlation measures the degree of similarity of the chromatograms' shapes, 1.0 being a perfect correlation.

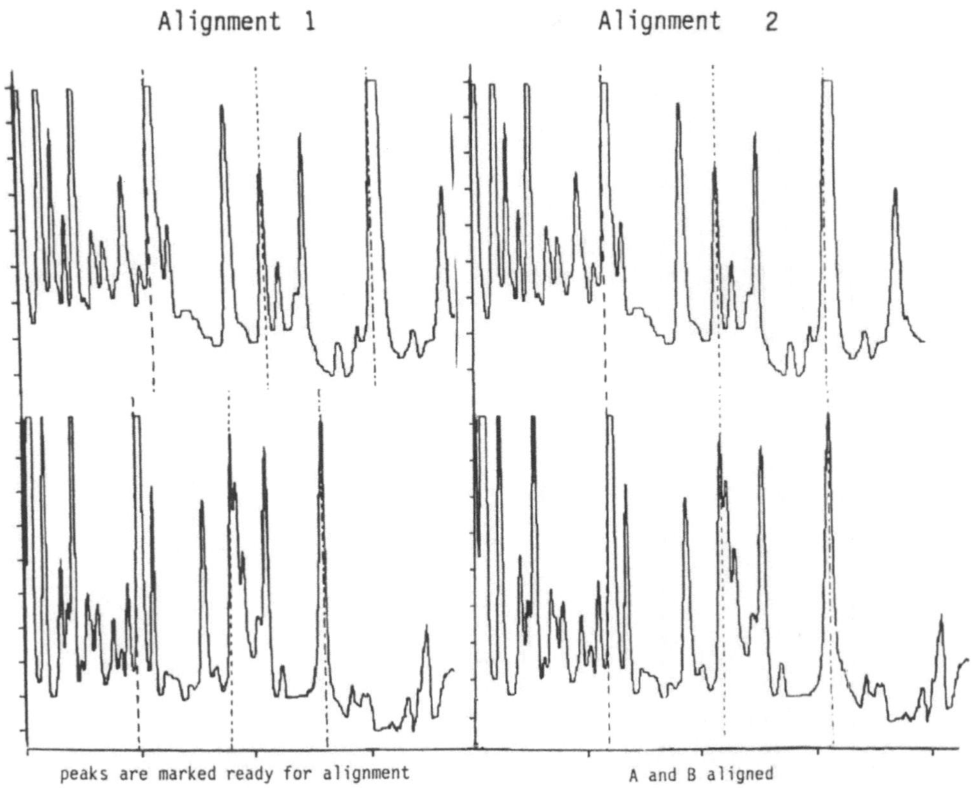

Fig. 3. Chromatograms of sweat volatiles from a pair of identical twins demonstrating the peak alignment facility to correct for variations in retention time between runs.

It is calculated as:

$$\frac{\int_{O}^{T} U_t V_t \, dt}{\left[\int_{O}^{T} U_t 2 \, dt \quad \int_{O}^{T} V_t 2 \, dt \right]^{0.5}}$$

where $U_t = B_t - \bar{B}$ and $V_t = A_t - \bar{A}$; $\bar{B}$ and $\bar{A}$ are the mean values of B_t and A_t, i.e., $\bar{B} = B_t \, dt/T$ and $\bar{A} = A_t \, dt/T$.

3. Euclidian distance summates the difference in height of the two chromatograms at frequent time intervals and is, therefore, a dissimilarity measure where 0.0 indicates a perfect match. Its use is particularly appropriate where a few large differences are anticipated.

It is calculated:

$$\left[\int_{O}^{T} W_t \, (B_t - A_t)^2 dt \right]^{0.5}$$

where W_t is the optional Gaussian weighting factor.

4. Box car or absolute distance is a similar measure to the Euclidian distance but gives the same weighting to a few large differences as to many small differences.

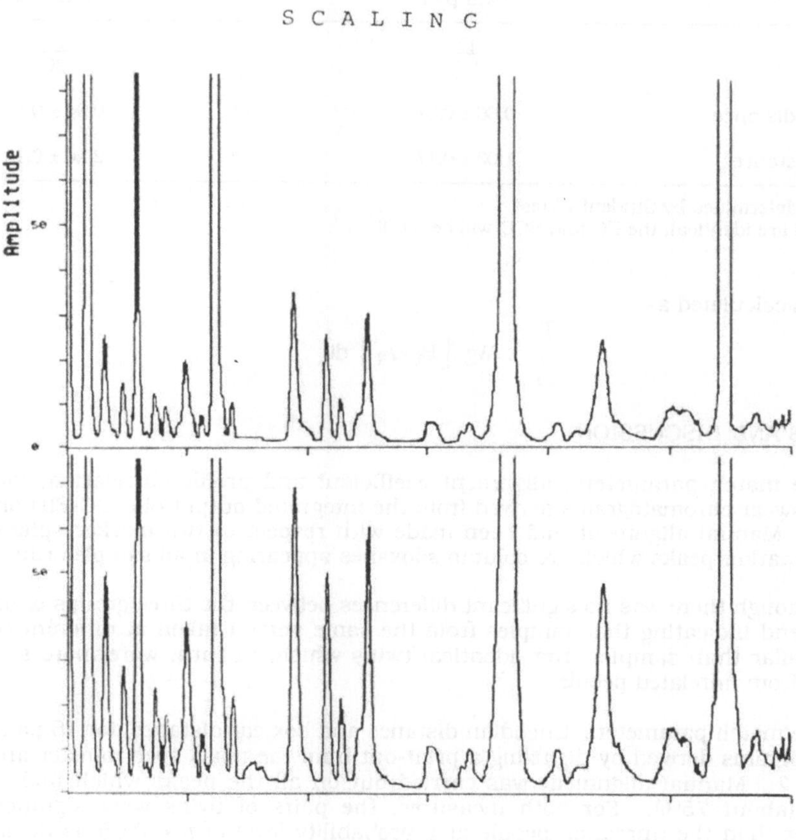

B expanded in 'y' direction to match peak heights in A.

Fig. 4. Chromatograms of sweat volatiles from a pair of identical twins demonstrating the scaling facility to correct for variation in sample size.

Table 1. Alignment coefficient and profile correlation for pairs of chromatograms derived from GC/FID

		mean ± standard of mean	
(n)	(5)	(6)	(7)
A and B =	Same person	Twin pair	Unrelated people
Common peaks	12/19	11/21	12/24
Alignment coefficient	0.64 ± 0.08	0.53 ± 0.04	0.51 ± 0.08
Profile correlation	0.52 ± 0.28	0.35 ± 0.15	0.33 ± 0.08

(1.00 represents perfect alignment)

Table 2. Euclidian distance and box car distance for pairs of chromatograms derived from digitised mass spectrometer reconstructed ion current print-outs

(n) A and B =	mean ± standard error of mean (6) Twin pair		(10) Unrelated people
Common peaks	17/27		15/26
Euclidian distance	0.66 ± 0.07	*	0.94 ± 0.06
Box car distance	1.03 ± 0.17	*	2.26 ± 0.19

* $p < 0.05$ determined by Student's t test
(If A and B are identical, the EC and BCD will be 0.00)

It is calculated as:

$$\int_{O}^{T} W_2 \left[B_t - A_t\right] dt$$

RESULTS AND DISCUSSION

The match parameters, alignment coefficient and profile correlation, for eighteen pairs of sweat chromatograms derived from the integrated output of a GC-FID are given in Table 1. Manual alignment had been made with respect to two marker spikes and two internal marker peaks which are column siloxanes appearing in all samples run.

Although there was no significant differences between the three groups of pairs, there was a trend indicating that samples from the same person taken at different times were more similar than samples from identical twins which, in turn, were more similar than samples from unrelated people.

The match parameters, Euclidian distance and box car distance, for 16 pairs of sweat chromatograms derived by digitizing a print-out from the mass spectrometer are provided in Table 2. Manual alignment was carried out on all the peaks which had a common identity (about 75%). For both measures, the pairs of twins were significantly less dissimilar than the unrelated people at a probability level of $p < 0.05$ as determined by Student's t test.

ACKNOWLEDGEMENTS

We should like to thank the Wellcome Trust for their generous donation of the Finnigan MAT mass spectrometer and Ian Walker, its technician.

REFERENCES

1. B. A. Sommerville, M. A. Green and D. J. Gee, Using chromatography and a dog to identify some of the compounds in human sweat which are under genetic influence, Proceedings of Chemical Signals in Vertebrates V, Oxford 1988, in press (1990).
2. R. J. Marshall, A. J. Bleasby, R. Turner and E. D. Cooper, A computer system for analysis of chromatographic data, *Chemom. Intell. Lab. Syst*, 1:285-295 (1987).

The abstracts provided here are for those papers and posters for which no manuscript was received for publication. They have been retyped but are otherwise printed essentially as received.

GAS CHROMATOGRAPHY-MASS SPECTROMETRY (GC-MS) AND MS-MS ANALYSES OF JUVENILE HORMONES AND METABOLITES OR ANALOGUES

B. Mauchamp, F. Couillaud*, A. Messeguer**, F. Sanchez**, H. Virelizier† and J. Einhorn ‡

Laboratoire Physiologie de l'Insecte, INRA, F-78026 Versailles, France
*Laboratoire Neuroendocrinologie, URA-CNRS 1138
Université de Bordeaux I, F-33405 Talence, France
**Department of Biological Organic Chemistry
CID (CSIC), J. Girona 18, 08034 Barcelona, Spain
†CEN Saclay, B.P. 91190 Gif-sur-Yvette, France, and
‡ Laboratoire Médiateurs Chimiques, INRA-CNRS
Magny-les-Hameaux, F-78470 St. Rémy-les-Chevreuse, France

Several years ago we developed a physico-chemical technique to detect and quantify juvenile hormones (JHs) of insects. This technique was based on gas chromatography-mass spectrometry (GC-MS) analysis of JHs without derivatization, unlike other methods simultaneously developed in several laboratories.

By GC analysis it is easy to separate the known JHs purified from whole insects or hemolymph by a complex procedure. When a mass spectrometer is coupled with a gas chromatograph, it is possible to detect minute amounts of chemicals mainly by using techniques such as multiple ion detection (MID) or selected ion monitoring (SIM). The possibilities of the technique were heavily dependent on the nature of fragmentation. We reported that the electron impact technique of fragmentation yielded a large number of fragments in the 50-100 mass range. Consequently, the molecular ion M^+ was virtually invisible. After several attempts we evidenced chemical ionization using ammonia as the reagent gas to be the best suited for analysis of JHs. Comparative studies allowed us to propose a general scheme for fragmentation, which was appropriate for analysis of analogues or derivatives. The technique was applied to fluorinated derivatives tested as chemicals acting on the JH biosynthetic pathway. The fluorinated analogues of methyl farnesoate were added to incubation medium containing isolated corpora allata (CA). The level of JH III synthesized was evaluated as well as the nature of the derivatives obtained from the fluorinated precursor. Some derivatives were detected in spite of the lack of references.

Using CI/MH3 we observed that protonated molecular ion MH^+ was always detectable. It suggests that these ions can be used to complete MS-MS analysis. In MS-MS all the components of a mixture are first ionized to obtain a set of molecular ions or specific fragments. These ions are then separated by the first mass spectrometer which plays the role of a particularly efficient chromatographic device. The structure of selected ions can be studied by fragmentation with the second mass spectrometer. Chemicals can be detected from a biological mixture without extraction.

We further developed this technique for analysis of a complex mixture of JH III and its metabolites in a CA incubation medium.

Comparison of MS-MS and GC-MS techniques is discussed.

BIOAMINES AND PUPAL DIAPAUSE IN *PIERIS BRASSICAE*

J. Puiroux, R. Moreau, L. Peypelut, J. Vieillemaringe and L. Gourdoux

Laboratoire de Neuroendocrinologie de l'Insecte
URA-CNRS 1138, Université de Bordeaux I
F-33405 Talence, France

According to the photoperiodic length, the post-embryonic development of *Pieris brassicae* can be direct (in the case of a long photoperiod) or interrupted at the pupal stage (under a short photoperiod): this phenomenon is called facultative diapause. Several studies have shown that the brain could be the direct receptor of the light duration variations. Biogenic amines are attempted to be involved in the perception of these variations and implicated in the regulation of the development, particularly in the diapause induction and termination.

High performance liquid chromatography (pump Beckman 114 m SDM and injector 210 A with an injection loop of 20 µl) with a dual electrochemical detection (HPLC-EICD) has been used to separate and estimate several monoamines. The brain and suboesophagal ganglion of direct and diapausing pupae of *Pieris* have been studied. The use of two different reverse-phase columns (Beckman ultrasphere C_{18} analytical 250 mm and Spherisorb ODS_2 C_{18} 250 mm with a precolumn ODS_2) and a dual electrode electrochemical detector (ESA model 5100A with the first electrode set at 0,35V and the second set at 0,6V) allowed the analysis of the following compounds: tyrosine (TYR), dopamine (DA), dihydroxyphenyl-acetic acid (DOPAC), *n*-acetyldopamine (NADA) and tryptophan (TRP), 5-hydroxy-tryptamine (5-HT), 5-hydroxyindolacetic acid (5-HIAA).

During the direct development, the levels of DA and NADA (one of its metabolites) in the brain were more important in 4 day old pupae and at the end of the pupal life (8th day). DOPAC (another metabolite of DA) showed no variation. 5-HT was observed at the higher level in 2-3 day old pupae but 5-HIAA was never detected. In the young diapausing pupae, the same variations of DA amount were observed even though a slight decrease of DOPAC and NADA was noted. 5-HT appeared more lately (4-5 days) and higher. In the old diapausing pupae, a marked increase in DA level can be pointed out, most particularly when the pupae were placed at low temperature (4°C). DOPAC and NADA exhibited less important differences. During diapause, 5-HT was reduced or absent.

The pupal diapause of *Pieris* could be characterized by higher dopaminergic activity and a lower serotoninergic activity.

PUTATIVE SEX PHEROMONE GLANDS IN THE *DACUS OLEAE* GMEL MALE (DIPTERA, TEPHRITIDAE)

M. Solinas, G. Nuzzaci* and L. De Marzo*

Agricultural Entomology Institute
Perugia University, 06100 Perugia, Italy, and
*Agricultural Entomology Institute
Bari University, 70126 Bari, Italy

The rectal sac of *D. oleae* males, like in other *Dacus* species, displays a secretory epithelium lining the latero-ventral and posterior walls as well as a special postero-ventral swelling. The epithelium produces an "oily substance" soon after adult emergence, but it becomes completely developed in the sexually mature males, when a special "odour" is produced in addition. Ultrastructural observations revealed a remarkable activity of the secretory cells both before and after sexual maturation, but the cell organelles involved in secretion (especially endoplasmis reticulum) were quite different in sexually immature males and mature ones. Laboratory trials with an olfactometer gave evidence that sexually mature males attract females by the "odour" at day time when the females are sexually receptive.

CEREBRAL BIOGENIC AMINES AND LARVAL DIAPAUSE IN THE EUROPEAN CORN BORER, *OSTRINIA NUBILALIS* Hbn

L. Peypelut, L. Lavenseau, J. Puiroux and J. Vieillemaringe

Laboratoire de Neuroendocrineologie de l'Insecte
URA-CNRS 1138, Université de Bordeaux I
F-33405 Talence, France

Catecholamine (dopamine (DA), N-acetyldopamine (NADA)) and indolamines (5-hydroxytryptamine (5-HT)) variations were studied in brains of non-diapausing, diapausing and diapause terminating fifth instar larvae of the European corn borer. The compounds were separated and identified by high performance liquid chromatography using a reverse phase ODS 4,6 mm x 25 cm spherical particle column and a mobile phase consisted of 50 mM tri-sodium citrate, 7% methanol adjusted to pH 4,1 with a citric acid solution. Catecholamines and indolamines levels were measured electrochemically using an ESA 5100 A coulometric detector system.

We demonstrated that:

1. In non-diapausing larvae (18L:6D, 25°C), the dopaminergic and serotoninergic activity increased at the beginning of the last instar (2-3 days) but decreased between the fifth and the seventh day.

2. In diapausing larvae (12L:12D, 6°C), the dopaminergic activity remained low throughout the entire diapause period. On the other hand, serotonin levels increased progressively during the same period. This suggests that serotonin may be involved in the control of the typical neurosecretory status of diapause.

3. During the first 3 days of the diapause termination (18L:6D, 25°C, water), no changes were observed in comparison with diapause. Conversely catecholamines and indolamines amounts increased sharply during the following 3 days, then decreased to levels observed in late non-diapausing larvae. Identical changes were observed in diapausing larvae subjected to a short heat stress (1 hour at 22°C) before being replaced under diapausing conditions. It strongly suggests that biogenic amines variations may be related to the modifications of the external conditions. This hypothesis is consistent with recent results obtained *in vitro*, after action of long day photoperiod on cultured brains.

The described techniques allowed us to show that the cerebral aminergic system is affected by some changes of the milieu and that it may act on the regulation of the diapause cycle.

INSECT NEUROPEPTIDE BIOSYNTHESIS STUDIED BY ANALYTICAL HIGH-PERFORMANCE LIQUID CHROMATOGRAPHY

M. O'Shea

Cell Biology Laboratory, Royal Holloway and Bedford
New College, University of London, Englesham Green
Egham, Surrey TW20 0EX, UK

Size exclusion and reverse phase high-performance liquid chromatography (HPLC) has been used to analyze the mode of biosynthesis of the locust adipokinetic hormones AKH I and AKH II. By performing pulse chase and pulse translation block experiments we have been able to purify and identify for structural characterization the precursors of the AKH hormones. This analysis has also led to the identification of novel peptide products generated during the enzymatic processing of the AKH precursors. These peptides are called AKH Precursor Related Peptides (APRPs) and they occur in three dimeric forms (two homodimers and one heterodimer) of two 28 residue peptides called the α and β chains. Our analysis of the precursors of AKH and APRPs by size exclusion and reverse phase HPLC showed that the AKH precursors were also dimers. The structures of the precursors and the APRPs will be given. In addition some of the pitfalls of using HPLC in combination with pulse-chase experiments to analyze peptide biosynthesis will be described.

CHEMICAL SIGNATURE PLASTICITY IN EXPERIMENTAL MIXED COLONIES OF THE ANTS *MANICA RUBIDA* (MYRMICINAE) AND *FORMICA SELYSI* (FORMICINAE)

C. Errard, A. G. Bagneres, J. L. Clement and C. Lange

Laboratoire d'Ethologie, Université Paris XIII
93430 Villefaneuse, France

Chemical analyses have showed that the closure of the sympatric societies of *Manica rubida* and *Formica selysi*, is correlated with cuticular hydrocarbons, each species having a species-specific mixture.

Hydrocarbon profiles were analysed by gas chromatography and mass spectrometry (GC-MS: EI and CI): saturated alkanes (n-alkanes, mono and dimethylalkanes) were predominant for *M. rubida*, and insaturated alkanes (monoenes and dienes) with a few saturated alkanes for *F. selysi*.

The analysis of cuticular hydrocarbons showed that when the two species are associated in artificial mixed colonies, each species acquired the cuticular hydrocarbons of the other. A new chemical signature was formed which was composed with the species specific cuticular hydrocarbons of the two species. A multivariate technique with a principal component analysis was employed to show these results clearly.

COMPOUND INDEX

SPECIES INDEX

SUBJECT INDEX

MIX
Papier aus verantwortungsvollen Quellen
Paper from responsible sources
FSC® C105338

If you have any concerns about our products,
you can contact us on
ProductSafety@springernature.com

In case Publisher is established outside the EU,
the EU authorized representative is:
Springer Nature Customer Service Center GmbH
Europaplatz 3, 69115 Heidelberg, Germany

Printed by Libri Plureos GmbH
in Hamburg, Germany